D0999767

# COMPREHENSIVE CHEMICAL KINETICS

# COMPREHENSIVE

# CHEMICAL KINETICS

EDITED BY

## C. H. BAMFORD

M.A., Ph.D., Sc.D. (Cantab.), F.R.I.C., F.R.S.

*Campbell-Brown Professor of Industrial Chemistry,*
*University of Liverpool*

AND

## C. F. H. TIPPER

Ph.D. (Bristol), D.Sc. (Edinburgh)

*Senior Lecturer in Physical Chemistry,*
*University of Liverpool*

VOLUME 13

REACTIONS OF AROMATIC COMPOUNDS

ELSEVIER PUBLISHING COMPANY
AMSTERDAM - LONDON - NEW YORK
1972

SETON HALL UNIVERSITY
SCIENCE LIBRARY
SOUTH ORANGE, N. J.

ELSEVIER PUBLISHING COMPANY
335 JAN VAN GALENSTRAAT
P. O. BOX 211, AMSTERDAM, THE NETHERLANDS

AMERICAN ELSEVIER PUBLISHING COMPANY, INC.
52 VANDERBILT AVENUE
NEW YORK, NEW YORK 10017

Science
QD
501
B3
V.13

LIBRARY OF CONGRESS CARD NUMBER 72-168910
ISBN 0-444-40937-8

WITH 8 ILLUSTRATIONS AND 277 TABLES

COPYRIGHT © 1972 BY ELSEVIER PUBLISHING COMPANY, AMSTERDAM
ALL RIGHTS RESERVED

NO PART OF THIS PUBLICATION MAY BE REPRODUCED,
STORED IN A RETRIEVAL SYSTEM, OR TRANSMITTED IN ANY FORM OR BY ANY MEANS,
ELECTRONIC, MECHANICAL, PHOTOCOPYING, RECORDING, OR OTHERWISE,
WITHOUT THE PRIOR WRITTEN PERMISSION OF THE PUBLISHER,
ELSEVIER PUBLISHING COMPANY, JAN VAN GALENSTRAAT 335, AMSTERDAM

PRINTED IN THE NETHERLANDS

COMPREHENSIVE CHEMICAL KINETICS

ADVISORY BOARD

Professor S. W. BENSON
Professor SIR FREDERICK DAINTON
Professor G. GEE
the late Professor P. GOLDFINGER
Professor G. S. HAMMOND
Professor W. JOST
Professor G. B. KISTIAKOWSKY
Professor V. N. KONDRATIEV
Professor K. J. LAIDLER
Professor M. MAGAT
Professor SIR HARRY MELVILLE
Professor G. NATTA
Professor R. G. W. NORRISH
Professor S. OKAMURA
Professor SIR ERIC RIDEAL
Professor N. N. SEMENOV
Professor Z. G. SZABÓ
Professor O. WICHTERLE

# Contributors to Volume 13

S. D. Ross        Research and Development Laboratories,
                  Sprague Electric Company,
                  North Adams, Mass., U.S.A.

R. Taylor         The Chemical Laboratory,
                  University of Sussex,
                  Falmer, Brighton, Sussex, England

D. L. H. Williams Chemistry Department,
                  University of Durham,
                  Durham City, England

# Preface

Section 4 deals almost exclusively with reactions recognised as organic in a traditional sense, but excluding (unless very relevant) those already considered in Sections 2 and 3 and biochemical systems. Also oxidations, *e.g.* of hydrocarbons, by molecular oxygen, polymerization reactions and fully heterogeneous processes are considered latter. The relationships of mechanism and kinetics, *e.g.* effects of structure of reactants and solvent, isotope effects, are fully discussed. Rate parameters of individual elementary steps, as well as of overall processes, are given if available. We have endeavoured, in conformity with our earlier policy, to organise this section according to the types of chemical transformation and with the minimum of recourse to mechanistic classification. Nevertheless it seemed desirable to divide up certain general processes on the basis of nucleophilic or electrophilic character.

In Volume 13 reactions of aromatic compounds, excluding homolytic processes due to attack of atoms and radicals (treated in a later volume), are covered. The first chapter on electrophilic substitution (nitration, sulphonation, halogenation, hydrogen exchange, etc.) constitutes the bulk of the text, and in the other two chapters nucleophilic substitution and rearrangement reactions are considered.

The Editors express their thanks for the continuing advice and support from members of the Advisory Board.

C. H. Bamford
C. F. H. Tipper

*Liverpool*
*February, 1972*

# Contents

*Chapter* 2 (S. D. ROSS)

# Nucleophilic aromatic substitution: the $S_{N2}$ mechanism · · · · · · · · · · · 407

*Chapter* 3 (D. L. H. WILLIAMS)

# Aromatic rearrangements · · · · · · · · · · · · · · · · · · · · · 433

# Index · · · · · · · · · · · · · · · · · · · · · · · · · 487

*Chapter* 1

# Kinetics of Electrophilic Aromatic Substitution

R. TAYLOR

## 1. Introduction

In this chapter on electrophilic aromatic substitution, the literature published prior to 1969 is exhaustively surveyed. The reactions are presented in the same sequence used in an earlier and related review[1], except that some newly discovered reactions are included, and others, for which no kinetic studies have been made, are omitted.

The vast majority of the kinetic detail is presented in tabular form. Amassing of data in this way has revealed a number of errors, to which attention is drawn, and also demonstrated the need for the expression of the rate data in common units. Accordingly, all units of rate coefficients in this section have been converted to $mole.l^{-1}.sec^{-1}$ for zeroth-order coefficients $(k_0)$, $sec^{-1}$ for first-order coefficients $(k_1)$, $l.mole^{-1}.sec^{-1}$ for second-order coefficients $(k_2)$, $l^2.mole^{-2}.sec^{-1}$ for third-order coefficients $(k_3)$, etc., and consequently *no further reference to units is made*. Likewise, energies and enthalpies of activation are all in kcal. $mole^{-1}$, and entropies of activation are in $cal.deg^{-1}mole^{-1}$. Where these latter parameters have been obtained over a temperature range which precludes the accuracy favoured by the authors, attention has been drawn to this and also to a few papers, mainly early ones, in which the units of the rate coefficients (and even the reaction orders) cannot be ascertained. In cases where a number of measurements have been made under the same conditions by the same workers, the average values of the observed rate coefficients are quoted. In many reactions much of the kinetic data has been obtained under competitive conditions such that rate coefficients are not available; in these cases the relative reactivities (usually relative to benzene) are quoted.

In general, kinetics studies have fallen into two classes: those aimed at determining the reaction mechanism, and those which utilise the reaction to determine substituent effects in electrophilic aromatic substitution. Paradoxically perhaps, where a reaction has received considerable attention under the former category, the kinetic complications which have necessitated these studies have usually precluded the use of the reaction in the latter category. The significant exception to this generalisation is the hydrogen-exchange reaction.

Very many different experimental techniques have been involved in determining rate coefficients. In earlier work, measurements of product yields, volume changes, and gas evolutions at various times, was common, but more recently, techniques involving gas chromatography, spectroscopy (UV, IR, NMR) and (for hydrogen exchange involving tritium) scintillation counting have been used. The use of these latter methods has meant, amongst other advantages, that the range of compounds which can be studied has been extended to those available in only milligram quantities.

## 2. General principles

### 2.1 INTERPRETATION OF THE RATE LAWS

It is possible to envisage a number of steps in an electrophilic aromatic substitution. These are

(*i*) Formation of the electrophile $E^+$ from an unpolarised species EZ.

(*ii*) Reaction of the electrophile or unpolarised species with the aromatic to give an intermediate in a step which may be fast

$$EZ + ArX \underset{k'_{-1}}{\overset{k'_1}{\rightleftharpoons}} Ar^+XE + Z^- \text{ (fast)} \tag{1}$$

or slow

$$EZ + ArX \underset{k'_{-1}}{\overset{k'_1}{\rightleftharpoons}} Ar^+XE + Z^- \text{ (slow)} \tag{1a}$$

(*iii*) Loss of the substituent being replaced in a step which may be slow

$$Ar^+XE \underset{k'_{-2}}{\overset{k'_2}{\rightleftharpoons}} ArE + X^+ \text{ (slow)} \tag{2}$$

or fast

$$Ar^+XE \underset{k'_{-2}}{\overset{k'_2}{\rightleftharpoons}} ArE + X^+ \text{ (fast)} \tag{2a}$$

or may, if slow, be catalysed by a base

$$Ar^+XE + B \underset{k'_{-2}}{\overset{k'_2}{\rightleftharpoons}} ArE + BX^+ \text{ (slow)} \tag{3}$$

(*iv*) Synchronous formation and cleavage of the ArE and ArX bonds, respectively.

$$E^+ + ArX + B \overset{k'_1}{\rightarrow} (E\ldots Ar\ldots X\ldots B)^+ \overset{k'_2}{\rightarrow} ArE + BX^+ \tag{4}$$

(*v*) As an alternative to any of the above we may have a rate-determining formation of a carbanion followed by rapid reaction of the anion with the electrophile,

$$ArH + Y^- \underset{k'_{-1}}{\overset{k'_1}{\rightleftharpoons}} Ar^- + HY \quad \text{(slow)}$$

$$Ar^- + EZ \underset{k'_{-2}}{\overset{k'_2}{\rightleftharpoons}} ArE + Z^- \quad \text{(fast)}$$

(5)

Kinetic studies have shown that the mechanism represented by a combination of reactions (1) or (1a) with (2), (2a), or (4), except of course (1) with (2a), is almost general for electrophilic substitution and is termed the $S_E2$ process; in certain circumstances the intermediate $ArXE^+$ is isolable. No example of the alternative $S_E3$ mechanism, reaction (4), has been discovered but two examples of the mechanism given by reaction (5) are now known. The first essentially irreversible (*i.e.* rate-determining) step in the reaction mechanism is the rate of the overall reaction and steps prior to this give equilibrium concentrations of intermediates. If, for the $S_E2$ mechanism, the formation of the electrophile is rate-determining then the reaction is zeroth-order in aromatic; if, however, formation of the electrophile is fast then a subsequent reaction with the aromatic will be rate-determining and the reaction will be first-order in aromatic. In one reaction the aromatic undergoing substitution also functions as the base according to reaction (3) and the reaction, therefore, is second-order in aromatic. The order may be anything up to third in the substituting reagent according to how many molecules of reagent are required to produce an electrophile sufficiently polar to substitute in the aromatic; in many cases the solvent acts as the polarising entity and hence does not appear in the kinetic equation, since it is in constant excess, but in other examples, a "catalyst" may be required to polarise the reagent and this then appears in the rate law.

Many electrophilic substitutions involve an acid (or a proton) as the electrophile and since the reaction is then acid-catalysed it is designated an $A\text{-}S_E2$ reaction provided that reaction between the aromatic and the acid takes place in a slow step. Variations of the $A\text{-}S_E2$ mechanism have, however, been envisaged in which the reaction between the acid and aromatic is very rapid so that the rate-determining step is loss of X from the conjugate acid of the aromatic (A-1 mechanism), or alternatively the loss of X may be base-catalysed in this rate-determining step (A-2 mechanism). Though both of these latter mechanisms have eventually been eliminated for all of the reactions so far investigated, the amount of effort which has gone into the search for them necessitates a short description of the kinetic consequences of their occurrence.

For the $A\text{-}S_E2$ mechanism we have reaction (1a), with EZ = HA, followed by reaction (2a) with E = H. Clearly, the kinetics follow

$$\text{Rate} = k'_1 [\text{ArX}][\text{HA}]$$

(6)

and since HA can be any acidic species present the reaction is subject to *general acid catalysis* and a linear relationship can be expected between $\log k_2$ and $\log C_{H^+}$, where $k_2$ is the second-order rate coefficient and $C_H^+$ is the total concentration of solvated $H^+$ ions.

For the A-1 mechanism, we have reaction (1) with EZ = HA followed by reaction (2) with E = H. The kinetics are governed by

$$\text{Rate} = k_2'[\text{Ar}^+\text{XH}] = \frac{k_2' k_1'}{k_{-1}'} \frac{[\text{ArX}][\text{HA}]}{[\text{A}^-]} \tag{7}$$

and if the ionisation of HA is rapid compared with the substitution then the rate expression is

$$\text{Rate} = \frac{k_1' k_2'}{k_{-1}' K_a} [\text{ArX}][\text{H}_3\text{O}^+] \tag{8}$$

where $K_a = [\text{H}_3\text{O}^+][\text{A}^-][\text{HA}]^{-1}$. The reaction is thus subject to *specific acid catalysis*.

For the A-2 mechanism, we have reaction (1) with EZ = HA followed by reaction (3) with E = H. The rate equation is

$$\text{Rate} = k_2'[\text{Ar}^+\text{XH}][\text{B}] = \frac{k_2' k_1'}{k_{-1}'} \frac{[\text{ArX}][\text{HA}][\text{B}]}{[\text{A}^-]} \tag{9}$$

and since the catalysing base B can and indeed is likely to be, the conjugate base $A^-$ of the catalysing acid, equation (9) reduces to

$$\text{Rate} = \frac{k_2' k_1'}{k_{-1}'} [\text{ArX}][\text{HA}] \tag{10}$$

and we again have *general acid catalysis* so that some other criterion is necessary to distinguish an A-2 and an A-$S_E$2 reaction.

The criterion which has been extensively employed to distinguish A-1 mechanisms from A-2 or A-$S_E$2 mechanism has been the linearity of plots of reaction rate coefficient *versus* the acidity function $h_0$. The acidity function (see Volume 2, p. 358) is a measure of the proton-donating ability of a medium (as measured by its tendency to protonate a base B) which is given by equation (11), *viz.*

$$h_0 = a_{H^+} \frac{f_B}{f_{BH^+}} \tag{11}$$

It can be shown[2] that the first-order rate coefficient for the A-1 mechanism $k_1$

is given by

$$k_1 = \frac{k_2'}{K_{Ar^+XH}} \frac{h_0 f_{BH} + f_{ArX}}{f_B f^{\ddagger}} \tag{12}$$

where $f^{\ddagger}$ is the activity coefficient of the transition state and $K_{Ar^+XH}$ is the acid ionization constant of $Ar^+XH$. Since by definition we have that the acidity function $H_0 = -\log h_0$ then

$$\log k_1 = -H_0 + \log \left\{\frac{f_{BH} + f_{ArX}}{f_B f^{\ddagger}}\right\} + \text{const.} \tag{13}$$

If the activity coefficient term on the right of equation (13) stays constant with changing medium, then a linear correlation of $\log k_1$ with $-H_0$ should follow for the A-1 mechanism. The above diagnosis of the A-1 mechanism is known as the Zucker–Hammett hypothesis[3].

However, the Zucker–Hammett hypothesis is now no longer regarded as a safe criterion of mechanism because a number of cases are now recorded where a reaction mechanism indicated by the hypothesis is different from that which is reliably indicated by other means. Thus in the racemisation of 1-phenylethanol[4], the acid-catalysed hydrolysis of the oxides of ethylene, propylene, and isobutylene[5, 6], the acid-catalysed hydrolysis of diethyl ether[6, 7], the acid-catalysed ring-opening of epichlorohydrin[6, 8], the bromination of acetone (which proceeds *via* initial acid-catalysed enolisation)[9] and the effect of acidity on the $^{18}O$ exchange and dehydration rates of tertiary alcohols[10], the Zucker–Hammett hypothesis predicts the A-1 mechanism, whereas there are sound indications from other data that the A-2 mechanism occurs. It has also been argued that it is not even a condition of a rate-determining reaction of $Ar^+XH$ that the rate should correlate with $h_0$[11].

The reason for these discrepancies may arise from the very important discovery that a single acidity function cannot describe the protonation of indicator bases of even the same charge type[12-19] and Katritzky *et al.*[19] have emphasised that this renders the attempt by Bunnett[20] to obtain a more satisfactory diagnosis of the A-1, A-2, and A-$S_E2$ reactions by the use of $w$ and $w^*$ values less reliable than it at one time appeared to be.

In deducing from the resulting kinetic equation the nature of the electrophile and how it is produced it is important to represent all the reagents present in terms of the species which they may produce. In this way it is possible to eliminate many negative or fractional orders in reagent and generally obtain a simpler kinetic equation. For example, the observed rate law in the uncatalyzed iodination of aniline can be written[21, 22] as

$$\text{Rate} = k[PhNH_3^+][H^+]^{-1}[I_3^-][I^-]^{-2} \tag{14}$$

but since $PhNH_3^+ \rightleftharpoons PhNH_2 + H^+$, $I_3^- \rightleftharpoons I^- + I_2$ and $I_2 \rightleftharpoons I^+ + I^-$, equation (14) reduces to

$$\text{Rate} = k_2[PhNH_2][I^+] \tag{15}$$

Fractional orders usually result when no single step in the reaction is solely rate-determining and intermediate kinetics result. They may also arise if the electrophile is produced by the dissociation of a reagent such that the species produced are not buffered.

The role of the solvent is sometimes difficult to interpret since, with an electrophile which is some species associated with a water molecule, a kinetic equation indistinguishable from that in which the solvent is not considered (in dilute aqueous solution) will be obtained. If the kinetic order falls to zero as the aromatic concentration is increased then the rate-determining step must be the formation of the electrophile $E^+$ from the species $E^+OH_2$. The alternative slow proton transfer, *i.e.* of $H^+$ to EOH can be identified by using heavy water. Because of the lower zero-point energy of the O–D bond the rate should be slower in $D_2O$ than in $H_2O$. On the other hand, since the equilibrium concentration of the conjugate acid of a solute is higher in $D_2O$ than in $H_2O$, then the concentration of $EOH_2{}^+$ will increase and the rate will increase if the rate-determining step occurs after the formation of $EOH_2{}^+$.

The effect upon the rate of varying the deuterium fraction in a mixed deuterium and protium-containing solvent has been used in attempts, now regarded as unsuccessful, to determine which of the A-1, A-2 or A-$S_E$2 mechanisms would be appropriate for a particular acid-catalysed electrophilic substitution. The Gross–Butler theory of solvent isotope effects[23] predicted, from an analysis of thermodynamic properties of the solvent, that the rate coefficient for a reaction involving a pre-equilibrium proton transfer would vary according to a function of the deuterium fraction of the solvent and which would contain a cubic term in this fraction. Thus the increase in rate with increasing deuterium fraction would be proportionately greater at higher deuterium contents, so that a curved plot of rate *versus* deuterium fraction would result for the A-1 mechanism. However, the theory assumed the presence of monohydrated protons and deuterons and ignored a medium effect, the free energy of transfer of the relevant species from one medium to another. When these are taken into account, different forms of the plot are obtained which, therefore, renders valueless the diagnostic value of the theory[24].

## 2.2 SUBSTRATE KINETIC ISOTOPE EFFECTS

Because the zero-point energy of carbon–hydrogen bonds decreases in the order

C–H > C–D > C–T, substitution in tritiated compounds requires a higher activation energy than substitution in deuterated compounds which in turn require a higher energy than protonated compounds. It is this factor which has helped to show both that the $S_E2$ mechanism is appropriate for electrophilic substitution and that the breaking of the carbon–hydrogen bond (when X = H) is usually a fast step. Melander[25] showed that tritiated benzene derivatives were nitrated and brominated at the same rate as the protonated compounds and consequently the carbon–hydrogen bond was not being broken in a rate-determining step, which ruled out the $S_E3$ mechanism. One is left with the $S_E2$ alternative, a consequence of which is the expectation that intermediates might be isolable and many examples of these have now been obtained[26]. A second consequence arises from the kinetic equation (16) for the $S_E2$ process, *viz.*

$$\text{Rate} = \frac{k_1'[\text{ArX}][\text{E}^+](k_2'/k_{-1}')[\text{B}]}{1+(k_2'/k_{-1}')[\text{B}]} \tag{16}$$

It follows that when the second step is fast, $k_2'/k_{-1}'[\text{B}]$ is large ($\gg 1$) and equation (16) reduces to

$$\text{Rate} = k_1'[\text{ArX}][\text{E}^+] \tag{17}$$

*i.e.*, the first step only determines the rate and no base catalysis will be observed. On the other hand if many more molecules of intermediate return to reactants than go to products then $k_{-1}' \gg k_2'$, $k_2'/k_{-1}'$ will be much less than unity, and the observed rate coefficient will be $k_1'k_2'/k_{-1}'$, *i.e.* equal to the equilibrium constant of the first step multiplied by the rate coefficient of the second step. Thus the reaction will show an isotope effect because the rate depends on $k_2'$, and also a dependence of rate on base concentration, which will only be a linear dependence if $k_2'[\text{B}]/k_{-1}' \ll 1$ and will be less than linear for all other cases.

The $S_E3$ mechanism gives the kinetic equation

$$\text{Rate} = k_1[\text{E}^+][\text{ArX}][\text{B}] \tag{18}$$

from which a linear dependence of rate on base concentration would be expected, and the observed kinetic isotope effect would be independent of the base catalysis. In fact, electrophilic substitutions show either no isotope effect, or one which depends on the base concentration; the two observations together with the isolation of intermediates provide conclusive proof for the $S_E2$ mechanism.

Kinetic isotope effects also show a dependence upon the reactivity of the electrophile. Thus some reactions, *e.g.* positive chlorination, show no isotope effect whereas others, *e.g.* sulphonation, do show an isotope effect. There are two ways of visualising the reasons for this and they are closely related. Very electronegative electrophiles will be very reactive such that the transition states for their reactions will be close to the ground state. Consequently, the C–H bond will be largely broken by the time the second transition state is reached and a negligible isotope effect results. Alternatively, consider the intermediate formed with a very electronegative electrophile; the strong electron-withdrawal will aid C–H bond-breaking thereby making the second step faster and less likely to be rate-determining. As a general rule then, reactions with small Hammett rho-factors, *e.g.* positive chlorination, do not give kinetic isotope effects, whereas those with larger rho-factors, *e.g.* molecular iodination, do give kinetic isotope effects.

This effect is observable within a series of very similar electrophiles. Zol-linger[27] found that in reactions of diazonium ions substituted with 4–Cl, 3–Cl, and 3–NO$_2$ substituents (*i.e.* the reactivity and electron-withdrawing power of the ion increased along the series) the respective kinetic isotope effects were 6.55, 5.48 and 4.78.

It is apparent from equation (16) that if $k'_{-1}$ becomes much larger than $k'_2$, the rate will depend upon $k'_2$ and so a kinetic isotope effect will be observed. Now $k'_{-1}$ will become large if there is steric hindrance to formation of the intermediate, and a number of examples are now known where an electrophile which normally gives no isotope effect, does so if formation of the intermediate is hindered.

Recently, the substrate kinetic isotope effects have been examined with regard to the postulate that they will be at a maximum when the transition state is most symmetrical[28]. It follows that this will arise when the aromatic and the base from which the electrophile is being transferred are of equal base strength. The reaction for which this situation is most easily studied is hydrogen exchange and variation in the value of $k_H/k_D$ with acid strength have been detected. Ideally, the value should be at a maximum when the Brönsted exponent for acid catalysis ($\alpha$) is 0.5 for the proton should then be approximately half transferred from catalysing acid to aromatic base in the transition state.

### 2.3 KINETIC EVALUATION OF COMPLEX FORMATION

Kinetic studies are of little value in attempting to determine the extent of complex formation in the reaction path of electrophilic substitution. The reasons for this have been adequately presented elsewhere[29] and the conclusions are that, unless the formation of the complex is rate-determining, the kinetic form is independent of complex formation. Further, the influence of complex formation on reaction rates only comes from the factors which lead in the first place to complex formation, and substituent effects are inadequate for showing the extent of complex formation though when they indicate similar effects on substitution and complex formation they provide evidence that the latter is a pathway of the former.

## 2.4 ENCOUNTER RATE AND DIFFUSION CONTROL

One facet of kinetic studies which must be considered is the fact that the observed reaction rate coefficients in first- and higher-order reactions are assumed to be related to the electronic structure of the molecule. However, recent work has shown that this assumption can be highly misleading if, in fact, the observed reaction rate is close to the encounter rate, *i.e.* reaction occurs at almost every collision and is limited only by the speed with which the reacting entities can diffuse through the medium; the reaction is then said to be subject to *diffusion control* (see Volume 2, Chapter 4). It is apparent that substituent effects derived from reaction rates measured under these conditions may or will be meaningless since the rate of substitution is already at or near the maximum possible.

Diffusion control can be particularly important in reactions in which two aromatic substances of differing reactivity are reacting with a deficiency of reagent. The more reactive aromatic will react first and since diffusion is slow compared with the rate of reaction it becomes impoverished in the reaction zone, and ensuing reaction will occur mainly with the less reactive aromatic which is now in large excess. The observed relative reaction rate then comes out to be less than it would otherwise be. It follows that this may also be true even when the aromatics are reacting at considerably less than the encounter rate.

## 2.5 ISOKINETIC TEMPERATURE RELATIONSHIPS

Application of the Hammett equation to electrophilic aromatic substitutions and other reactions is only valid if the entropy of activation is constant for each compound studied, or if the difference in entropy is proportional to the difference in activation energy or enthalpy[30]. Consequently, it has become the practice where the former condition does not apply, to plot $\Delta H^{\ddagger}$ against $\Delta S^{\ddagger}$ and regard the reasonable scatter of points about a straight line as a satisfactory indication that an isokinetic temperature relationship exists. The isokinetic temperature (the slope of this plot) is the temperature at which the rate coefficients for all of the compounds are the same and hence the point through which all the Arrhenius plots must pass. Unfortunately, it has not been fully appreciated that $\Delta S^{\ddagger}$ is a linear function of $\Delta H^{\ddagger}$ so that any collection of lines through a point on an Arrhenius plot will give a perfect correlation, and for a reaction of small $\rho$-factor we have almost this situation, so that an excellent correlation is unavoidable. Peterson[31] has also drawn attention to this by showing the excellent correlation which can arise from a hypothetical set of data which has only a small range of rate coefficients. It should be noted that his isokinetic plot is *not* a result of accidental compensation of errors, for if the given rate coefficients are modified such that all of the Arrhenius lines pass through a point, a perfect isokinetic relationship is obtained, *i.e.* a *perfect* correlation can *only* be obtained if a true isokinetic relationship exists. For reaction of small $\rho$-factor therefore, a very high correlation coefficient is required for the existence of this relationship to be proved, but for reactions of higher $\rho$-factor a reasonable scatter would be an acceptable experimental error.

## 3. Reactions with electrophilic nitrogen

### 3.1 NITRATION

The kinetics of nitration have been studied fairly intensively over a number of years and show that the reaction can be represented by the general steps

$$ArH + NO_2^+X^- \overset{slow}{\rightleftharpoons} ArHNO_2^+ + X^- \qquad (19)$$

$$ArHNO_2^+ \overset{fast}{\rightarrow} ArNO_2 + H^+ \qquad (20)$$

though examples have now been found where the second step is partially rate-determining leading to a kinetic isotope effect[32]. This occurs in the nitration of 2-substituted derivatives of 1,3,5-tri-t-butylbenzene in nitric acid–sulphuric acid–nitromethane which gives $k_H/k_D = 2.8$ (2-NO$_2$, 40 °C); 3.8 (2-Me, 0°C), 2.1 (2-F, $-10$ and $+21$ °C) whereas under the same conditions 1,3,5-tri-$t$-butyl-benzene gives no isotope effect[33]. These isotope effects were determined from product analysis and a kinetic investigation of the former compound gave $k_1(H) = 2.95 \times 10^{-5}$ and $k_1(D) = 1.65 \times 10^{-5}$ in nitromethane containing 10.8 $M$ HNO$_3$ and 0.47 $M$ H$_2$SO$_4$, which yields $k_H : k_D = 1.8$. Presumably the greater steric hindrance of the substituted compounds is sufficient to cause $k'_{-1}$ to become sufficiently fast relative to $k'_2$ for the isotope effect to become measurable. In all other cases examined however, no primary kinetic isotope effect has been observed[34].

The outstanding problem is to decide how much, if any, association exists between NO$_2^+$ and X$^-$ in the generally rate-determining step of the reaction. Kinetic studies tend to indicate the presence of different electrophiles under different conditions whereas the derived partial rate factors are closely similar and therefore indicate one electrophile common to most, if not all, nitrating agents. The more electron-attracting is X$^-$, the more easily is NO$_2^+$ displaced from it and hence a reactivity sequence should be

$$NO_2^+ > NO_2OH_2^+ > NO_2NO_3 > NO_2OAc > NO_2OH$$

and this is observed for nitrating mixtures thought to contain these species[35].

### 3.1.1 Nitration by nitric acid in aqueous mineral acid solution

Possible nitrating entities in these media are the nitronium ion (the concentration of which can be measured spectroscopically), molecular nitric acid or the nitricacidium ion. Nitric acid is not the nitrating species because dramatic changes in the reaction rate are observed in the nitration of 2-phenylethylsulphonic acid in aqueous nitric acid on addition of perchloric acid (Table 1). Also, when nitric

TABLE 1

NITRATION OF Ph·CH$_2$·CH$_2$·SO$_3$H IN AQUEOUS HClO$_4$–HNO$_3$ AT AN HNO$_3$ : H$_2$O RATIO
OF 18 : 82 MOLE % AT 30.1 °C

| HClO$_4$ (Mole %) | 7.8 | 8.3 | 9.6 | 10.0 | 10.3 | 10.6 | 10.7 | 10.9 |
|---|---|---|---|---|---|---|---|---|
| $10^4 k_1$ | 1.16 | 2.68 | 7.48 | 11.3 | 14.5 | 21.7 | 27.0 | 30.0 |

acid is added to aqueous solutions of perchloric acid, the rate increases non-linearly with the increase in concentration of nitric acid, and the point at which a more rapid increase occurs depends upon the concentration of aqueous perchloric

TABLE 2

NITRATION OF Ph·CH$_2$·CH$_2$SO$_3$H IN AQUEOUS HClO$_4$–HNO$_3$ AT AN HClO$_4$ : H$_2$O RATIO
OF 9.2 : 90.8 MOLE % AT 30.1 °C

| HNO$_3$ (Mole %) | 15.7 | 18.4 | 20.7 | 21.9 | 22.2 | 23.2 | 25.5 |
|---|---|---|---|---|---|---|---|
| $10^4 k_1$ | 0.99 | 2.33 | 6.33 | 9.88 | 11.7 | 18.7 | 53.7 |

AS ABOVE AT A RATIO OF 12.8 : 87.2 MOLE %

| HNO$_3$ (Mole %) | 12.2 | 13.3 | 14.1 | 15.1 | 15.7 | 16.5 | 16.7 |
|---|---|---|---|---|---|---|---|
| $10^4 k_1$ | 4.32 | 6.73 | 9.33 | 15.2 | 19.7 | 28.7 | 33.0 |

acid (Table 2). In the neighbourhood of such a threshold, sodium perchlorate accelerated whilst sodium nitrate and ammonium nitrate retarded the reaction. Thus addition of approximately 2 mole % of sodium nitrate to a medium of approximate composition H$_2$O : HNO$_3$ : HClO$_4$ of 72 : 18 : 10 mole % caused about 20 % rate decrease, whilst addition of approximately 1 mole % of sodium perchlorate caused about 50 % rate increase. Activation energies for a range of media (determined over a 20° range) were in the range 17–20 with log A being 9.5–11.5, but individual rate coefficients were given for nitration in one medium only[36].

Kinetic studies have also eliminated the possibility that nitricacidium ion formed *via* the equilibrium

$$HNO_3 + H_2SO_4 \overset{fast}{\rightleftharpoons} H_2NO_3^+ + HSO_4^- \tag{21}$$

is the nitration species by showing that the heterolysis represented by equilibrium (22)

$$H_2NO_3^+ \overset{slow}{\rightleftharpoons} NO_2^+ + H_2O \qquad (22)$$

occurs prior to nitration. A very reactive aromatic will react with the nitronium ion as fast as it is formed giving kinetics zeroth-order in aromatic, and this is partially observed in nitration of mesitylene- and isodurene-$\alpha$-sulphonic acids in aqueous nitric acid or aqueous nitric acid–perchloric acid[37], the reaction following a mixture of zeroth- and first-order kinetics. Hence there must be a rate-determining step prior to attack of the electrophilic on the aromatic; this cannot be protonation of nitric acid which is obviously very fast and must almost certainly, therefore, be the heterolysis step. Consistent with this was the observation that the kinetic order in aromatic in nitration of 2-phenylethylsulphonic acid decreased as the concentration of aromatic was increased. Further supporting evidence is the fact that the rate of $^{18}O$ exchange between nitric acid and water is closely similar to the rates of nitration under zeroth-order conditions[37]. Calculations show that in nitric acid of approximately 40 mole % concentration, for zeroth-order kinetics to be followed to about 90% of reaction, then the reactivity of the aromatic towards nitric acid would have to be 20,000 times that of water. The reactivities relative to water for some compounds were found to be 2-phenylethylsulphonic acid $\approx$ 2-nitro-N-methylaniline (100–250), benzylsulphonic acid (20–45), 4-nitrophenol (10–13), N-methyl-N,2,4-trinitroaniline (0.04).

### 3.1.2 Nitration by nitric acid in concentrated strong mineral acid

Nitration in strong sulphuric acid gives second-order kinetics, viz.

$$\text{Rate} = k_2[\text{ArH}][\text{HNO}_3] \qquad (23)$$

for a fairly wide range of deactivated aromatics and the measured second-order rate coefficients are in the range 0.00005–30.0 depending upon the substituent, sulphuric acid concentration (70.8–107.8 wt. %) and temperature (15–120 °C)[38,39]. Compounds studied were anthraquinone and its 1,2-benzo, 1- and 2-hydroxy, 1- and 2-chloro and 2-carboxy derivatives, dinitromesitylene, 4,6-dinitro-$m$-xylene, 2,4-dinitrotoluene, 4-nitrotoluene, 2-, 3-, and 4-chloronitrobenzenes, nitrobenzene, 2,4-dinitroanisole, 2,4-dinitrophenol, benzoic and benzenesulphonic acids, and trimethylanilinium ion (unsubstituted, and with 4-methyl and 4-chloro substituents). In the absence of added mineral acid, i.e. constant excess of nitric acid, the kinetic form reduces as expected to

$$\text{Rate} = k_1[\text{ArH}] \qquad (24)$$

as shown by kinetic studies with fluorobenzene, 4-chloronitrobenzene, and 1-nitroanthraquinone[40]. Both rate equations are consistent with a rate-determining attack of nitronium ion on the aromatic since nitric acid is completely ionised in the presence of mineral acid and appreciably so by itself[41] (about 4 %).

Measured activation energies, which are not independent of temperature nor of the acid concentration, vary between 13.3 and 24.2, show a minimum at the acid concentration giving the maximum rate and these fairly low energies for such unreactive substrates are consistent with a highly reactive electrophile.

The rates increase up to a maximum at about 90 wt. % sulphuric acid (this point varies slightly according to the aromatic reactivity) and the increase with increasing acid concentration is consistent with the increase in the concentration of nitronium ions. The occurrence of a maximum indicates an opposing factor and is thought[42] to be partly due to protonation of the aromatic (most of the measured compounds contain the group >X=0) but since it also occurs for $PhNMe_3^+$, medium effects must be involved, $i.e.$ the activities of the species present varies, whilst the concentrations remain the same. The kinetic equation for reaction of nitronium ion with an aromatic is

$$\text{Rate} = k[\text{NO}_2^+][\text{ArH}](f_{\text{NO}_2^+} f_{\text{ArH}}/f^\ddagger) \tag{25}$$

where $f^\ddagger$ is the activity coefficient of the transition state. It has been argued[40, 43] that the ratio $f_{\text{NO}_2^+}/f^\ddagger$ should be minimum in anhydrous sulphuric acid, so that the decrease in rate towards oleum then follows. Further, evidence that $f_{\text{ArH}}$ is greater in aqueous media[44] has been confirmed by measurements on some nitro compounds which show that the decrease in the second-order rate coefficient $(k_2)$ towards oleum almost exactly parallels the decrease in $f_{\text{ArH}}$, strongly suggesting that this latter effect is responsible for the rate decrease[39]. It should be noted that in the region > 100 wt. % sulphuric acid $H_2SO_4$ ($i.e.$ free $SO_3$ present) the behaviour of $PhNMe_3^+$ and nitro compounds is different (the former shows a minimum in rate at ~ 100 wt. % sulphuric acid) and this points to protonation being a factor in affecting the rates of the latter.

Second-order rate coefficients have been determined for the nitration of more reactive aromatics by nitric acid in a range of sulphuric acid media well below the region in which a maximum of rate is observed[45-47]. The range of rates (quoted as $10^3k_2$) obtained at 25 °C at the acid concentrations (wt. %, in parenthesis) by Deno and Stein[45] were anisole, 30–0.0004 (62.8–42.3); benzene, 26–0.0004 (68.0–51.1); fluorobenzene, 13–0.0004 (70.0–54.0); chlorobenzene, 47–0.0007 (72.0–57.0); bromobenzene, 45–0.0007 (72.0–57.0); benzonitrile, 9–0.1 (85.0–80.0), and by Tillett[46], 3-nitrotoluene, 10,000–0.23 (87.3–72.3). It should be noted, however, that all these values with the exception of the last are in error since they are, in fact, (pseudo) first-order coefficients; the true values are up to three times as large (see ref. 47). Tillett found that 2- and 4-nitrotoluenes reacted 2.23 and

1.14 times faster than the 3-compound in 97 wt. % sulphuric acid, which agreed with a previous determination with 82.5 wt. % acid by Westheimer and Kharasch[38], and also that nitrobenzene is $6 \times 10^{-8}$ times less reactive than benzene, leading to the partial rate factors: $f_o^{NO_2} 1.22 \times 10^{-8}$; $f_m^{NO_2} 16.5 \times 10^{-8}$; $f_p^{NO_2} 0.5 \times 10^{-8}$. However, this assumes an identical dependence of rate upon the acid composition for both compounds, which is now known to be incorrect[47], and also uses the incorrect rate coefficient for benzene.

It has been pointed out that since the rate coefficient of encounter of nitronium ion with aromatics is of the order of $10^9$ l.mole$^{-1}$.sec$^{-1}$, when allowance is made for the concentration of nitronium ions in these weaker sulphuric acid media, the calculated rate coefficient comes out to be close to some of the above observed rate coefficients[47]. Thus in 68 wt. % sulphuric acid the encounter rate coefficient is $6 \times 10^8$ l.mole$^{-1}$.sec$^{-1}$ and the stoichiometric rate coefficient for nitration of benzene is $0.042^{47}$ (cf. 45). Since $[NO_2^+] = [HNO_3] \times 10^{-8}$ in this medium, the second-order rate coefficient for reaction of nitronium ion with benzene is $4.2 \times 10^6$, i.e. within a factor of 100 of the encounter rate. Clearly, rates derived for compounds of greater reactivity than benzene may not be meaningful as they will be encounter-controlled. This has been confirmed in a study of the nitration of a range of fairly reactive aromatics at 25 °C, the second-order coefficients $(10^3 k_2)$ in 68.3 wt. % sulphuric acid being benzene, 58.5; toluene, 1000; diphenyl, 920; p-xylene, 2,200; o-xylene, 2.200; m-xylene, 2,200; mesitylene, 2,100; naphthalene, 1,600; 2-methylnaphthalene, 1,600; 1-methoxynaphthalene, 2,000; phenol, 1,400. For some of these compounds rates were determined over a range of sulphuric acid concentrations (in parentheses) as follows: benzene, $3.7 \times 10^7$–1.27 (82.2–63.2); toluene, $1.2 \times 10^7$–21 (80.0–63.2); mesitylene, $1.8 \times 10^7$–1.8(80.0–56.3); naphthalene, 9,300–6.0 (70.6–60.8); and p-xylene 24 at 60.7 wt. % acid. Plots of logarithms of rates against $-(H_R + \log a_{H_2O})$ were curves which converged towards higher acidity so that it is clear that relative reactivities determined by overlap techniques cannot be satisfactory. Determination of the Arrhenius parameters for some of the compounds in 67.1 wt. % sulphuric acid gave the following results: benzene, $E_a = 18.0$, $\log A = 11.4$ (25–44 °C), $E_a = 13.9$, $\log A = 9.5$ (73.2 wt. % acid, 2–25°); mesitylene, $E_a = 18.0$, $\log A = 10.1$ (25–53°); naphthalene, $E_a = 15.5$, $\log A = 11.0$ (25–44°). It follows that if the reaction is encounter controlled, the values of the activation energy must be almost entirely attributed to the high endothermicity of the dissociation of nitric acid to nitronium ion in these media.

The above difference in reactivity of benzene and toluene is close to that observed in a less viscous system so that the results for compounds of the reactivity of benzene (and perhaps toluene) or less in these media are meaningful, but not for compounds which are normally more reactive. This conclusion was confirmed for nitration in nitric acid–perchloric acid of a range of compounds which normally are fairly reactive. The derived second-order rate coefficients $(10k_2)$ for nitration in 61.05 wt. % perchloric acid at 25 °C were: benzene, 0.83; toluene, 16; *p*-xylene, 70; mesitylene, 65; naphthalene, 22; 2-methylnaphthalene 47; 1-methoxynaphthalene, 73; 1-naphthol, 70; phenol, 26; *m*-cresol, 49; thiophen, 43, and this shows the similarity of rate for the latter compounds, contrary to the usual expectation. Rate coefficients $(10 k_2)$ were determined for some compounds in a range of perchloric acid media (concentration in parentheses) as follows: benzene, 39–0.0138 (64.4–57.1); mesitylene 65–1.6 (61.1–57.1), naphthalene, 22–0.38 (61.1–57.1); phenol, 26–0.63 (61.1–57.1). Interestingly, the ratios of the observed coefficients for mesitylene and benzene were quantitatively related to the viscosities of the solutions as follows: 80 in 61.1 wt. % $HClO_4$ ($\eta = 0.03$ P), 40 in 68.3 wt. % $H_2SO_4$ ($\eta = 0.09$), 6 in 80.0 wt. % $H_2SO_4$ ($\eta = 0.18$). In view of the very large differences in reactivity observed in nitration of some reactive aromatics by nitric acid in acetic anhydride, it was argued that these differences cannot be accounted for by viscosity differences alone and indicate that the electrophile is not the nitronium ion in this medium.

The significance of encounter control has been emphasised by Ridd *et al.*[48, 49], who have investigated the rates of nitration of a range of compounds containing nitrogen which can be protonated. For the heterocyclics, quinoline, pyrazoline, and imidazole, second-order kinetics were established and rate coefficients determined over a range of sulphuric acid concentrations as given in Table 3. For each compound the rate passed through a maximum at *ca.* 90 wt. % sulphuric acid and the form of each curve was very similar in the common acid range. Furthermore, for quinoline the rate increased again in very strong acid and this would not be expected for reaction of the neutral species. Hence reaction for all three compounds was assumed to take place in the protonated species as indicated also by the large negative entropies of activation. The same conclusion was reached by considering the encounter rates for the neutral species which would be about $10^{-6}$ for quinoline, $10^{-8}$ for imidazole and $10^{-3.5}$ for pyrazole. For the first two compounds this is much less than the observed rates (extrapolated to 25 °C), and for pyrazole it is slightly higher. However, when allowance is made for the fact that the activation energy of the reaction will not be zero, it becomes clear that reaction cannot occur with the neutral species. From this work, partial rate factors for the 5 and 8 positions of the quinolinium ion $(7.5 \times 10^{-8})$, the 4 position of the pyrazolinium ion $(2.1 \times 10^{-10})$ and for the 4,5 position of the imidazolinium ion $(3.0 \times 10^{-9})$ were determined[49].

The conclusion that the nitration of quinoline in sulphuric acid takes place *via* the conjugate acid has been confirmed by Moodie *et al.*[50], who measured the rates of nitration of a wide range of heterocyclic compounds in nitric acid–sulphuric acid mixtures at a range of temperatures. A summary of the second-order rate coefficients and Arrhenius parameters is given in Table 4. From an analysis of the shapes of the plots of log $k_2$ *versus* sulphuric acid acidity (or some function of this), it was concluded that all of the compounds starred in Table 4

TABLE 3

VALUES OF $10^3 k_2$ FOR NITRATION OF ArH IN $x$ WT. % SULPHURIC ACID

| x | ArH | | |
| --- | --- | --- | --- |
| | Quinoline (25.27 °C) | Pyrazole (80.8 °C) | Imidazole (25 °C) |
| 77.45 | 0.01 | | |
| 83.75 | | 7.85 | |
| 83.80 | | | 0.258 |
| 86.8 | | | 1.67 |
| 87.5 | 33.5 | 5.22 | |
| 89.5 | 50.3 | | |
| 89.6 | | | 3.15 |
| 90.2 | 51.4 | | |
| 90.4 | | 6.15 | |
| 90.9 | | | 2.73 |
| 91.0 | 43.5 | | |
| 91.9 | 40.8 | | |
| 91.95 | | 5.1 | |
| 93.65 | 30.0 | | |
| 93.85 | | | 1.14 |
| 93.95 | | 4.4 | |
| 94.9 | 24.5 | | |
| 95.0 | | 3.93 | |
| 95.7 | 19.9 | | |
| 95.9 | | 2.68 | |
| 96.1 | | | 0.742 |
| 97.95 | 9.44[a] | | |
| 98.7 | | 1.32[b] | |
| 98.9 | 8.22 | | |
| 99.3 | | | 0.307 |
| 101.5 | 10.7 | | |

[a] Respective rates at 31.12 and 45.05 °C were 21.3 and 45.3 giving $E_a = 15.0$, $\log A = 9.0$ and $\Delta S^{\ddagger} = -19.3$.
[b] Respective rates at 54.9 and 100.0 °C were 0.187 and 4.48 giving $E_a = 17.3$, $\log A = 7.8$ and $\Delta S^{\ddagger} = -24.8$.

nitrated as the positive ion. These conclusions were in many cases reinforced by consideration of the relative reactivities of the unsubstituted molecule (*e.g.* isoquinoline) and its derivative (*e.g.* N-methyl isoquinoline); in this example, the latter compound must react as the quaternary ion and the similarity in the rate of nitration of the former compound means that it must also nitrate as the conjugate acid. In the case of cinnoline-2-oxide it appeared that the 5- and 8-nitro derivatives are obtained by nitration of the cation and the 6-nitro derivative mainly from nitration of the free base. Thus the proportion of the 6-compound formed relative to the 5- and 8-compounds decreases with increasing concentration of sulphuric acid. Likewise with quinoline-1-oxide, the 5- and 8-nitro compounds are derived from the conjugate acid and the 4-compound from free base. For pyridine-

TABLE 4

RATE COEFFICIENTS AND ARRHENIUS PARAMETERS FOR NITRATION OF
HETEROCYCLICS IN NITRIC ACID–SULPHURIC ACID[50]

| Compound | $H_2SO_4$ (Wt. %) | Temp. (°C) | $10^6 k_2$ | $E_a$ | log A |
|---|---|---|---|---|---|
| Quinoline* | 81.3 | 25 | 167 | 16.4 | 8.3 |
| N-Methylquinolinium ion* | 81.3 | 25 | 44 | 18.0 | 8.8 |
| Isoquinoline* | 81.3 | 25 | 2,320 | 14.1 | 7.7 |
| N-Methylisoquinolinium ion* | 81.3 | 25 | 2,580 | 14.9 | 8.4 |
| Pyridine-N-oxide | 87.9 | 25 | 0.0366 | 23.5 | 9.8 |
| N-Methoxypyridinium ion* | 87.9 | 125 | <1 | | |
| 2,6-Lutidine-N-oxide | 78.2 | 25 | 0.0231 | 26.5 | 11.8 |
| | 81.4 | 25 | 0.148 | 23.9 | 10.7 |
| | 84.3 | 25 | 0.607 | 23.2 | 10.8 |
| | 87.7 | 25 | 1.01 | 22.9 | 10.8 |
| | 92.5 | 25 | 1.41 | 22.7 | 10.8 |
| | 97.8 | 25 | 0.143 | 24.6 | 11.2 |
| Isoquinoline-N-oxide* | 83.1 | 25 | 3,300 | 14.7 | 8.3 |
| N-Methoxyisoquinolinium ion* | 83.1 | 25 | 1,074 | 15.5 | 8.4 |
| Cinnoline* | 76.1 | 80 | 11.8 | 24.4 | 10.15 |
| | 81.1 | 80 | 321 | 20.6 | 9.2 |
| 2(N)-Methylcinnolinium ion* | 81.2 | 80 | 735 | 19.1 | 8.7 |
| 2(N)-Methoxycinnolinium ion* | 81.5 | 80 | 974 | | |
| Cinnoline-2(N)-oxide* | 81.2 | 80 | 222 | | |
| N-Methoxyquinolinium ion* | 82.0 | 25 | 15.9 | | |
| | 72.5 | 65 | 1.1 | | |
| Quinoline-N-oxide* | 55.6 | 25 | 0.0174 | 29.0 | 13.5 |
| | 60.1 | 25 | 0.100 | 26.6 | 12.5 |
| | 66.9 | 25 | 0.603 | 25.0 | 12.1 |
| | 82.0 | 25 | 236 | 19.7 | 10.8 |
| Quinoline-N- (4-nitration only) | 82.0 | 0 | 4.9 | 22.3 | 12.5 |
| | | 25 | 158 | | |
| Quinoline-N- (5+8-nitration only) | 82.0 | 0 | 6.2 | 16.2 | 7.8 |
| | | 25 | 78.5 | | |

* See text.

and 2,6-lutidine-N-oxides, the mechanism of nitration is uncertain for although the protonated species seem not be to involved, the experimentally determined log A factors differ from those which can be calculated, by 6–8 units; the problem here is thus similar to that noted below for nitration of substituted anilines (see Table 8, p. 26). The log A values also fail to show that the difference in reactivity of quinoline and its N-methyl derivative arise from the steric hindrance to 8-substitution which was proposed as the reason for this difference. Partial rate factors were derived from this work as $0.36 \times 10^{-6}$ for the 5- and 8-positions of quinoline, $9.0 \times 10^{-6}$ and $1.0 \times 10^{-6}$ for the 5- and 8-positions of isoquinoline respectively; the former values seriously disagree with those obtained by Ridd et al.[48,49] given above (p. 15).

The nitration of some heterocyclic compounds by nitric acid in sulphuric acid has been studied by Katritzky et al.[50a-d] and the results are exactly as expected in that electron-supplying substituents in the ring favour reaction on the conjugate acid whereas electron-withdrawing substituents produce reaction on the free base. Rate coefficients and the kinetic parameters for nitration of pyridine derivatives (and some benzene analogues)[50a] are given in Table 4a.

A comparison of the second-order rate coefficients for nitration of 2,4,6-trimethylpyridine and 1,2,4,6-tetramethylpyridinium ion (both at the 3-position) shows similarity of profile in the common acidity region and a rapidly increasing rate with acidity for the trimethyl compound at acidities below 90 wt. % (where the usual maximum is obtained). These two pieces of evidence show reaction to occur on the conjugate acid as also indicated by the large negative entropy of activation. Surprisingly, the tetramethyl compound is *less* reactive than the trimethyl compound so maybe this is an example of steric hindrance to solvation. Calculation of the encounter rate also showed that reaction on the free base was unlikely.

The data for the 2,4- and 2,6-dimethoxypyridines (nitration at the 3 position) indicates reaction on the conjugate acid both by the rate–acidity profile for the former compound and by the large negative entropies for both. For dimethoxy-benzene, and 2,4-dimethoxynitrobenzene plots of $\log k_2$ *versus* $-[H_R + \log a_{H_2O}]$ gave near unit slopes as did the literature data[45] for anisole, and this follows from the nitric acid molecule existing in 60–84 wt. % sulphuric acid as the mono-hydrate[50]. For 2,6-dimethoxy-3-nitropyridine, reaction on the free base might be expected in view of the $pK_a$ value of $-3.86$, and comparison of the rate coefficients at $> 90$ wt. % with those of 2,6-dimethoxypyridine show the former to decrease much more rapidly with increasing acidity which is the expected behaviour for reaction of the free base. Correction of the observed rate coefficients to take account of the true concentration of free base present, by means of the expression $H_0 = pK_a + \log ([B]/[BH^+])$, gave the rates $k_2$ (calc.) in the Table; these then showed a rate acidity profile similar to that of the 2,6-dimethoxy compound. The entropy of activation derived from the calculated rate coefficients also indicated reaction on the free base since it was positive. However, in view of the calculation involved this is perhaps a foregone conclusion and cannot be regarded as positive evidence; the observed rate coefficients gave an entropy of activation of $-10.9$.

Nitration of 3,5-dimethoxypyridine and 3,5-dimethoxy-2-nitropyridine takes place at the 6 position as expected. For the former compound the rate–acidity profile shows reaction to clearly occur on the conjugate acid, whereas the latter only has a small increase in rate coefficient between $H_0 - 7.8$ and $-9.0$, suggesting reaction on the free base. Correction of the observed rate coefficients as before gives a rate–acidity profile similar to that of 3,5-dimethoxypyridine showing that the reaction takes place upon the free base of the nitro compound. The corrected

## TABLE 4a

RATE COEFFICIENTS AND KINETIC PARAMETERS FOR REACTION OF ArH IN $HNO_3$–$H_2SO_4$[50a]

*ArH = 2,4,6-Trimethylpyridine(P) and 1,2,4,6-Tetramethylpyridinium ion(N) at 101.3 °C*

| $H_2SO_4(Wt.\%)$ | 86.03 | 87.04 | 90.70 | 91.20 | 92.27 | 93.51 | 95.15 | 98.05 |
|---|---|---|---|---|---|---|---|---|
| $10^4 k_2(P)$ | 1.50 | 4.62 | 9.89[a] | 11.3 | 10.8 | 9.62 | 6.20 | 3.01[b] |

| $H_2SO_4(Wt.\%)$ | 100.0 | 103.0 | 107.6 | 124.6 | 126.2 | 137.7 | 147.4 |
|---|---|---|---|---|---|---|---|
| $10^4 k_2(P)$ | 1.46 | 1.57 | | 1.59 | | 1.30 | |
| $10^4 k_2(N)$ | | | 1.43 | | 1.45 | | 1.24 |

[a, b] $E_a = 17.7$ and 20.3, $\log A = 7.4$ and 8.4, $\Delta S^\ddagger = -26.1$ and $-20.5$, respectively, from data at other temperatures.

*ArH = 1,3-Dimethoxybenzene (B) and 2,6-Dimethoxypyridine (P) at 22.7 °C*

| $H_2SO_4(Wt.\%)$ | 57.67 | 58.27 | 58.93 | 61.83 | 62.59 | 63.08 | 64.24 |
|---|---|---|---|---|---|---|---|
| $10^3 k_2(B)$ | 1.49 | 1.99 | 3.74 | 19.1 | 32.7 | 35.2 | 69.0 |

| $H_2SO_4(Wt.\%)$ | 86.12 | 89.80 | 90.4 | 94.19 | 98.48 |
|---|---|---|---|---|---|
| $10^3 k_2(P)$ | 6.37 | 34.2 | 33.4[c] | 26.8 | 17.4 |

[c] This result at 22.5 °C. Rates at other temperatures give $E_a = 13.1$, $\log A = 8.95$ and $\Delta S^\ddagger = -18.2$. For the 2,4-dimethoxy compound these parameters were 15.5, 9.5, and $-15.8$, respectively, in 89.83 wt. % acid, the rate coefficient at 25.0 °C being $1.30 \times 10^{-3}$.

*ArH = 2,4-Dimethoxynitrobenzene (B) and 2,6-Dimethoxy-3-nitropyridine (P) at 31 °C*

| $H_2SO_4(Wt.\%)$ | 67.47 | 69.13 | 69.62 | 71.02 | 72.83 |
|---|---|---|---|---|---|
| $10^3 k_2(B)$ | 19.7 | 57.1 | 119 | 229 | 985 |

| $H_2SO_4(Wt.\%)$ | 86.02 | 89.83 | 94.07 | 96.01 | 98.34 |
|---|---|---|---|---|---|
| $10^5 k_2(P)$ | 77.7 | 110 | 24.0 | 9.14 | 1.83 |
| $k_2(calc.)$[e] | 8.51 | 41.4[d] | 47.8 | 36.4 | 19.6 |

[d] $E_a = 19.7$, $\log A = 15.9$, and $\Delta S^\ddagger = +12.4$ from data at other temperatures.
[e] See text.

*ArH = 3,5-Dimethoxypyridine(P) at 25 °C and 3,5-Dimethoxy-2-nitropyridine at 51 °C*

| $-H_0$ | 7.79 | 8.06 | 8.45 | 8.88[f] | 9.36 | 9.90 | 10.36 |
|---|---|---|---|---|---|---|---|
| $10^3 k_2(P)$ | 5.55 | 48.3 | 202 | 1,050 | 719 | 565 | 390 |

| $-H_0$ | 7.87 | 7.97 | 8.28 | 8.82 | 8.91[g] | 9.45 | 9.78 | 9.98 |
|---|---|---|---|---|---|---|---|---|
| $10^5 k_2(N)$ | 6.51 | 8.67 | 17.6 | 26.6 | 24.2 | 21.2 | 7.93 | 5.13 |
| $10^5 k_2(N)(calc.)$[h] | 4.72 | 7.63 | 30.2 | 139 | 153 | 423 | 316 | 310 |

[f] Other data using acid of approximately this concentration gave $E_a = 12.4$, $\log A = 8.8$, and $\Delta S^\ddagger = -20.0$.
[g] As[f], $E_a = 22.7$, and from the calculated data $\log A = 17.5$ and $\Delta S^\ddagger = +15.9$.
[h] See text.

*ArH = 2-Phenylpyridine at 25 °C*

| $H_2SO_4(Wt.\%)$ | 77.1 | 78.0 | 78.6 | 79.2 | 80.9 |
|---|---|---|---|---|---|
| $k_2$ | 0.068[i] | 0.27 | 0.47 | 0.60 | 2.7 |

[i] Wrongly reported as 0.68.

entropy of activation also comes out positive and indicates (as before) that reaction is on the free base.

For 2,6-dichloropyridine, rates (for nitration at the 3 position) were measured in media containing high proportions of nitric acid so that examination of rate *versus* sulphuric acid acidity plots was considered to be insufficiently rigorous. In a medium of 81.5 % sulphuric acid, 11.6 % nitric acid and 5.9 % water, rate coefficients ($10 k_2$) at 74.7, 86.0, 95.0 and 104.4 °C were 8.6, 32.3, 80.0, and 194 respectively, giving $E_a = 27.1$ and $\Delta S^{\ddagger} \sim +15$, so that reaction on the free base is indicated. For 1,3-dichlorobenzene the corresponding parameters were 17.5 and $-7.5$ in a medium containing 98.27 wt. % sulphuric acid. The possibility existed, however, that this kinetic data was for reaction of 2,4-dichlorobenzenesulphonic acid, and the difference in the entropies suggests a marked difference in solvation for the two molecules, and this would be expected in view of the large steric hindrance in the sulphonic acid.

From this work the deactivation of pyridine to benzene was estimated as about $10^7$. The partial rate factors for nitration of the 3 position of pyridine and the corresponding pyridinium ion were $10^{1.7-2.5}$ and $10^{20}$ respectively. 2-Phenylpyridine was evaluated as $4.9 \times 10^{-5}$ times less reactive than benzene.

A kinetic study of the electrophilic substitution of pyridine-N-oxides has also been carried out[5b, c]. Rate–acidity dependencies were unfortunately given in graphical form only and the rate parameters (determined mostly over a 30 °C range) are given in Table 4b. There is considerable confusion in Tables 3 and 5 of the original paper, where the rate coefficients are labelled as referring to the free base. In fact the rate coefficients for the first three substituted compounds in

TABLE 4b

RATE PARAMETERS FOR REACTION OF PYRIDINE-N-OXIDES WITH $HNO_3$–$H_2SO_4$

| Compound | Position of substitution | $H_2SO_4$ (Wt. %) | $E_a$ | $\log A^b$ | $\Delta S^{\ddagger}$ | $\Delta S^{\ddagger}$ (calc.)$^b$ |
|---|---|---|---|---|---|---|
| Pyridine-N-oxide$^a$ | 4 | 87.9 | 23.5 | | $-15.5$ | |
| 3,5-Me$_2$-Pyridine-N-oxide | 4 | 91.8 | 25.3 | 19.0 | $-15.1$ | $+26.3$ |
| 2,6-Cl$_2$-Pyridine-N-oxide | 4 | 82.4 | 26.9 | 13.5 | $-12.5$ | $+1.5$ |
| 2,6-Cl$_2$-Pyridine-N-oxide | 4 | 87.9 | 25.7 | 14.2 | $-9.6$ | $+4.6$ |
| 2,6-Cl$_2$-Pyridine-N-oxide | 4 | 94.5 | 23.5 | 13.8 | $-14.2$ | $+2.7$ |
| 3,5-Cl$_2$-Pyridine-N-oxide | 4 | 90.8 | 27.85 | 14.0 | $-7.2$ | $+3.4^c$ |
| 2,4,6-(MeO)$_3$-Pyridine-N-oxide | 3 | 81.0 | 16.9 | | $-17.2$ | |
| 2,6-(MeO)$_2$-Pyridine-N-oxide | 3 | 84.8 | 14.5 | | $-26.1$ | |
| 4-MeO-2,6-Me$_2$-Pyridine-N-oxide | 3 | 90.3 | 16.6 | | $-18.5$ | |
| 3,5-(MeO)$_2$-Pyridine-N-oxide | 2 | 82.3 | 20.3 | | $-12.3$ | |
| 3,5-(MeO)$_2$-2-NO$_2$-Pyridine-N-oxide | 2 | 82.0 | 14.55 | | $-29.0$ | |
| 1,3,5-(MeO)$_3$-Pyridinium perchlorate | 2 | 82.3 | 23.8 | | $-13.0$ | |

$^a$ See Table 4.
$^b$ Calculated for reaction of the free base.
$^c$ Misreported as $+15.1$.

the Tables (the 3,5-dimethyl, 2,6- and 3,5-dichloro compounds) seem to be the experimentally observed values corrected for the true concentration of free base present, since the rate–acidity profiles indicated the free bases to be the reacting species. The remaining rates in the Tables refer to reaction of the conjugate acids which are the reacting forms of the compounds indicated. The activation energies given in Table 4b appear to be those experimentally observed, whereas the log $A$ values are corrected for the concentration of the reacting species present. The entropies of activation are those obtained before and after correction of the rate for the base–conjugate acid equilibrium as before; the latter parameters came out to be positive as required for a reaction on the free base, and by contrast to the experimentally observed entropies of activation which are negative; these conclusions were supported by calculations of the encounter rate. For nitration at the 2 and 3 positions, the rate–acidity profiles, and entropies of activation show reaction to occur on the conjugate acid.

The partial rate factor for nitration of pyridine-N-oxide in the 4 position was estimated as $4 \times 10^{-6}$ which is, therefore, close to that found for the 3 position of pyridine, and 2-phenylpyridine-N-oxide was evaluated as $2 \times 10^{-4}$ times less reactive than benzene from rate measurements in 74.7–78.6 wt. % acid at 25 °C.

Kinetic studies of the nitration of pyridones have also been carried out[50d], and some representative kinetic data are given in Table 4c. For the media of high acidity, the rate–acidity profiles for 4-pyridone and 4-methoxypyridine are similar and similar also to that of 2,4,6-trimethylpyridinium ion thereby indicating reaction on the conjugate acid of the former compound as shown also by the large negative entropy of activation. However, in addition, the rate coefficients for the former two compounds were similar, and similar also to that of 1-methyl-4-pyridone ($10^5 k_2$ at 109 °C in 93.8 wt. % acid = 636) so that it is clear that 4-pyridone reacts as the diprotonated form, being protonated at both nitrogen and oxygen. The rate–acidity profile for nitration of 2-methoxy-3-methylpyridine is also similar to these compounds and suggests reaction on the conjugate acid, as expected for a base with these electron-supplying substituents.

By contrast, the rate–acidity profiles for nitration of 3-nitro-4-pyridone, 3- and 5-methyl-2-pyridone and 1,5-dimethyl-2-pyridone resemble each other and differ from the above-indicated reaction upon the free base, and correction of the observed rates to allow for the concentration of free base actually present gave rate–acidity profiles of the expected form; the corrected entropies of activation then turned out to be positive. Furthermore, if the logarithms of the corrected rate coefficients obtained in media of low acidity were plotted against $H_R + \log a_{H_2O}$, then slopes of near unity were obtained (see above, p. 18), but not otherwise. A similar result was obtained from the nitration data for 4-pyridone in media of low acidity suggesting that here it reacts as the free base. A further test which was applied was to calculate the concentration of nitronium ions in the various media and to correct the observed rate coefficients for this the logarithms of these coeffi-

TABLE 4c

RATE COEFFICIENTS AND PARAMETERS FOR REACTION OF PYRIDONES WITH HNO$_3$–H$_2$SO$_4$

$ArH = 4$-Pyridone[a]

| $H_2SO_4(Wt. \%)$ | 85.18 | 86.53 | 88.22 | 89.94 | 92.80 | 94.91[b] | 96.62 | 98.58 |
|---|---|---|---|---|---|---|---|---|
| $10^5k_2(86\ °C)$ | 7.78 | 19.1 | 43.6 | 70.8 | 65.3 | 38.9 | 26.5 | 15.2 |

| $H_2SO_4(Wt. \%)$ | 85.21 | 87.00 | 87.87 | 89.84 | 91.66 | 94.56 | 96.36 | 98.58 |
|---|---|---|---|---|---|---|---|---|
| $10^5k_2\ (110\ °C)$ | 60.0 | 120 | 189 | 364 | 444 | 242 | 165 | 73.5 |

| $H_2SO_4(Wt. \%)$ | 79.21 | 80.54 | 82.05 | 83.09 | 84.34 | 85.43 |
|---|---|---|---|---|---|---|
| $10^5k_2\ (133.5\ °C)$ | 20.5 | 33.6 | 75.0 | 97.5 | 171 | 296 |

| $H_2SO_4(Wt. \%)$ | 72.76 | 74.58 | 76.44 | 78.47 | 80.55 |
|---|---|---|---|---|---|
| $10^5k_2\ (157.5\ °C)$ | 11.2 | 20.5 | 39.7 | 73.7 | 180 |

[a] The temperature referred to as 110 °C in Table 3 of ref. 50d should in fact read 86 °C.
[b] In this medium (approx.) $E_a = 19.3$ and $\Delta S^{\ddagger} = -20.3$.

$ArH = 3$-Nitro-4-pyridone

| $H_2SO_4(Wt. \%)$ | 83.01 | 85.72 | 89.95 | 91.80 | 93.69 | 93.89 | 95.64 | 97.70 |
|---|---|---|---|---|---|---|---|---|
| $10^5k_2\ (157\ °C)$ | 10.8 | 26.4 | 46.3 | 47.0 | 32.8 | 27.6 | 13.4 | 5.60 |

$ArH = 4$-Methoxypyridine[c]

| $H_2SO_4(Wt. \%)$ | 85.86 | 87.78 | 90.10 | 91.31 | 93.06 | 95.49 | 96.28 | 97.79 |
|---|---|---|---|---|---|---|---|---|
| $10^5k_2\ (100\ °C)$ | 15.5 | 52.1 | 115 | 115 | 108 | 79.6 | 69.8 | 48.4 |

[c] Wrongly referred to as 4-methoxypyridone in Table 5 of ref. 50d.

cients being plotted against acidity. When this is done, compounds known to act *via* a species whose concentration does not significantly change with acidity gave a horizontal line, whereas 5- and 3-methyl-2-pyridones and also 4-pyridone in media of low acidity gave straight lines with a negative slope, again indicating reaction on the free base the concentration of which decreases with increasing acidity; it is probably unnecessary to apply this technique, however, for it seems superfluous to the evidence of the slope of the original rate–acidity profile. Analogies to the change in mechanism for nitration of 4-pyridone from reaction on the free base at low acidity to reaction on the conjugate acid at high acidity

Table 4c (continued)

*ArH = 3-Methyl-2-pyridone*

| $H_2SO_4(Wt.\%)$ | 78.28 | 81.83 | 84.11 | 86.61 | 88.19 | 89.07 | 90.07 | 93.91[d] | 97.80 |
|---|---|---|---|---|---|---|---|---|---|
| $10^3k_2$ (31 °C) | 18.4 | 59.7 | 135 | 502 | 504 | 521 | 335 | 569 | 6.45 |

| $H_2SO_4(Wt.\%)$ | 75.00 | 76.59 | 77.00 | 78.14 | 79.45 | 80.92 |
|---|---|---|---|---|---|---|
| $10^3k_2$ (34.5 °C) | 6.08 | 10.7 | 13.2 | 17.8 | 23.6 | 46.5 |

| $H_2SO_4(Wt.\%)$ | 61.79 | 65.53 | 67.66 | 69.94 | 71.78 | 73.01 | 75.24 | 76.67 |
|---|---|---|---|---|---|---|---|---|
| $10^4k_2$ (44.2 °C) | 0.94 | 4.10 | 7.85 | 23.8 | 57.4 | 75.3 | 197 | 289 |

[d] In this medium (approx.) $E_a = 14.1$, $\Delta S^{\ddagger}$ (free base) = 7.8, $\Delta S^{\ddagger}$ (conjugate acid) = −14.5.

*ArH = 5-Methyl-2-pyridone*

| $H_2SO_4(Wt.\%)$ | 84.12 | 87.35 | 88.20 | 89.14 | 89.95 | 90.77 | 92.76 | 94.81[e] | 96.77 |
|---|---|---|---|---|---|---|---|---|---|
| $10^3k_2$ (31.5 °C) | 33.4 | 74.2 | 107 | 103 | 117 | 70.0 | 28.8 | 16.2 | 6.66 |

| $H_2SO_4(Wt.\%)$ | 74.11 | 75.32 | 76.81 | 77.49 | 78.13 | 79.17 | 80.96 |
|---|---|---|---|---|---|---|---|
| $10^4\,k_2$ (35 °C) | 8.30 | 11.5 | 21.8 | 34.7 | 36.2 | 53.1 | 118 |

[e] In this medium (approx.) $E_a = 19.2$, $\Delta S^{\ddagger}$ (free base) = 29.1, $\Delta S^{\ddagger}$ (conjugate acid) = −13.6.

*ArH = 1,5-Dimethyl-2-pyridone*

| $H_2SO_4(Wt.\%)$ | 76.00 | 82.51 | 86.02 | 89.08 | 90.70 | 92.49 | 94.08 | 95.79 | 97.86 |
|---|---|---|---|---|---|---|---|---|---|
| $10^3k_2$ (29 °C) | 6.02 | 63.2 | 137 | 157 | 154 | 106 | 49.7 | 21.3 | 6.21 |

| $H_2SO_4(Wt.\%)$ | 71.42 | 74.42 | 75.92 | 77.53 | 78.39 |
|---|---|---|---|---|---|
| $10^3k_2$ (39.5 °C) | 0.836 | 3.05 | 5.89 | 12.9 | 17.0 |

*ArH = 3-Methyl-2-methoxypyridine*

| $H_2SO_4(Wt.\%)$ | 85.63 | 87.31 | 88.74 | 90.57 | 92.16[f] | 94.24 | 95.84 |
|---|---|---|---|---|---|---|---|
| $10^5k_2$ (25.9 °C) | 3.79 | 11.2 | 16.5 | 10.4 | 15.6 | 11.3 | 6.85 |

[f] In this medium (approx.) $E_a = 15.5$ and $\Delta S^{\ddagger} = -22.6$.

*References pp. 388–406*

are commonly observed in aromatic hydrogen exchange (see pp. 226–238). From this work the deactivation of a $NH^+$ group was estimated as $10^{13}$.

Ridd *et al.*[48] have studied the nitration of aniline by nitric acid in 82.0–100.0 wt. % sulphuric acid, and the second-order rate coefficients were separated (from product analysis) into those appropriate for *ortho*, *meta*, and *para* substitution (Table 5).

TABLE 5

DEPENDENCE OF ANILINIUM ION NITRATION RATE AT 25 °C ON $H_2SO_4$ CONCENTRA TION[48]

| $H_2SO_4$ (Wt. %) | $10^2 k_2$ | $10^2 k_2$ | | |
|---|---|---|---|---|
| | | *ortho*[a] | *meta* | *para* |
| 82.0 | 1.2 | 0.031 | 0.22 | 0.74 |
| 84.0 | 7.9 | 0.20 | 1.4 | 4.7 |
| 84.9 | 16 | 0.32 | 3.0 | 9.7 |
| 86.5 | 66 | 1.3 | 13 | 38 |
| 87.5 | 110 | 1.6 | 23 | 60 |
| 88.5 | 200 | 3.0 | 45.5 | 100 |
| 89.4 | 261 | | 58.7 | 136 |
| 89.5 | 260 | 3.9 | 63 | 130 |
| 92.4 | 201 | | 53.3 | 94.5 |
| 94.8 | 147 | | 41.9 | 63.2 |
| 96.4 | 108 | | 31.3 | 45.4 |
| 98.0 | 66.8 | | 20.7 | 25.4 |
| 100.0 | 65.5 | | 21.0 | 23.6 |

[a] The percentage of *ortho* substitution was not determined under all conditions.

Firstly, there is a very similar dependence of the rate upon acidity for nitration of all three positions [the slopes of log rate *versus* wt. % $H_2SO_4$ plots below 87 % acid are 0.33 (*o*), 0.38 (*m*), and 0.36 (*p*)], and since it is argued that the substantial *meta* nitration will occur *via* the anilinium ion, then *ortho* and *para* nitration must occur *via* the same species. Furthermore, the rates show signs of passing through a minimum at *ca.* 100 wt. % acid and this is inconsistent with reaction on a neutral species. The dependence of rate upon acidity is also very similar to that obtained by Gillespie and Norton[38] for nitration of the phenyltrimethylammonium ion, which again argues for reaction on the protonated species. Calculations of the encounter rate coefficient produce the same conclusion, for this is approximately $2.5 \times 10^8$ l.mole$^{-1}$.sec$^{-1}$. in 98 wt. % sulphuric acid. The concentration of the free base is $10^{-15}$ times that of the stoichiometric concentration of aniline so that stoichiometric rate coefficient for reaction of the free base should be $1.5 \times 10^7$, which is $10^6$ times less than observed; a similar calculation for 85 wt. % acid produces the same conclusion. Rate coefficients ($10^2 k_2$) for nitration at 25 °C of 4-chloroanilinium ion and 4-chlorophenyltrimethylammonium ion were determined as 23.1 and 0.58 (in 91.0 wt. % $H_2SO_4$), 5.6 and 0.15 (in 99.5 wt. %

$H_2SO_4$). For the former compound the chloro substituent should increase the amount of free amine in solution and hence speed up the rate so that a significant difference in the relative rates in the two acid media should be observed if reaction occurs on the free amine; the relative rates were, however, 40 and 37.

Hartshorn and Ridd[48] showed that there is a negligible solvent isotope effect on nitrating anilinium ions in sulphuric acid and deuterated sulphuric acid (*cf.* an earlier less accurate determination by Brickman and Ridd[48]). The absence of a solvent isotope effect also argues against reaction on the free base because the free base concentration would be lower by a factor of about four in the deuterium-containing medium. Consequently, the differences in the rate coefficients in Table 6

TABLE 6

RATE COEFFICIENTS FOR NITRATION OF ANILINIUM IONS IN $H_2SO_4$ AND OF THE CORRESPONDING N-DEUTERATED IONS IN $D_2SO_4$ AT 25 °C[48]

| $H_2O$ or $D_2O$ (mole %) | ArH | $10^2 k_2$ | $[k_2(m)/k_2(p)]_H / [k_2(m)/k_2(p)]_D$ |
|---|---|---|---|
| 4.8 | $PhNH_3{}^+$ | 44 | 1.36 |
| 4.8 | $PhND_3{}^+$ | 33 | |
| 4.8 | $PhNHMe_2{}^+$ | 3.2 | 1.13 |
| 4.8 | $PhNDMe_2{}^+$ | 3.0 | |
| 47.4 | $PhNH_3{}^+$ | 35 | 1.57 |
| 47.4 | $PhND_3{}^+$ | 23 | |
| 53.0 | $PhNH_3{}^+$ | 2.8 | 1.64 |
| 53.0 | $PhND_3{}^+$ | 1.1 | |

arise from the substrate only, and the difference in the substrate isotope for the methylated and unmethylated compounds implies a contribution from *each* hydrogen, so that the effect cannot be a primary one resulting from the partial transfer of one proton to a base in the rate-determining transition state. The fact that N–H hydrogen exchange in these media takes place much less readily than nitration[51] further confirms that nitration must occur on the conjugate acid of the amine. The difference in the isotope effect for *meta* and *para* nitration (calculated as 1.08 and 1.18 per N–H hydrogen atom for *meta* and *para* substitution) was considered to arise from the same cause that gives a variation of *para* : *meta* nitration rate with sulphuric acid strength, namely a differential salt $(H_3O^+HSO_4^-)$ effect, though the manner in which this operates was not clear. The ratio of *para* : *meta* substitution rate increases regularly from 100 wt. % sulphuric acid to 84.5 wt. % acid, this latter concentration being that at which the molar concentrations of water and sulphuric acid are equal and also that at which the concentration of the salt $H_3O^+HSO_4^-$ becomes relatively constant in the direction of decreasing acid strength.

Extension of this work to the kinetics of nitration of N-methyl- and N, N-dimethylaniline[48] gave the rate coefficients given in Table 7 together with those

### TABLE 7

DEPENDENCE OF NITRATION RATES AT 25 °C UPON $H_2SO_4$ CONCENTRATION[38,48]

| PhNH₂Me⁺ | | | | PhNHMe₂⁺ | | | | PhNMe₃⁺ | | |
|---|---|---|---|---|---|---|---|---|---|---|
| $H_2SO_4$ (Wt. %) | $10^2k_2$ | $10^2k_2$ (m) | $10^2k_2$ (p) | $H_2SO_4$ (Wt. %) | $10^2k_2$ | $10^2k_2$ (m) | $10^2k_2$ (p) | $H_2SO_4$ (Wt. %) | $10^2k_2$ (m) | $10^2k_2$ (p) |
| 90.85 | 74.7 | 22.8 | 29.1 | 91.1 | 13.5 | 4.99 | 3.51 | 98.0 | 0.55 | 0.14 |
| 96.2 | 31.7 | 10.7 | 10.3 | 96.2 | 5.92 | 2.25 | 1.42 | | | |
| 99.9 | 12.7 | 4.46 | 3.77 | 100 | 2.68 | 1.06 | 0.56 | | | |

calculated from the data of Gillespie and Norton[38]. The dependence of the rates upon the sulphuric acid concentration again indicated that reaction occurs on the conjugate acids. N-Methylation reduced the rate of nitration at both *meta* and *para* positions, the latter being the most affected. It was concluded that N–H hyperconjugation was not responsible since the rate coefficients $(10^2k_2)$ for nitration at 25 °C of benzylammonium ion and benzyltrimethylammonium ion, in which the N–H bonds cannot conjugate with the ring, were 158 and 2.52 (with 78.7 wt. % $H_2SO_4$), 372 and 6.0 (with 80.05 wt. % $H_2SO_4$); the constant relative rate of 62 indicated that here also the free base was not involved. The rate differences were, therefore, attributed to the difference in solvation of protonated and methylated nitrogen poles.

From the above work, partial rate factors were determined as follows (all values $\times 10^{-8}$): $f_o^{NH_3^+}$, 19; $f_m^{NH_3^+}$, 138; $f_p^{NH_3^+}$ 451 (with 82.0 wt. % $H_2SO_4$); $f_o^{NH_3^+}$, 4.3; $f_m^{NH_3^+}$, 173; $f_p^{NH_3^+}$, 213; $f_m^{NH_2Me^+}$, 57; $f_p^{NH_2Me^+}$ 49; $f_m^{NHMe_2^+}$, 12.3; $f_p^{NHMe_2^+}$, 7.1; $f_m^{NMe_3^+}$, 4.2; $f_p^{NMe_3^+}$, 1.0 (all with 98.0 wt. % $H_2SO_4$) and all of these values are subject to the possibility for error noted above (p. 14).

Hartshorn and Ridd[48] have also measured the variation of rate coefficient with acidity for nitration of 4-nitroaniline and 2-chloro-4-nitroaniline (Table 8).

### TABLE 8

DEPENDENCE OF NITRATION RATES AT 25 °C UPON THE $H_2SO_4$ CONCENTRATION[48]

| $H_2SO_4$ (Wt. %) | | | | | | | | | |
|---|---|---|---|---|---|---|---|---|---|
| 84.0 | 84.7 | 86.5 | 87.5 | 90.1 | 93.8 | 95.4 | 98.0 | 99.2 | 100 |
| $10^2k_2$ (4-nitroaniline) | | | | | | | | | |
| 0.123 | a | 0.47 | 0.785 | 1.20ᵃ | 0.505 | 0.313 | 0.104ᵃ | | 0.018 |
| $10^2k_2$ (2-chloro-4-nitroaniline) | | | | | | | | | |
| | 15.3 | 45.5 | 63.5 | 72.1 | 16.0ᵇ | 8.20 | 1.80ᵇ | 1.08 | |

ᵃ Rate coefficients determined between 24.9 and 39.95 °C gave $E_a$ and log $A$ values of 22.4, 13.8 (84.7 %), 19.2, 12.2 (90.1 %), and 18.2, 10.4 (98.0 %) respectively.
ᵇ Rate coefficients determined between 25.0 and 39.95 °C gave $E_a$ and log $A$ values of 19.6, 13.6 (93.8 %) and 19.1, 12.3 (98.0 %) respectively.

TABLE 9

RATE COEFFICIENTS FOR NITRATION AT 25 °C OF 4-NITRO- AND 2-CHLORO-4-NITRO-
ANILINE IN $H_2SO_4$ AND $D_2SO_4$ AT THE SAME COMPOSITION IN MOLE % WATER[48]

| ArH | Water (Mole %) | $10^3 k_2(H)$ | $10^3 k_2(D)$ | $k_2(H)/k_2(D)$ |
|---|---|---|---|---|
| 4-Nitroaniline | 19.4 | 2.63 | 1.30 | 2.02 |
| | 45.7 | 5.52 | 1.43 | 3.66 |
| 2-Chloro-4-nitro- | 15.0 | 36.2 | 19.0 | 1.90 |
| aniline | 30.9 | 316 | 157 | 2.00 |
| | 365 | 640 | 220 | 2.90 |
| | 38.7 | 758 | 240 | 3.16 |
| | 44.8 | 512 | 157 | 3.28 |

The log rate *versus* acid strength curve for the latter compound is of the exact form expected for reactions of the free base, whilst that of the former compound is intermediate between this form and that obtained for the nitration of aniline and phenyltrimethylammonium ion, *i.e.* compounds which react as positive species. That these compounds react mainly or entirely *via* the free base is also indicated by the comparison of the rate coefficients in Table 8 with those in Table 5, from which it can be seen that the nitro substituent here only deactivates weakly, whilst the chloro substitutent appears to activate. In addition, both compounds show a solvent isotope effect (Table 9), the rate coefficients being lower for the deuterium-containing media, as expected since the free base concentration will be lower in these.

A problem is associated with the Arrhenius parameters given in Table 8, in that the calculated stoichiometric rate coefficients for reaction of the free amines with nitronium ions are of the order of $10^8$, which is only slightly less than that for the encounter rate ($10^{8.4}$). This would mean a reaction of very low activation energy contrary to observation. To overcome this difficulty Hartshorn and Ridd[48] suggest that approach of nitronium ions to the conjugate acid of the amine facilitates proton loss to a base (perhaps by only a field effect), and this is followed by rate-determining attack of nitronium ion on the hydrogen-bonded amine-base complex. However, since the concentration of the conjugate acid is much higher than that of the free base, this seems to worsen the situation, unless this facilitation of proton loss has a very high activation energy.

Nitration by nitric acid in sulphuric acid has also been by Modro and Ridd[52] in a kinetic study of the mechanism by which the substituent effects of positive poles are transmitted in electrophilic substitution. The rate coefficients for nitration of the compounds $Ph(CH_2)_n NMe_3^+$ ($n = 0–3$) given in Table 10 show that insertion of methylene groups causes a substantial decrease in deactivation by the $NMe_3^+$ group as expected. Since analysis of this effect is complicated by the superimposed activation by the introduced alkyl group, the reactivities of the

## TABLE 10

RATE COEFFICIENTS (AT 25.1 °C) AND KINETIC PARAMETERS FOR NITRATION OF THE IONS $Ph(CH_2)_n NMe_3^+$ IN $H_2SO_4$ MEDIA

| $H_2SO_4$ (Wt. %) | $10^3 k_2$ | | |
| --- | --- | --- | --- |
| | $n = 1$ | $n = 2$ | $n = 3$ |
| 61.0 | | | 0.482 |
| 63.4 | | 0.195 | 2.31 |
| 68.3 | | 8.23 | 105 |
| 72.7 | | 357 | |
| 74.5 | 0.56 | | |
| 76.5 | 2.93 | 8,040 | |
| 79.3 | 34.5 | | |
| 81.6 | 283 | | |
| $\Delta H^{\ddagger}$ | 18.3 ⎱ 70.0% acid | 16.2 ⎱ 70.0% acid | 16.2[a] ⎱ 68.3% acid    15.8 ⎱ 68.3% acid |
| $\Delta S^{\ddagger}$ | −18.2 ⎰ acid | −10.4 ⎰ acid | −14.0[a] ⎰ acid    −10.1 ⎰ acid |

[a] Determined over a 12 °C range only; all other values refer to a range $> 20$ °C.
For $4\text{-}MeC_6H_4CH_2NMe_3^+$, $\Delta H^{\ddagger} = 17.8$ and $\Delta S^{\ddagger} = -9.6$ in 69.7 wt. acid.

*meta* positions were derived since these would be least susceptible to alkyl group substituent effects. It appeared that the resultant modification of the $NMe_3^+$ substituent effect was best accommodated in terms of an electrostatic interaction between the nitrogen pole and the charge in the transition state. The primary dependence of rate upon entropy of activation was considered consistent with the mode of substituent effect transmission, though the fact that the rate enhancement of 360 at 25 °C arising from substituting a 4-methyl substituent in the benzyltrimethylammonium ion derives from the entropy of activation is difficult to explain satisfactorily. This rate enhancement is greater than for a methyl substituent in benzene, but this is expected since the transition state for nitration of these unreactive compounds will be shifted towards products, and consequently will be more highly charged.

Ridd *et al.*[53] have also measured the deactivating effects of the quaternary ions $PhXMe_3^+$ in nitric acid–sulphuric acid, and the second-order rate coefficients at 25 °C are given in Table 11 together with the partial rate factors, those for the antimony compound being obtained by extrapolation. The greater reactivity with increasing atomic weight of X accords with expected decrease in the $-I$ effect, and consistently significant *ortho* substitution appears for the antimony compound. From the ratio of $\log f_p : \log f_m$ it can be seen that *para* substitution becomes increasingly unfavourable with increasing weight of X and this accords with an increasing $-M$ effect (which must operate mainly at the *para* position as indicated by the significant *ortho* substitution; however, this accords with modern theory[54]).

The large activating effect of a methyl group upon the reactivity of aromatic quaternary nitrogen compounds noted above has been confirmed by Utley and Vaughan[55], who have measured the second-order rate coefficients and the

TABLE 11

RATE COEFFICIENTS AND PARTIAL RATE FACTORS FOR NITRATION OF $PhXMe_3^+$ IN SULPHURIC ACID AT 25 °C

| X | $H_2SO_4$ (Wt. %) | $10^3k_2$ | $10^8f_o$ | $10^8f_m$ | $10^8f_p$ | $\log f_p : \log f_m$ |
|---|---|---|---|---|---|---|
| N | 98.7 | 10.4 | | 4.67 | 1.15 | 1.08 |
| P | 98.7 | 50.1 | | 24.7 | $\sim 1.0$ | 1.21 |
| As | 98.7 | 400 | | 194 | $<16.2$ | 1.19 |
| Sb | 82.15 | 360 | 840 | 9,140 | 1,050 | 1.23 |
| Sb | 81.6 | 192 | | | | |
| Sb | 80.7 | 76.5 | | | | |
| Sb | 79.3 | 25.8 | | | | |
| Sb | 76.5 | 1.83 | | | | |
| Sb | 75.95 | 1.15 | | | | |

Arrhenius parameters for nitration of 4-alkylphenyltrimethylammonium ions in nitric acid–sulphuric acid mixtures (Table 12). It was argued that the observed Baker–Nathan order of alkyl substituent effect was, in fact, the result of a steric effect superimposed upon an inductive order. However, a number of assumptions were involved in this deduction, and these render the conclusion less reliable than one would like; it would be useful to have the thermodynamic parameters for nitration of the methyl substituted compound in particular, in order to compare with the data for the $t$-butyl compound, though experimental difficulties may preclude this. It would not be surprising if a true Baker–Nathan order were observed because it is observed for all other electrophilic substitutions in this medium[1].

Although the nitrating agent in nitration by nitric acid is less clear than when

TABLE 12

RATE COEFFICIENTS ($10^3k_2$) AND ARRHENIUS PARAMETERS FOR NITRATION OF $4\text{-}RC_6H_4NMe_3^+$ [55]

| $H_2SO_4$ (Wt. %) | Temp. (°C) | R | | | | |
|---|---|---|---|---|---|---|
| | | H | Me | Et | i-Pr | t-Bu |
| 98 | $-4^a$ | 0.835 | 2,610 | 2,470 | 1,030 | 213 |
| 98 | 4.8 | | | | | 599 |
| 98 | 10.2 | | | | | 853 |
| 98 | 15.0 | 4.79 | | | | 1,105 |
| 98 | 25.0 | 12.9 | | | | |
| 98 | 40.1 | 45.4 | | | | |
| 82 | 15.1 | 0.093 | 218 | 190 | 88.0 | 17.2 |
| $E_a$ | | 16.0 | | | | 9.45 |
| $\log A$ | | 11.6 | | | | 9.02 |

g Reaction solution was 0.1 M in $K_2SO_4$.

a stronger mineral acid is present, the kinetic evidence all points to nitronium ion as the effective entity. Addition of sulphuric acid (in the nitration of 1-nitro-anthraquinone at 20 °C) of nitrate ions (in the nitration of nitrobenzene at 0 °C) and of water (in the nitration of nitrobenzene at −13.3 °C) caused an increase, a decrease, and a decrease in the rate, respectively, without altering the kinetic form. In the first two cases there is initially a small change in rate, and, above $\sim 0.1$ $M$, the rate changes almost linearly with change in concentration of the added reagent, indicating that both rate changes derive from a similar source. Since the equilibrium for the formation of nitronium ions from sulphuric acid is

$$2\,HNO_3 \rightleftharpoons NO_2^+ + NO_3^- + H_2O \tag{26}$$

it follows from this and from the equilibrium

$$NO_3^- + H_2SO_4 \rightleftharpoons HNO_3 + HSO_4^- \tag{27}$$

that the effect of the added reagents would be as observed if nitronium ion is the effective reagent.

Some interesting results have been obtained by Akand and Wyatt[56] for the effect of added non-electrolytes upon the rates of nitration of benzenesulphonic acid and benzoic acid (as benzoic acidium ion in this medium) by nitric acid in sulphuric acid. Division of the rate coefficients obtained in the presence of non-electrolyte by the concentration of benzenesulphonic acid gave rate coefficients which were, however, dependent upon the sulphonic acid concentration e.g. $k_2$ was 0.183 at 0.075 molal, 0.078 at 0.25 molal and 0.166 at 0.75 molal (at 25 °C). With a constant concentration of non-electrolyte (sulphonic acid +, for example, 2, 4, 6-trinitrotoluene) the rate coefficients were then independent of the initial concentration of sulphonic acid and only dependent upon the total concentration of non-electrolyte. For nitration of benzoic acid a very much smaller effect was observed; nitromethane and sulphuryl chloride had a similar effect upon the rate of nitration of benzenesulphonic acid. No explanation was offered for the pheno-menon.

### 3.1.3 Nitration by nitric acid in organic solvents other than acetic anhydride

The kinetics of nitration by nitric acid in nitromethane and acetic acid have been studied in detail[40], and interpretation of the data here is perhaps made a little more difficult by the fact that spectroscopic and conductimetric analyses give no evidence for the existence of the nitronium ion in these media[57].

The kinetics obtained are very similar to those with aqueous nitric acid–sulphuric acid. Thus reactive aromatics, i.e. reactivity of toluene and above, give zeroth-order kinetics, benzene gives zeroth-order kinetics in nitromethane and mixed order in acetic acid, whilst the less reactive halogen-substituted aromatics mostly give first-order kinetics with both media[40, 58]. This is nicely illustrated for nitration in acetic acid by the average rate coefficients given in Table 13. The effect of variable nitrous acid concentration (always below 0.015 $M$) is omitted, since when this is below 0.1 $M$ a minor retarding effect is observed according to the rate law $1/k_{0(1)} = a_{0(1)} + b_{0(1)}[HNO_2]^{\frac{1}{2}}$, and this effect does not alter the conclusions. Above 0.1 $M$, nitrous acid causes a marked anticatalytic effect, e.g. 0.2 $M$ nitrous acid reduced the rate $(10^5 k_1)$ for nitration of chlorobenzene by 10 $M$ nitric acid in 99.8 % acetic acid at 20 °C from 325 to 15. This anticatalytic effect was attributed to ionisation by the nitric acid of the oxides of nitrogen which are present to give nitrates which reduce the nitronium ion concentration. The data in Table 13 show clearly the change over from first- to zeroth-order kinetics

TABLE 13

RATE COEFFICIENTS FOR NITRATION OF ArH BY $HNO_3$ IN AcOH[40, 58]

| ArH | [$HNO_3$] (M) | Temp. (°C) | $10^5 k_0$ | $10^5 k_1$ |
|---|---|---|---|---|
| Benzene | 12.4 | 10.7 | 97 | |
| Toluene | 12.4 | 10.7 | 103 | |
| Toluene | 7.0 | 20 | 6.31 | |
| Ethylbenzene | 7.0 | 20 | 5.95 | |
| p-Xylene | 7.0(5.0) | 20 | 5.93(0.433) | |
| Mesitylene | 7.0(5.0) | 20 | 5.58(0.417) | |
| Benzene | 7.0(8.0) | 20 | | 4.02(119) |
| Fluorobenzene | 9.0(8.0) | 20 | | 107(28) |
| Iodobenzene | 9.0(8.0) | 20 | | 137(28) |
| Chlorobenzene | 9.0 | 20 | | 52 |
| Bromobenzene | 9.0 | 20 | | 42 |
| Chlorobenzene | 10.0 | 20 | | 365(0.05 M PhCl)–281(0.30 M PhCl) |
| Benzene | 9.7 | 20 | | 167(0.10 M PhH)–75(0.34 M PhH) |
| o-Dichlorobenzene | 10.5 | 20 | | 25.7 |
| m-Dichlorobenzene | 10.5 | 20 | | 49.8 |
| p-Dichlorobenzene | 10.5(12.2) | 20 | | 20.0(223) |
| Ethylbenzoate | 10.5 | 20 | | 15.5 |
| 1,2,4-Trichlorobenzene | 12.2 | 20 | | 18.0 |

as the nitric acid concentration is increased, and the acid strength at which this change occurs is higher the less reactive the aromatic; thus chlorobenzene shows less departure from first-order behaviour with 10 $M$ acid than does benzene with 9.7 $M$ acid.

In general, nitrations in nitromethane show a greater tendency towards zeroth-order kinetics, so that, for example, whereas benzene gives first-order kinetics on nitration by 7.0 $M$ nitric acid in acetic acid (Table 13), in nitromethane the reaction is zeroth-order $(10^5 k_0 = 17.2)$. Hence first-order kinetics are not attained unless a relatively unreactive aromatic, $p$-dichlorobenzene is used. The rate coefficients for nitration in nitromethane are also much more susceptible to the concentration of nitrous acid as the data in Table 14 show,

TABLE 14

RATE COEFFICIENTS FOR NITRATION OF ArH BY $HNO_3$ IN $MeNO_2$[40, 58]

| ArH | [ArH] (M) | [HNO₃] (M) | [HNO₂] (M) | Temp. (°C) | $10^5 k_0$ | $10^5 k_1$ |
|---|---|---|---|---|---|---|
| Toluene | 0.09 | 7.0 | 0.448 | −6 | 0.64 | |
| Toluene | 0.09 | 7.0 | 0.0038 | −6 | 7.78 | |
| $p$-Dichlorobenzene | 0.10 | 8.5 | 0.174 | 20 | | 45 |
| $p$-Dichlorobenzene | 0.10 | 8.5 | 0.024 | 20 | | 367 |
| $p$-Dichlorobenzene | 0.10 | 8.5 | 0.0018 | 20 | | 649 |
| $p$-Dichlorobenzene | 0.17 | 8.5 | 0.0018 | 20 | | 617 |
| $p$-Dichlorobenzene | 0.10 | 11.5 | 0.0269 | 20 | | 2470 |
| 1,2,4-Trichlorobenzene | 0.10 | 8.5 | 0.0018 | 20 | | 55.4 |

the variation of rate with nitrous acid concentration following the equation for nitration in acetic acid above, but with different parameters $a$ and $b$. Comparison of Tables 13 and 14 also reveals that nitration is faster in nitromethane than in acetic acid.

It is clear from the above that the appropriate reactions are (21a)

$$HNO_3 + HNO_3 \stackrel{fast}{\rightleftharpoons} H_2NO_3^+ + NO_3^- \tag{21a}$$

and (22) (p. 12), the nitronium ion being the electrophile. Additional evidence to support this comes from the fact that whereas addition of ionised salts usually speeds up reaction rates, addition of nitrates dramatically slowed it down without altering the kinetic form (0.01 $M$ $KNO_3$ approximately halved the zeroth- and first-order rate of nitration by nitric acid in nitromethane whereas 0.01 $M$ $KClO_4$ caused a small rate increase), and urea nitrate decreased the first-order rates for nitration in acetic acid and thus the pre-equilibrium step must be affected[40]. It also follows that a stronger acid will increase the protonation of nitric acid and speed up the rate, and consistently sulphuric acid increased the rate; addition of only 0.04 $M$ sulphuric acid caused a 60-fold increase in the zeroth-order rate of nitration of benzene by 3 $M$ nitric acid in nitromethane, and from the data[40] it can be deduced that addition of the same molar quantity of nitric acid would produce a far smaller rate increase.

Further confirmation of the proposed mechanism is provided by the fact that the reaction rates in acetic acid and nitromethane are little affected by the addition of small amounts of water, but when larger amounts (*ca.* 5 %) are added (to the acetic acid medium) competition between water and the aromatic for nitronium ion causes a change over to a first-order dependence on aromatic concentration.

The kinetic effect of increased pressure is also in agreement with the proposed mechanism. A pressure of 2000 atm increased the first-order rates of nitration of toluene in acetic acid at 20 °C and in nitromethane at 0 °C by a factor of about 2, and increased the rates of the zeroth-order nitrations of $p$-dichlorobenzene in nitromethane at 0 °C and of chlorobenzene and benzene in acetic acid at 0 °C by a factor of about 5[59]. The products of the equilibrium (21a) have a smaller volume than the reactants and hence an increase in pressure speeds up the rate by increasing the formation of $H_2NO_3^+$. Likewise, the heterolysis of the nitric acidium ion in equilibrium (22) and the reaction of the nitronium ion with the aromatic are processes both of which have a volume decrease, consequently the first-order reactions are also speeded up and to a greater extent than the zeroth-order reactions.

As might be expected, an increase in pressure reduces the selectivity of the nitronium ion (since the reactivity is effectively increased) and also increases the amount of *ortho* substitution; this has been shown from the relative rates of nitration of benzene and $t$-butylbenzene in acetic acid at 45 °C[60].

Finally, it is customary to compare the partial rate factors obtained under different conditions to indicate the reactivity of the electrophile. Unfortunately, the medium, nitric acid in sulphuric acid, in which the nitronium ion is most clearly established as the electrophile, is such a poor solvent for aromatics that meaningful competitive nitrations are impossible and kinetic studies are hampered by the difficulties noted above. However, since the isomer distribution is a function of the rate factors, inspection of these distributions (Table 15) shows a very

TABLE 15

ISOMER DISTRIBUTIONS FOR NITRATION OF TOLUENE AND $t$-BUTYLBENZENE

| Condition | Temp. (°C) | Toluene | | | t-Butylbenzene | | | Ref. |
|---|---|---|---|---|---|---|---|---|
| | | $o$ | $m$ | $p$ | $o$ | $m$ | $p$ | |
| $HNO_3/H_2SO_4$ | 25 | 56.4 | 4.8 | 38.4 | 15.8 | 11.5 | 72.7 | 61, 62 |
| $HNO_3/AcOH$ | 25 | 56.9 | 2.8 | 40.3 | | | | 61 |
| $HNO_3/aq.AcOH$ | 45 | 56.5 | 3.5 | 40.0 | 12.0 | 8.5 | 79.5 | 63 |
| $HNO_3/MeNO_2$ | 25 | 61.5 | 3.1 | 35.4 | 12.2 | 8.2 | 79.6 | 61, 64 |
| $HNO_3/CF_3COOH$ | 25 | 61.6 | 2.6 | 35.8 | | | | 65 |

consistent pattern indicative of a common electrophile under each condition; those conditions which facilitate determination of partial rate factors also yield closely similar values for these factors.

Relatively few kinetic studies have been carried out using nitric acid (or nitric acid–sulphuric acid) in organic solvents (other than acetic anhydride), with the object of determining relative reactivities and hence partial rate factors. Some rates relative to benzene (mostly determined by competition methods) are summarized in Table 16; in addition, the relative reactivity of toluene to $t$-butylbenzene was found to be 1.56 (HNO$_3$, 90 % aq. AcOH, 45 °C[63]) and 1.4 (HNO$_3$, MeNO$_2$, 25 °C)[64], and of fluoranthene relative to naphthalene, 5.5 (HNO$_3$, AcOH, 50 °C)[68]. From the data in Table 16, Olah *et al.*[61] argued that only in concentrated solutions of nitric acid–sulphuric acid, and these acids in organic solvents (and where the nitronium ion can be spectroscopically observed), does nitration occur *via* pre-formed nitronium ions and this gives rise to very low substrate selectivity. Under

TABLE 16

REACTIVITIES RELATIVE TO BENZENE FOR NITRATION BY HNO$_3$ AND HNO$_3$–H$_2$SO$_4$
IN SOME ORGANIC SOLVENTS[62−67]

| Acid | Solvent | Temp. (°C) | ArH | $k_{rel}$ |
|---|---|---|---|---|
| HNO$_3$ | CF$_3$CO$_2$H | 25 | Toluene | 28 |
| | AcOH | 25 | Toluene | 28.8 |
| | AcOH | 25 | $p$-Xylene, mesitylene | >1,000 |
| | AcOH | 24 | 1-Methylnaphthalene | 28 |
| | AcOH | 24 | 2-Methylnaphthalene | 26 |
| | AcOH | ? | 1-Methoxynaphthalene | $2.6 \times 10^4$ |
| | AcOH | ? | 2-Methoxynaphthalene | $3.2 \times 10^3$ |
| | MeNO$_2$ | 30 | Toluene | 21 |
| | MeNO$_2$ | 25 | Iodobenzene | 0.22 |
| | MeNO$_2$ | 25 | Toluene | 26.4 |
| | MeNO$_2$ | 25 | Ethylbenzene | 22.6 |
| | MeNO$_2$ | 25 | $p$-Xylene, mesitylene | >1,000 |
| | MeNO$_2$ | 24 | 1-Methylnaphthalene | 18.0 |
| | MeNO$_2$ | 24 | 2-Methylnaphthalene | 18.5 |
| | (CH$_2$)$_4$SO$_2$ | 25 | Toluene | 17 |
| | (CH$_2$)$_4$SO$_2$ | 25 | Mesitylene | >1,000 |
| HNO$_3$–H$_2$SO$_4$ | AcOH | 25 | Toluene | 21[a], 2.13[b] |
| | AcOH | 25 | $p$-Xylene | >500[a] |
| | AcOH | 25 | Mesitylene | >1,000[a] |
| | AcOH | 25 | $m$-Xylene | 1.27[b] |
| | AcOH | 24 | 1-Methylnaphthalene | 2.4[b] |
| | AcOH | 24 | 2-Methylnaphthalene | 2.5[b] |
| | AcOH | ? | 2-Methoxynaphthalene | 6.0[b] |
| | (CH$_2$)$_4$SO$_2$ | 25 | Toluene | 28[a], 1.60[b] |
| | (CH$_2$)$_4$SO$_2$ | 25 | Ethylbenzene | 24[a], 1.35[b] |
| | (CH$_2$)$_4$SO$_2$ | 25 | Isopropylbenzene | 13.8[a] |
| | (CH$_2$)$_4$SO$_2$ | 25 | $p$-Xylene | >500 |
| | (CH$_2$)$_4$SO$_2$ | 25 | Mesitylene | >1,000[a], 0.33[b] |
| | (CH$_2$)$_4$SO$_2$ | 25 | $o$-Xylene | 0.9[b] |
| | (CH$_2$)$_4$SO$_2$ | 25 | $m$-Xylene | 1.1[b] |

[a] Data refers to a 30 % solution of acid.
[b] Data refers to a 75 % solution of acid.

all other conditions nitration involves "the formation of the active nitrating agent as the rate-determining step, and the interaction of aromatic substrates with the nitrating agent involves not free $NO_2^+$, but a weaker electrophilic precursor of it, probably $H_2NO_3^+$". However, this fails to explain the differing reactivities of substrates in the media in which they consider $NO_2^+$ is not present.

A kinetic study of nitration by nitric acid in carbon tetrachloride has been briefly reported and is of interest because of the third-order dependence of rate upon nitric acid concentration, for nitration of N-methyl-N-nitrosoaniline. This is believed to arise from equilibria (28) and (29) below, which give rise to a nitrosating species and nitration is achieved through subsequent oxidation of the nitrosated aromatic[69].

$$PhNMe.NO + HNO_3 \rightleftharpoons PhNHMe + N_2O_4 \qquad (28)$$

$$N_2O_4 + 2\, HNO_3 \rightleftharpoons NO^+ 2\, HNO_3.NO_3^- \qquad (29)$$

### 3.1.4 Nitration by nitric acid in acetic anhydride

The nature of the electrophile in this nitrating mixture is still not wholly agreed upon; whereas kinetic evidence can be interpreted as consistent with nitration by nitronium ion, the fact that substituents with lone pairs of electrons or $\pi$-electrons give markedly different ortho : para ratios from other nitrating mixtures is usually conceded to be consistent with the electrophile being something other than the nitronium ion. The balance of evidence at present is in favour of protonated acetyl nitrate being the electrophile.

Kinetic studies with benzene in acetic anhydride containing 0.4–2 M nitric acid at 25 °C show the reaction to be first-order in benzene and approximately second-order in nitric acid; this falls to first-order in nitric acid on addition of sulphuric acid, which also increases the first-order rate coefficient (first-order in benzene) from $4.5 \times 10^{-4}$ to $6.1 \times 10^{-4}$. By contrast the addition of as little as 0.001 M sodium nitrate reduced the rate to $0.9 \times 10^{-4}$ without affecting the kinetic order[70]. These results were, therefore, interpreted as nitration by nitronium ion via equilibria (21a) and (22).

Bordwell and Garbisch[71] contested this conclusion since they found that nitric acid in acetic anhydride prepared at $-10$ °C contained a much less effective nitrating species (the nitric acid could be recovered quantitatively) than when mixed at 25 °C and cooled to $-10$ °C (the nitric acid being then mostly unrecoverable). Further, these latter solutions reacted with alkenes to give predominantly cis addition products (nitro-acetates), which indicates association of the nitronium ion with some other species. It has been argued[72] that this does not necessarily follow, since nitration of aromatics may involve a different species than does addition to alkenes. This argument seems unlikely, however, since it has been shown that nitration of biphenyl is much slower in nitrating mixtures prepared at low temperatures compared to those prepared at 25 °C and then cooled[73]. Further, at −40 °C the latter solutions solidify, almost certainly due to the presence of acetic acid formed through hydrolysis of acetic anhydride by the aqueous nitric acid; the former solutions solidify after standing for 24 h, and nitration by these media is then rapid. It seems very probable then that the addition reagent which Bordwell and Garbisch found to be formed by reaction of the reagents is the same one giving aromatic nitration, and they proposed that the effective nitrating species was protonated acetyl nitrate formed according to equilibria (30) and (31), *viz.*

$$Ac_2O + HNO_3 \rightleftharpoons AcONO_2 + HOAc \qquad (30)$$

$$AcONO_2 + HNO_3 \text{ (or } H_2SO_4) \rightleftharpoons AcOHNO_2^+ + NO_3^- \text{ (or } HSO_4^-) \qquad (31)$$

this explains the observed kinetic and catalytic effects (the reaction was also retarded by acetate ion due, it is believed, to proton abstraction from the proto-nated species)[71].

A further observation is the fact that differences in rates of nitration between the reagents prepared at different temperatures tended to zero as the water con-centration of the added nitric acid was decreased to zero[73]. It has been argued that, since the acid-catalysed hydrolysis of acetic anhydride must be very rapid at 25 °C and removes water which initially competes with acetic anhydride and acetyl nitrate for protons, this removal permits equilibria (30) and (31) to be displaced towards products. The more anhydrous the nitric acid, the less important is this initial hydrolysis of the acetic anhydride and so the difference in the ni-trating power of the differently prepared mixtures becomes less. When reagents are mixed at low temperatures, the hydrolysis of the anhydride is very slow, but once this is accomplished, formation of the protonated acetyl nitrate and subsequent nitration is rapid as observed[73].

It should be noted that protonated acetyl nitrate can be regarded as a nitronium ion solvated by acetic acid[74] (I)

(I)

and this has been postulated as a species present in nitration in acetic acid[75]. Further, Sparks[74] has shown that nitration of chlorobenzene by nitric acid in acetic acid gives *ortho* : *para* ratios intermediate between those obtained with nitric acid–sulphuric acid, and nitric acid in acetic anhydride; this result has been confirmed for nitration of biphenyl (Table 17)[73].

Acetoxylation has been found to accompany the nitration of *o*- and *m*-xylene

TABLE 17

VARIATION OF *ortho*: *para*-RATIO WITH NITRATING CONDITION[73, 74]

| Condition | PhCl | PhPh |
|---|---|---|
| $HNO_3/H_2SO_4$ | 0.54 | 0.7 |
| $HNO_3/AcOH(+H_2SO_4$ for PhCl) | 0.35 | 1.4 |
| $HNO_3/Ac_2O$ | 0.27 | 2.0 |

in nitric acid–acetic anhydride, both reactions giving kinetics approximately zeroth-order in aromatic, acetoxylation being zeroth-order overall, and nitration being third- and second-order in nitric acid in the absence and presence, respectively, of acetic acid. The ratio of acetoxylation to nitration remained constant for a 330-fold variation in rate and was independent of the rate increase and decrease brought about by added sulphuric acid and lithium nitrate, respectively (Table 18)[76].

TABLE 18

RATE COEFFICIENTS FOR ACETOXYLATION (A, $k_0$) AND NITRATION (N, $k_3$) BY $HNO_3$ IN $Ac_2O$ AT 25 °C[74]

| $10^2[HNO_3]$ (M) | [o-Xylene] (M) | $10^6 k_0$ | $10^3 k_3$ | A/N |
|---|---|---|---|---|
| 7.12 | 0.344 | 5.44 | 15.1 | 0.71 |
| 5.96 | 0.337 | 3.06 | 14.5 | 0.71 |
| 2.12 | 0.393 | 0.117 | 12.3 | |
| 2.07 | 0.486 | 0.106 | 12.0 | 0.73 |
| 4.60 | 0.029 | 0.82 | 8.4 | |
| 4.50 | 0.114(0.369 M AcOH) | 5.9 | 7.0 | |
| 3.91 | 0.319(45 × 10$^{-6}$ M H$_2$SO$_4$) | 15.8 | 264 | 0.73 |
| 6.64 | 0.682(8.4 × 10$^{-4}$ M LiNO$_3$ +0.298 M AcOH) | 1.43 | 0.89 | 0.72 |
| | [m-Xylene] (M) | | | |
| 4.50 | 0.111(0.369 M AcOH) | 6.58 | 7.9 | 0.034 |

The constancy of the acetoxylation : nitration ratio and zeroth reaction order in aromatic suggested that both reactions were brought about by a common species (protonated acetyl nitrate) the formation of which was rate-determining according to equilibria (30) and (31), with the latter being a slow step. Since the total rate of acetoxylation plus nitration is given by

$$\text{Rate} = k[AcONO_2][HNO_3] \tag{32}$$

from equilibrium (30) we obtain

$$\text{Rate} = \frac{k[\text{AcONO}_2]^2[\text{AcOH}]}{K[\text{Ac}_2\text{O}]} \tag{33}$$

Since nitration produces acetic acid, the concentration of this as well as of acetyl nitrate can be shown to depend upon the nitric acid concentration giving kinetics third-order in nitric acid (3.16 actually observed). It follows that in the presence of acetic acid the order in nitric acid should fall to 2 (2.31 observed). Likewise, in the presence of added sulphuric acid, from equilibrium (31) it follows that the order in nitric acid should fall, the observed order in this being 1.4 and 1.7 in added sulphuric acid. The retardation by added nitrate was attributed to competition by this ion for protonated acetyl nitrate, *viz.*

$$\text{AcONO}_2\text{H}^+\text{A}^- + \text{LiNO}_3 \rightarrow \text{AcNO}_2 + \text{HNO}_3 + \text{LiA}^- \tag{34}$$
$$\text{or}$$
$$\text{AcOH} + \text{N}_2\text{O}_5 + \text{LiA}^-$$

though when $\text{A}^-$ is $\text{NO}_3^-$ it is hard to understand why this should follow. It was argued that an increase in the concentration of aromatic should increase its effectiveness as a competitor for protonated acetyl nitrate so that the zeroth-order rate dependence on aromatic concentration should not be preserved in the presence of added nitrate; evidence was presented substantiating this proposal. Finally, the extent of acetoxylation appeared to decrease with increase in the reactivity of the aromatic (Table 18) which is to be expected if the slow rate-determining formation of the nitrating and acetoxylating precursor is short-circuited by the reaction of a highly reactive aromatic with a less reactive species (which nitrates only)[76].

An alternative to the above mechanism is that acetoxylation is an addition–elimination process involving $\text{NO}_2^+$ and $\text{OAc}^-$, leading to nitro and acetoxy products[77], and it follows that this process would be less likely to occur, for example, with mesitylene and significantly perhaps, experiments seeking acetoxylation in mesitylene have failed[78]; on the other hand, mesitylene is a very reactive substrate so it could be that an alternative nitrating species is involved here.

TABLE 19

*Ortho* : *para* RATIOS FOR NITRATION OF SOME AROMATICS

| Compound | $HNO_3$–$Ac_2O$ | $HNO_3$–$H_2SO_4$ |
|---|---|---|
| Biphenyl[73] | 2.0 | 0.7 |
| Cyclopropylbenzene[81] | 4.7 | 2.1 |
| Methyl phenethyl ether[79] | 1.83 | 0.54 |
| Anisole[82] | 2.54 | 0.46 |
| Acetanilide[83] | 2.28 | 0.25 |

Further, if the addition–elimination mechanism is correct, then one should observe acetoxylation by nitric acid in acetic acid; none has been reported.

Finally, dinitrogen pentoxide has been proposed as the nitrating species in nitric acid–acetic anhydride which gives rise to enhanced *ortho* substitution because of the similar *ortho* : *para* values obtained in nitration of aryl ethers by nitric acid in acetic anhydride and by dinitrogen pentoxide in acetonitrile[79], though since the rate of nitration of benzene[70] and *p*-terphenyl[80] by nitric acid–acetic anhydride is depressed three- to five-fold by the addition of about 0.001–0.0002 *M* nitrate ion, it seems improbable that what is likely to be a small amount of nitration through dinitrogen pentoxide could cause such dramatic changes in *ortho* : *para* value. These difficulties have been overcome by the proposal[73] that all nitrating species of the type $NO_2X$ can give rise to enhanced *ortho* substitution by the same mechanism proposed[79] for dinitrogen pentoxide (involving nucleophilic displacement of $X^-$ from $NO_2X$ by lone pairs or $\pi$-electrons, see Table 19) and this, therefore, accommodates protonated acetyl nitrate[73].

TABLE 20

REACTIVITIES RELATIVE TO BENZENE FOR NITRATION BY $HNO_3$ IN $Ac_2O$

| ArH | Temp. (°C) | $k_{rel}$ | $A_rH$ | Temp. (°C) | $k_{rel}$ |
|---|---|---|---|---|---|
| Toluene[85] | 0(35) | 27(23) | Naphthalene[94] | 0 | 400 |
| Fluorobenzene[86, 87] | 18(25) | 0.15(0.14) | Phenanthrene[94] | 0 | 415 |
| Chlorobenzene[86] | 18 | 0.033 | Pyrene[94] | 0 | 10,600 |
| Bromobenzene[86] | 18 | 0.030 | Triphenylene[94] | 0 | 1,840 |
| Iodobenzene[86, 66] | 18(25) | 0.18(0.13) | Chrysene[94] | 0 | 1,580 |
| Ethyl benzoate[88] | 18 | 0.0037 | Perylene[94] | 0 | 60,000 |
| Cinnamic acid[89] | 25 | 0.111 | Benzopyrene | 0 | 25,600 |
| Ethyl phenylacetate[90, 91] | 25(45) | 3.66, 3.86 (3.52) | Anthanthrene[94] | 0 | 116,000 |
| Benzylchloride[90, 91] | 25 | 0.184, 0.302 | Coronene[94] | 23 | 2,760 |
| Benzyl methyl ether[90] | 25 | 1.70 | Diphenylmethane[94] | 25 | 19.5 |
| Benzyl cyanide[90] | 25 | 0.486 | Fluorene[94] | 25 | 101 |
| Phenyl nitromethane[90] | 25 | 0.171 | Diphenylether[94] | 25 | 157 |
| *t*-Butylbenzene[90] | 25 | 14.9 | Dibenzofuran[94] | 25 | 78.5 |
| Phenyltrimethylsilane[92] | *ca.*−10 | 1.64 | Diphenylamine[94] | 25 | 740,000 |
| Biphenyl[93, 94] | 0 | 39–42, 16.2 | Carbazole[94] | 25 | 37,000 |
| 2-Nitrobiphenyl[93] | 0 | 0.33 | 1-Methylnaphthalene[67] | 24 | 2.2[b] |
| 3-Nitrobiphenyl[93] | 0 | 1.03 | 2-Methylnaphthalene[67] | 24 | 2.6[b] |
| 4-Nitrobiphenyl[93] | 0 | 0.31 | 1-Methoxynaphthalene[67] | 24 | 22[b] |
| Ethylbenzene[87] | 0 | 0.85[a] | 2-Methoxynaphthalene | 24 | 10[b] |
| *i*-Propylbenzene[87] | 0 | 0.65[a] | Fluoranthrene[68] | 0 | 3.0[b] |
| *t*-Butylbenzene[87, 63] | 0(25) | 0.56, (0.50)[a] | 1,3,5-Triphenylbenzene[95] | 0 | 1.36[b] |
| | | | 10,9-Borazarophenanthrene[96] | 0 | $9.37 \times 10^{5c}$ |
| | | | | | $20.6 \times 10^{5d}$ |

[a] Rates relative to toluene.
[b] Rates relative to naphthalene.
[c] Partial rate factor for 8 position.
[d] Partial rate factor for 6 position.

It has been suggested in defence of the nitronium ion hypothesis that, in the media giving these high ratios, prior attachment of the ion to the lone pairs occurs, followed by rearrangement to give the *ortho*-substituted product[84], but that this is prevented in nitric acid–sulphuric acid due to hydrogen bonding. This can be discounted, however, by the results for biphenyl and cyclopropylbenzene where H-bonding is unlikely, and it has also been shown that the substituent effects (in terms of the *meta* : *para* values) are not altered for ethers under different nitrating conditions[79] as would be expected as a result of this hydrogen-bonding.

In summary then, the kinetics and related data are most consistent with protonated acetyl nitrate as the reagent in this medium. It is unfortunate that there is doubt as to the nature of the electrophile, as this medium combines high reactivity with good solvent properties, which has made it popular for studying substituent effects in nitration. Some relative reactivities (mostly obtained under competition conditions) are given in Table 20.

### 3.1.5 Nitration by dinitrogen pentoxide in organic solvents

Dilute solutions of dinitrogen pentoxide in carbon tetrachloride, nitromethane, acetonitrile, chloroform, and mixtures of the first two solvents give approximately second-order kinetics, *viz.*

$$\text{Rate} = k_2[\text{ArH}][\text{N}_2\text{O}_5] \tag{35}$$

Some typical results are given in Table 21 and Table 22 shows the average rate coefficients (in some cases corrected for temperature and solvent differences) obtained for a range of compounds in carbon tetrachloride[97].

Evidence in favour of the molecular nitrating species is the relative insensitivity of rate to solvent effects, rates differing in nitromethane by only a factor of 6

TABLE 21

RATE COEFFICIENTS FOR NITRATION OF $PhCO_2Et$ BY $N_2O_5$ IN $CCl_4$ UNDER FIRST-ORDER CONDITIONS (EXCESS AROMATIC) AT 15 °C [97]

| $10[PhCO_2Et]$ (M) | $10^2[N_2O_5]$ (M) | $10^4 k_1$ | $10^4 k_2$ |
|---|---|---|---|
| 5.46 | 4.10 | 10.2 | 18.7 |
| 5.78 | 3.29 | 10.5 | 18.2 |
| 6.92 | 5.29 | 12.2 | 17.7 |
| 8.02 | 5.29 | 12.6 | 17.0 |
| 8.71 | 3.19 | 14.5 | 16.7 |
| 10.76 | 5.29 | 16.3 | 15.3 |
| 10.93 | 4.10 | 17.4 | 15.9 |
| 13.52 | 5.29 | 19.5 | 14.4 |

TABLE 22
SECOND-ORDER RATE COEFFICIENTS FOR NITRATION OF ArH BY $N_2O_5$
IN $CCl_4$ AT 15 °C[97]

| $ArH$ | $10^4 k_2$ | $ArH$ | $10^4 k_2$ | $ArH$ | $10^4 k_2$ |
|---|---|---|---|---|---|
| 1,2-$C_6H_4Cl_2$ | 8.5 | 1,4-$C_6H_4Br_2$ | 5.7 | $PhCO_2Bu^t$ | 30.2 |
| 1,3-$C_6H_4Cl_2$ | 12.1 | 1,2,4-$C_6H_3Cl_3$ | 2.5 | 4-$MeC_6H_4CO_2Me$ | 21.8 |
| 1,4-$C_6H_4Cl_2$ | 5.1 | $PhCO_2Me$ | 18.2 | 3-$ClC_6H_4CO_2Et$ | 8.7 |
| 1,4-$C_6H_4ClBr$ | 5.5 | $PhCO_2Et$ | 21.7 | 1,2-$C_6H_4(CO_2Et)_2$ | 3.0 |

from those obtained in carbon tetrachloride. The possibility that nitration was occurring *via* the nitronium ion or the ion pair $NO_2^+NO_3^-$ was considered to be eliminated by the fact that added salts, including nitrates, caused rate acceleration. Thus addition of *ca.* 0.004 $M$ tetraethylammonium picrate, hydrogen sulphate, and nitrate produced rate increases of 280, 33, and 10 % respectively. Difficulty in drawing conclusions from these quantitative differences arose in view of the change in kinetic form produced by the added salts; the accelerative effect was less at the end of a run than at the beginning.

The reaction was shown to be catalysed by nitric acid the rate law being

$$\text{Rate} = k_2[\text{ArH}][\text{N}_2\text{O}_5] + k_{2+n}[\text{ArH}][\text{N}_2\text{O}_5][\text{HNO}_3]^n \qquad (34)$$

where $n$ is at least 2 and some results for conditions where $n = 2$ are given in Table 23. Since nitric acid is produced in nitration by dinitrogen pentoxide, these nitrations suffer autocatalysis. The high order in nitric acid means that auto-catalysis will be more important at higher reactant concentrations, and since the activation energy of the catalysed reaction is less than that of the uncatalysed reaction it follows that the former predominates at lower temperatures. The catalytic effect of nitric acid (and also sulphuric acid) and salts was considered to arise from the induced ionisation of dinitrogen pentoxide to give nitronium

TABLE 23
EFFECT OF ADDED $HNO_3$ ON CALCULATED AND OBSERVED SECOND-ORDER RATE
COEFFICIENTS FOR NITRATION OF 1,4-$C_6H_4Cl_2$ (0.6 $M$) BY $N_2O_5$ (0.04 $M$) IN $CCl_4$ AT
20 °C[97]

| $[HNO_3]$ (M) | $10^4 k_2$ | $11 + 80[HNO_3]^2$ |
|---|---|---|
| 0 | 10.1 | 11.0 |
| 0.071 | 16.1 | 15.0 |
| 0.080 | 16.6 | 16.1 |
| 0.174 | 33.1 | 35.5 |
| 0.181 | 37.5 | 37.2 |
| 0.251 | 61.1 | 61.5 |

ions. Consistent with this was the observation that the *ortho* : *para* ratio in nitration of chlorobenzene in the catalysed reaction was similar (27 % *ortho*, 73 % *para*) to that obtained in nitric acid–sulphuric acid[86] (33 % *ortho*, 67 % *para*) but quite different to that obtained in the uncatalysed reaction[98] (43 % para)[†].

### 3.1.6  Nitration by acyl nitrates in organic solvents

Ingold *et al.*[99] showed that rates of nitration by benzoyl nitrate in carbon tetrachloride were depressed by the addition of benzoic anhydride. Thus addition of 0.035 $M$ of the anhydride to a solution 2.36 $M$ in benzene and 0.030 $M$ in nitrate decreased the first-order rate coefficient from $44 \times 10^{-4}$ to $20 \times 10^{-4}$. This is consistent with nitration by dinitrogen pentoxide formed *via* equilibrium (37), *viz.*

$$2\,PhCOONO_2 \rightleftharpoons (PhCO)_2O + N_2O_5 \qquad (37)$$

The equilibrium is apparently acid catalysed, for the pure anhydride had no effect unless benzoic acid was present.

Further evidence for this mechanism comes from the fact that a solution 0.305 $M$ in dinitrogen pentoxide and 0.0346 $M$ in anhydride, *i.e.* equivalent to the conditions above, gave almost the same rate, much slower than that obtained in the absence of the anhydride; addition of further amounts of anhydride produced a regular decrease in the rate.

No satisfactory analysis has been made with acetyl nitrate but the qualitative retarding effect of addition of acetic anhydride[100] has been likewise interpreted as resulting from the preformation of dinitrogen pentoxide.

### 3.1.7  The kinetic importance of nitrous acid

Nitrous acid can have both a catalytic and an anticatalytic effect on aromatic nitration, the former being appropriate in dilute (*ca.* 6 $M$) nitric acid solutions and the latter appropriate to more concentrated solutions.

#### (a)  The anticatalytic effect

Ingold *et al.*[40] showed that nitrous acid retards both zeroth- and first-order

---

† This latter result is a little surprising in that acyl nitrates[74,86] and nitric acid in acetic anhydride[66], both of which usually give very similar *ortho* : *para* values to dinitrogen pentoxide, are reported as giving *lower* ratios than nitric acid–sulphuric acid. Nevertheless, the nitration of fluorobenzene by dinitrogen pentoxide in carbon tetrachloride also gives a higher *ortho* : *para* ratio (0.39) than nitric acid in sulphuric acid (0.14)[74].

nitrations without changing the kinetic order, and this applies to nitration in acetic acid and nitromethane and in aqueous and concentrated nitric acid solutions without organic solvent. At low nitrous acid concentrations, and in the absence of water, the retardation is proportional to $[HNO_2]^{\frac{1}{2}}$ and at higher concentrations, or in the presence of water, it is proportional to $[HNO_2]^{\frac{3}{2}}$. The measured rate decreases are ten–twentyfold for a hundredfold increase in nitrous acid concentration.

In nitric acid, nitrous acid exists mainly as dinitrogen tetroxide which gives rise to nitrosonium and nitrate ions; the change in the dependence on nitrous acid concentration was originally explained in terms of an increase in the proportion of dinitrogen tetroxide and correspondingly nitrate ions. Thus the half order dependence was attributed to nitrate ion and the three-halves order to nitrite ion[40]. A more recent explanation[101] attributes the variation entirely to the effect of nitrate ion. It follows from the equilibrium

$$2\,HNO_3 \rightleftharpoons NO_2^+ + NO_3^- + H_2O \tag{38}$$

that in pure nitric acid $[NO_2^+][NO_3^-][H_2O]$ = constant and since $[NO_2^+]$ = $[H_2O]$ it follows that $[NO_2^+] \propto 1/[NO_3^-]^{\frac{1}{2}}$. Thus addition of nitrous acid and hence nitrate ion causes the observed reduction in nitration rate. In media containing water, the concentration of this is effectively constant so that $[NO_2^+] \propto 1/[NO_3^-]$, and in addition equilibrium (38) is suppressed, so that the nitration rate is much more sensitive to the concentration of added nitrate ion.

The effect of nitrous acid on nitration in nitromethane and acetic acid is also attributed to the effect of nitrate ions even though the ionisation of the dinitrogen tetroxide is much less in these solvents. As noted above (p. 31), the anticatalytic effect of nitrous acid is not governed by $k^{-1} = a+b[HNO_3]^{\frac{1}{2}}$ at nitrous acid concentrations above 0.1 $M$.

In sulphuric acid, nitrous acid has a catalytic effect since dinitrogen tetroxide reacts with sulphuric acid to give nitronium ions according to equilibrium (39)[102].

$$N_2O_4 + 3\,H_2SO_4 \rightleftharpoons NO^+ + NO_2^+ + H_3O^+ + 3\,HSO_4^- \tag{39}$$

### (b) The catalytic effect

Amines and phenols can be nitrated by dilute ($< 5\,M$) solutions of nitric acid in water or acetic acid provided nitrous acid is present[103]. A mixture of products is frequently obtained and the *ortho* : *para* ratio can vary widely according to the nitrous acid concentration[104, 105], thereby indicating a different electrophile under these conditions.

Kinetic studies have been made on the nitration of anisole, 4-chloroanisole, 4-nitrophenol and mesitylene[105]. For anisole the kinetics are complicated by demethylation; for 4-chloroanisole, nitration by 4–10 $M$ nitric acid in glacial acetic acid showed that, at nitric acid concentrations up to 6 $M$, the kinetic form was

$$\text{Rate} = k_2[\text{ArOH}][\text{HNO}_2] \tag{40}$$

For a given concentration of nitrous acid, the first-order rate coefficients increase as the sixth to seventh power of the nitric acid concentration and this catalysed nitration is presumed to be nitrosation followed by oxidation, reactions (41) and (42), *viz.*

$$\text{ArH} + \text{HNO}_2 \rightarrow \text{ArNO} + \text{H}_2\text{O} \tag{41}$$

$$\text{ArNO} + \text{HNO}_3 \rightarrow \text{ArNO}_2 + \text{HNO}_2 \tag{42}$$

As the concentration of nitric acid is increased the kinetic form changes and in 10 $M$ nitric acid the reaction is zeroth-order and anti-catalysed by nitrous acid (as described above) presumably due to nitronium ion nitration.

For 4-nitrophenol (studied in the range 1.4–10 $M$ nitric acid) the first-order rate coefficients (at constant nitrous acid concentration) decrease approximately 50 % as the nitric acid concentration is increased from 2 $M$ to 5 $M$ but increase considerably as it is further increased to 10 $M$; the increase is greater the lower the fixed concentration of nitrous acid and is attributed to the catalysed reaction. The rate decrease was attributed to superimposition upon the normal catalysis noted above for 4-chloroanisole of the effect of lowering of the concentration of the highly reactive phenoxide ion as the acidity was increased. In 10 $M$ nitric acid the anti-catalysed reaction was again observed.

Mesitylene was studied using the range 5–7 $M$ nitric acid, and when the nitrous acid concentration is small ($< 0.014$ $M$) nitronium ion nitration appears to occur, giving zeroth-order kinetics weakly retarded by nitrous acid. At rather higher nitrous acid concentrations the reaction is catalysed by nitrous acid and the kinetics go over to first-order (at constant nitrous acid concentration).

### 3.1.8 Nitration by nitronium salts in organic solvents

Aromatics can be very readily nitrated with nitronium salts such as $\text{NO}_2^+\text{BF}_4^-$, $\text{NO}_2^+\text{ClO}_4^-$, $\text{NO}_2^+\text{PF}_6^-$, $\text{NO}_2\text{AsF}_6^-$ and $\text{NO}_2^+\text{SbF}_6^-$ [106], and in tetramethylene sulphone and nitromethane as solvents it is found that the larger the anion, the slower the rate of substitution into mesitylene, a hindered aromatic, so that this indicates that the nitrating species is an ion pair[107]. Table 24 gives the relative rates of nitration of aromatics relative to benzene by nitronium borofluoride in tetramethylene sulphone, the corresponding values obtained under flow conditions in tetramethylene sulphone and nitromethane, and the range of relative reactivities

## TABLE 24

SPREAD OF RATES RELATIVE TO BENZENE FOR NITRATION OF ArH BY $NO_2X$ (SEE TEXT) AT 25 °C[67,107,109]

| ArH | $NO_2BF_4-$ $(CH_2)_4SO_2$ | $NO_2X-$ $(CH_2)_4SO_2$ | $NO_2BF_4-$ $MeNO_2$ | $NO_2X-$ $MeNO_2$ |
|---|---|---|---|---|
| Toluene | 1.67(2.07)[a] | 1.40–1.67 | 2.50[a] | 0.91–1.19 |
| Ethylbenzene | 1.60 | | | |
| n-Propylbenzene | 1.46 | | | |
| i-Propylbenzene | 1.32 | | | |
| n-Butylbenzene | 1.39 | | | |
| t-Butylbenzene | 1.18 | | | |
| o-Xylene | 1.75 | 1.10–1.75 | | |
| m-Xylene | 1.65(2.68)[a] | 1.23–1.71 | 6.54 | |
| p-Xylene | 1.96 | 1.60–1.96 | | |
| Mesitylene | 2.71(3.50)[a] | 0.90–3.25 | 10.4 | |
| Fluorobenzene | 0.45 | 0.39–0.54 | | |
| Chlorobenzene | 0.14 | 0.12–0.21 | | |
| Bromobenzene | 0.12 | | | |
| Iodobenzene | 0.28 | | | |
| Biphenyl | 2.08 | | | |
| Naphthalene | 2.05[b] | | | |
| Phenanthrene | 1.40 | | | |
| 1-Methylnaphthalene | 1.4[c] | | | |
| 2-Methylnaphthalene | 2.2[c] | | | |

[a] Under flow conditions.
[b] Also given as 2.55 in the same paper.
[c] Rates relative to naphthalene.

obtained using different nitronium salts[67,107-109]; the latter data shows the particular sensitivity of mesitylene to the nature of the nitronium salt.

Kinetic studies on the nitration of nitrobenzene by nitronium borofluoride in the polar solvents sulphuric acid, methane–sulphuric acid, and acetonitrile show the reaction to be first-order in both nitronium salt and aromatic[110]. With the first two solvents, the rate coefficients are similar for nitration by nitric acid and by the nitronium salts, indicating a common nitrating entity. With aceto-nitrile the rate coefficients are very much lower, consistent with a much lower concentration of free nitronium ions in this medium and thus with the nitronium salts existing as ion pairs in organic solvents (see Table 25).

The kinetic complication in nitration by nitronium salts in organic solvents is that although the product isomer distributions are similar to those obtained in nitration by other nitrating media (e.g. in Table 15) the relative reactivities of aromatic substrates is very much less; for example, the ratio $k_{PhCH_3}/k_{PhH}$ is 1.67 for nitration by nitronium borofluoride in tetramethylene sulphone, cf. 27 for most other nitrations[61], and it has been proposed[108] that this arises because the rate-determining step is formation of an orientated π-complex since the relative reactivities of alkyl- and halogeno-benzenes correlate with π-complex stabilities

TABLE 25

RATE COEFFICIENTS $(10^2 k_2)$ FOR NITRATION OF $PhNO_2$[110]

| Solvent | Temp. (°C) | $NO_2{}^+ BF_4{}^-$ | $HNO_3$ |
|---|---|---|---|
| $H_2SO_4$ | 23 | 0.70 | 0.77 |
| $CH_3SO_3H$ | 25, 24 | 1.84 | 2.0 |
| $CH_3CN$ | 21 | 0.18 | $<3 \times 10^{-5}$ |

of the aromatics and that this $\pi$-complex goes over to a $\sigma$-complex in the product forming step so that the usual product orientation is observed. As noted above (p. 34) it has been proposed that only in strong acid solutions and with nitronium salts is nitronium ion the electrophile and that the higher substrate selectivity otherwise observed is due to the electrophile being a weak precursor of the nitronium ion[61].

There is, however, the possibility that the low relative reactivities of the aromatic substrates arises from imperfect mixing *i.e.* the reaction is subject to diffusion control. Thus if two substrates of unlike reactivity are in the vicinity of the reagent molecules and one of the substrates reacts very quickly and becomes sufficiently impoverished as to be unable to compete with the remaining large number of less reactive molecules, these will then react before further of the more reactive molecules can diffuse to the area of reaction. Tolgyesi[111] has claimed that the selectivity of nitronium salt nitrations becomes the same as that observed with other nitrating agents when improved mixing is employed. Relevant points which argue for and against this are:

(*a*) The reaction rate is very fast indeed ($t_\frac{1}{2} = 10^{-3}$ sec), but this is stated to be many orders of magnitude less than the speed at which diffusion control would be expected[109].

(*b*) The isomer distribution is normal even when the low relative reactivity of substrates is observed. If reaction occurred at every collision then the statistical isomer distribution (40 % *ortho*, 40 % *meta*, and 20 % *para* for monoalkylbenzenes for example) would be expected, but this is not the case.

(*c*) Nevertheless, nitration of toluene by nitronium borofluoride in nitromethane and in sulpholane gives the same isomer distribution but different reactivities relative to benzene, the values being 1.19 and 1.67 respectively[107], and this strongly suggests a mixing problem.

(*d*) Use of flow system[109, 112] changes the relative rates (see Table 24) which leads one to the same conclusion.

(*e*) Dinitration occurs which clearly demonstrates imperfect mixing but the extent of this is very small[109].

(*f*) Nitration of dibenzyl gives mainly dinitro products (nitro group in each ring) and unchanged dibenzyl; this shows clearly that diffusion control operates[113].

(*g*) In tetramethylene sulphone, nitrobenzene and benzotrifluoride react $10^{-3}$–$10^{-4}$ times slower than benzene which is of the order normally expected. This has been attributed to the fact that these molecules unlike alkyl aromatics are not π-bases, which is a not unreasonable explanation[113].

(*h*) Use of higher dilutions of nitronium salts and better mixing causes the relative reaction rate for toluene to benzene to increase to as much as 6.77[111]. This has been attributed to reaction of impurities in tetramethylene sulphone with the nitronium salt to give reagents to which can nitrate *via* the normal mechanism, but it may also be due to mixing control[112].

(*i*) The nitration of pentamethylbenzene by nitronium borofluoride in tetramethylene sulphone at 25 °C has been found to occur *via* the *fast* formation of an addition complex which then forms the product in a *slow* step[114].

A small isotope effect has been observed in nitration of benzene by nitronium borofluoride in tetramethylene sulphone at 30 °C ($k_H/k_D = 0.86$) and this has been attributed to a secondary effect of the change in hybridisation from $sp^2$ to $sp^3$ of the ring carbon during the course of the reaction[109]. However, naphthalene gives an isotope effect of 1.15 under the same conditions, and anthracene a value of 2.6[115]. It does not seem at all clear why these relatively unhindered and normally more reactive molecules should give rise to an isotope effect when benzene does not.

Finally, nitrations with nitronium borofluoride of aromatics containing lone pairs of electrons or π-electrons give rise to high *ortho* : *para* ratios[73, 116], *unlike* nitrations by nitric acid–sulphuric acid, and for biphenyl at least the temperature dependence of the *ortho* : *para* ratio was very similar to that for dinitrogen pentoxide in acetonitrile and for nitric acid in acetic anhydride. The proposal was, therefore, made that nucleophilic displacement of the $BF_4^-$ anion by the lone pairs or π-electrons occurs in exactly the same way as for other nitrating species of the type $NO_2X$. This implies, of course, that nitration by nitronium salts is different in mechanism to nitrations by media containing nitronium ions, contrary to the views expressed above, and is similar to many nitrations in organic solvents. Like other nitrating species believed to be of the form $NO_2X$, nitronium borofluoride gives rise to the same *ortho* : *para* ratios in the nitration of toluene as do solutions containing nitronium ions.

## 3.2 NITROSATION

The kinetics of aromatic nitrosation at ring carbon have received little attention. The first attempt to determine the nature of the electrophile was made by Ingold *et al.*[117], who measured the rates of the nitrous acid-catalysed nitration of 4-chloroanisole by nitric acid in acetic acid which proceeds *via* initial nitrosation of the aromatic ring. Assuming that the electrophiles are the nitrosonium ion and dinitrogen tetroxide, the rate equation is

$$\text{Rate} = k_2^*[\text{ArH}][\text{N}_2\text{O}_4] + k_2^{**}[\text{ArH}][\text{NO}^+] \tag{43}$$

Since Goulden and Millen showed[102] that dinitrogen tetroxide is slightly ionised according to the equilibrium

$$\text{N}_2\text{O}_4 \rightleftharpoons \text{NO}^+ + \text{NO}_3^- \tag{44}$$

the rate expression becomes

$$\text{Rate} = k_2 \,[\text{ArH}][\text{N}_2\text{O}_4] \tag{45}$$

where $k_2 = k_2^* + K k_2^{**}[\text{NO}_3^-]^{-1}$.

The constancy of the second-order rate coefficients $(k_2)$ in the presence of tetra-alkyl-ammonium nitrates (Table 26) indicates the validity of the assumptions.

TABLE 26

EFFECT OF ADDED NITRATES ON THE RATE OF NITRATION *via* NITROSATION OF
4-ClC$_6$H$_4$OMe BY 4.44 $M$ HNO$_3$ IN HOAc AT 25 °C[117]

| $10^2[HNO_2]$ ($M$) | $10_3[NMe_4{}^+NO_3{}^-]$ ($M$) | $10^3\,k_2$ |
|---|---|---|
| 7.5 | | 34.9 |
| 5.8 | 7.0 | 29.9 |
| 5.0 | 9.7 | 27.0 |
| 2.4 | 12.6 | 25.9 |
| 5.8 | 16.7 | 23.5 |
| 6.2 | 20.8 | 21.4 |
| 6.6 | 23.4 | 21.0 |
| 4.8 | 80.0 | 15.4 |
| 2.5 | 24.2ᵃ | 19.8 |
| 2.1 | 11.5ᵃ | 25.2 |

ᵃ For NEt$_4{}^+$NO$_3{}^-$.

From an analysis of the results, the values of $k_2^*$ and $k_2^{**}$ were determined as $13 \times 10^{-3}$ and $132 \times 10^{-3}$ respectively, which implies, not unreasonably, that NO$^+$ is a more reactive electrophile than N$_2$O$_4$; the nitrosation rate was also relatively independent of the water concentration.

The kinetics of nitrosation of phenol in aqueous mineral acid have been studied in some detail. Suzawa *et al.*[118] showed that, with 0.01 $M$ nitrous acid at 0° a second-order rate coefficient of 0.00148 was obtained and that this was increased to 0.00225 by the addition of hydrochloric acid to pH 1.3. Morrison and Turney[119] studied the effect of varying the perchloric acid concentration upon the rate of nitrosation of phenol by sodium nitrate at 0.6 °C. They confirmed that the rate equation is

$$\text{Rate} = k_2 \, [\text{ArH}][\text{HNO}_2] \tag{46}$$

the rate increasing with acidity (Table 27), and the linear correlation of log rate *versus* the acidity function $H_0$ (at least up to 6 $M$ acid) was interpreted to mean that the electrophile was the nitrousacidium ion $H_2NO_2^+$, rather than the nitrosonium ion $NO^+$. The same conclusion was reached by Schmid *et al.*[120] from a study of the nitrosation of phenol in 70 % aq. sulphuric acid. They also found that, at constant ionic strength, the rate was independent of the hydrogen ion concentration in the range $10^{-5}$–$10^{-1}$ $M$, and the logarithm of the rate was proportional to the ionic strength. Methanol caused a decrease in the rate due to conversion of nitrous acid to its methyl ester; the activation energy for nitrosation of phenol was found to be 15.3.

The conclusions from the foregoing studies with phenol have been challenged by Challis and Lawson[121], who find that rates of nitrosation (shown graphically) pass through a maximum at about 8 $M$ perchloric acid, and also that the reaction shows a large primary kinetic isotope effect at 0.7 °C (Table 27). Hence loss of a

TABLE 27

RATE COEFFICIENTS FOR NITROSATION OF $C_6H_5OH$ AND $[4\text{-}^2H]\text{-}C_6H_4OH$ IN $HClO_4$[121]

| $[HClO_4]$ (M) | $10^3 \, k_2^H$ (0.7 °C) | $10^3 \, k_2^D$ (0.7 °C) | $k_2^H/k_2^D$ | $10^3 \, k_2^H$ (0.6 °C) |
|---|---|---|---|---|
| 0.105 | 1.35 | 0.35 | 3.9 | |
| 1.67 | 3.02 | 0.717 | 4.2 | |
| 2.25 | 4.36 | 1.07 | 4.1 | |
| 2.44 | | | | 2.5 |
| 3.64 | | | | 5.5 |
| 3.86 | 21.7 | 6.5 | 3.3 | |
| 4.13 | 30.8 | 8.66 | 3.6 | |
| 4.87 | 87.5 | 25.7 | 3.4 | 160 |
| 5.47 | | | | 425 |
| 5.89 | 550 | 157 | 3.6 | |
| 6.09 | | | | 950 |
| 6.68 | | | | 1,580 |
| 7.30 | | | | 2,000 |
| 8.55 | | | | 2,580 |

proton from the intermediate would seem to be rate-determining, and it was suggested that this loss may be either spontaneous (at low acidities) or acid-catalysed (at high acidities). The decrease in rate at acidities > 8 $M$ indicated that the protonated intermediate is in equilibrium with the reactants under these conditions.

Clearly, the previously observed acidity dependencies cannot be satisfactorily diagnostic of the nitrousacidium ion being the electrophile, and, significantly perhaps, the rate of N-nitrosation of benzamide follows the $h_R$ acidity function, which indicates that the nitrosonium ion is the electrophile[122].

A large primary isotope effect $k_H/k_D = 3.6$ had also been found earlier by Ibne-Rasa[122a] in the nitrosation of 2,6-dibromophenol in the 4 position which was also shown to be base-catalysed. These values are not unexpected in view of the isotope effect found with diazonium coupling which involves a similarly unreactive electrophile, so that the rate-determining transition state will be displaced well towards products. Furthermore, the intermediate will have a quinonoid structure and will, therefore, be of low energy; consequently, the energy barrier for the second step of the reaction will be high.

### 3.3 DIAZONIUM COUPLING

Conant and Peterson[123] made the first kinetic study of the coupling of diazonium ions with aromatics, and measured the rates of reaction of diazotised aniline and its 2-MeO, 4-Me, 4-Br and 4-SO$_3$H derivatives with the sodium salts of 4-hydroxybenzenesulphonic acid, 2-naphthol-3,6-disulphonic acid, 1-naphthol-3,8-disulphonic acid, and 1-naphthol-4-sulphonic acid. The reaction was second-order, *viz.*

$$\text{Rate} = k_2 \, [\text{ArN}_2\text{OH}][\text{Ar}'\text{OH}] \tag{47}$$

though at pH > 9 a departure from second-order kinetics was observed. It was found that the logarithm of the reaction rate increased as the pH of the buffered solution according to

$$\log k_2 = \log k_2^0 + \text{pH} \tag{48}$$

and some values of $k_2^0$ obtained at 15 °C with a solution of ionic strength 0.24 were: *o*-anisidine + 2-naphthol-3,6-disulphonate, $1.04 \times 10^{-8}$; *o*-anisidine + 1-naphthol-4-sulphonate, $2.4 \times 10^{-8}$; sulphanilic acid + 2-naphthol-3,6-disulphonate, $84 \times 10^{-8}$; sulphanilic acid + 1-naphthol-4-sulphonate, $590 \times 10^{-8}$. These results indicate steric hindrance since the reactivity sequence for substitution at the 1 and 2 positions of naphthalene is the reverse of that expected, and this is also indicated by the different relative reactivities obtained with the two (diazotised) amines and the fact that 2-naphthol-6,8-disulphonate was found to be much less reactive than 2-naphthol-3,6-disulphonate; activation energies (determined only between 15 and 25 °C) were all approximately 15.0. In general, the reactivities of the diazotised amines (XC$_6$H$_4$NH$_2$) followed the sequence (X =)

4-SO$_3$H = 4-Br > H > 4-Me > 2-MeO, and the mechanism was formulated as

$$ArN_2X + OH^- \overset{\text{fast}}{\rightleftharpoons} ArN_2OH + X^- \tag{49}$$

$$ArN_2OH + Ar'OH \overset{\text{slow}}{\rightleftharpoons} ArN_2Ar' + H_2O \tag{50}$$

it being argued that equilibrium concentration of the diazohydroxide would be a function of hydroxide ion concentration and hence pH.

From a comparison of the model curves of the variation in log rate *versus* pH expected for reaction of possible diazonium species with possible amine and phenolic species (Figs. 1 and 2) with those obtained (Table 28) for the reaction of

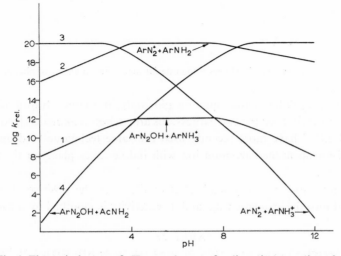

Fig. 1. Theoretical curve of pH *versus* log rate for diazonium coupling of amines.

1-naphthylamine-4-sulphonic acid with diazotised sulphanilic acid and of 1-naphthylamine-8-sulphonic acid with diazotised aniline, Wistar and Bartlett[124] found that the pairs of species giving rise to either curve 1 or 2 (Fig. 1) were indicated. However, it is inconceivable that reaction was occurring on the protonated amine since the observations that *meta* rather than *ortho*, *para* substitution occurs, and that only the most reactive aromatics couple with diazonium compounds (ArNH$_3^+$ being very unreactive in other electrophilic substitutions), are both inconsistent with this mechanism. Hence reaction was interpreted as occurring between ArN$_2^+$ and Ar'NH$_2$ and by analogy, between ArN$_2^+$ and Ar'O$^-$. This latter combination would give the log rate *versus* pH plot 6 (Fig. 2) with a unit slope as observed by Conant and Peterson[123], and also the same kinetic equation that they observed, *viz.*

$$\text{Rate} = k_2[ArN_2^+][Ar'O^-][H_2O] \tag{51}$$

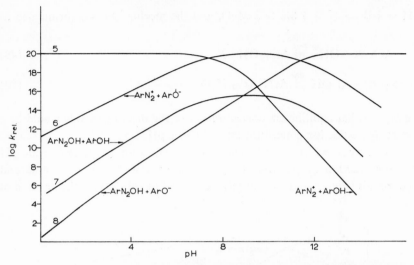

Fig. 2. Theoretical curve of pH versus log rate for diazonium coupling of phenols.

since equation (51) differs from equation (46) only in a concentration of water which is in constant excess; the two expressions are, therefore, kinetically identical. Binks and Ridd[125] have more recently used the same technique to deduce that reaction of 4-nitrobenzenediazonium ion with indole takes place on the neutral indole molecule.

The diazonium ion is a very weak electrophile since the positive charge can be delocalised into the aromatic ring, and its reactivity is modified by substituents

TABLE 28

RATE COEFFICIENTS FOR REACTION OF DIAZOTISED $Ar'NH_2$ WITH $ArH$[124, 125]

| $Ar'NH_2$ = Sulphanilic acid $ArH$ = 1-Naphthylamine-4-sulphonic acid | | Aniline 1-Naphthylamine-8-sulphonic acid | | 4-Nitroaniline Indole | |
|---|---|---|---|---|---|
| $pH$ | $\log k_2{}^a$ | $pH$ | $\log k_2{}^a$ | $pH$ | $k_2{}^b$ |
| 5.18 | 0.41 | 6.25 | 1.70 | 9.20 | 4.50 |
| 4.51 | 0.40 | 4.95 | 1.52 | 8.60 | 5.20 |
| 3.49 | 0.45 | 4.32 | 1.04 | 7.01 | 3.90 |
| 3.43 | 0.01 | 3.49 | 0.81 | 6.78 | 3.16 |
| 2.04 | −0.18 | 2.43 | −0.29 | 6.68 | 3.10 |
| | | | | 6.17 | 2.04 |
| | | | | 5.69 | 2.08 |
| | | | | 5.21 | 2.20 |
| | | | | 4.87 | 2.18 |
| | | | | 4.55 | 2.24 |
| | | | | 3.78 | 2.26 |

$^a$ Calculated assuming the original values were in $l.mole^{-1}.min^{-1}$. At 25 °C.
$^b$ At 0 °C.

in the aromatic ring to the extent that 2,4,6-trinitrobenzenediazonium ion couples with mesitylene which is less reactive than anisole[126], whereas the unsubstituted ion will not even substitute in the anisole. This additionally proves that Ar'O$^-$ rather than Ar'OH must be the reactive entity, since benzenediazonium ion does substitute in phenol, and phenol and anisole usually have comparable reactivities in electrophilic substitution.

Some diazonium couplings are subject to base catalysis and in these cases kinetic isotope effects are observed but since the rate of catalysed reaction is not linearly related to the base concentration (see p. 7), the $S_E3$ mechanism is ruled out and the $S_E2$ mechanism must operate, viz.

$$ ArN_2^+ + Ar'H \underset{k'_{-1}}{\overset{k'_1}{\rightleftarrows}} Ar\overset{+}{\underset{N_2Ar}{\diagdown}}{}^H \xrightarrow[k'_2]{+B} ArN_2Ar' + BH^+ \tag{52} $$

The kinetic isotope effect will thus be observed when $k'_2, k_{-1}$ is small; this will be occasioned by low reactivity of the reagents, and steric hindrance to reaction.

Kinetic isotope effects have not been observed in the reaction of 1-naphthol-4-sulphonic-2-$d$ acid with 2-methoxydiazobenzene[127], imidazole-2,4,5-$d_3$ with 4-diazobenzene sulphonic acid[128], or indole-3-$d$ with 4-nitrodiazobenzene[125], nor has base catalysis been observed in those cases where it has been measured; in each of these reactions one or both of the reagents is relatively reactive.

For coupling with 2-naphthol-6,8-disulphonic-1-$d$-acid the isotope effects ($k_H/k_D$) varied with the substituent in the benzenediazonium ion as follows: 4-Cl (6.55); 3-Cl (5.48); 4-NO$_2$ (4.78), i.e. the reactivity of the ion was increased so that $k_{-1}$ correspondingly decreased. Base catalysis was observed[127, 129], and there was a free energy relationship between this catalytic effect and the basicity of pyridine, 3- and 4-picoline. However, for 2-picoline and 2,6-lutidine, the catalysis was 3 times and 10 times less than expected from their basicities showing that, in this particular proton transfer, steric hindrance is important.

A further effect of steric hindrance in the transition state was shown by comparing the isotope effects obtained in the reactions of the sulphonic acids (II–V)

|  | (II) | (III) | (IV) | (V) |
|---|---|---|---|---|
| $k_H/k_D$ | 1.04 | 3.10 | 6.55 | 6.2 |

with 4-chlorodiazobenzene (the arrow indicates the principal position of substitution)[130]. The isotope effect clearly decreases along with the steric hindrance at the position of substitution. In addition, substitution at the 2 position of (III) occurs more readily than at the 4 position, consequently the 2 : 4 ratio is higher when deuterium is present. The reason is that reaction at the 4 position is more strongly base-catalysed, consequently the kinetic isotope effect for 4 substitution is also larger.

The coupling of 4-chlorodiazobenzene with 1,3,5-trimethoxybenzene-2-$d$ gives rise to an inverse isotope effect ($k_H/k_T = 0.88$) coupled with base catalysis[131]. Since these two observations are seemingly irreconcilable the latter has been attributed to a salt effect, and the inverse isotope effect is considered to be secondary, arising from hybridisation changes at the position of substitution on going through the transition state.

Recently, a kinetic study has been made of the substitution of diazotised sulphanilic acid in the 2 position of 4-substituted phenols under first-order conditions (phenol in excess) in aqueous buffer solutions at 0 °C[131a]. A rough Hammett correlation existed between reaction rates and $\sigma_m$ values, with $\rho$ about -3.8; however, the point for the methoxy substituent deviated by two orders of magnitude and no explanation was available for this. The unexpectedly low $\rho$-factor was attributed to the high reactivities of the aromatic substrates, so that the transition state would be nearer to the ground state than for reaction of monosubstituted benzene derivatives.

## 4. Reactions with electrophilic oxygen and sulphur

### 4.1 HYDROXYLATION AND RELATED REACTIONS

Aromatic compounds can be hydroxylated, benzoxylated and acetoxylated and the isomer distributions and substituent effects indicate that the reaction is an electrophilic substitution[132]. Very little kinetic work has been carried out so that the nature of the electrophile is in some doubt.

#### 4.1.1 Hydroxylation

In hydroxylation, quinones are usually obtained since the initial hydroxyl product is further oxidised. Kinetic studies on the hydroxylation of 1,3,5-trimethoxybenzene with perbenzoic acid gave second-order rate coefficients (Table 29) which remained fairly constant for a wide variation in concentration of aromatic and acid thus indicating that the rate-determining step is bimolecular[133]. The variation was considered to be within the rather large experimental error for the reaction which was very fast and, therefore, studied at low temperature ($-12.4$ °C). Since more than one mole of acid per mole of aromatic was eventually consumed, the mechanism was formulated as

TABLE 29

RATE COEFFICIENTS FOR REACTION OF $PhCO_3H$ WITH 1,3,5-TRIMETHOXYBENZENE AT $-12.4\,°C$[133]

| $10^2[PhCO_3H]$ (M) | $10^2[C_6H_3(OMe)_3]$ (M) | $10^3k_2$ |
|---|---|---|
| 5.64 | 12.0 | 2.72 |
| 5.68 | 9.05 | 2.73 |
| 5.71 | 17.5 | 2.27 |
| 5.68 | 17.7 | 2.58 |
| 7.19 | 7.5 | 2.30 |
| 7.16 | 22.6 | 2.33 |
| 7.17 | 29.8 | 2.08 |
| 7.16 | 28.9 | 1.93 |
| 9.41 | 19.7 | 2.40 |
| 5.04 | 31.8 | 2.40 |

$$(53)$$

Consistent with this mechanism is the observation that trifluoroperoxyacetic acid is reported to be the most effective peracid in aromatic oxidation[134]; the great stability of the trifluoroacetate anion causes it to be an excellent leaving group so that heterolysis to give hydroxyl cations $OH^+$ occurs most readily.

### 4.1.2 Benzoxylation

The first kinetic studies on benzoxylation showed that the reaction of dibenzoyl peroxide with m-cresol was first-order in peroxide[135]. A more extensive study of the reaction of dibenzoyl peroxide with phenol showed that the reaction was not free-radical, was speeded up by electron-supplying groups (though also to a much smaller extent by electron-withdrawing groups) and retarded by bulky ortho-substituents, was neither acid- nor base-catalysed, gave a small isotope effect $k_H/k_D = 1.32\pm0.03$, and followed the rate equation[136]

$$\text{Rate} = k_2\,[\text{phenol}][\text{peroxide}] \qquad (54)$$

Second-order rate coefficients are given in Table 30. The predominance of ortho-

TABLE 30

RATE COEFFICIENTS FOR REACTION OF 0.1 $M$ (PhCOO)$_2$ WITH 3.75 $M$ ArOH IN PhH AT 30 °C

| Substituent in Ar | $10^7 k_2$ | Substituent in Ar | $10^7 k_2$ | Substituent in Ar | $10^7 k_2$ |
|---|---|---|---|---|---|
| 2,4,6-Bu$^t_3$ | <0.1 | 4-Cl | 28.5 | 2-MeO | 77 |
| 2-Me-4,6-Bu$^t_2$ | <0.1 | 2,6-Pr$^i_2$ | 29.9 | 2,6-Me$_2$ | 78 |
| H | 5.7 | 4-Br | 35.7 | 4-Me | 107 |
| 3-Cl | 7.2 | 3-EtO | 45.2 | 2,6-(MeO)$_2$ | 267 |
| 3-Me | 27.4 | 2,4-Me$_2$ | 48.5 | 4-MeO | 9020 |

substitution, the kinetic isotope effect, and the large negative entropy for the reaction ($-26.7$ for 4-methylphenol and $-22.3$ for 4-methoxyphenol, $\Delta H^{\ddagger}$ being 17.4 and 15.3 respectively, 0–40 °C) lead to the proposed mechanism

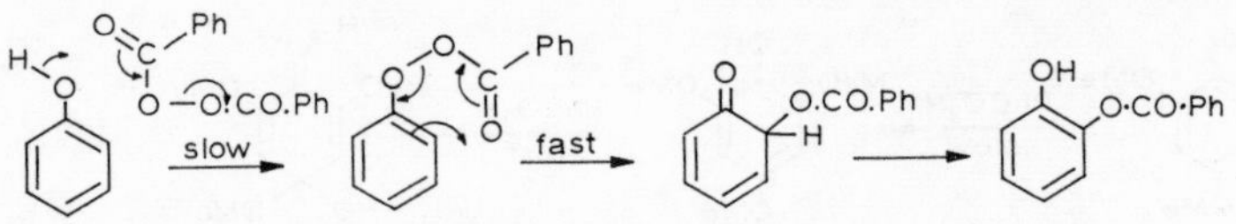

(55)

### 4.1.3 Acetoxylation

Acetoxylation is found to accompany nitration of fairly reactive aromatics by nitric acid in acetic anhydride and gives rise to zeroth-order kinetics[76]. The electrophile is believed to be protonated acetyl nitrate the formation of which is rate-determining, hence the kinetic order (see p. 37). Acetoxylation can also accompany halogenation by positive halogenating agents in acetic acid solvent, especially in the presence of sodium acetate[137], but no kinetic studies have been carried out.

## 4.2 SULPHONATION

Early studies showed that sulphonation, reaction (56)

$$ArH + SO_3(+H_2O) \rightleftarrows ArSO_3H(+H_2O)$$   (56)

involved an electrophilic reagent[138, 139], but kinetic studies have been hampered by the fact that the reaction is reversible, the mechanism may differ according to the strength of sulphuric acid, and be different again in aprotic media, and the reaction products (which can isomerise) are difficult to isolate. In addition, water is produced during the reaction with sulphuric acid, hence rate coefficients decrease during a run unless the acid is in large excess.

Four main mechanisms for sulphonation have been proposed and these may be represented by the following combinations of the steps below: (57), (59), (60); (58), (59), (60); (57), (61), (62); (58), (61), (62).

$$ArH + SO_3(H_2SO_4) \rightleftharpoons Ar{\overset{+}{<}}{\overset{H}{\underset{SO_3^-}{}}} \quad (+H_2SO_4) \tag{57, 57a}$$

$$ArH + H_3SO_4^+ \rightleftharpoons Ar{<}{\overset{H}{\underset{SO_3^-}{}}} + H_3O^+ \tag{58}$$

$$Ar{\overset{+}{<}}{\overset{H}{\underset{SO_3^-}{}}} + H^+(H_2O) \rightleftharpoons Ar{\overset{+}{<}}{\overset{H}{\underset{SO_3H}{}}} \quad (H_2O) \tag{59, 59a}$$

$$Ar{<}{\overset{H}{\underset{SO_3H}{}}} + HSO_4^- \rightleftharpoons ArSO_3H + H_2SO_4 \tag{60}$$

$$Ar{\overset{+}{<}}{\overset{H}{\underset{SO_3^-}{}}} + HSO_4^- \rightleftharpoons ArSO_3^- + H_2SO_4 \tag{61}$$

$$ArSO_3^- + H^+(H_2O) \rightleftharpoons ArSO_3H + H_2O \tag{62, 62a}$$

In addition a number of the species involved may be solvated giving rise to the alternative equilibria (57a) etc., so that clearly to unravel the nature of the appropriate reacting species, the sequence in which they are involved, and the rate-determining step is a formidable kinetic task. It is therefore not surprising that the number of mechanisms proposed, and the number of papers published on the subject, has tended to be comparable. The problem is heightened by the fact that there is a relatively large number of species present in sulphuric acid to consider, and any equilibrium which produces a molecule of water then requires a further molecule of sulphuric acid to ionise the water *via* the equilibrium

$$H_2O + H_2SO_4 \rightleftharpoons H_3O^+ + HSO_4^- \tag{63}$$

Hence equilibria leading to possible sulphonating species are

$$2\,H_2SO_4 \rightleftharpoons H_3SO_4^+ + HSO_4^- \tag{64}$$

$$2\,H_2SO_4 \rightleftharpoons SO_3 + H_3O^+ + HSO_4^- \tag{65}$$

$$3\,H_2SO_4 \rightleftharpoons HSO_3^+ + H_3O^+ + 2\,HSO_4^- \tag{66}$$

$$4\,H_2SO_4 \rightleftharpoons S_2O_6 + 2\,H_3O^+ + 2\,HSO_4^- \tag{67}$$

$$H_2SO_4 + SO_3 \rightleftharpoons H_2SO_4(SO_3) \text{ or } H_2S_2O_7 \tag{68}$$

Suppose $H_3SO_4^+$ is the sulphonating species. Then according to equilibrium (68), $K = [H_2SO_4^+][HSO_4^-]/[H_2SO_4]^2$, and if the concentration of unionised sulphuric acid is maintained approximately constant then $[H_3SO_4^+] \propto 1/[HSO_4^-]$, *i.e.* inversely proportional to the concentration of added water by virtue of (63).

Likewise, water will affect the sulphonation rates by $SO_3$, $HSO_3^+$, and $S_2O_6$ according to the inverse second, third, and fourth powers of its concentration, respectively.

It should be noted that of the four main mechanisms described above, reversal of the first two suggests that the rate-determining step in desulphonation takes place upon the sulphonic acid, whereas in the latter two mechanisms, it will take place upon the anhydride.

### 4.2.1 Sulphonation in non- or part-aqueous media

Most research workers have employed an aqueous sulphuric acid solution of the aromatic, but a few have used other solvent systems and sulphonating agents. Principal among these are Hinshelwood et al.[139], who measured the rates of sulphonation of aromatics by sulphur trioxide in nitrobenzene at temperatures between 0 and 100 °C, a minimum of about 40 °C being employed for each compound. The initial reaction rate was given by

$$\text{Rate} = k_3 \, [\text{ArH}][\text{SO}_3]^2 \tag{69}$$

and the reaction was studied under second-order conditions (aromatic in excess), a correction being applied for the reaction with nitrobenzene for those compounds for which this was significant; the derived third-order rate coefficients at $40 \pm 0.1$ °C are given in Table 31. The second order in sulphur trioxide was interpreted to

TABLE 31

RATE COEFFICIENTS AND ARRHENIUS PARAMETERS FOR REACTION OF $SO_3$ WITH ArH IN $PhNO_2$ AT $40 \pm 0.1$ °C[139]

| ArH | $k_3$ | $E_a$ | log A |
|---|---|---|---|
| Benzene | 40.8, 48.8 | 4.8, 5.5 | ca. 5.24 |
| 4-Nitroanisole | 6.29 | 4.3 | ca. 5.24 |
| 1-Nitronaphthalene | 3.27 | 7.9 | ca. 5.24 |
| Chlorobenzene | 2.40 | 7.7 | 5.14 |
| Bromobenzene | 2.10 | 7.8 | ca. 5.24 |
| 1,3-Dichlorobenzene | $4.36 \times 10^{-2}$ | 9.2 | ca. 5.24 |
| 4-Nitrotoluene | $9.53 \times 10^{-4}$ | 11.0 | 4.79 |
| Nitrobenzene | $7.85 \times 10^{-6}$ | 11.4 | 3.91 |

mean that the electrophile is either $S_2O_6$, or $SO_3$ with a second molecule acting as a base (catalysing the removal of the proton) in a rate-determining step. An alternative possibility that a molecule of sulphur trioxide was interacting with the aromatic substituent was ruled out by the constancy of the kinetic form with change in the aromatic. Rate coefficients decreased during a run but this was not due to water as none is produced under these conditions, and arose from the formation of a complex between sulphur trioxide and the sulphonic acid product; this is thus analogous to solvation of sulphur trioxide by sulphuric acid[140]. The reaction rates were controlled by activation energies except for nitrobenzene and 4-nitroanisole, the latter only being considered anomalous; the third-order rate for nitrobenzene may be in considerable error in view of the method of derivation. The low values of the activation energies are consistent with the highly polar environment and high reactivity of the sulphur trioxide electrophile.

The view that the second molecule of sulphur trioxide may act as a base in a rate-determining step must be considered alongside the rather small isotope effects observed by Cerfontain et al.[141], in the sulphonation of benzene by sulphur trioxide in nitrobenzene ($k_H : k_D = 1.35 \pm 0.16$) and by sulphur trioxide alone ($k_H : k_D = 1.16$). These workers consider that one molecule of sulphur trioxide attacks the aromatic ring to give the usual $\sigma$-complex and a second molecule of sulphur trioxide then adds to the incipient sulphonate group, followed by a rapid intramolecular proton shift to give the arylpyrosulphonic acid which is in equilibrium with the usual sulphonic acid product and sulphur trioxide according to the scheme[142]

$$\text{ArH} + \text{SO}_3 \underset{\text{slow}}{\rightleftarrows} \text{Ar}^{+}\!\!\begin{smallmatrix}\nearrow\text{H}\\\searrow\text{SO}_3^-\end{smallmatrix} \underset{\text{slow}}{\overset{\text{SO}_3}{\rightleftarrows}} \text{Ar}^{+}\!\!\begin{smallmatrix}\nearrow\text{H}\\\searrow\text{SO}_2\text{-O}\end{smallmatrix}\!\!\nearrow^{\text{SO}_3^-} \overset{\text{fast}}{\longrightarrow} \text{ArSO}_2\text{OSO}_3\text{H} \tag{70}$$
$$\big\updownarrow$$
$$\text{ArSO}_3\text{H} + \text{SO}_3$$

This mechanism does not, however, explain the fall-off in rate coefficients observed by Hinshelwood et al.

The kinetics of the sulphonation of chlorobenzene and 1,4-dichlorobenzene by sulphur trioxide in nitromethane[143] follow the rate expression (69) and this dependence has been attributed to mechanism (70). (The rate of sulphonation of anisole in dioxan also depends on the first power of the sulphur trioxide concentration, but the kinetic equation includes a third term which is a complex function of the concentration of sulphur trioxide, sulphuric acid impurity, and sulphonic acid product[144].) For chlorobenzene, rate coefficients ($10^2 k_3$) at $-28$, $-13.5$, and 3.5 °C were 9, 32, and 60 respectively giving $\Delta H^{\ddagger} = 7.5$ and $\Delta S^{\ddagger} = -32$, and the low negative entropy is consistent with the postulated second slow step of the reaction (and would also be consistent with $S_2O_6$ as the electrophile). For dichlorobenzene, third-order coefficients at 20 °C were $43 \times 10^{-7}$ and $36 \times 10^{-7}$ for the protium- and 91.5 % deuterium-containing substrates, which gives $k_H : k_D = 1.23$. Benzene, which reacted too fast for the kinetic order to be determined gave an isotope effect of 1.34 at 20 °C by the competition method, and these small effects were argued to be consistent with the proposed mechanism (70).

With trichlorofluoromethane as solvent, the kinetics followed

$$\text{Rate} = k_2[\text{ArH}][\text{SO}_3] \tag{71}$$

and since the reaction product is again the pyrosulphonic acid, the mechanism was considered to be a modified scheme (70) in which the second step is fast[143]. It could be equally likely that the first intermediate gives the sulphonic acid *via* a proton shift and that this is in equilibrium with the pyrosulphonic acid as before. Dichlorobenzene gave second-order coefficients $(10^4 k_2)$ at $-46.7$, $-28$ and $-10$ °C as 3.8, 21.8, and 225 respectively ($\Delta H^{\ddagger} = 13.1 \pm 1.8$, $\Delta S^{\ddagger} = -16 \pm 7$), and a kinetic isotope effect $k_H : k_D$ of $1.08 \pm 0.17$ at $-28$ °C. At $-35$ °C, benzene gave a value of 1.23 and these low values are not inconsistent with either of the mechanistic proposals. The competitively determined relative reactivities of toluene and benzene at $-18$, 0, and 21 °C were 232–177, 157–122, and 125–63 in nitromethane, and 51–23, 53–24, and 36–21 in trichlorofluoromethane so that there is a roughly decreasing selectivity with temperature which, however, was not paralleled by a change in isomer selectivity. The lower reactivity and higher selectivity in nitromethane was tentatively interpreted in terms of complex formation between sulphur trioxide and nitromethane, *viz.* equilibrium (72)

$$\text{CH}_3\text{NO}_2 + \text{SO}_3 \rightleftharpoons \text{CH}_3 \cdot \overset{+}{\text{N}} \overset{\displaystyle \nwarrow^{O}}{\underset{\displaystyle O-SO_3^-}{}} \tag{72}$$

though the selectivity change would seem to be a natural consequence of the postulated second slow step of the reaction mechanism and one can anticipate that sulphonation of very reactive aromatics by sulphur trioxide in nitromethane will follow the kinetic equation (71).

Hinshelwood *et al.*[145] measured the rates of sulphonation of a wide range of aromatics by sulphuric acid in nitrobenzene, at temperatures between 5 and 100 °C (Table 32), and in particular the effect of adding up to 0.012 $M$ water was determined. The reaction followed the complex rate law

$$\text{Rate} = k_2 [\text{ArH}][\text{H}_2\text{SO}_4](x+a)^{-\frac{1}{3}} - k_1[\text{H}_2\text{SO}_4](1+bx)^{-1} \tag{73}$$

where $a$ and $b$ are constants and $x$ is the water formed in the reaction, and it was concluded from this work that $\text{HSO}_3^+$ was the electrophile, and this conclusion must be considered alongside the fact that activation energies were generally higher than for sulphonation of the same compounds by sulphur trioxide in nitrobenzene (compare Tables 31 and 32). The rates of reaction were here also controlled essentially by the activation energies, though a precise correlation in log rate with activation energies (which are probably very accurate in view of the 45 °C temperature range employed) was not obtained so that entropy factors are important and this is generally true of sulphonation in view of the large electrophile.

Further details relating to the mechanism of sulphonation under these con-

TABLE 32

RATE COEFFICIENTS (INITIAL) AND ACTIVATION ENERGIES FOR REACTION OF $H_2SO_4$
WITH ArH IN $PhNO_2$ AT 40 °C[145]

| ArH | $10^6 k_2$ | $E_a$ | ArH | $10^6 k_2$ | $E_a$ |
|---|---|---|---|---|---|
| Naphthalene | 141.3 | 6.1 | 4-Chlorotoluene | 17.1 | 7.4 |
| m-Xylene | 116.8 | 6.4 | 1,3-Dichlorobenzene | 6.70 | 9.5 |
| Toluene | 78.7 | 6.8 | 4-Nitrotoluene | 3.26 | 9.8 |
| 1-Nitronaphthalene | 26.1 | 8.4 | 1,4-Dibromobenzene | 1.01 | 9.7 |
| Benzene | 15.5 | 7.5 | 1,4-Dichlorobenzene | 0.98 | 9.6 |
| Chlorobenzene | 10.6 | 8.9 | 1,2,4-Trichlorobenzene | 0.73 | 9.9 |
| Bromobenzene | 9.55 | 8.9 | Nitrobenzene | 0.24 | 11.1 |

ditions are sparse, but it may be relevant that the sulphonation of [4-$^3$H]- and [4-$^2$H]-bromobenzene by 110–112 wt. % sulphuric acid in nitrobenzene at temperatures between 0 and 50 °C gave kinetic isotope effects $k_H : k_T$ and $k_H : k_D$ of 1.7–2.5 and 1.3–1.7 respectively[146]. These results are not affected by hydrogen exchange, which does not occur under these conditions, so we may conclude that the breaking of the carbon–hydrogen bond is rate-determining to a small extent.

Eaborn and Taylor[147] measured first-order rate coefficients for sulphonation of some aromatics in mixtures of trifluoroacetic acid–aqueous sulphuric acid, as sulphonation proved to be a troublesome side reaction accompanying hydrogen exchange in these media. They introduced a technique which has been found useful by later workers and makes use of the high solubility of sulphonic acids in

TABLE 33

RATE COEFFICIENTS FOR REACTION OF ArH WITH AQ. $H_2SO_4$–$CF_3COOH$ AT 25 °C[147]

| ArH | $CF_3COOH$ (mole %) | $H_2O$ (mole %) | $H_2SO_4$ (mole %) | $10^7 k_1$ |
|---|---|---|---|---|
| Toluene | 82.36 | 3.17 | 14.47 | 9,550 |
| Toluene | 85.94 | 6.22 | 7.84 | 188 |
| Toluene | 93.60 | 3.95 | 2.45 | 25.8 |
| Toluene | 95.31 | 2.21 | 2.48 | 26 |
| Benzene | 95.31 | 2.21 | 2.48 | <0.15 |
| Benzene | 83.40 | 1.98 | 14.62 | 63 |
| Chlorobenzene | 83.40 | 1.98 | 14.62 | 2.5 |
| Naphthalene | 83.40 | 1.98 | 14.62 | 104 |
| Biphenyl | 83.40 | 1.98 | 14.62 | 48 |
| Biphenyl | 85.08 | 6.42 | 8.50 | 341 |

aqueous solution. The concentration of unsulphonated aromatic in the hexane-extracted alkali-washed reaction products was determined spectrophotometrically and the first-order rate coefficients thus obtained decreased during a run due to the increasing concentration of the water by-product as shown by the fact that this decrease was greater, the lower the initial concentration of water; some initial rate coefficients are given in Table 33. For a given concentration of sulphuric acid, sulphonation in these media was much faster than in the aqueous acid (*cf.* the data below).

### 4.2.2 Sulphonation in aqueous sulphuric acid

The first kinetic study appears to have been that of Martinsen[148], who found that the sulphonation of 4-nitrotoluene in 99.4–100.54 wt. % sulphuric acid was first-order in aromatic and apparently zeroth-order in sulphur trioxide, the rate being very susceptible to the water concentration. By contrast, Ioffe[149] considered the reaction to be first-order in both aromatic and sulphur trioxide, but the experimental data of both workers was inconclusive. The first-order dependence upon aromatic concentration was confirmed by Pinnow[150], who determined the equilibrium concentrations of quinol and quinolsulphonic acid after reacting mixtures of these with 40–70 wt. % sulphuric acid at temperatures between 50 and 100 °C; the first-order rate coefficients for sulphonation and desulphonation are given in Tables 34 and 35. The logarithms of the rate coefficients for sulphonation

TABLE 34

RATE COEFFICIENTS FOR SULPHONATION(S) AND DESULPHONATION (D) OF QUINOL AND QUINOLSULPHONIC ACID, RESPECTIVELY, AT 100 °C[150]

| $H_2SO_4$ (Wt. %) | $10^5 k_1(S)$ | $10^5 k_1(D)$ | $H_2SO_4$ (Wt. %) | $10^5 k_1(S)$ | $10^5 k_1(D)$ |
|---|---|---|---|---|---|
| 38.9 |      | 0.83 | 58.4  | 9.75  | 14.6 |
| 50.9 | 1.17 | 4.50 | 62.9  | 35.6  | 24.5 |
| 53.2 | 2.00 | 6.00 | 62.9ᵃ | 46.6  | 29.4 |
| 53.6 | 2.17 | 6.00 | 62.9ᵇ | 49.8  | 27.5 |
| 56.0 | 4.50 | 9.67 | 63.0  | 38.7  | 26.2 |
| 56.1 | 4.50 | 9.00 | 65.0  | 61.7  | 35.8 |
| 56.3 | 4.67 | 10.5 | 66.2  | 112.5 | 43.3 |
| 58.3 | 9.17 | 13.2 | 69.4  | 305   | 55.0 |

ᵃ Containing 0.213 $M$ $K_2SO_4$.
ᵇ Containing 0.49 $M$ $NaSO_4$.

were linearly proportional to the molarity of sulphuric acid. This work demonstrated what has been subsequently confirmed many times, namely that rates of desulphonation increase less readily with increasing acid concentration than do rates of sulphonation, and one would indeed expect that desulphonation would

TABLE 35

RATE COEFFICIENTS FOR SULPHONATION (S) AND DESULPHONATION (D) OF QUINOL
AND QUINOLSULPHONIC ACID, RESPECTIVELY, AT VARIOUS TEMPERATURES[150]

| $H_2SO_4$ (wt. %) | 80 °C | | 70 °C | | 60 °C | | 50 °C | |
|---|---|---|---|---|---|---|---|---|
| | $10^5k_1(S)$ | $10^5k_1(D)$ | $10^5k_1(S)$ | $10^5k_1(D)$ | $10^5k_1(S)$ | $10^5k_1(D)$ | $10^5k_1(S)$ | $10^5k_1(D)$ |
| 65.0 | 7.58 | 3.78 | | | 0.665 | 0.29 | | |
| 67.9 | 22.8 | 8.09 | 6.67 | 2.03 | 2.01 | 0.575 | | |
| 69.8 | 35.5 | 9.50 | 11.2 | 2.40 | 3.64 | 0.80 | | |
| 71.1 | 59.5 | 13.0 | | | 6.42 | 1.13 | | |
| 73.1 | 133 | 15.2 | | | 13.6 | 1.25 | | |
| 75.45 | | | | | 33.5 | 2.00 | 9.27 | 0.40 |
| 77.4 | | | 196 | 6.73 | 65.0 | 2.33 | 19.7 | 0.50 |

be favoured by more aqueous media since this will cause equilibrium (56) (p. 56) to lie increasingly to the left. The variation in sulphonation and desulphonation rates with temperature show clearly that the activation energy for sulphonation is less than that for desulphonation in this medium and that this difference is greater the more concentrated the acid; again this is the intuitive expectation.

A badly reported kinetic study was that of Lauer and Oda[151], who measured the rates of sulphonation of anthraquinone in 89.5–119.9 wt. % sulphuric acid at temperatures between 130 and 180 °C. First-order rate coefficients were initially reported (without units but probably in min$^{-1}$) for a range of temperatures[151a]; subsequently[151b] another set were given with neither units nor temperatures of measurement, and both sets appeared to be quite unrelated. The latter values (which may be in min$^{-1}$ and appropriate to a temperature of 100 °C, obtained by extrapolation from data at other temperatures) are given in Table 36 and are of little value except to show that there is an unusual dependence of rate on acid concentration; an incorrect account of this dependence was given by Cowdrey and Davies[152]. From an erroneous interpretation of the heats of activation it was concluded that molecular sulphuric acid was the sulphonating species, sulphur

TABLE 36

RATE COEFFICIENTS AT AN UNKNOWN TEMPERATURE AND ACTIVATION ENERGIES
FOR REACTION OF $H_2SO_4$ WITH ANTHRAQUINONE[151]

| $H_2SO_4$ (wt. %) | $10^6k_1$ (min$^{-1}$?) | $E_a$ | $H_2SO_4$ (wt. %) | $10^6k_1$ (min$^{-1}$?) | $E_a$ |
|---|---|---|---|---|---|
| 89.5 | 0.14 | 37.9 | 100.3 | 2.9 | 23.1 |
| 93.3 | 0.37 | 35.5 | 100.7 | 5.3 | 23.9 |
| 95.8 | 0.56 | 33.5 | 101.1 | 1.0 | 28.5 |
| 98.3 | 1.6 | 32.5 | 101.9 | 3.7 | 26.6 |
| 99.2 | 2.9 | 32.2 | 104.5 | 9.0 | 22.0 |
| 100.0 | 44 | 28.5 | | | |

trioxide aiding reaction only through its ability to combine with the water by-product. The decrease in rate in $> 100$ wt. % acid was attributed to sulphur trioxide being present in an inactive form in this acid range. Protonation of the ketone is also likely to be an important factor, and, significantly, the reaction gives extremely high activation energies; the pattern of reaction rate *versus* acid strength is also similar to that found for nitration of carbonyl compounds in nitric acid–sulphuric acid and for that reaction, protonation was considered to be a contributing factor (see p. 13). An investigation of the effect upon the rate of adding up to 3 $M$ bisulphate revealed very small rate and activation energy increases with 89.5 and 95.8 wt. % acid, but with 100.7 wt. % acid the rate decreased by about 50 % though the activation energy increased[153]; all that one may safely conclude from this is that the breaking of the C–H bond is not appreciably base-catalysed.

Lauer and Irie[154] also reported two sets of unitless rate coefficients for the sulphonation of 1,8-benzanthrone in 80.6–99.0 wt. % acid. First-order rate coefficients (believed to be in min$^{-1}$ and extrapolated to *ca*. 95 °C from rates obtained at temperatures up to 170 °C) and activation energies relating to acid strengths (in parentheses) were; 12, 30.1 (91 %); 110, 26.6 (95.6 %), 2500, 24.8 (99 %). With 80.6 wt. % acid an extremely high activation energy of 47.4 was obtained, and this is the highest recorded for an electrophilic substitution. However, rate measurements were made at 150 and 170 °C only, so the value may be subject to considerable error. Rates were unfortunately not determined for this compound in $> 100$ wt. % acid, for it would have been useful to see if the abnormal dependence of the rate upon acid concentration was repeated here. Assuming that the units of the rate coefficients are the same for both compounds, this work indicated that 1,9-benzanthrone is more reactive than anthraquinone.

In a study of the sulphonation and desulphonation of naphthalene Lanz[155] measured the proportion of naphthalene converted to sulphonic acids in its reaction with 51.4–94.4 wt. % acid at temperatures ranging from 60–180 °C, and from this data rough sulphonation rates may be deduced, but no conclusion concerning the mechanism was reached.

Cowdrey and Davis[152] extended the kinetic studies of Martinsen[148] to include a wider range of acid and found that the rate of sulphonation of 4-nitrotoluene in 100.22–101.24 wt. % sulphuric acid was given by equation (71), but in 92–99 wt. % acid the rate was given by

$$\text{Rate} = k_2 \, [\text{ArH}][\text{H}_2\text{O}]^{-1}[\text{H}_2\text{O} + \text{NaHSO}_4]^{-1} \tag{74}$$

*i.e.* the rate varied inversely as $[\text{H}_2\text{O}]^2$, thereby indicating that sulphur trioxide or its solvate with sulphuric acid was the electrophile. Their conclusion that sulphuric acid would be the proton-removing base is subject to criticism because it was assumed that the concentration of unionised sulphuric acid remains constant in the 92–99 wt. % acid range, whereas later work[156] has shown that this is not so. It was also assumed that the activity coefficients of the species indicated in equation (74) are independent of the acidity of the medium. In addition, a later analysis[157] has shown that the bisulphate anion is the most effective base present.

A significant difference in activation energies was found between the reactions carried out in media containing free water and containing free sulphur trioxide. With $< 100$ wt. % acid, $E_a = 27.4$ (the rate at 117 °C should be 0.45 and not as reported) whereas with $> 100$ wt. % acid $E_a = 18.0$; this large difference is probably partly due to experimental error (see below). Rate coefficients calculated according to equation (71) for sulphonation in oleum showed signs of increasing at high sulphur trioxide concentration, especially at low temperature, and this indicated the existence of an additional reaction of higher order in sulphur trioxide; the average second-order coefficient at 25 °C $(8.8 \times 10^{-5})$ was identical to that obtained by Martinsen[148] at the same temperature.

Brand et al.[158] made extensive measurements of the rates of sulphonation of a range of unreactive aromatics by oleum containing up to 41 % of sulphur trioxide at temperatures between 0 and 45 °C, and found the reaction to be first-order in aromatic. First-order rate coefficients at 25 °C are given in Table 37, some of the

TABLE 37

RATE COEFFICIENTS $(10^5 k_1)$ FOR REACTION OF ArH WITH $H_2SO_4$ AT 25 °C[158]

| $H_2SO_4$ (wt. %) | X in $XC_6H_4NMe_3{}^+$ | | | | | Y in $YC_6H_4NO_2$ | |
|---|---|---|---|---|---|---|---|
| | 4-Me | H | 4-F | 4-Cl | 4-Br | 4-Me | H |
| 100.5 | 142 | | | | | 4.7 | |
| 100.9 | 503 | 1.24 | | | | 9.0 | |
| 101.3 | 1,130 | 2.83 | | | | 14.2 | 0.51 |
| 101.8 | 1,700 | 7.1 | | | | 23.5 | 0.85 |
| 102.6 | 5,400 | 21.9 | | | | 43.9 | 1.7 |
| 104.2 | | 129 | | | 1.6 | 100 | 4.2 |
| 104.7 | | 220 | 3.5 | 2.42 | 2.53 | 118 | 5.3 |
| 105.9 | | 665 | 10.2 | 7.56 | 7.72 | | 9.6 |
| 106.9 | | 1,870 | 24.7 | 22.0 | 21.0 | | 12.0 |
| 107.4 | | 3,330 | 38.2 | 33.2 | 35.3 | | 15.2 |
| 108.9 | | | 83.4 | 75.2 | 84.6 | | |

values being obtained by interpolation from data at other acid concentrations. For the sulphonation of 4-nitrotoluene, equation (71) was again obeyed and the second-order rate coefficient $(7.3 \times 10^{-5})$ at low (free) sulphur trioxide concentrations was in very good agreement with the earlier measurements[148, 152], and was confirmed as increasing at higher sulphur trioxide concentrations, e.g. a value of $18.9 \times 10^{-5}$ was obtained in 103.0 wt. % acid. Surprisingly, 4-nitrotoluene was found to be much less reactive than the phenyltrimethylammonium ion, and this was attributed to protonation of the former so that reaction occurred only through the small equilibrium concentration of free base; the true coefficients for reaction of the free base are up to 135 times greater than the observed coefficients. (In the earlier publication a much larger factor was indicated, due to the inaccuracy of the available $H_0$ values, and clearly there is a large margin of error in the rate coefficients derived for the free base.) The dependence of rate upon the acid concentration for the nitro compounds was markedly different from that for quaternary ammonium compounds unless a correction for reaction on the free base was applied. When this was done, the slopes of the log $k$ *versus* $-H_0 + \log P_{SO_3}$ plots ($P_{SO_3}$ is the partial pressure of sulphur trioxide) were about 1.2 and 1.0 for the ammonium and nitro compounds, respectively. Consequently, the relative reactivities given in Table 38 are medium-dependent and are only approximate. The

TABLE 38

RELATIVE RATES AND ARRHENIUS PARAMETERS FOR REACTION OF $XC_6H_4NMe^+$ AND $YC_6H_4NO_2$ WITH $H_2SO_4$ AT 25 °C[157]

| X | Y | $k_{rel}$ | $E_a$ | log $A$ |
|------|------|-------|------|------|
| 4-Me |      | 400   | 13.3 | 7.3  |
| H    |      | 1     | 15.3 | 7.8  |
| 4-F  |      | 0.014 |      |      |
| 4-Cl |      | 0.010 |      |      |
| 4-Br |      | 0.011 |      |      |
|      | 4-Me | 85    | 19.5 | 10.8 |
|      | H    | 0.4   |      |      |

relative rates for the nitro compounds refer to reaction on the free base, whereas the $E_a$ and $A$ values are for the protonated base and the very magnitude of these latter values compared to those for the quaternary ions is a further indication that the nitro group is protonated. The activation energy for 4-nitrotoluene is greater than that obtained by Cowdrey and Davis[152], and is almost certainly more accurate in view of the large (45 °C) temperature range employed. The Arrhenius parameters were independent (within experimental error) of the acid concentration.

In this work, Brand and Horning[158] showed that the rate of sulphonation of phenyltrimethylammonium ion was linearly related to the calculated concentration of protonated sulphur trioxide $HSO_3^+$, indicating it to be the electrophile. Added sulphate anions reduced the rate for 4-nitrotoluene in direct proportion to their concentration, and this followed from the equilibrium

$$SO_4^{2-} + H_2S_2O_7 \rightleftharpoons HS_2O_7^- + HSO_4^- \tag{75}$$

in which the bisulphate anion is produced and this reduces the concentration of $HSO_3^+$ (or $SO_3$) according to equilibria (66) and (65), respectively. Subsequently, Brand et al.[159] showed that $HSO_3^+$ was less likely to be the electrophile since sulphonation rates for the $p$-tolyltrimethylammonium ion were slower in $D_2S_2O_7$ than in $H_2S_2O_7$ ($k_H : k_D \approx 1.35$ at 25 °C), whereas indicator measurements showed the concentration of $DSO_3^+$ in $D_2SO_4$ to be greater than $HSO_3^+$ in $H_2SO_4$[159]. For the sulphonation of nitrobenzene and aryltrimethylammonium ions in oleum at 25 °C, a substrate isotope effect $k_H : k_D$ of 1.6–2.1 was obtained which was independent of the acid composition. For these substrates acid-catalysed hydrogen exchange was sufficiently slow not to affect the validity of these observations. In view of this isotope effect and the correlation with $H_0$ indicating a rate-determining proton transfer, they proposed that for sulphonation in oleum the mechanism consisted of reactions (57), (59), and (60), (though $HSO_4^-$) was not specified as the base in the latter step) and reactions (59) and (60) were rate-determining. They also showed that a mechanism proposed by Gold and Satchell[160] (who studied the sulphonation of benzene in 77.5–87.6 wt. % sulphuric acid) was unacceptable for explaining the kinetics with 101–109 wt. % acid*, since the Brönsted rate equation predicts a linear correlation of rate *versus* activity of sulphur trioxide for this mechanism, contrary to observation. The mechanism required that $SO_3$, or less likely but by no means impossibly $HSO_3^+$, forms a $\pi$-complex with the aromatic in a fast step, followed by slow interchange of $SO_3$ and H and then a fast removal of the $\pi$-bonded hydrogen to the $SO_3^-$ substituent. This mechanism was thus analogous to the no-longer-accepted mechanism for acid-catalysed hydrogen exchange (see p. 198).

From a study of the sulphonation of arylsulphonic acids in 93.3–109.8 wt. % sulphuric acid at 25 °C, Cerfontain[161] concluded that the Brand mechanism for sulphonation in oleum was an acceptable one, and the slopes of the log $k$ *versus* $-H_0 + \log P_{SO_3}$ plots were all about 1.2 as for the arylammonium compounds. In the region of < 100 wt. % acid, plots of log $k$ *versus* log $[SO_3]$ were linear with slopes of about 0.8 and this was considered to be in agreement with a mechanism consisting of reactions (57), (61) and (62), the former being the slow step; an intramolecular proton shift was considered to be an acceptable alternative to the latter two steps. Additional evidence regarded as favourable to the concept of two mechanisms was the fact that first-order rate coefficients showed a very dramatic increase with increasing acid strength in the region of 100 wt. % acid (Table 39)[161, 162]. A similar change in rate with change in acid concentration in the region of about 100 wt. % had previously been observed by Kilpatrick and Meyer[163] in the sulphonation of mesitylenesulphonic acid at 12.5 °C (Table 40), and the inconsistencies in data in the Table are due to the difficulty in determining the acid concentration in the range employed; a similar difficulty was found by Cerfontain. It will be seen that the rate coefficients for mesitylenesulphonic acid in Tables 39 and 40 differ by orders of magnitude and this reflects the experimental difficulties. Since the activity of water decreases very markedly in this region (a

---

* Medium contains free anhydride.

*References pp. 388–406*

TABLE 39

RATE COEFFICIENTS ($10^5 k_1$) FOR REACTION OF $ArSO_3H$ WITH $H_2SO_4$ AT 25 °C[161,162]

| $H_2SO_4$ (wt. %) | Substituent in Ar | | | | | | | | | |
|---|---|---|---|---|---|---|---|---|---|---|
| | H | 3-Me | 2-Me | 4-Me | 2,3-Me₂ | 3,4-Me₂ | 2,4-Me₂ | 2,4,6-Me₃ | 2,3,4-Me₃ | 1,3,4-Me₃ |
| 93.3 | | | 0.0082ᵇ | | | | | | | |
| 95.9 | | | 0.0152ᵇ | | | | | | | |
| 97.4 | | | | 0.0018 | | | | | | |
| 98.8 | | | 0.071 | | | | | | | |
| 99.0 | | | 0.18 | 0.0072 | | | | | | |
| 99.6 | | | 1.0 | 0.26 | 5.9 | 0.69 | 41 | 1.1 | 68 | 1.1 |
| 100.0 | | | 49 | | | | | | | |
| 100.05 | | | 1,070 | 84 | | | | | | |
| 100.15 | 0.074 | 0.81 | 1,600 | 140 | | | | | | |
| 100.4 | 0.76 | 6.3 | | | | | | | | |
| 100.5 | | | 6,700 | 330 | | | | | | |
| 100.8 | 45 | 44 | | | | | | | | |
| 101.4 | | | | 2,200 | | | | | | |
| 101.7 | 8.5ᵃ | 100 | | | | | | | | |
| 103.3 | 56 | 470 | | 10,500 | | | | | | |
| 105.4 | 272 | 2,690 | | | | | | | | |
| 109.8 | 5,600 | | | | | | | | | |

ᵃ Rate coefficients at 36.5, 45.4, 54.5, and 64.1 °C were 28.3, 58, 122, and 304, giving $E_a = 18.0$.
ᵇ Rate coefficients at 65.0 °C in 93.3 and 95.9 wt. % acid were 0.70 and 0.98 giving $E_a$ values of 22.3 and 20.9 respectively.

factor of $10^3$ from 97.4 to 100 wt. % acid) the results were seen as consistent with the kinetic equation (76) (p. 70), which they found to describe the rate of sulphonation of benzene in more aqueous media.

As in the sulphonation of nitro compounds, the observed rate coefficients are not the true rate coefficients for reaction of the neutral substance since protonation occurs: the true rate coefficients are thus greater and increasingly so the higher the acid concentration, *e.g.* in 100.2 and 109.8 wt. % acid the coefficients for reaction of the unprotonated benzenesulphonic acid are 0.162 and 826,000, respectively, which are 2.2 and 147 times greater than observed. From the corrected rate coefficients, average partial rate factors ($f_m^{Me}$, 14.5; $f_o^{Me}$, 3,260; $f_p^{Me}$, 37,000) may be calculated and these extremely high values are consistent with reaction upon highly deactivated substrates; there was a large variation in the individual values,

TABLE 40

FIRST-ORDER RATE COEFFICIENTS FOR REACTION OF $2,4,6\text{-}Me_3C_6H_2SO_3H$ WITH $H_2SO_4$ AT 12.5 °C[163]

| $H_2SO_4$ (wt. %) | 98.20 | 98.40 | 98.48 | 98.52 | 98.56 | 98.64 | 98.84 | 98.90 |
|---|---|---|---|---|---|---|---|---|
| $10^5 k_1$ | 3.0 | 2.4 | 3.9 | 8.5 | 2.8 | 46 | 65 | 41 |

*e.g.* the *meta* values varied randomly from 11.5 to 17.8 and this undoubtedly indicates the magnitude of the experimental error.

Some of the most extensive kinetic studies of sulphonation have been carried out using benzene and alkylbenzenes. For benzene, first-order rate coefficients obtained by various workers[157, 160, 164, 165] are given in Table 41, and in order to facilitate presentation and assimilation some of the coefficients (those italicised) have been obtained by interpolation from plots of log rate *versus* acid strength in the original data. It is immediately apparent from the Table that there are considerable differences in the rate for reaction at 25 °C. (Note that in ref. 165, the authors plot of their own data does not correspond with the data given in their table, which appears to be totally in error.) There may be a number of factors responsible for these differences: firstly, the extreme sensitivity of the reaction to acid concentration means that a small error in estimating the acid concentration leads to a large error in the rate coefficient, and the preparation of standard

TABLE 41

RATE COEFFICIENTS ($10^6 k_1$) FOR REACTION OF PhH WITH $H_2SO_4$

| $H_2SO_4$ (wt. %) | Temperature (°C) | | | | | |
|---|---|---|---|---|---|---|
| | 5 | 12.3 | 25ᵃ | 25ᵇ | 25ᶜ | 45 |
| 77.6 | | | | 0.26 | 0.061 | 0.86 |
| 79.2 | | | | 0.75 | *0.18* | |
| 81.9 | 0.088 | | | 5.5 | *1.55* | 20.3 |
| 82.9 | | | | *11.0* | 3.3 | |
| 83.4 | | | | *15.5* | *5.12* | |
| 84.0 | | | | *22.9* | *8.12* | 1,049 |
| 85.1 | | | | *57* | 20.0 | |
| 85.7 | | | *46* | 76 | *35* | |
| 86.5 | | 25.8 | *87* | 132 | *63.1* | |
| 87.2 | 11.9 | | *145* | 214 | *115* | 1,150 |
| 87.7 | | 66.8 | *209* | 320 | *182* | |
| 88.3 | | 108 | *330* | | 289 | |
| 88.7 | | 153 | *448* | | *400* | |
| 89.1 | | | *590* | | *500* | 3,960 |
| 89.4 | | 266 | *740* | | 590 | |
| 89.7 | | 317 | *922* | | 708 | |
| 91.0 | 205 | | *2,370* | | 1,744 | 12,000 |
| 91.2 | | 878 | *2,760* | | 2,111 | |
| 91.4 | | | *3,160* | | 2,200 | |
| 93.4 | 1,010 | | | | 8,060 | |
| 96.3 | 5,000 | | | | 57,200 | |
| 97.6 | 10,820 | | | | | |
| 98.8 | 38,500 | | | | | |
| 99.2 | 103,500 | | | | | |

ᵃ Ref. 164.
ᵇ Ref. 160.
ᶜ Ref. 151.

solutions for titration calls for considerable care in view of the hygroscopic nature of the acid. Secondly, the half-lives of the reactions recorded in Table 41 range from 6.5 sec to 17 weeks. For reactions of the former speed, quite exceptional skill is needed to obtain accurate results, whilst for reactions of the latter speed, extreme precautions are necessary to prevent loss of benzene from these solutions (which happen to involve the poorest solvents in the range) and anything less than a totally glass-sealed vessel is probably inadequate; loss of benzene will lead to rate coefficients which are too high. Thirdly, use of vessels with an appreciable vapour space will lead to rate coefficients which are too low unless the vessel is shaken very rapidly so that a rapid equilibrium is established between vapour and solution; this equilibrium must, of course, be more rapid than the rate of sulphonation.

Considerable differences in activation energies were also obtained, those reported by Kilpatrick et al.[165] being 21.8 (85.5 %) and 12.0 (91.5 %) and those by Cerfontain et al.[157] being 25.2 (77.6 %), 24.8 wrongly reported as 23.8 (81.9 %), 20.0 (87.2 %), 17.8 (91.0 %), 17.6 (93.4 %), and 18.0 (95.9 %). The former set of values were obtained only over a 13 °C range and the value of 12.0 is unquestionably wrong by a large margin; both sets of data show the usual sulphonation pattern of decreasing activation energy with increasing acid strength.

Different mechanisms have been advanced to account for the dependence of rate upon concentration, and since the experimental data does not agree it is no surprise that the mechanistic conclusions are different. Gold and Satchell[160] advanced the mechanism which has been described above (p. 67) and which is no longer acceptable. Kilpatrick et al.[165] found the rate coefficients to be approximately correlated by means of the equation

$$\text{Rate} = k_3[\text{ArH}][\text{H}_2\text{SO}_4]_{\text{sp}}^2/a_{\text{H}_2\text{O}} \tag{76}$$

where $[\text{H}_2\text{SO}_4]_{\text{sp}}$ is the concentration of the molecular sulphuric acid species present, and it was argued that this equation was consistent with either $\text{H}_2\text{S}_2\text{O}_7$, $\text{SO}_3$ or $\text{HSO}_3^+$ being the electrophile and $\text{H}_2\text{S}_2\text{O}_7$, $\text{H}_2\text{SO}_4$ or $\text{HSO}_4^-$ being the corresponding proton-removing bases. For the rate law to be obeyed, the addition of the electrophile has to be rate-determining for the first possibility, whereas the removal of the proton has to be rate-determining for the latter two alternatives. Cerfontain et al.[157] examined the correlation of log $k$ with the logarithms of the concentrations of the various species present in sulphuric acid and since their measurements covered a wider range of acid concentrations than the other two studies, the conclusions should be the most valid. Correlations with the concentration of $\text{SO}_3$ or $\text{H}_2\text{S}_2\text{O}_7$ were the most satisfactory linear, the slopes being approximately 0.7 and 1.0, respectively. Addition of up to 1.57 $M$ bisulphate ion to 98.8 wt. % acid at 5 °C was found to decrease the rate in direct proportion to the total bisulphate ion concentration, which follows from equilibrium (65) if $\text{SO}_3$

(or $H_2S_2O_7$) are the electrophiles. The removal of a proton is not then appreciably base-catalysed and is unlikely to take place in a predominantly rate-determining step, though this cannot be confirmed by kinetic isotope studies since hydrogen exchange of benzene is more rapid than sulphonation in these media[166]. It was argued that $H_2S_2O_7$ was less likely to be the sulphonating species in oleum since plots of log $k$ *versus* log $[H_2S_2O_7]$ and log $[H_2S_2O_7] - H_0$ for benzenesulphonic acid had slopes of 0.7–0.4 and 0.5–0.1, respectively, and, therefore, it was unlikely to be the species in aqueous sulphuric acid either. However, this argument loses force through the subsequent conclusion that different mechanisms apply in aqueous acid and oleum[161] (see p. 77).

The kinetics of the sulphonation of alkylbenzenes have been examined and the first-order rate coefficients are gathered together in Table 42[147a, 164, 167, 168];

TABLE 42

RATE COEFFICIENTS ($10^6 k_1$) FOR REACTION OF PhR WITH $H_2SO_4$

| $H_2SO_4$ (wt. %) | R = Me | | | | R = Et | R = i-Pr | R = t-Bu | | |
|---|---|---|---|---|---|---|---|---|---|
| | 5 °C | 25 °C$^a$ | 25 °C$^b$ | 45 °C | 25 °C | 25 °C | 5 °C | 25 °C | 35 °C |
| 70.0 | | | 0.045 | | | | | | |
| 72.4 | | | *0.195* | 3.04 | 0.17 | | | 0.062 | |
| 74.8 | | | 0.93 | | | | | | |
| 75.3 | | 1.76 | *1.23* | | | 0.56 | | 0.43 | |
| 77.5 | 0.295 | | | | | | 0.099 | 2.86 | 14.8 |
| 77.6 | | 8.50 | 4.63 | 79.5 | 4.3 | 3.25 | | 3.22 | |
| 79.8 | | 35.9 | 27.2 | | 25.2 | 20.9 | 0.63 | 17.2 | 77 |
| 81.14 | | 97.6 | 67.5 | | | | | | |
| 81.9 | 9.3 | *170* | *113* | 1,410 | | | | 66 | |
| 82.9 | | *355* | 224 | | 180 | | | | |
| 84.0 | | | | | | | 18.6 | 205 | 690 |
| 84.1 | 53.3 | *800* | *520* | 5,650 | | 267 | | | |
| 84.3 | | | 523 | | 524 | | | | |
| 85.8 | | | | | | | 88 | 520 | 1,520 |
| 86.1 | | | | 16,700 | 1,304 | 1,100 | | | |
| 87.2 | 426 | | 3,680 | 43,000 | 3,030 | 1,390 | | 680 | |
| 89.1 | 1,720 | | 13,900 | 116,000 | | 1,580 | | 1,450 | |
| 91.0 | 3,210 | | 53,000 | | | | | 5,200 | |
| 93.4 | 12,600 | | | | | | | | |
| 95.9 | 44,700 | | | | | | | | |

$^a$ Ref. 147a.
$^b$ Ref. 167.

italicised rate coefficients are those obtained by interpolation from graphs of log rates *versus* acid strength, and are intended to facilitate assimilation of the data.

It can be seen from the Table that there is reasonable agreement between the data for toluene obtained by the radioactive labelling[147a], and spectrophotometric[167] methods, but as in the case of benzene sulphonation, the rate coefficients obtained by Cerfontain et al.[167] are the lowest values which may or may not be significant; the differences in rate coefficients are equivalent to an error of 0.5 % in estimating the concentration of sulphuric acid.

Rate coefficients for the alkylbenzenes increase with acidity in the order toluene > ethylbenzene > $i$-propylbenzene ≈ $t$-butylbenzene, and this is not a simple selectivity effect because the order is also that of decreasing reactivity and for a selectivity effect it should be the converse. Steric hindrance is undoubtedly a major factor and if there is a change in the electrophile with acidity, such that its size increases with increasing acidity, the observed results would follow. This is also indicated by the change in activation energy with acidity which decreases regularly for toluene from 23.1 (81.9 %) to 18.5 (89.1 %) and for $t$-butylbenzene from 28.5 (77.5 %) to 16.0 (89.1 %). Since the greater decrease in activation energy for $t$-butylbenzene is not compensated for by an increase in relative reactivity but rather the reverse, there must be significant changes in activation entropy. In fact the sulphonation of both toluene and $t$-butylbenzene has a considerably more negative activation entropy in the stronger acid and the difference in activation entropy from that for sulphonation in the weaker acid is greater for $t$-butylbenzene, suggesting a change in the size of the electrophile.

Comparison of the toluene and benzene sulphonation data shows the relative reactivities to be very medium dependent, the $k_{rel}$ values being stated to vary from 30 (91 wt. %) to 110 (79 wt. %)[167]. This seems intuitively to be too large, and, indeed, careful examination of the authors' data shows the latter value to be not greater than 100 and it is probably about 85. The range of values disagrees with the constant value of 31 ± 1.5 found by Eaborn and Taylor[147a] over a limited acid range, but since this value was obtained by comparison of their toluene data with the benzene data of Gold and Satchell[160] it cannot be regarded as other than very approximate. The decrease in reactivity is not merely due to the possible effect of decreased *ortho* substitution, for the *para* partial rate factors ($f_p^{Me}$) also decreased as follows: 301 (81.9 %), 150 (85.5 %), 85 (89.1 %)[167]. Interestingly, the *para* partial rate factors for sulphonation of *tert.*-butylbenzene (18, 89 %; 53, 86.3 %)[169, 170] are considerably less than for toluene (84; 140) so that sulphonation follows the hyperconjugative order as do most electrophilic substitutions, carried out in sulphuric acid. Furthermore, the *para* : *meta* substitution ratio for sulphonation of $t$-butylbenzene decreases regularly towards 100 wt. % acid and then increases again, suggesting that a change in mechanism occurs at this point[169] as described above (p. 67).

Kilpatrick and Meyer[163] measured the first-order rates of sulphonation of some alkylbenzenes by sulphuric acid at 12.3 °C. Rate coefficients were given graphically only and the parameters for the linear log $(k \times 10^6)$ *versus* $-a+b$ [H$_2$SO$_4$]$_{st}$ plots (where the stoichiometric acid concentration is expressed in mole.l$^{-1}$) were as follows, the acid range examined being given in parentheses:

toluene, 17.013, 1.3385 (15.5–14.6); *p*-xylene, 16.765, 1.3600 (15.2–14.4); *o*-xylene, 16.637, 1.2168 (15.2–14.2); 1,2,4-trimethylbenzene, 12.933, 1.1361 (14.4–13.4); *m*-xylene, 12.781, 1.1286 (14.4–13.6); 1,2,3-trimethylbenzene, 11.216, 1.0296 (14.4–13.4). For mesitylene[171], rates of sulphonation, governed by the

TABLE 43

RATE COEFFICIENTS $(10^5 k_1)$ FOR SULPHONATION OF MESITYLENE AND DESULPHONA-
TION OF MESITYLENE SULPHONIC ACID AT 12.3 °C[171]

| $H_2SO_4$ (wt. %) | Sulphonation | Desulphonation |
|---|---|---|
| 72.0 | 3.67 | 1.89 |
| 73.9 | 10.5 | 2.81 |
| 75.5 | 38.0 | 3.61 |
| 76.2 | 52.3 | 4.25 |
| 77.3 | 107 | 5.24 |
| 77.8 | 147 | 5.06 |

parameters 11.671, 1.100 (13.5–12.0) were comparable with rates of desulphonation (Table 43) and this no doubt arises from steric acceleration favouring the latter. Thus reaction of mesitylene with sulphuric acid does not go to complete formation of the sulphonic acid as in the case of the other hydrocarbons studied. The data in Table 43 show that as noted above (p. 62), rates of sulphonation increase more rapidly with increasing acid concentration than do rates of desulphonation.

Since the dependence of the rate of benzene sulphonation on the acid concentration was given by $\log (k \times 10^6) = -19.373 + 1.379 [H_2SO_4]_{st}$ relative rates of sulphonation in 82.7 wt. % acid at 12.3 °C can be calculated as follows: benzene, 1.0; toluene, 53; *p*-xylene, 195; *o*-xylene, 210; 1,2,4-trimethylbenzene, 690; *m*-xylene, 740; 1,2,3-trimethylbenzene, 935; mesitylene, 3,710. The relative reactivity of benzene to toluene is thus within the range found by Cerfontain *et al.*[167], and for the most part the dependence of rate acidity decreases with increasing reactivity. However, the dependence for *p*-xylene was very much greater than for *o*-xylene even though they possess very similar reactivities in the acid range studied, and analysis of the data of Kaandorp[172] and Prinsen[162] (Table 44) confirms this result. Interpolation of the data of the Dutch group to the temperature used by Kilpatrick *et al.*[171] again shows that the former workers obtain lower rate coefficients. The relative reactivities also differ, being as follows: toluene, 85 (incorrectly quoted as 108); *p*-xylene, 265; *o*-xylene, 387; *m*-xylene, 1,663. Much of the discrepancy, however, seems to arise from the difference in the reactivities of benzene since the reactivities relative to toluene are closely similar; at no concentration studied, however, did Cerfontain *et al.*[168] find the relative reactivities of *o*- and *p*-xylene actually to reverse, in contrast to Kilpatrick and Meyer[163].

TABLE 44

FIRST-ORDER RATE COEFFICIENTS $(10^6 k_1)$ FOR REACTION OF ArH WITH $H_2SO_4$[162, 172]

| $H_2SO_4$ (wt. %) | p-Xylene | | | o-Xylene 25 °C | m-Xylene 25 °C |
|---|---|---|---|---|---|
| | 5 °C | 25 °C | 35 °C | | |
| 70.9 | | | | 0.40 | |
| 72.3 | 0.027 | 0.45 | 1.56 | 0.93ᵃ | 5.4 |
| 75.2 | | | | | 33 |
| 77.5 | 0.96 | 18.5 | 64.2 | 29 | |
| 78.1 | | 26 | | 38 | 163 |
| 79.8 | | | | 119 | 488 |
| 81.8 | 22.9 | 342 | 1,020 | 500 | |
| 84.8 | | | | | 8,060 |
| 86.1 | | 5,500 | | 7,400 | |
| 87.0 | 770 | 12,600 | 43,100 | | |
| 88.9 | 1,295 | 22,500 | 71,200 | | |

ᵃ Italicised data is interpolated from values at other acidities.

An abnormal dependence of rate upon acidity for the sulphonation of p-xylene was found by Leitman et al.[173], who measured the rates of sulphonation of o-, m-, and p-xylenes and ethyltoluenes, and also of ethylbenzene in 70–90 wt. % acid between 0 and 140 °C. The logarithms of rates were again linearly related to the acid concentration, the xylenes were 2–4 times as reactive as the ethyltoluenes, and in the main, activation energies decreased with increasing acidity. The reactions were studied under pseudo-homogeneous conditions, the speed of mixing of the heterogeneous reaction mixtures being in the range where variation of speed did not alter the speed of sulphonation, thereby indicating that a homogeneous reaction was taking place. Since an equilibrium was established between dissolved and undissolved aromatic, zeroth-order conditions prevailed. Whilst at any particular temperature, the reactivities of o- and m-xylene relative to ethylbenzene were relatively independent of an increase of acid concentration, the reactivity of p-xylene increased dramatically. Contrary to the usual observations, the relative reactivities *increased* with increasing temperature; this seems unlikely and may indicate that the reaction being followed was not truly homogeneous. The reason for the different acidity dependence of p-xylene is not known but seems well established. A possible factor is the existence of two mechanisms for sulphonation with differing steric requirements (see below).

Rates of sulphonation of naphthalene by aqueous sulphuric acid have been measured[174], the first-order coefficients $(10^5 k_1)$ at 25 °C (wt. %) acid in parentheses) being as follows: 13.1 (79.0), 38.7 (80.6), 57.8 (81.1), 122 (82.3) and 217 (83.4). The reactivity is thus slightly lower than that of toluene, and relative to benzene decreases from 80 (79.0 wt. %) to 43 (83.4 wt. %). The results are thus very different to those obtained with media composed mainly of trifluoroacetic acid (Table 33), and, once again, reactions carried out in this acid give a much greater spread of rates than otherwise, presumably due to its poorer solvating power. The ratio $\log f_1^{naph} : \log f_2^{naph}$ (where $f$ is the partial rate factor) was found to remain constant (1.57–1.64) over the acid range studied and to conform very well to the linear free energy analysis of substitution in naphthalene given by Taylor and Smith[175]. This is very surprising in view of the extreme steric hindrance found in sulphonation so that clearly an unexplained factor is operating here.

The kinetics of the sulphonation of chlorobenzene have been examined by Kort and Cerfontain[176] in order to obtain further information regarding the sulphonation mechanism in very concentrated aqueous sulphuric acid media. First-order rate coefficients (obtained at assorted temperatures) are given in Table 45 and plots of the logarithms of the *para* partial rate *coefficient* (used in

TABLE 45

RATE COEFFICIENTS FOR REACTION OF PhCl WITH $H_2SO_4$[176]

| $H_2SO_4$ (wt. %) | Temp. (° C) | $10^5 k_1$ |
|---|---|---|
| 83.4 | 25.0 | 0.034 |
| 87.9 | 25.4 | 1.34 |
| 91.9 | 25.0 | 26.4 |
| 94.7 | 25.0 | 158 |
| 95.7 | 25.5 | 268 |
| 96.6 | 25.0 | 506 |
| 98.2 | 25.9 | 1,950 |
| 98.5 | 26.5 | 2,240 |
| 98.9 | 26.0 | 4,000 |
| 99.2 | 26.8 | 4,620 |
| 99.6 | 25.7 | 10,500 |

order to avoid any complications from steric effects) against $\log a_{H_2S_2O_7}$ were more satisfactorily linear (slope 0.93–0.6) than those against $\log a_{SO_3}$ (slope 1.34–0.6). (Note that the scale of the abscissa was in error for this plot by three units throughout.) The earlier proposal of Cerfontain *et al.*[157] that $SO_3$ rather than $H_2S_2O_7$ was the electrophile was retracted since it was unsoundly based as argued above (p. 71). The departure from linearity at high acidity (> 98 wt. % acid) was thought to be due to the proton-removal step becoming rate-determining. If this were catalysed by bisulphate anion, then the decreasing concentration of this with increasing acidity would cause the rate to tend towards a decrease with increasing acidity. This proposal was supported by the observation of a kinetic isotope effect $k_H : k_D = 1.25$ (95 wt. %) which increased to about 2.5 (97.5 wt. %), *i.e.* the proton removal becomes increasingly rate-determining at higher acidity. Under the reaction conditions hydrogen exchange occurred and a method was given for determining the true isotope effect for sulphonation in presence of this exchange.

From this work a relative reactivity for chlorobenzene to benzene of 0.065 is obtained and this may be compared with a value of 0.34 obtained by Kilpatrick and Meyer[163] for bromobenzene at 12.3 °C; this is rather a large difference in view of the fact that there is very little *ortho* sulphonation of either substrate.

Recently, much of the previous kinetic data on sulphonation has been re-examined by Kort and Cerfontain[177], who have concluded that in aqueous sulphuric acid, two mechanisms are involved. Plots of logarithms of rate coefficients for sulphonation of benzene, some alkyl- and halogeno-benzenes and the xylenes against logarithms of the activities of $H_2S_2O_7$ and of $H_3SO_4^+$ are of approximately unit slope for the former with $> 85$ wt. % acid and for the latter with $< 80$ wt. % acid. It was, therefore, concluded that in the weaker acid media the electrophile is $H_3SO_4^+$, the reaction of which with the aromatic is the rate-determining step. The base-catalysing removal of the proton in the subsequent fast step was assumed to be the bisulphate ion in preference to $H_2SO_4$, since the former is the stronger base and in less than 92 wt. % acid it is present in higher concentration. From a further analysis of the data it was concluded that $H_3SO_4^+$ is a weaker electrophile than $H_2S_2O_7$, the respective rho factors being $-9.3$ and $-6.1$, and the former also gives the lowest *ortho* : *para* ratios, since the transition state is nearer to the Wheland intermediate so that the carbon–sulphur bond is shorter in this state leading to greater hindrance. (Electronic considerations would also predict that the reaction with the larger rho factor should have the smaller *ortho* : *para* ratio[178], but the variation in the present circumstances seems to be too great to be accounted for by this alone.)

In conclusion, it is appropriate to summarise the current views on the various mechanisms of sulphonation.

(*a*) The electrophile is predominantly $H_3SO_4^+$ in $< 80$ wt. % acid, changes to $H_2S_2O_7$ in $> 85$ wt. % acid (the exact concentration at which change over from reaction predominantly *via* a particular electrophile to reaction *via* the other depends upon the reactivity of the aromatic), and in oleum it is $SO_3$. This is reasonable since we progress from $SO_3$ solvated by $H_3O^+$, then by $H_2SO_4$, through to the poorly solvated or unsolvated species and the point at which the concentration of $H_2SO_4$ becomes greater than $H_2O^+$ is *ca.* 91 wt. % acid[179]; the 'unsolvated' $SO_3$ will be very much more reactive, hence the dramatic increase in rate at 100 wt. % acid.

(*b*) Attachment of $H_3SO_4^+$ to the aromatic is *rate-determining*, as is the attachment of $H_2S_2O_7$ except in media $> 97$ wt. %. This latter change is not unexpected since the concentration of this species is approaching a maximum. In oleum, the very high reactivity of the 'unsolvated' $SO_3$ causes the attachment of this to the aromatic to be *fast*. The lower reactivity of $H_2SO_4^+$ relative to $H_2S_2O_7$ causes the activation energy for the first rate-determining step to be higher, consequently observed activation energies decrease towards higher acidity.

(c) Removal of the aromatic proton in the weaker acid media is *fast*, and this is reasonable in view of the high concentration of highly basic $HSO_4^-$. This condition persists until the concentration of $HSO_4^-$ becomes very small, the predominant base being then $H_2SO_4$, and thus in media > 97 wt. % acid, removal of the aromatic proton becomes *rate-determining*. In oleum this transfer may be rate-determining.

(d) The attachment of the proton to the sulphonate group in the intermediate is *fast* in aqueous acid, but probably rate-determining in oleum. The latter transfer was concluded to be rate-determining in view of the dependence of rate upon the acidity function; this dependence is not now regarded as a rigorous diagnosis of mechanism so the situation in oleum is the least certain.

(e) In terms of equilibria (57)–(62) we have the following sequences.
   < 85 wt. % acid; (58) slow, (61) fast, (62a) fast:
   90–97 wt. % acid; (57a) slow, (61) fast, (62a) fast.
   97–100 wt. % acid; (57a) fast, (61) slow, (62a) fast.
   > 100 wt. % acid; (57) fast (59) slow?, (60) slow?, base not specified, or (57) fast, (61) slow?, base not specified, (62) slow?

## 4.3 SULPHONYLATION

Sulphonylation, *e.g.*

$$ArH + RSO_2Cl \xrightarrow{AlCl_3} ArSO_2R$$

is a kinetically complex reaction because the reaction order and rate-determining step depend upon the reactivity of the aromatic substrate in the medium in which the reaction is carried out.

The mechanism was first studied by Olivier[180-183], who measured the rates of 4-bromobenzenesulphonylation of a variety of aromatic compounds[180], of benzene using various sulphonyl chlorides[181] and mixtures of sulphonyl chlorides[182], and of benzene and chlorobenzene using benzenesulphonyl chloride as solvent[183], all at 30 °C. The reaction rates were proportional to the concentration of aluminium halide provided that this was not in excess; a small excess caused a very large increase in the rate, whereas an excess of sulphonyl chloride did not. The sulphone reaction product was shown to be complexed in 1:1 ratio with aluminium chloride, for the catalytic effect of the latter was inhibited in the presence of an equimolar quantity of the former[180]. The electrophilic nature of the reaction followed from the relative rates of 4-bromobenzenesulphonylation of some aromatics thus: nitrobenzene (0.0); bromobenzene, 10.2; chlorobenzene, 8.02; benzene, 11.1; toluene, 41.3. Electron supply in the aryl ring of the sulphonyl chloride also caused a rate increase, for the relative rates for substituents were[181]: 4-Me, 64.6; H, 21.2; 4-I, 14.0; 4-Br, 11.1; 4-Cl, 10.6; 3-NO$_2$, 1.36. The reaction rates were sensitive to small concentrations of impurities, so that purification of benzene gave an increase in the reaction rate[180]. Kinetic studies using benzenesulphonyl chloride as solvent showed that 1 mole of aluminium chloride catalysed 1 mole of sulphonyl chloride only[183].

The next and only other major kinetic study was carried out by Jensen and Brown[184], who used aluminium chloride as catalyst, nitrobenzene as solvent, benzene- and p-toluene-sulphonyl chlorides as sulphonylating agents and benzene, chlorobenzene, alkyl- and polyalkylbenzenes as aromatic substrates[184].

The reaction of p-toluenesulphonyl chloride with toluene at 25 °C gave ditolyl sulphone and third-order kinetics, viz.

$$\text{Rate} = k_3[\text{AlCl}_3][\text{PhCH}_3][\text{4-CH}_3\text{C}_6\text{H}_4\text{SO}_2\text{Cl}] \tag{77}$$

The reaction was first order in sulphonyl chloride, and changing the initial concentration of this did not alter the value of $k_3$ (Table 46). However, increasing the initial concentrations of aluminium chloride and toluene caused a rate decrease and *vice versa* (Table 46).

TABLE 46

RATE COEFFICIENTS FOR THE AlCl$_3$-CATALYSED REACTION OF 4-MeC$_6$H$_4$SO$_2$Cl WITH PhCH$_3$ IN PhNO$_2$ AT 25 °C[184]

| 10[PhCH$_3$] (M) | 10[ArSO$_2$Cl] (M) | 10[AlCl$_3$] (M) | 10$^4k_3$ |
|---|---|---|---|
| 3.17 | 3.17 | 3.17 | 13.2 |
| 3.17 | 1.57 | 3.17 | 12.2 |
| 3.17 | 4.73 | 3.17 | 21.8 |
| 3.17 | 3.17 | 5.25 | 9.91 |
| 3.17 | 3.17 | 1.71 | 15.9 |
| 5.20 | 3.17 | 3.17 | 9.95 |
| 1.03 | 3.17 | 3.17 | 17.5 |

The reaction of p-toluenesulphonyl chloride with chlorobenzene at 60 °C also gave a third-order law, being again first-order in sulphonyl chloride and subject to a rate decrease when the initial concentration of aluminium chloride was increased, but in this case the reaction was clearly first-order in aromatic since increasing or decreasing the initial concentration caused only a slight rate decrease or increase, respectively (Table 47) and this was attributed to a solvent effect. The decrease in the rate with increase in aluminium chloride concentration was here so large that the reaction order with respect to this appeared to be almost zero, *i.e.* $k_3[\text{AlCl}_3]_{\text{initial}} = $ constant.

TABLE 47

RATE COEFFICIENTS FOR THE $AlCl_3$-CATALYSED REACTION OF $4\text{-}MeC_6H_4SO_2Cl$ WITH PhCl IN $PhNO_2$ AT 60 °C[184]

| $10[PhCl]$ (M) | $10[ArSO_2Cl]$ (M) | $10[AlCl_3]$ (M) | $10^4k_3$ |
|---|---|---|---|
| 3.05 | 3.05 | 3.05 | 2.02 |
| 3.05 | 8.21 | 3.05 | 1.80 |
| 3.05 | 3.05 | 8.05 | 0.942 |
| 3.05 | 3.05 | 1.19 | 4.48 |
| 1.17 | 3.05 | 3.05 | 2.23 |
| 8.31 | 3.05 | 3.05 | 1.80 |

The reaction of benzenesulphonyl chloride instead of $p$-toluenesulphonyl chloride gave much the same kinetic results except that reaction rates were approximately four times slower, which agrees well with the result obtained by Olivier. For the reaction with toluene the activation energy and entropy values (determined over a 25 °C range) were 19.6 and $-10.6$ ($\log A = 10.9$) respectively, and this latter value seems to indicate a highly ordered transition state. Reaction of benzenesulphonyl chloride with other alkyl aromatics also gave third-order kinetics and the Baker–Nathan order of reactivity, $k_{rel}$ values being as follows: benzene, 1.0; toluene, 9.0; ethylbenzene, 6.8; $i$-propylbenzene, 4.8; $t$-butylbenzene, 3.5. However, with more reactive alkyl aromatics, the increase in rate per alkyl group fell off, suggesting that the reaction becomes independent of structure and, therefore, concentration. (For example, $m$-xylene was only 2.4 times as reactive as toluene and only 1.2 times as reactive as $o$-xylene which has the same number of reaction sites but is usually much less reactive.) Thus mesitylene gave second-order kinetics, *viz.*

$$\text{Rate} = k_2[AlCl_3][PhSO_2Cl] \tag{78}$$

and compounds of lesser reactivity gave kinetics intermediate between second and third order suggesting the intervention of at least one equilibrium prior to the reaction with the aromatic.

The order of reaction with respect to aluminium chloride was ill-defined. Since nitrobenzene and aluminium chloride form a 1 : 1 complex which exhibits the simple monomeric molecular weight in nitrobenzene solution[185], and since even in solutions of aluminium chloride and benzoyl chloride in nitrobenzene the aluminium chloride is preferentially associated with the solvent[184], then in nitrobenzene solutions of aluminium chloride and benzenesulphonyl chloride the lesser basicity of the latter relative to benzoyl chloride means that the aluminium chloride must be mainly associated with nitrobenzene and in equilibrium with aluminium chloride associated with the sulphonyl chloride[184]. By analogy with benzoylation, where the aluminium chloride aids the ionisation which forms the benzoyl cation, the effect in sulphonylation was assumed to be similar and the mechanism was, therefore, proposed as

$$Ar \cdot SO_2Cl + AlCl_3 PhNO_2 \underset{k'_{-1}}{\overset{k'_1}{\rightleftharpoons}} ArSO_2Cl \cdot AlCl_3 + PhNO_2 \qquad (79)$$

$$ArSO_2Cl \cdot AlCl_3 \underset{k'_{-2}}{\overset{k'_2}{\rightleftharpoons}} ArSO_2^+ AlCl_4^- \qquad (80)$$

$$ArH + ArSO_2^+ AlCl_4^- \overset{k'_3}{\rightarrow} ArSO_2Ar \cdot AlCl_3 + HCl \qquad (81)$$

This mechanism is consistent with all the observations except the variation in rate with initial aluminium chloride concentration. With very reactive aromatics the ionisation step (80) is rate-determining, leading to second-order kinetics, but with less reactive aromatics the ionisation is fast compared with the subsequent reaction of the ionised complex with the aromatic (81), so that this latter then becomes rate-determining.

The reaction is less selective than the related benzoylation reaction ($f_p^{Me} = 30.2$, cf. 626), thereby indicating a greater charge on the electrophile; this is in complete agreement with the greater ease of nuclophilic substitution of sulphonic acids and derivatives compared to carboxylic acids and derivatives and may be rationalized from a consideration of resonance structures. The effect of substituents on the reactivity of the sulphonyl chloride follows from the effect of stabilizing the arylsulphonium ion formed in the ionisation step (81) or from the effect on the pre-equilibrium step (79).

The above conclusion as to which step is rate-determining must be treated with caution because the substrate kinetic isotope effect $k_H : k_D$ for the reaction has been determined as $0.96 \pm 0.04$ in nitrobenzene, $0.94 \pm 0.07$ in nitromethane[141, 184] and 1.91–2.44 in dichloromethane[186]. The latter values were obtained for the reaction of benzene and benzene-$d_6$ with $p$-toluenesulphonyl chloride which gave clean third-order kinetics, the rate coefficients $10^4 k_{H(D)}$ at 0, 7 and 15.5 °C being 1.18, (0.48); 1.92, (1.01); 4.02, (1.93). Comparison of these values with those in Table 46 (p. 78) shows that the reaction of benzene in dichloromethane takes place at a similar rate to that of toluene in nitrobenzene so that the former medium appears to enhance reaction. Presumably, therefore, the equilibrium leading up to the rate-determining step must be more rapidly established in dichloromethane so that the step involving breaking of the carbon–hydrogen bond becomes rate-determining; this will still give third-order kinetics. In this study values of the activation energy 11.9 (H), 14.0 (D) and entropy −34.6 (H), −28.9 (D) are probably subject to considerable error, but certainly confirm that the reaction involves a highly negative activation entropy.

Jensen and Brown[184] have also reinvestigated the kinetics of benzenesulphonylation using benzenesulphonyl chloride as solvent. They found that for toluene,

TABLE 48

RATE COEFFICIENTS FOR THE $AlCl_3$-CATALYZED REACTION OF $PhSO_2Cl$ WITH $PhCH_3$ IN $PhSO_2Cl$ AT 25 °C[184]

| $10[PhCH_3]$ (M) | $10[AlCl_3]$ (M) | $10^3 k_2$ |
|---|---|---|
| 3.10 | 3.10 | 2.07 |
| 3.10 | 1.55 | 2.08 |
| 3.10 | 4.74 | 2.02 |
| 1.56 | 3.12 | 2.12 |
| 6.13 | 3.08 | 1.82 |

TABLE 49

RATE COEFFICIENTS FOR THE $AlCl_3$-CATALYSED REACTION OF $PhSO_2Cl$ WITH PhCl IN $PhSO_2Cl$ AT 25 °C[184]

| $10[PhCl]$ (M) | $10[AlCl_3]$ (M) | $k_{\frac{3}{2}}$ |
|---|---|---|
| 4.55 | 4.58 | 2.31 |
| 4.53 | 2.06 | 2.53 |
| 4.53 | 9.99 | 2.12 |
| 2.18 | 4.52 | 2.60 |
| 9.93 | 4.53 | 1.74 |

satisfactory second-order kinetics, equation (82),

$$\text{Rate} = k_2[AlCl_3][PhCH_3] \qquad (82)$$

were obtained (Table 48), whereas with chlorobenzene the order was three-halves (Table 49), *viz.*,

$$\text{Rate} = k_{\frac{3}{2}}[AlCl_3]^{\frac{1}{2}}[PhCl] \qquad (83)$$

They also showed that if Olivier's data[183] for this system was treated as three-halves-order instead of second-order as he had assumed, then the initial rate coefficients in his kinetic runs were satisfactorily consistent, contrary to his observation (Table 50). It was necessary to use the initial values since Olivier found a decrease in the rate coefficients with time, contrary to the observations of Jensen and Brown, who assumed, therefore, that Olivier's reagent may have contained impurities.

In arriving at the mechanism of sulphonylation under these conditions account was taken of the facts that (a) Olivier[183] had shown that aluminium chloride and benzenesulphonyl chloride forms a 1:1 complex so that the rate equations (82) and (83) become (84) and (85), respectively, *viz.*

TABLE 50

RATE COEFFICIENTS FOR THE $AlCl_3$-CATALYZED REACTION OF $PhSO_2Cl$ WITH PhCl
IN $PhSO_2Cl$ AT 30 °C[183, 184]

| 10[PhCl] (M) | 10[AlCl₃] (M) | 10²k₂ | 10²k₃⁄₂ |
|---|---|---|---|
| 3.86 | 2.00 | 9.95 | 4.46 |
| 3.13 | 3.00 | 8.15 | 4.47 |
| 7.24 | 3.00 | 7.56 | 4.14 |
| 8.27 | 3.00 | 6.94 | 3.79 |
| 3.53 | 5.00 | 6.46 | 4.56 |
| 7.84 | 5.00 | 5.72 | 4.03 |
| 7.89 | 5.00 | 5.74 | 4.05 |
| 8.00 | 8.00 | 4.32 | 3.85 |
| 15.7 | 8.00 | 3.61 | 3.22 |

$$\text{Rate} = k_2[PhSO_2Cl\cdot AlCl_3][PhCH_3] \tag{84}$$

$$\text{Rate} = k_{\frac{3}{2}}[PhSO_2Cl\cdot AlCl_3]^{\frac{1}{2}}[PhCl] \tag{85}$$

and (b) the addition of cyclohexane to a particular reaction mixture caused a
very large rate decrease, hence at least one rate-determining step in the reaction
must involve separation of charge. The proposal was, therefore, made that, in
the reaction with chlorobenzene, the addition compound which first forms dis-
sociates, viz.

$$Ph\cdot SO_2^+AlCl_4^- \rightleftharpoons PhSO_2^+ + AlCl_4^- \tag{86}$$

before reacting with the aromatic in the rate-determining step

$$ArH + PhSO_2^+ \rightleftharpoons PhSO_2Ar^+H \tag{87}$$

and this is followed by a fast proton loss

$$PhSO_2Ar^+H + AlCl_4^- \rightarrow PhArSO_2AlCl_3 + HCl \tag{88}$$

consistent with the absence of a significant isotope effect, $k_H: k_D = 0.86 \pm 0.06$[141].
Now since $[PhSO_2^+] = [AlCl_4^-] = (K_{86}[PhSO_2^+AlCl_4^-])^{\frac{1}{2}}$, then if equilibrium
(87) is rate-determining the kinetic equation (85) follows. For toluene the direct
reaction of the aromatic with the undissociated ion-pair would give the required
kinetic form but this is inconsistent with the effect of changing the medium
polarity, so it is assumed that toluene reacts with the ion-pair in the rate-determining
step to give the required kinetic equation (84). The fall-off in rate coefficients
when large concentrations of both chlorobenzene and toluene were used follows

from a solvent effect analogous to that obtained on adding cyclohexane, and the greater effect with chlorobenzene than with toluene is consistent with the former reaction proceeding through dissociated ions rather than with ion-pairs and hence to be more susceptible to a change in the dielectric constant of the medium.

The benzenesulphonylation of benzene did not give a single kinetic form. Satisfactory second-order rate coefficients could be obtained provided that the initial benzene concentration was not small, when a small increase in the calculated rate coefficients as the reaction proceeds is observed. However, the values of the second-order rate coefficients are not independent of the initial concentration of reactants (Table 51). It was concluded that benzene reacts by both mechanisms;

TABLE 51

RATE COEFFICIENTS FOR THE $AlCl_3$-CATALYSED REACTION OF $PhSO_2Cl$ WITH PhH IN $PhSO_2Cl$ AT 25 °C[184]

| $10[PhH]$ | $10[AlCl_3]$ | $10^4 k_2$ |
|-----------|--------------|-----------|
| 5.90 | 5.46 | 2.51 |
| 5.53 | 1.73 | 2.83 |
| 1.63 | 5.38 | 3.47 |
| 1.90 | 5.48 | 3.02 |

at low concentrations the mechanism involving ion-pair dissociation predominates, but as the concentration of benzene increases the attack on the ion-pair prior to dissociation will obviously become more important and so the kinetic order increases as the benzene concentration increases.

The very large effect of adding an amount of aluminium chloride greater than stoichiometric as noted by Olivier has been attributed[187] to an as yet unspecified property of the dimer which differs from that of the monomer.

Finally, it is interesting to note that the similarity between benzenesulphonylation and benzoylation also shows up in the $\log f_o : \log f_p$ ratios for toluene which are $0.56^{184}$ and $0.54^{188}$, respectively, indicating a similar size of electrophile in each case.

## 5. Electrophilic halogenation

### 5.1 POSITIVE HALOGENATION

The addition of mineral acids to hypohalous acids produces a large increase in the rate at which these latter acids halogenate and reaction under these conditions is usually referred to as "positive halogenation" which has been subjected to intensive kinetic studies. Whilst there is ample evidence supporting the existence of positive chlorinating and brominating species, the arguments regarding positive iodination are somewhat controversial since the kinetic data can also be interpreted in favour of molecular iodine as electrophile.

### 5.1.1 Positive bromination

Positive bromination was first observed by Shilov and Kaniaev[189] who found that the bromination of sodium anisole-$m$-sulphonate by bromine-free hypobromous acid was accelerated by the addition of nitric or sulphuric acids, and was governed by the kinetic equation

$$\text{Rate} = k_3[\text{ArH}][\text{HOBr}][\text{H}^+] \tag{89}$$

The choice of conditions is fairly critical in studying positive bromination for it has been shown[190] that the rate of hypobromous acid decomposition is given by $k_4[\text{HOBr}]^3[\text{OH}^-]$, hence dilute solutions of hypobromous acid and the presence of strong mineral acid are the most satisfactory if the kinetics are not to become too complicated. With these conditions, Wilson and Soper[191] investigated the bromination of benzene and 2-nitroanisole by hypobromous acid and found that the reaction rate was very slow but increased rapidly with increasing [$\text{H}^+$] and they therefore proposed that the brominating species was $\text{H}_2\text{OBr}^+$.

Derbyshire and Waters[192] measured the rates of bromination of sodium toluene-$m$-sulphonate (in water) and of benzoic acid (in aqueous acetic acid) by hypobromous acid with sulphuric or perchloric acids as catalysts, all at 21.5 °C. No bromination occurred in the absence of mineral acid and the reaction was strictly first-order in aromatic and in hypobromous acid. The function of the catalyst was considered to be the formation of a positive brominating species, according to the equilibrium

$$\text{H}^+ + \text{HOBr} \rightleftharpoons \text{H}_2\text{O} + \text{Br}^+ \text{ (or } \text{H}_2\text{OBr}^+\text{)} \tag{90}$$

Addition of perchlorate ion had little kinetic effect, but addition of chloride ion decreased the rate and complicated the kinetics, probably through intervention of the equilibrium

$$\text{H}_2\text{OBr}^+ + \text{Cl}^- \rightleftharpoons \text{BrCl} + \text{H}_2\text{O} \tag{91}$$

For the bromination of benzoic acid, the rate increased more rapidly than [$\text{H}^+$] at high mineral acid concentrations and apparently depended on the acidity function $h_0$ rather than [$\text{H}^+$]. Neutral salts such as disodium hydrogen phosphate, lithium and sodium sulphates also increased the rate due either to a positive salt effect, or to the effect of these salts on the acidity function. Measurement of the effect of added acetate ion revealed a rate decrease, which, however, was not as large as would be expected if the bromination was carried out by $H_2OBr^+$ rather than $Br^+$, but the difference in rate was attributed to a possible contrary positive salt effect. It was also suggested that in aqueous acetic acid, protonated bromine acetate $AcOHBr^+$ might be a significant positive brominating species. By analysing the effect of bromination of sodium-4-anisate by hypobromous acid at 25 °C, in phosphate buffers between pH 7–8, hypobromous acid was shown to be 2000 times less reactive than molecular bromine (produced by addition of bromide ion).

The positive bromination of aromatics ethers was first studied by Bradfield et al.[193] and by Branch and Jones[194]. The reaction of hypobromous acid in 75 % aqueous acetic acid with benzyl 4-nitrophenyl ether and 4-nitrophenetole at 20 °C was very rapid and approximately second-order[193]. The value of $k_2/[H^+]$ remained constant in the $[H^+]$ range 0.005–0.090 $M$ for the effect of added mineral acids on the bromination of 4-nitroanisole and 4-nitrophenetole (at 19.8 °C)[194]. The variation in reaction rate with the percentage of acetic acid in the medium was also studied and showed a large increase in the 0–10 % range with a levelling off at approximately 25 % acetic acid (Table 52); this was attributed

TABLE 52

EFFECT OF HOAc CONCENTRATION ON REACTION OF $4\text{-}NO_2C_6H_4OMe$ WITH HOBr AT 19.8 °C[194]

| HOAc (wt. %) | 0 | 12 | 25 | 50 | 65 | 75 | 85 |
|---|---|---|---|---|---|---|---|
| $10k_2$ | 0.21 | 1.83 | 1.94 | 1.89 | 1.83 | 1.62 | 1.34 |

to the change in the hydrogen ion concentration in the medium. At high acetic acid concentrations the rate decreased and this was attributed to the decrease in the dielectric constant of the medium. In 75 wt. % acetic acid at 19.8 °C, rate coefficients ($10\,k_2$) for reaction of 4-nitrophenetole and benzyl 4-nitrophenyl ether were 3.28 and 1.16 respectively; with the former compound rate coefficients decreased during a run unless highly purified hypobromous acid was used, and this was attributed to molecular bromine impurity. For the latter aromatic, rates increased during a run, due probably to the formation of 4-nitrophenol.

Aqueous dioxan (50 %) has been used as a medium for bromination with acidified hypobromous acid and de la Mare and Harvey[195] showed that, with perchloric acid as catalyst, the bromination of toluene followed the usual kinetic equation (89). At 25 °C, in ca. 0.0013 $M$ hypobromous acid, the average value of $k_2/[H^+]$ for toluene (0.008–0.15 $M$) was 21.7 and for benzene (0.0011–0.016 $M$) was 0.60, so that $k_{toluene}/k_{benzene}$ was 36.2. The bromination of t-butylbenzene[196] and biphenyl[197] gave $k_2/[H^+] = 7.25$ and 7.52, and hence relative rates of 12.1, and 12.6, respectively, both compounds giving the same kinetic equation.

De la Mare and Hilton[198] measured the rates at 25 °C of bromination of benzene, benzoic acid, phthalic acid, 2-nitrobenzoic acid, trimethylanilinium perchlorate and nitrobenzene by hypobromous acid with sulphuric or perchloric acids as catalysts, in some cases in aqueous dioxan, in an attempt to discover if $Br^+$ or $H_2OBr^+$ was the appropriate brominating species since the logarithm of the rates should then follow the acidity functions $H_0$ or $H_R$ ($J_0$) respectively. The results, however, were inconclusive[†], and relative rates of bromination were determined (see Table 53).

TABLE 53

RELATIVE RATES OF REACTION OF ArH WITH HOBr IN $H_2O$ OR $H_2O$–DIOXAN AT 25 °C[198]

| ArH | Catalysing acid | |
|---|---|---|
| | Perchloric | Sulphuric |
| Benzene | 1 | |
| Benzoic acid | $7.5 \times 10^{-3}$ | $7.5 \times 10^{-3}$ |
| Phthalic acid | $4.9 \times 10^{-5}$ | |
| 2-Nitrobenzoic acid | $1.1 \times 10^{-7}$ | |
| Nitrobenzene | $1.6 \times 10^{-5}$ | $7.9 \times 10^{-6}$ |
| Trimethylanilinium perchlorate | $1.6 \times 10^{-5}$ | $4.0 \times 10^{-6}$ |

De la Mare and Maxwell[199] measured the rate of bromination of biphenyl by hypobromous acid in 75 % aqueous acetic acid, in some cases catalysed by perchloric acid, at temperatures between $-3.78$ and $+20.1$ °C. They showed that whereas when mineral acid is present the brominating species is $Br^+$ (or a solvate), in the absence of mineral acid it is BrOAc which is a highly reactive brominating species giving $E_a = 7.9$ (this value is only approximate since it also includes a contribution from bromination by HOBr), and the appropriate kinetic equation is then

$$\text{Rate} = k_2[\text{ArH}][\text{BrOAc}] \tag{92}$$

At 0.9 °C the rate of bromination of biphenyl relative to benzene was approximately 1,270, compared to 26.9 in the presence of mineral acid, and this latter value is fairly close to that obtained with 50 % aqueous dioxan. The possibility that the positive brominating species might be protonated bromine acetate, $AcOHBr^+$, was considered a likely one since the reaction rate is faster in aqueous acetic acid than in water, but this latter effect might be an environmental one since bromination by acidified hypobromous acid is slower in 50 % aqueous dioxan than in

---

[†] Had a "conclusive" result been obtained it would now probably be regarded as unsatisfactory in view of the demise of the Zucker-Hammett hypothesis (see pp. 4–5).

water[198]. The *ortho* : *para* ratio was 0.13 in the absence, and 0.57 in the presence of mineral acid; the difference was attributed to the differing sizes of the brominating species, but, in fact, most of this difference arises from a selectivity effect since the $\log f_o$ : $\log f_p$ values are 0.75 and 0.85, respectively[200].

Measurement of the kinetic isotope effect in the perchloric acid-catalysed bromination of benzene by hypobromous acid at 25 °C gave a value of $k_H/k_D = 1.0$, indicating that breaking of the carbon–hydrogen bond was not rate-determining[201]. However, in the bromination of 1,3,5-tri-*t*-butylbenzene by bromine in 50 % aqueous dioxan containing perchloric acid and silver perchlorate (to suppress molecular bromine) a very large isotope effect ($k_H/k_D \approx 10$) was obtained and attributed to steric hindrance retarding conversion of the intermediate into the bromo product *i.e.* the ratio of $k_2'/k_{-1}'$ is small[33] (see p. 8).

Protonated hypobromous acid or protonated acetyl hyprobromite have been suggested as probable brominating species in the bromination of benzene by potassium bromate in aqueous acetic acid in the presence of sulphuric acid at 60 °C. The reaction did not occur in the absence of mineral acid, appeared to be facilitated by electron supply and was approximately second-order overall, being first-order in benzene and in bromate ion, though incursion of a minor third-order reaction (second-order in benzene) was indicated. With 50 % aqueous acetic acid and 3 $M$ sulphuric acid the activation energy and entropy were 22.3 and $-4.0$ respectively[200a]. Hypobromous acid was believed to be formed from bromate ion *via* protonation and loss of oxygen, with the electrophiles derived from this in the usual way. The reaction second-order in benzene was postulated as arising from attack on benzene of a benzene-protonated acetyl hypobromite $\pi$-complex, which seems rather unlikely.

### 5.1.2 Positive chlorination

Kinetic studies using acidified hypochlorous acid are rather more complicated than these with hypobromous acid. Much higher concentrations of mineral acid are necessary so that the activities of the reacting entities do not correspond closely to their molecular concentrations, and the kinetic order of reaction varies according to the acid concentration and the reactivity of the aromatic.

Derbyshire and Waters[202] carried out the first kinetic study, and showed that the chlorination of sodium toluene-*m*-sulphonate by hypochlorous acid at 21.5 °C was catalysed more strongly by sulphuric acid than by perchloric acid and that the rate was increased by addition of chloride ion. A more extensive examination by de la Mare *et al.*[203] of the rate of chlorination of the more reactive compounds, anisole, phenol, and *p*-dimethoxybenzene by hypochlorous acid catalysed by perchloric acid, and with added silver perchlorate to suppress the formation of $Cl_2$ and $Cl_2O$ (which would occur in the presence of $Cl^-$ and $ClO^-$, respectively), showed that the reaction rate was proportional to hypochlorous acid concentration, was dependent upon the acidity of the medium but was independent of the concentration of the aromatic, being the same $(10^4 k_1 = 1.43)$ for anisole $(0.033–0.01\ M)$, phenol $(0.003–0.08\ M)$ and $p$-dimethoxybenzene. The kinetic order was not, however, entirely clear, but, by using lower concentrations of hypochlorous acid $(< 0.001\ M)$, the same workers found that for anisole (in the range $0.004–0.01\ M$) clear kinetic orders could be obtained and the appropriate equation was

$$\text{Rate} = k_1[\text{HOCl}] + k_2[\text{HOCl}][\text{H}^+] \tag{93}$$

where $k_1 = 1.25 \times 10^{-4}$ and $k_2 = 1.18 \times 10^{-3}$ at 25 °C, the reaction giving approximately the same rate, at 25 °C in aqueous dioxan. At higher aromatic concentrations, and with the latter solvent, the kinetic form changed to

$$\text{Rate} = k_1[\text{HOCl}] + k_2[\text{HOCl}][\text{H}^+] + k_3[\text{HOCl}][\text{H}^+][\text{ArH}] \tag{94}$$

and this was confirmed by using mesitylene, phenol, and methyl $p$- and $m$-tolyl ethers. The terms in this equation were interpreted as representing fission of HOCl to give $\text{Cl}^+$, of $\text{H}_2\text{OCl}^+$ to give $\text{Cl}^+$, and attack of the aromatic on $\text{H}_2\text{OCl}^+$, respectively, and it follows from this that the latter term becomes significant at higher aromatic concentrations. It also follows that as the concentration is decreased, the point at which the rate falls off as the reaction becomes independent of aromatic concentration should be lower the more reactive the aromatic, *i.e.* the most reactive aromatics remain effective in competing with $\text{OH}^-$ and $\text{H}_2\text{O}$ for $\text{Cl}^+$ down to a lower concentration. This was observed, the concentrations at which a significant decrease in rate occurred being 0.05 $M$ for methyl $p$-tolyl ether, 0.005 $M$ for anisole and 0.0015 $M$ for phenol. For the most reactive compounds, *e.g.* methyl $m$-tolyl ether, equation (94) applied at all concentrations.

The alternative interpretation of the second term in equation (94), *i.e.* that it represents a slow proton transfer to hypochlorous acid, has been ruled out on two grounds: firstly, the rate of this proton transfer has been measured and found to be much higher than the first-order chlorination rate[204], and, secondly, the third term should disappear when the reactivity of the aromatic is very high since then the reaction responsible for the second term would be dominant[203].

It follows from the above that, in the reactions of fairly unreactive aromatics, the formation of $\text{Cl}^+$ (either from HOCl or $\text{H}_2\text{OCl}^+$) will be relatively fast compared with the subsequent reaction of this ion with the aromatic so that the kinetics will be governed mainly by the third term in equation (94). Hence de la Mare *et al.*[204a] found the rate of chlorination of benzene and toluene by acidified hypochlorous acid to depend on the concentration and nature of the aromatic and to increase with hydrogen ion concentration though (as in the case of positive bromination) the dependence at higher acidities was more closely governed by the acidity function $h_0$. At 25 °C, values of $10^4 k_2/[H^+]$ were found to be 250 and 4.2 for toluene and benzene, respectively, this latter reacting about 3 times faster than sodium toluene $m$-sulphonate (at 21.5 °C).

Thus chlorination by acidified hypochlorous acid seems to indicate $Cl^+$ as the chlorinating species, whereas there is no clear indication from bromination by hypobromous acid whether $Br^+$ or $H_2OBr^+$ is responsible. Evidence which further supports $Cl^+$ as the chlorinating species comes from the measurement of the rates at 25 °C of perchloric acid-catalysed chlorination of methyl $p$-tolyl ether by hypochlorous acid in 62.5 % aqueous dioxan. In media containing $H_2O$ the rate was only 52 % of that in media containing $D_2O$, and this accords with the higher acid dissociation constants of protium- than deuterium-containing substrates appropriate to formation of $Cl^+$ being rate-determining; if formation of $H_2OCl^+$ was rate-determining the opposite effect would have been expected because of the lower zero point energy of $D_3O^+$ compared with $H_3O^+$ [205]. Unfortunately, thermodynamic calculations seem to show that neither $Cl^+$ nor $H_2OCl^+$ can be the chlorinating species, but the possibility exists that these calculations are in error[206].

Kinetics studies of acid-catalysed chlorination by hypochlorous acid in aqueous acetic acid have been carried out, and the mechanism of the reactions depends upon the strength of the acetic acid and the reactivity of the aromatic. Different groups of workers have also obtained different kinetic results. Stanley and Shorter[207] studied the chlorination of anisic acid by hypochlorous acid in 70 % aqueous acetic acid at 20 °C, and found the reaction rate to be apparently independent of the hydrogen ion concentration because added perchloric acid and sodium perchlorate of similar molar concentration (below 0.05 $M$, however) both produced similar and small rate increases. The kinetics were complicated, initial rates being proportional to aromatic concentration up to 0.01 $M$, but less so thereafter, and described by

$$\text{Initial rate} = A[\text{ArH}][\text{HOCl}]/(1 + B[\text{HOCl}]) \qquad (95)$$

where $A$ and $B$ are constants found subsequently to depend on the medium and aromatic. The apparent absence of catalysis by hydrogen ion was regarded as showing the concentration of free hypochlorous acid to be small in 75 % aqueous acetic acid, and that the chlorinating species was chlorine acetate, the formation and subsequent reaction of which was not catalysed by hydrogen ion. Kinetic studies with 4-substituted ethoxy-, $n$- and $i$-propoxy-, $n$- and $i$-butoxy-, $n$- and $i$-pentyloxy- and $n$-hexyloxy-benzoic acids, 4-chloroanisole, and methyl $p$-anisate confirmed these experimental observations. The chlorination rate increased with increasing size of the alkyl chain, and there appeared to be an equilibrium to form an "active" form of the ethers from the bulk form, the equilibrium seeming to depend on the structure of the ether.

In contrast to these results, Stanley and Shorter[207] found the catalytic effect of perchloric acid to be large in 40 % aqueous acetic acid, the difference in behaviour from 75 % acid being attributed to the presence of a much higher concentration of hypochlorous acid in the more aqueous medium.

Rather different experimental results were obtained by de la Mare *et al.*[208, 209], who studied chlorination by hypochlorous acid in 51, 75 and 98 % aqueous acetic acid. With the latter medium, the chlorination of anisole or *m*-xylene (at an unspecified temperature) was independent of the concentration of aromatic, and catalysed by perchloric acid to a much greater extent than an equimolar amount of lithium perchlorate; the reaction was also catalysed by the base, sodium acetate. The reactive species was postulated as chlorine acetate produced as in equilibrium (96)

$$HOCl + HOAc \rightleftharpoons ClOAc + H_2O \tag{96}$$

or *via* the base-catalysed equilibrium (97)

$$AcO^- + HOCl \rightleftharpoons ClOAc + OH^- \tag{97}$$

or *via* the acid-catalysed equilibria (98) and (99)

$$HOCl + H^+ \rightleftharpoons H_2OCl^+ \tag{98}$$

$$H_2OCl^+ + HOAc \rightleftharpoons AcOHCl^+ + H_2O \tag{99}$$

which produce protonated chlorine acetate so that in the presence of mineral acid either $Cl^+$ or $AcOHCl^+$ can be the electrophile. The reaction with toluene was slower and second-order kinetics, obtained, *viz.*

$$Rate = k_2[HOCl][ArH] \tag{100}$$

so it was proposed that an equilibrium concentration of chlorine acetate is present which chlorinates in a fast step with reactive aromatics and in a rate-determining step with less reactive aromatics[208].

Kinetic studies at 25 °C showed that for benzene, toluene, *o*-, *m*-, and *p*-xylene, *t*-butylbenzene, mesitylene, 4-chloroanisole, and *p*-anisic acid in 51 and 75 % aqueous acetic acid addition of small amounts of perchloric acid had only a slight effect on the reaction rate which followed equation (100). At higher concentrations of perchloric acid (up to 0.4 $M$) the rate rose linearly with acid concentration, and more rapidly thereafter so that the kinetic form in high acid concentration was

$$Rate = k_2[ArH][HOCl] + k_3[ArH][HOCl]f[H^+] \tag{101}$$

With 77 % aqueous acetic acid, the rates were found to be more affected by added perchloric acid than by sodium perchlorate (but only at higher concentrations than those used by Stanley and Shorter[207], which accounts for the failure of these workers to observe acid catalysis, but their observation of kinetic orders in hypochlorous acid of less than one remains unaccounted for). The difference in the effect of the added electrolyte increased with concentration, and the rates of the acid-catalysed reaction reached a maximum in *ca.* 50 % aqueous acetic acid, passed through a minimum at *ca.* 90 % aqueous acetic acid and rose very rapidly thereafter. The faster chlorination in 50 % acid than in water was, therefore, considered consistent with chlorination by $AcOHCl^+$, which is subject to an increasing solvent effect in the direction of less aqueous media (hence the minimum in 90 % acid), and a third factor operates, *viz.* that in pure acetic acid the bulk source of chlorine is chlorine acetate rather than HOCl and causes the rapid rise in rate towards the anhydrous medium. The relative rates of the acid-catalysed (acidity $\geqslant 0.49$ $M$) chlorination of some aromatics in 76 % aqueous acetic acid at 25 °C were found to be: toluene, 69; benzene, 1; chlorobenzene, 0.097; benzoic acid, 0.004. Some of these kinetic observations were confirmed in a study of the chlorination of diphenylmethane; in the presence of 0.030 $M$ perchloric acid, second-order rate coefficients were obtained at 25 °C as follows[209a]: 0.161 (98 vol. % aqueous acetic acid); *ca.* 0.078 (75 vol. % acid), and, in the latter solvent in the presence of 0.50 $M$ perchloric acid, diphenylmethane was approximately 30 times more reactive than benzene.

In the absence of added mineral acid, the effective chlorinating species was concluded to be chlorine acetate. Like the catalysed chlorination, the rate of chlorination (of toluene) falls rapidly on changing the solvent from anhydrous to 98 % aqueous acetic acid, passes through a shallow minimum and thence to a maximum in 50 % aqueous acid; this was thus attributed to a combination of the decrease in concentration of chlorine acetate as water is added and a solvent effect. By correcting for the change in concentration of chlorine acetate in the different media it was shown that the reaction rate increases as the water content of the media increases.

The kinetics of chlorination by other species believed to be positively charged have been investigated. In the reaction of diethylchloramine with phenol, 2- and 4-chlorophenol, and *p*-cresol, Brown and Soper[210] found that, in buffered aqueous solutions at 25 °C and with excess phenol, the reaction was first-order in diethyl-chloramine. In the pH range 3.5–7.5 each phenol gave a constant (though different) reaction rate and the relative reactivities were 2-Cl > H > 4-Me > 4-Cl, which is the relative order of the ionisation constants of these phenols so that this was interpreted as meaning that the phenoxide ion was the reacting species. Since good second-order kinetics were obtained, the rate being expressed by

$$\text{Rate} = k_2[\text{PhOH}][\text{R}_2\text{NCl}] \tag{102}$$

this implies that the diethylchloroammonium cation $R_2NHCl^+$ was the halo-genating species, and evidence in favour of this rather than $Cl^+$ was the constancy of rate in the 3.5–7.5 pH range since the formation of the latter ion from the former should be dependent upon the pH. The reactivity of $R_2NHCl^+$ was determined as $10^3$ times greater than that of HOCl.

Carr and England[211] investigated the kinetics of the hydrochloric acid-catalysed chlorination of phenol by N-chloro-succinimide, -acetamide, and -morpholine, and found that the latter compound gave third-order kinetics, viz.

$$\text{Rate} = k_3[\text{ArOH}][\text{H}^+][\text{R}_2\text{NCl}] \tag{103}$$

The reaction rate (at 25 °C) was identical if perchloric acid was used instead of hydrochloric acid, so that chloride ion plays no part in the reaction contrary to observations with other N-chloro compounds which produce molecular chlorine as the chlorinating species; the addition of chloride ions also failed to produce a change in rate. The addition of a large excess of morpholine produced no change in the specific rate so that, although this basic reaction by-product picks up a proton thereby reducing the hydroxonium ion concentration, the conjugate acid is formed after the rate-determining step of the reaction which is, therefore, not subject to general acid-catalysis by the morpholinium cation. The mechanism of the reaction was, therefore, envisaged as

$$R_2NCl + H_3O^+ \rightleftharpoons R_2NHCl^+ + H_2O \tag{104}$$

$$R_2NHCl^+ + ArOH \rightleftharpoons ClAr'OH + R_2NH + H^+ \tag{105}$$

### 5.1.3 Positive iodination

The kinetics of iodination under conditions in which a positive species is be-lieved to be formed have been investigated less extensively than for positive bromination and chlorination. Cofman[212] carried out the first investigation and found that iodine only iodinates phenol in alkaline solution, so he assumed that the active iodinating agent was hypoiodous acid. Soper and Smith[213] investigated this reaction further and showed that the reaction rate was proportional to $[I^-]^{-2}$ which they took to indicate hypoiodous acid, and not molecular iodine as the iodinating species this latter requiring a rate proportional to $[I^-]^{-1}$. At constant $[I^-]$ they found that the iodinating rate was proportional to a power of $[H^+]$ between $-2$ and $-1$, the latter appropriate if phenol is the reacting species and the former if it is the phenoxide ion; hence they assumed both species were reac-ting. Subsequently, Painter and Soper[214] found that, in buffer solution at 25 °C, the iodination rate was proportional to $[H^+]^{-1}[I^-]^{-2}$, indicating a reaction between HOI and PhOH, though this is, of course, kinetically indistinguishable from the more likely reaction between $I^+$ and $PhO^-$. In addition, they observed general acid-catalysis by acetic acid in a solution at constant pH, and this catalytic effect was proportional to $[H^+]^{-2}$. They, therefore, considered that iodine acetate, formed as in the equilibrium

$$AcOH + I_2 \rightleftharpoons AcOI + I^- + H^+ \tag{106}$$

was reacting with phenoxide ion[214]. In summary, the observed rate coefficients can be expressed as the sum of the catalytic constants according to equation (107), *viz.*

$$k_{obs.} = k_0[H^+]^{-1} + k_{cat.}[H^+]^{-2} \text{ [general acid]} \tag{107}$$

which is equivalent to the equation

$$k_{obs.} = k_0[OH^-] + k_{cat.}[OH^-]^2 \text{ [general base]} \tag{108}$$

Berliner[215] subsequently investigated the kinetics of iodination of aniline by iodine in aqueous buffer solutions (pH 5.6–6.1) at 25 °C, and found the reaction to be second-order overall and first-order in each reagent. It was subject to general base-catalysis (which was not a salt effect since the ionic strength of the medium was maintained constant by addition of sodium chloride), and the reaction rate was independent of $[H^+]$ and proportional to $[I^-]^{-2}$ so that the kinetics were interpreted as involving reaction of $I^+$ with $PhNH_2$, which is also indistinguishable from the reaction between HOI and $PhNH_3^+$. However, *para* substitution occurred, thereby ruling out the latter reactants since positive poles are *meta* directing; hence by analogy the reaction of phenols with iodine is considered to involve $I^+$ and $PhO^-$. The observed rate coefficient in this case can be expressed as

$$k_{obs.} = k_0 + k_{cat.} \text{ [general base]} \tag{109}$$

which is similar to that obtained with phenols except that in equation (108) the concentration of hydroxide ion occurs in both terms which suggests that the ion serves a specific purpose, namely, the formation of phenoxide ion[216]. A further investigation of the rates of iodination of phenol at 25 °C and 35 °C (in buffer solutions, pH 5.6–6.1), and of aniline at 35 °C by iodine in water, gave rate coefficients for the uncatalyzed reaction from which rate coefficients for the reaction of phenol and phenoxide ion could be obtained and compared with the aniline data. Relative rates of $PhO^-$ : $PhNH_2$ : $PhOH$ as $9.2 \times 10^9$ : $3.7 \times 10^5$ : 1 were obtained with corresponding activation energies of 19.4, 23.9 and 25.2, and since this is the reactivity order expected for an electrophilic substitution it was argued that this supported the assumption that phenoxide ion is involved rather than phenol[216].

The interpretation of the above data on iodination has been questioned by Buss and Taylor[217], and by Grovenstein *et al.*[218, 219]. The former workers studied the iodination of 2,4-dichlorophenol at about 25 °C using a stirred flow reactor, the advantages of which are that once a steady state has been reached there is no change in the concentration of the reactive species in the reactor with time and the rate of reaction is simply a product of extent of reaction multiplied by the reciprocal of the contact time; hence it is possible to use unbuffered solutions and low iodide ion concentrations. They found general catalysis by the base component of added phosphate buffers and the observed rate coefficients varied with $[H^+]$ according to

$$k_{obs.} = k/(K_i + [H^+]) \tag{110}$$

where $K_i$ is the ionisation constant of phenol, and with $[I^-]$ according to

$$k_{obs.} = k/[I^-](K + [I^-]) \tag{111}$$

where $K$ is the dissociation constant for the $I_3^-$ complex. The results were considered to be compatible with iodination of phenoxide ion by $I_2$, HOI, or $I^+$ ($H_2OI^+$) the latter being the least likely.

Grovenstein and Kilby[218] showed that the kinetic isotope effect $k_H/k_D$ is 3.97 for the iodination 2,4,6-trideuterophenol by iodine in aqueous buffer at 25 °C, and this is in accord with the base catalysis described above. However, this large isotope effect means that the intermediate is in fairly rapid equilibrium with the reactants, so that it is difficult to determine from kinetic studies which iodinating species is involved. Thus it might be positive iodine, equilibria (112), (113), (115)

$$I_2 + H_2O \rightleftharpoons H_2OI^+ + I^- \tag{112}$$

$$H_2OI^+ + ArH \underset{k'_{-1}}{\overset{k'_1}{\rightleftharpoons}} ArHI^+ + H_2O \tag{113}$$

or molecular iodine, equilibria (114), (115)

$$I_2 + ArH \underset{k'_{-1}}{\overset{k'_1}{\rightleftharpoons}} ArHI^+ + I^- \tag{114}$$

$$ArHI^+ \overset{k'_2}{\rightleftharpoons} ArI + H^+ \tag{115}$$

Subsequently, Grovenstein and Aprahamian[219] investigated the iodination of 4-nitrophenol and its 2,6-dideuterated derivative, by iodine in aqueous solution at 50 °C. Contrary to the observations of Soper *et al.*[213, 214] they found that the iodination rate was proportional to $[I^-]^{-1}$ at high concentration and independent of it at low concentration, but like these workers they found that the rate was proportional to $[H^+]^{-1}$ though this order became greater than minus one at low $[I^-]$. This implies, therefore[214], that the iodinating species is molecular iodine; no reason for the disagreement in the two studies is available though the latter study was more extensive. Also, by using the technique evolved by Berliner[220] for investigating the variation in the isotope effect with change in the concentration of iodide ion, they found that $k_H/k_D$ fell from *ca.* 5.5 at $[I^-] =$ 0.006 *M* to 2.3 at 0.00001 *M*. Now the mechanism represented by equilibria (112), (113) and (115) should not give rise to a kinetic isotope effect which depends upon $[I^-]$ since the latter appears only in a pre-equilibrium whereas the mechanisms given by equilibria (114) and (115) should give the dependence. Hence it would appear that iodine in water may iodinate as molecular iodine rather than as a positive species.

A positive iodinating species has been indicated in the investigation of the iodination of 4-chloroaniline with iodine monochloride in water in the presence of excess hydrogen ion (to prevent rapid hydrolysis of iodine monochloride) and chloride ion (to minimize the kinetic effect of combination of iodine monochloride with the small amount of chloride ion produced during the reaction) and between 11.5 °C and 35 °C[221]. The reaction is second-order overall and first-order in each reagent, and the rate is proportional to $[H^+]^{-1}$ and $[Cl^-]^{-1}$ (after allowing for complexing of ICl with $Cl^-$) which is consistent with the scheme represented by equilibria (116) and (117), *viz.*

$$ICl + H_2O \rightleftharpoons H_2OI^+ + Cl^- \tag{116}$$

$$H_2OI^+ + ArNH_2 \rightleftharpoons IAr'NH_2 + H^+ + H_2O \tag{117}$$

Additional evidence supporting the mechanism involving the free amine is the observation that substitution occurred *ortho* to the $NH_2$ group, rather than *meta* which would occur if the nitrogen were protonated. The values for $E_a$, log $A$, and $\Delta S^{\ddagger}$ were found to be 16.5, 10.8, and $-10.9$, respectively; the highly negative activation entropy is a composite value for the iodination step and the pre-equilibria and was considered to additionally support the mechanism in which ions are formed from molecules. Subsequently, the kinetics of iodination of 2,4-dichlorophenol and anisole were examined, the observations were similar and it was concluded that the phenol reacted as the corresponding phenoxide ion[222]. The activation energies were 13.8 and 14.7, the log $A$ factors 8.5 and 9.3, and the entropies $-21.5$ and $-17.8$, respectively. The relative reactivities of 2,4-dichlorophenol (as phenoxide ion): 4-chloroaniline: anisole were $8.4 \times 10^6 : 2.8 \times 10^3 : 1$.

Although this kinetic work appears to show that the iodinating species is $I^+$

or $H_2OI^+$, Berliner[220] subsequently found an isotope effect, $k_H/k_D = 3.8$, in the iodination of 2,4,6-trideuteroanisole by iodine monochloride in water containing perchloric acid and chloride ions at 25 °C, and this was independent of $[H^+]$ between 0.1 and 0.0001 $M$ and $[Cl^-]$ between 0.3–0.9 $M$. The activation energy (determined between 18 °C and 38 °C) was 15.5, log $A$ was 9.37, $\Delta S^{\ddagger}$ $-17.6$, and there was a solvent isotope effect, $k_{H_2O}/k_{D_2O} = 1.40$. By the same reasoning applied (above) to the dependence of the isotope effect on the concentration of iodide ion in iodination by iodine it was argued that iodination by iodine monochloride followed the mechanisms given by equilibria (116), (113) and (115). However, it proved difficult to determine kinetics accurately below $[Cl^-] = 0.01$ $M$ because the reaction becomes too fast, but at this lowest concentration no change in the isotope effect was apparent. This, therefore, favours the scheme involving formation of a positive iodinating species, but does not disprove that involving molecular iodine monochloride, equilibrium (118)

$$ICl + ArH \rightleftharpoons ArHI^+ + Cl^- \tag{118}$$

followed by equilibrium (115) because the chloride ion concentration necessary to reverse (118) may be very low indeed[220].

A positive iodinating species was postulated to account for the kinetics and isotope effect observed in the iodination of some amines by iodine in aqueous potassium iodide (in some cases in the presence of acetate, lactate, or phosphate ion). The isotope effects ($k_H/k_D$ values in parenthesis) for these compounds studied were: 2,4,6-trideutero-$m$-dimethylaminobenzenesulphonate ion, 25 °C (1.0); 2,4,6-trideutero-$m$-dimethylbenzoate ion, 30 °C (1.4); 2,4,6-trideuterodimethylaniline, 30 °C, lactate (3.0); 2,4,6-trideuteromethylaniline, 25 °C, acetate (3.2); 2,4,6-trideuteroaniline, 25 °C (3.5), phosphate (4.0); 2,4,6-trideuterometanilate ion, 35 °C (2.0); 2,4,6-trideutero-$m$-aminobenzoate ion, 30 °C (4.8), phosphate (3.0); 2,6-dideutero-1-dimethylaminobenzene-4-sulphonate ion, 25 °C, phosphate (1.0); 4-deutero-1-dimethylaminobenzene-3-sulphonate ion, 25 °C, phosphate (1.0). The kinetics of these reactions was given by

$$\text{Rate} = k_3[ArNR_2]^2[\text{"Iodine"}] + k_3[ArNR_2][\text{"Iodine"}][B] \tag{119}$$

where B is the buffer base, and since the first term was observed regardless of the occurrence of an isotope effect, the second molecule of aromatic amine seemed not to be involved in removing the displaced aromatic proton and was assumed therefore to be involved in forming some other iodinating species, especially as it can be shown that the concentration of $I^+$ in the medium would be very low ($10^{-14}$–$10^{-15}$ $M$). This species was thought to be $ArNR_2I^+$ and likewise the base B could be involved in forming the species $BI^+$ [223]. However, Vainshtein and Shilov[224] later reported that in the iodination of the anions of $m$- and $p$-dimethylaminobenzenesulphonic acid, $m$-dimethylaminobenzoic acid and benzenesulphonic acid in buffer solution, isotope effects $(k_H/k_D)$ of 3, 1.8−1, and 1 were observed, respectively, and the isotope effects decreased with an increase in the basicity and concentration of the basic component of the buffer and *vice versa*. The reaction rates which were said to be governed by

$$\text{Rate} = k_3[\text{ArNR}_2][\text{I}^+][\text{B}] + k_3[\text{ArNR}_2][\text{I}_2][\text{B}] \tag{120}$$

were increased by an increase in the basicity and concentration of the basic component of the buffer. The base appears therefore to be employed in breaking the carbon–hydrogen bond and not in forming the iodinating species so that the earlier evidence in favour of this species being positive seems unsatisfactory and equation (120) suspect.

Grimison and Ridd[225] also suggested the preformation of a positive iodinating species as a means of explaining the results obtained in the iodination by iodine in aqueous potassium iodide, of imidazole at 25 °C. The kinetics of this reaction follow equation (121)

$$\text{Rate} = k[\text{ArH}][\text{I}_3^-][\text{H}^+]^{-1}[\text{I}^-]^{-2} + k(\text{cat})[\text{ArH}]^2[\text{I}_3^-][\text{H}^+]^{-1}[\text{I}^-]^{-2} \tag{121}$$

which is equivalent to

$$\text{Rate} = k_2[\text{Ar}^-][\text{H}^+] + k_3(\text{cat})[\text{Ar}^-][\text{I}^+][\text{ArH}] \tag{122}$$

The values of $k_H/k_D$ for the uncatalysed and catalysed reactions were 4.36 and 4.47 respectively, yet the isotope effect is not necessarily diminished on reducing the concentration of iodide ion to zero and by the arguments elaborated above (p. 95) this implies that molecular iodine is not the iodinating species and that this species is formed in some pre-equilibrium, the function of the base being to form the species and not to remove the proton. This argument assumes, as does the previous discussion of the effect of iodide ion concentration on isotope effects, that a minute concentration of $\text{I}^-$ is insufficient to compete effectively with the reaction involving proton loss.

### 5.2 MOLECULAR HALOGENATION

There have been rather more kinetic studies of molecular halogenation than of positive halogenation and apart from the problems with iodination noted above, interpretation of the kinetics has presented few difficulties.

### 5.2.1 Molecular chlorination

(a) *Molecular chlorine as electrophile*

Orton and King[226] made the first kinetic study of chlorination under conditions believed to involve molecular chlorine as the electrophile; they measured the effects of 2- and 4-alkyl, 2- and 4-alkoxy, 2,3- and 3,4-benzo, and polymethyl substituents on the rates of chlorination of a range of N-acylanilides derived from benzoic and aliphatic acids, in acetic acid containing 0.14 % water at 16 °C. The reaction was first-order in anilide. Chlorine was produced by adding hydrogen chloride to acetylchloramino-2,4-dichloroacetanilide (which is itself not significantly chlorinated under the reaction conditions) and thus produces chlorine *via* the equilibrium

$$ArNClAc + HCl \rightleftharpoons ArNHAc + Cl_2 \tag{123}$$

which is established virtually instantaneously. Subsequent chlorination of the aromatic produces hydrogen chloride so that the reaction proceeds with a constant concentration of chlorine maintained, and apparent first-order kinetics are obtained. However, the rate is dependent on the initial concentration of added acid and hence of chlorine so that the kinetic equation for the reaction is

$$\text{Rate} = k_1[\text{ArH}][\text{Cl}_2] \tag{124}$$

Addition of up to a tenfold molar excess of hydrogen chloride did not appreciably alter the reaction rate. Orton and Bradfield[227] obtained the same kinetic form for the chlorination of formanilide, acetanilide, benzanilide, and benzenesulphonanilide in 99 % aqueous acetic acid at 20 °C; reaction rates were higher than previously obtained with the less aqueous medium, and this medium effect has been subsequently found to be general.

The first kinetic study using chlorine in acetic acid was carried out by Orton *et al.*[228], who measured rates of chlorination of halogen- and methyl-substituted anilides in 40 % aqueous acetic acid at 18 °C and also the relative rates of C- and N-chlorination, thereby confirming that chlorine is first (and rapidly) produced in the reaction of N-chloro-N-acylanilides with hydrogen chloride in acetic acid to give ring-chlorinated products. Subsequently, Bradfield and Jones[229, 230] measured the rates of chlorination by chlorine in 99 % aqueous acetic acid at 20 °C of a wide variety of aromatic ethers (2- and 4-ROArX, where X = Cl, COOH, or NO$_2$, and R = alkyl, benzyl and substituted benzyl)[229] and some anilides, anisic acid, and ethers (in some cases at 30 °C as well)[230] all of which have kinetics of the form of equation (124). This form was also found by de la Mare and Robertson[231], who measured the rates of chlorination of benzene and naphthalene in an attempt to evaluate the inconsistencies in the rates, heats of activation, and catalytic effect of hydrogen chloride reported for these compounds by Lauer and Oda[232]. Using 99 % aqueous acetic acid they found that the heat of activation for benzene, the least reactive compound, was greater (17, determined at 24.0 and 34.2 °C) than for naphthalene (7.5, determined at 15.3 and 24.0 °C), the respective second-order rate coefficients being $1.48 \times 10^{-6}$, $3.9 \times 10^{-6}$, 0.07 and 0.108. The effect of hydrogen chloride was not specific (as assumed by Lauer and Oda) since sulphuric acid had the same effect, and lithium chloride and sodium acetate also had small catalytic effects. The relative rates of chlorination of alkyl-benzenes RPh at 24 °C was: R = H (0.29), Me (100), Et (84), $i$-Pr (51) and $t$-Bu (32), showing that the hyperconjugative order of electron release operated. The relative rates of chlorination of polymethylbenzenes under the same conditions was found to be: toluene (1.0), $p$-xylene (6.3), $o$-xylene (13.4), $m$-xylene ($1.25 \times 10^3$), mesitylene ($5.3 \times 10^5$) and pentamethylbenzene ($2.25 \times 10^6$), indicating a fairly selective and, therefore, unreactive electrophile, entirely consistent with this being molecular chlorine. The catalytic effect of perchloric acid on the chlorination of $m$-xylene in acetic acid (probably 99 % aqueous) was subsequently measured and found to be small[233], and these results together with the observation that the catalytic acceleration by added electrolytes was in the order $HClO_4$ > $LiCl \approx HCl$ > NaOAc (which is closely related to the degree of ionisation of the electrolyte) was interpreted by Robertson[234] to show that chlorination by $Cl^+$, AcOCl, or $AcOHCl^+$ was not involved. If $Cl^+$ was involved a very large anticatalytic effect of $Cl^-$ would have been observed. Likewise, since AcOCl and $AcOHCl^+$ are produced $via$ equilibria (125) and (126), respectively, $viz.$

$$Cl_2 + AcOH \rightleftharpoons AcOCl + H^+ + Cl^- \tag{125}$$

$$Cl_2 + AcOH \rightleftharpoons AcOHCl^+ + Cl^- \tag{126}$$

addition of $Cl^-$ would displace these equilibria to the left and alter the rate contrary to observation. These conclusions were confirmed by de la Mare et al.[203], who prepared chlorine acetate and found that it was immediately decomposed by chloride ions and strongly catalysed by mineral acids in its chlorination reactions. The small catalysis by electrolytes and the more rapid reaction in the more aqueous media follows from the greater polarity of the transition state relative to the ground state involving neutral reagents.

Equation (124) has been found to describe the kinetics of molecular chlorination in other solvents. Roberts and Soper[235] studied the chlorination of acetyl-anthranilic acid, 4-chlorophenetole, 4-cresyl methyl ether, 4-chloroacetanilide, and N-methylacetanilide in the solvents hexane, chloroform, dichloroethane, and nitrobenzene, as well as in acetic acid (Table 54). Activation energies were derived from these values, but little significance can be attached to these in view of the 10 °C temperature range used in determining them. Surprisingly, the relative reactivities of the compounds studied appears to change according to the solvent

TABLE 54

SECOND-ORDER RATE COEFFICIENTS ($10^4 k_2$) FOR REACTION OF ArH WITH $Cl_2$ IN VARIOUS SOLVENTS[235]

| ArH | Hexane | | Chloroform | | Dichloroethane | | Nitrobenzene | | 99 % aq. acetic acid | |
|---|---|---|---|---|---|---|---|---|---|---|
| | 0 °C | 20 °C | 20 °C | 30 °C | 20 °C | 30 °C | 20 °C | 30 °C | 20 °C | 30 °C |
| Acetylanthranilic acid | | | 40.5 | 55.7 | 424 | 509 | 900 | 1,230 | 164 | 303 |
| 4-Chlorophenetole | | | 3.50 | | 49.7 | 55.0 | 29.4 | 43.4 | 317 | 580 |
| 4-Cresylmethyl ether | 2.63 | 2.63 | | | | | | | 324[a] | |
| 4-Chloroacetanilide | | | | | | | 40.4 | 59.1 | 59.1 | 116 |
| N-Methylacetanilide | | | | | | | 5.73 | 8.41 | 75.9 | 1,620 |

[a] Also 280 at 16 °C and 384 at 26 °C.

employed and this suggests solvent substrate interactions, though which solvents and which substrates are anomolous cannot be ascertained from the limited data.

Dewar and Mole[236] derived second-order rate coefficients for chlorination at 25 °C of benzene $(6 \times 10^{-7})$, diphenyl $(6.9 \times 10^{-4})$, naphthalene $(6.3 \times 10^{-2})$, phenanthrene $(2.9 \times 10^{-1})$ and triphenylene $(2.2 \times 10^{-2})$ in Analar acetic acid and of diphenyl $(9 \times 10^{-7})$, naphthalene $(1.9 \times 10^{-4})$, phenanthrene $(1.3 \times 10^{-3})$,

TABLE 55

SECOND-ORDER RATE COEFFICIENTS FOR REACTION OF ArH WITH $Cl_2$ IN VARIOUS SOLVENTS[237]

| Compound | Solvent | Temp. (°C) | $10^4 k_2$ | $E_a$ | $\Delta S^{\ddagger}$ |
|---|---|---|---|---|---|
| Toluene | Acetic anhydride | 25.0 | 3.40 | 9.1 | −46 |
| | | 45.4 | 8.90 | | |
| Toluene | Acetic acid | 25.2 | 8.00 | 13.0 | −30 |
| | | 45.4 | 34.6 | | |
| m-Xylene | Acetic acid | 25.9 | 3,980 | 9.4 | −45 |
| | | 43.2 | 9,460 | | |
| Toluene | Acetonitrile | 25.0 | 13.6 | 7.9 | −50 |
| | | 45.4 | 31.9 | | |
| Benzene | Acetonitrile | 25.0 | 0.0091 | | |
| p-Xylene | Acetonitrile | 25.0 | 99.0 | 7.7 | −31 |
| | | 45.4 | 233 | | |
| m-Xylene | Acetonitrile | 25.1 | 16,160 | 4.6 | −43 |
| | | 42.2 | 22,850 | | |
| Toluene | Nitromethane | 1.5 | 52.0 | 5.6 | −44 |
| | | 25.0 | 109 | | |
| Benzene | Nitromethane | 25.0 | 0.068 | | |
| Benzene | Trifluoroacetic acid | 25.0 | 68.0 | 11.4 | −32 |
| | | 45.4 | 242 | | |

TABLE 56
SECOND-ORDER RATE COEFFICIENTS ($10^3k_2$) FOR REACTIONS OF ArH WITH $Cl_2$ IN MeCN–MeNO$_2$ AT 25.4 °C[238]

| ArH | Nitromethane (mole %) | | | | |
|---|---|---|---|---|---|
| | 100 | 74.5 | 49.5 | 24.6 | 0 |
| Toluene | 13.6 | 6.94 | 4.15 | 2.65 | 1.80 |
| 2-Chloronaphthalene | 123 | 73.4 | 44.3 | 32.9 | 16.0 |

and pyrene $(7.1 \times 10^{-1})$ in a mixture of carbon tetrachloride and acetic acid (3 : 1 by volume), the reactions being studied under first-order conditions (excess of hydrocarbon)[236]; these results again show the high selectivity of the electrophile. Andrews and Keefer[237] measured the rates and kinetic parameters for chlorination of benzene, toluene, and m-xylene in a variety of solvents at 25 °C (Table 55), and showed that the rate varies with solvent as 1,2-dichloroethane $\ll$ acetic anhydride $\approx$ acetic acid $\approx$ acetonitrile $<$ nitromethane $\ll$ trifluoroacetic acid, and the activation energies decrease along this series, that for chlorination in dichloroethane being much higher to the extent that it would not occur unless hydrogen chloride was added, the rate being very sensitive to the hydrogen chloride concentration. The rate is thus not simply a function of the dielectric constant of the medium, and the results for acetic and trifluoroacetic acids show that hydroxylic solvents are good at solvating, electrostatically, the activated complex; consistent with this, they appear to give rise to less negative activation entropies though all of the entropy values seem to be exceptionally low even for formation of a positively charged transition state from neutral reagents. Similar results were obtained for the chlorination of toluene and 2-chloronaphthalene in mixtures of acetonitrile and nitromethane (Table 56) from which it is apparent that the rate increases as the proportion of the latter solvent increases; since both solvents have the same dielectric constant, nitromethane must be better able to solvate electrostatically the activated complex, i.e. it aids breaking of the Cl–Cl bond[238].

TABLE 57
SECOND-ORDER RATE COEFFICIENTS ($10^4k_2$) FOR REACTION OF ArH WITH $Cl_2$ IN VARIOUS SOLVENTS AT 25 °C[239]

| ArH | Solvent | | | | |
|---|---|---|---|---|---|
| | MeNO$_2$ | PhNO$_2$ | MeCN | Ac$_2$O | PhCl |
| Benzene | 0.054 | | | | |
| Toluene | 131 | 45 | 15 | 2.44 | 0.013 |
| t-Butylbenzene | 30.7 | 9.58 | 5.8 | | 0.0036 |
| p-Xylene | 723 | | 110 | | |

The kinetics of the chlorination of some alkylbenzenes in a range of solvents has been studied by Stock and Himoe[239], who again found second-order rate coefficients as given in Table 57. Although the range of rates varies by a factor of $10^4$, there was no marked change in the toluene: $t$-butylbenzene reactivity ratio, and it was, therefore, concluded that the Baker–Nathan order is produced by a polar rather .than a solvent effect.

Finally, a 1:1 mixture of acetic and propionic acids containing 2 % of water has been used in order to study the rates of chlorination of polyalkylbenzenes at low temperatures. Second-order rate coefficients were obtained and the values are recorded in Table 58 together with the energies and entropies of activation (which are given with the errors for 95 % confidence limits) from which it was concluded

TABLE 58

SECOND-ORDER RATE COEFFICIENTS $k_2$ AND KINETIC PARAMETERS FOR REACTION OF ArH WITH $Cl_2$ IN ACETIC ACID–PROPIONIC ACID[240]

| ArH | Temp. (°C)[a] | | | | | | $E_a$ | $\Delta S^\ddagger$ |
|---|---|---|---|---|---|---|---|---|
| | 30 | 20 | 10.1 | 4.9 | 0 | −10 | | |
| $m$-Xylene | 0.268 | 0.162 | 0.089 (9.8) | | 0.0491 (0.2) | | 9.4±0.2 | −34.3±1 |
| Mesitylene | 36.6 | 24.8 | 19.8 | | 12.3 | 7.62 (9.9) | 6.1±0.4 | −35.6±1.5 |
| 1,2,4,5-Tetramethyl-benzene | 2.24 | 1.36 | 0.735 | | 0.406 | 0.249 (10.2) | 8.8±0.5 | −31.4±2 |
| 1,2,3,5-Tetramethyl-benzene | 62.8 | 53.4 | 37.4 | | 28.3 | 15.9(9.8) 17.6 | 5.4±0.5 | −35.6±1 |
| Pentamethyl-benzene | 173 | 137 | 127 | 115 | 85.0 | 61.0 (9.4, 10.0) | 4.1±0.6 | −36.8±2.5 |

[a] Temperatures other than those given in the column headings are in parentheses.

that the rates of chlorination of these compounds are governed by the activation energies rather than entropies[240].

Most of the kinetic studies of molecular chlorination have, however, been carried out using acetic acid (usually with some water content) as solvent. Jones et al.[241–251] measured the rates of chlorination of a very wide range of ethers and anilides in 99 % aqueous acetic acid at 20 °C in order to evaluate the relative directive effects of substituents in electrophilic substitution and the compounds are summarised in Table 59. For some of these compounds rates were measured at a range of temperatures and showed that these were controlled by activation energies rather than by pre-exponential factors[244, 251]; the average activation energies for a series of ethers with variable A₁ groups (Table 60) showed the additivity effect of substituents upon the activation energies.

TABLE 59

ANILIDES AND ETHERS FOR WHICH RATE COEFFICIENTS HAVE BEEN MEASURED FOR
CHLORINATION IN 99.5 % AQUEOUS ACETIC ACID AT 20 °C

| Compound | R | X | Ref. |
|---|---|---|---|
| 2- and 4-$ROC_6H_4X$ | H, alkyl, substituted benzyl | Hal, phenyl sulphonate, ethyl carbonate, carboxylate | 241, 242 |
| 1,2,4-$ROC_6H_3X_2$ | Alkyl, benzyl | Hal, Me, $NO_2$ | 241, 243 |
| 1,2,4-$RC_6H_3X_2$ | OMe, NHAc | Cl, H, Me | 244 |
| 2- and 4-$NHRC_6H_4X$ | $R'CO$, $R'SO_2$ | Hal, $CO_2H$, $NO_2$ | 242, 245, 246 |
| 2- and 4-$ROC_6H_4X$ | Alkyl, benzyl | Hal, $CO_2H$, Me | 244, 247 |
| 4-$ROC_6H_4COC_6H_4X$(3') or (4') | Alkyl | H, Cl | 248 |
| 4-$ROC_6H_4COC_6H_4X$(4') | Alkyl | H, hal, $NO_2$ | 248 |
| 4-$ROC_6H_4COC_6H_4X$ | Alkyl | 2'-, 3'-, or 4'-Me, hal, or $NO_2$ | 248 |
| ROX | Alkyl, substituted benzyl | 2,4-dichlorophenyl, 2,4-dichloro-3,5-dimethyl-phenyl, 4-chloro-3,5-di-methyl-2-nitrophenyl | 243 |
| 3-$ROC_6H_4X$ | Alkyl, substituted benzyl | Cl, $CO_2H$, $NO_2$ | 249 |
| $ROC_6H_3X_2$ | Alkyl, substituted benzyl | 2,5-$Cl_2$, 3,5-$Cl_2$, or 2-Me-5-$NO_2$ | 249 |
| $ROC_6H_3X_2$[a] | Substituted benzyl | 4-Me-2-Br($NO_2$) 2-$i$-Pr-3-Me | 250 |

[a] Rates for some of these compounds in 99.5 % aqueous acetic acid were measured at 16 °C.

Various groups of workers have obtained second-order rate coefficients for the chlorination of alkylbenzenes in acetic acid and the data of de la Mare and Robertson[231] (24 °C, 99.9 % aqueous acetic acid), Brown and Stock[252] (25 °C, 99.87 % aqueous acetic acid), Keefer and Andrews[253] (25.2 °C, "purified" acetic acid), de la Mare and Hassan[254] (25 °C, water content ≯ 0.05 %), and Bacciocchi and Illuminati[255] (30 °C, "purified" acetic acid) are gathered in Table 61. The disagreement between the sets of data is not simply an effect arising from the different concentration of water in the acetic acid because the spread of rates is

TABLE 60

ACTIVATION ENERGIES FOR REACTION OF ROAr
WITH $Cl_2$ IN $HOAc$[244]

| Substituent in Ar | $E_a$ |
|---|---|
| 4-Cl | 11.2 |
| 2-Cl, 4-Cl | 14.2 |
| 2-Cl, 4-Cl, 3,5-$Me_2$ | 9.5 |
| 2-Br, 4-Me | 9.7 |
| 2-Me, 4-Br | 10.2 |
| 2-$i$-Pr, 3-Me | 7.7 |

TABLE 61

SECOND-ORDER RATE COEFFICIENTS $(10^3 k_2)$ FOR REACTION OF ArH WITH $Cl_2$ IN HOAc

| ArH | Ref. 231 | Ref. 252 | Ref. 253 | Ref. 254 | Ref. 255 |
|---|---|---|---|---|---|
| Benzene | | 0.00154 | 0.0069 | 0.0013 | |
| Toluene | 0.47 | 0.53 | 0.65 | 0.485 | |
| t-Butylbenzene | | 0.135 | | | |
| p-Di-t-butylbenzene | | 0.462[a] | | | |
| p-t-Butyltoluene | | 2.28[a] | | | |
| p-Xylene | 3.05 | 3.20 | 4.22 | | |
| o-Xylene | 6.35 | 3.23 | 7.9 | | |
| m-Xylene | | 285 | 337 | | |
| Mesitylene | | | | | 67,300 |
| 1,2,4,5-Tetramethyl benzene | | | 4,190 | | 2,010 |
| 1,2,3,5-Tetramethyl-benzene | | | | | 124,000 |
| Pentamethylbenzene | | | | | 311,000 |
| 2-Nitromesitylene | | | | | 0.1435 |
| 3-Nitro-1,2,4,5-tetra-methylbenzene | | | | | 0.00456 |
| 4-Nitro-1,2,3,5-tetra-methylbenzene | | | | | 0.272 |

[a] Includes the rate of chlorodebutylation.

very inconsistent as well as the data for o- and p-xylene; the rate coefficients for benzene may be compared with those obtained by Dewar and Mole[236] (0.0006) and Mason[256] (0.0012). The data for the nitropolyalkylbenzenes showed that the nitro group deactivated less than expected from analysis of its effect in benzene alone and thus was interpreted as showing that there was steric hindrance to resonance between the nitro group and the aromatic ring. Alternative explanations are possible since the transition state for the more reactive alkylbenzenes would be nearer to the ground state and hence the destabilising effect of a substituent would be smaller[257].

De la Mare and Hassan[254] obtained second-order rate coefficients (in parenthesis) for the following: 4-methylacetanilide (1.53), 2-methylacetanilide (0.193), 2,6-dimethylacetanilide (0.0118), acetanilide (0.93), 4-acetamidodiphenyl (0.248) and 1,4-diacetamidobenzene (0.231); the results for the acetanilides demonstrated the effect of steric hindrance to coplanarity thereby inhibiting resonance of the nitrogen lone pair with the aromatic ring. The rate coefficients for chlorination of 3-chloroacetanilide (0.215), 4-chloroacetanilide (0.010) 3-nitroacetanilide $(6.7 \times 10^{-5})$ and phenyl benzoate $(3.2 \times 10^{-6})$ have also been measured[258, 261].

Mason[256] has measured the second-order rate coefficients and Arrhenius parameters for the chlorination of benzene, biphenyl, naphthalene, and phenanthrene in acetic acid (containing 0.05 % water) and these are given in Table 62.

TABLE 62

SECOND-ORDER RATE COEFFICIENTS $(10^4k_2)$ AND ARRHENIUS PARAMETERS FOR REACTION OF ArH WITH $Cl_2$ IN HOAc[256]

| ArH | Temp. (°C) | | | | $E_a$ | log A |
|-----|-----|-----|-----|-----|-----|-----|
|     | 15 | 20 | 25 | 30 | | |
| Benzene | $3.8 \times 10^{-3}$ | $6.8 \times 10^{-3}$ | $1.2 \times 10^{-2}$ | $2.2 \times 10^{-1}$ | 20.0 | 9.0 |
| Biphenyl | 3.3 | 5.23 | 8.62 | 12.9 | 14.8 | 7.8 |
| Naphthalene | 370 | 521 | 730 | 1,050 | 11.5 | 7.3 |
| Phenanthrene | 3,170 | 4,500 | 5,910 | 7,940 | 10.5 | 7.4 |

Significantly, the pre-exponential factors decrease with increasing reactivity, and this suggests that the Wheland intermediate is more nearly formed in the transition state, the more reactive the compound. Or, considered another way, the position along the reaction co-ordinate at which a given amount of carbon–halogen bond formation occurs is nearer to the ground state the more reactive the compound.

Stock and Baker[259] measured the relative rates of chlorination of a number of halogenated aromatics in acetic acid containing 20.8 $M$ $H_2O$ and 1.2 $M$ HCl at 25 °C and the values of the second-order rate coefficients $(10^3k_2)$ are as follows: $p$-xylene (11,450), benzene (4.98), fluorobenzene (3.68), chlorobenzene (0.489), bromobenzene (0.362), 2-chlorotoluene (3.43), 3-chlorotoluene (191), 4-chloro-toluene (2.47), 4-fluorotoluene (9.70), 4-bromotoluene (2.47). Increasing the concentration of the aromatic, however, caused, in some cases, a decrease in the rate coefficients; thus an increase in the concentration of chlorobenzene from 0.1 $M$ to 0.2 $M$ caused a 20 % decrease in rate coefficient, whereas with 4-chloro- and 4-bromo-toluene, no such change was observed.

De la Mare et al.[260] measured the rates of chlorination of biphenyl, a wide range of its methyl derivatives, and anisole in acetic acid at 25 °C. Second-order rate coefficients $(10^4k_2)$ were: biphenyl (6.40), 2-methylbiphenyl (3.20), 3-methyl-biphenyl (820), 4-methylbiphenyl (30.0), 2.2'-dimethylbiphenyl (4.40), 3.3'-dimethylbiphenyl (2,630), 4,4'-dimethylbiphenyl (70.0), 2,6'-dimethylbiphenyl (1,130), 3,4,3',4'-tetramethylbiphenyl (19,300), anisole $(12.5 \times 10^4)$, and these results showed very clearly the effect of steric inhibition of resonance between the phenyl rings through the presence of ortho methyl groups[260]. Similar (but rather more emphatic) results were obtained[262] in chlorination of the $t$-butyl derivatives for which the corresponding rate coefficients were: 2-$t$-butylbiphenyl (1.0) 4-$t$-butylbiphenyl (25.7), 2,2'-di-$t$-butylbiphenyl (1.8), 4,4'-di-$t$-butylbiphenyl (70.0).

The rates of chlorination of bridged biphenyls have also been measured and show the effect of coplanarity on reactivity. Second-order rate coefficients $(10^4k_2)$ at 25 °C were: fluorene (1,700), 9,10-dihydrophenanthrene (170), 1,2 : 3,4-di-benzocyclohepta-1,3-diene (9.70), 5-methyl-1,2 : 3,4-dibenzocyclohepta-1,3-diene

(11.5), 6,7-dimethyl-1,2 : 3,4-dibenzocyclohepta-1,3-diene (6.80)[263], N-acetyl-carbazole (720), N-acetyldiphenylamine (183), N-acetyl-2-chlorocarbazole (135), N-acetyl-3-chlorocarbazole (95.0); carbazole, diphenylamine and triphenylamine reacted so rapidly that no chlorine could be detected after one minute, hence the rate coefficients were estimated as $> 170$[264].

Despite the relative simplicity of the kinetics of molecular chlorination, there has so far been only one measurement of the rate coefficient with a heterocyclic compound and the need for more work in this area is indicated. Marino[265] found that chlorination of thiophene by chlorine in acetic acid at 25 °C gave the second-order rate coefficient of $10.0 \pm 1.5$, so that thiophene is $1.7 \times 10^9$ times as reactive as benzene in this reaction and this large rate spread is clearly consistent with the neutral and hence relatively unreactive electrophile.

Second-order rate coefficients have been obtained for chlorination of alkyl-benzenes in acetic acid solutions (containing up to 27.6 $M$ of water) at temperatures between 0 and 35 °C, and enthalpies and entropies of activation (determined over 25 °C range) are given in Table 63 for the substitution at the position indicated[266].

TABLE 63

KINETIC PARAMETERS FOR THE CHLORINATION OF ALKYLBENZENES IN ACETIC ACID[266]

| Position of substitution | Anhydrous acid | | 4.11 M H$_2$O+1.11 M HCl | | 27.6 M H$_2$O+1.18 M HCl | |
|---|---|---|---|---|---|---|
| | $\Delta H^{\ddagger}$ | $\Delta S^{\ddagger}$ | $\Delta H^{\ddagger}$ | $\Delta S^{\ddagger}$ | $\Delta H^{\ddagger}$ | $\Delta S^{\ddagger}$ |
| Benzene (1) | 17 | −30 | 14.7 | −30.3 | 13.0 | −27.6 |
| Toluene (2) | 13 | −30 | 11.4 | −29 | 8.9 | −29.0 |
| Toluene (4) | 13 | −31.6 | 10.6 | −30.9 | 8.8 | −29.0 |
| t-Butylbenzene (2) | | | 12.7 | −28.7 | 10.5 | −27.7 |
| t-Butylbenzene (4) | | | 12.0 | −27.6 | 10.3 | −25.3 |

It was concluded that the variations in rate are due to variations in activation enthalpy rather than entropy, and since the rates of substitution rates at the *para* positions of toluene and *t*-butylbenzene varied by only 4 % for a change in reactivity of 6,430, it was concluded that the Baker–Nathan reactivity order does not arise from a solvent effect (*cf*. Table 57).

Kinetic studies have been carried out using the 1 : 1-complex iodobenzene dichloride as a source of molecular chlorine. In acetic acid solutions, the dissociation of this complex is slower than the rate of halogenation of reactive aromatics such as mesitylene or pentamethylbenzene, consequently the rate of chlorination of these is independent of the aromatic concentration. Thus at 25.2 °C first-order chlorination rate coefficients were obtained, being approximately $0.2 \times 10^{-3}$ whilst the first-order dissociation rate coefficient was $0.16 \times 10^{-3}$; from measurements at 25.2 and 45.6 °C the corresponding activation energies were 20.1 (pentamethylbenzene) and 20.2[253]. Prior dissociation of this complex in chlorination in carbon tetrachloride catalysed by trifluoroacetic acid, does not however occur, for the rate of chlorination of durene is dependent on the aromatic concentration and at 25 °C the average rate coefficient $k_0$ ($k_0 = -\mathrm{d}\ln[\mathrm{PhICl_2}]/\mathrm{d}t$) was $4.7 \times 10^{-3}$ sec$^{-1}$ compared with $7 \times 10^{-5}$ sec$^{-1}$ for dissociation[267]. The order with respect to the catalyst was a minimum of 2.2 and indicated that the acid dimer was involved and probably increases the polarity of the iodobenzene dichloride through hydrogen bonding.

There have been two determinations of the kinetic isotope effect in uncatalysed molecular chlorination. The first of these compared the rates of chlorination of 3-bromo-1,2,4,5-tetramethylbenzene with its 6-deuterated isomer in acetic acid at 30 °C; the respective second-order rate coefficients were 0.098 and 0.107, so that $k_H/k_D = 0.92$[268]. A similar inverse isotope effect was observed in the chlorination of naphthalene ($k_2 = 0.051$) and its octadeuterated isomer ($k_2 = 0.060$) in acetic acid at 25 °C, giving $k_H/k_D = 0.85$[269]. These effects were probably a balance of a normal and small isotope effect for the step involving breaking of the carbon–hydrogen bond together with an inverse isotope effect arising from the change in hybridisation $sp^2$ to $sp^3$ in going through the rate-determining transition state of the reaction.

### (b) Molecular chlorine acetate as electrophile

Kinetic studies with chlorine acetate as a molecular chlorinating species have been carried out. The kinetic form for chlorination of toluene by hypochlorous acid in aqueous acetic acid is given by

$$\text{Rate} = k_2[\text{ArH}][\text{HOCl}] \tag{127}$$

but since hypochlorous acid is a very ineffective chlorinating species, and could not possibly react with toluene under the conditions employed, the reacting species was concluded to be molecular chlorine acetate. From the equilibrium constant for the formation of chlorine acetate from hypochlorous acid in acetic acid, and from the variation in the rate of chlorination of toluene by mixtures of hypochlorous acid in aqueous acetic acid at 25 °C, the true rate of chlorination by chlorine acetate was deduced and compared with the rate of chlorination by molecular chlorine (Table 64)[209]. It appears then that the reactivity sequence for molecular electrophiles is ClOAc > ClCl > ClOH, whereas from the electron affinity of X in ClX one would have expected the sequence ClCl > ClOAc > ClOH; consequently, it was proposed that the transition state for chlorination by chlorine acetate involves a 6-centre cyclic structure (VI) in which the Cl–O bond fission is aided by intramolecular hydrogen bonding.

(VI)

TABLE 64

COMPARISON OF RATE COEFFICIENTS FOR REACTION OF $Cl_2$ AND ClOAc WITH PhMe
IN AQUEOUS HOAc AT 25 °C[209]

| Solvent (% AcOH) | | 100 | 98 | 95 | 86 | 76 | 51 |
|---|---|---|---|---|---|---|---|
| $k_2$(ClOAc) | ca. | 0.28 | 0.57 | 2.67 | 8.35 | 21.6 | 60.6 |

| Solvent (% AcOH) | 99.87 | 98.7 | 95 | 90 | 85 | 76 |
|---|---|---|---|---|---|---|
| $10^3 k_2$($Cl_2$) | 0.53 | 0.98 | 7.17 | 23.9 | 83.5 | 275 |

Relative rates and partial rate factors have been determined for the chlorination of some aromatics by chlorine acetate in 76 % aqueous acetic acid at 25 °C [209]; these are given in Table 65. The spread of rates is, therefore, smaller than is found with molecular chlorine and this is entirely consistent with the lower reactivity of the latter reagent.

TABLE 65

RELATIVE RATES AND PARTIAL RATE FACTORS FOR REACTION OF ArH WITH ClOAc
IN HOAc AT 25 °C[209]

| ArH | $k_{rel}$ | ArH | $k_{rel}$ | ArH | $k_{rel}$ |
|---|---|---|---|---|---|
| Benzene | 1.0 | o-Xylene | 580 | 4-Choroanisole | 560 |
| Toluene | 145[a] | p-Xylene | 690 | 4-Methoxybenzoic | 490 |
| t-Butylbenzene | 62.5[b] | m-Xylene | 12,500 | acid | |

[a] $f_2^{Me} = 306, f_4^{Me} = 237$.
[b] $f_4^{t-Bu} = 158$.

The rates of chlorination of benzene, biphenyl, and diphenylmethane by chlorine acetate in 98 % aqueous acetic acid at 25 °C have also been determined and the second-order rate coefficients are 0.00118, 0.0364, and 0.0311, respectively[209a, 270]. The variation in rate with change in water content of the acetic acid was the same as that previously observed[209] for toluene, and thus in ca. 75 % aqueous acid the coefficients were 0.00073, 0.027 and 0.0241; however, elsewhere in ref. 209a a 4-fold decrease in rate coefficient for diphenylmethane was claimed to accompany the same increase in water content of the medium.

(c) *Chlorination by a molecular species with a catalyst*

In the molecular chlorination described above, the breaking of the Cl–Cl bond is facilitated by the polarity of the solvent, which in addition may itself, by interaction with the negative end of the $Cl^{\delta+}$–$Cl^{\delta-}$ dipole, further aid the breaking of this bond in the transition state of the reaction. It is evident, therefore, that Lewis acids with their facility for accepting a pair of electrons will greatly aid this bond-breaking so that chlorination in the presence of these acids should be accelerated, and this is indeed observed. Though there have been numerous qualitative observations of this over the past seventy years or so, few quantitative studies have been carried out.

Schonken *et al.*[271], found that the chlorination of aromatics by chlorine (in the absence of solvent and, therefore, with excess in aromatic) was catalysed by hydrogen chloride and followed the rate expression

$$\text{Rate} = k_2[\text{Cl}_2][\text{HCl}] \tag{128}$$

In the hydrogen chloride-catalysed chlorination of toluene, the addition of solvents generally produced a rate change consistent with their polarities. Thus nitrobenzene caused a large rate increase, benzene and carbon tetrachloride caused rate decreases and acetic acid caused the rate to pass through a maximum at approximately 35 % by volume, probably due to the formation of dimers at high concentration. The effectiveness of di- and trichloroacetic acids as catalysts was also examined. Using the latter as catalyst for the chlorination of toluene, it was found that nitrobenzene increased the rate, and benzene reduced it as before, but in this case acetic acid caused a large rate decrease, probably due to the formation of mixed dimers thereby reducing the effectiveness of the chloroacid catalyst. The selectivity of the chlorinating species between *p*-xylene and chloro-*p*-xylene was much higher for the use of chlorinated acids as catalysts than with hydrogen chloride, thereby pointing to a more polar chlorinating species in the latter case.

Keefer and Andrews[253] measured the rates of chlorination of alkylbenzenes by chlorine catalysed by zinc chloride in acetic acid in solvent. They analysed the data in terms of a catalysed and an uncatalysed reaction (equation 129 and Table 66)

$$\text{Rate} = (k_2 + k_3[\text{ZnCl}_2])[\text{ArH}][\text{Cl}_2] \tag{129}$$

and the reaction appears, therefore, to be approximately first-order in catalyst. However, measurement of activation energies indicated that (for toluene) these were the same for the catalysed and uncatalysed reactions (13.4) so that there is a possibility that the "catalysis" might be only a salt effect.

The effect of catalysis by trifluoroacetic acid on chlorination in carbon tetrachloride has also been determined[272]. For 1,2,4,5-tetramethylbenzene, with low concentrations of catalyst, the order in catalyst is three-halves, but for toluene (which requires a higher concentration) the order is mixed three- and five-halves; the indication is, therefore, that a minimum of three catalyst monomers (or one monomer and one dimer) are necessary. Since trifluoroacetic acid is very likely to be dimeric in carbon tetrachloride, the concentration of monomer is pro-

TABLE 66

RATE COEFFICIENTS FOR THE ZnCl$_2$-CATALYSED REACTION OF Cl$_2$ WITH ArH IN HOAc AT 25 °C[253]

| ArH | [ZnCl$_2$] (M) | $10^3k_{2(obs)}$ | $10^3k_3$ |
|---|---|---|---|
| Benzene | 0.400 | 0.072 | 0.16 |
| Toluene | 0.400 | 8.4 | 19.5 |
| | 0.200 | 3.49 | 14.2 |
| | 0.100 | 1.8 | 12.0 |
| | 0.196 (45.2 °C) | 14.9 | 60.2 |
| o-Xylene | 0.200 | 33.1 | 128 |
| | 0.100 | 19.7 | 122 |
| m-Xylene | 0.207 | 920 | 2,820 |
| | 0.103 | 632 | 2,860 |
| p-Xylene | 0.200 | 29.2 | 124 |
| | 0.100 | 12.7 | 92 |
| | 0.050 | 8.0 | 91 |
| 1,2,4,5-Tetramethyl-benzene | 0.103 | 4,660 | |

portional to (CF$_3$COOH)$^{\frac{1}{2}}$. The addition of acetic acid produced a large rate decrease (as observed with chloroacetic acids as catalysts, above) and this was attributed to the probable formation of mixed dimers. The catalysis by trifluoroacetic acid can be attributed to its strong electron-attracting effect which facilitates breaking of the halogen–halogen bond, thus lowering the energy of the transition state.

Catalysis by hydrogen chloride or iodine monochloride in chlorination in carbon tetrachloride has also been examined. For the chlorination of pentamethylbenzene, the reaction was first-order in both aromatic and chlorine and either three-halves, or mixed first- and second-order in hydrogen chloride, but iodine monochloride was more effective as a catalyst and the chlorination of mesitylene was first-order in iodine monochloride; the activation energy for this latter reaction (determined from data at 1.2 and 25.0 °C) was only 0.4 [273].

Trifluoroacetic acid has been examined as a solvent and chlorination of benzene in this is first-order in aromatic and chlorine, but for benzene a higher activation energy (11.4, determined from data at 25.0 and 45.4 °C) was obtained than for chlorination in carbon tetrachloride; this unexpected result was attributed to an increase in desolvation energy of the reactants[273].

Catalysis by stannic chloride in the chlorination of alkylbenzenes in the absence of solvent has been shown to be first-order in catalyst so that the kinetic equation was[274]

$$\text{Rate} = k_2[\text{Cl}_2][\text{SnCl}_4] \tag{130}$$

Second-order rate coefficients at 20 °C were as follows: benzene (0.0133), toluene (1.26), p-xylene (62) o-xylene (70.2), and i-propylbenzene (0.395). It is difficult to evaluable the quantitative significance of this data, however, since there must be a solvent correction factor to each rate, arising from the differing polarities of the media. That this can be significant can be seen from the data in Table 67,

TABLE 67

RELATIVE REACTIVITIES OF ArH IN CHLORINATION BY $Cl_2$ WITH AND WITHOUT CATALYSTS AT 25 °C[275]

| ArH | No catalyst | FeCl₃ cat./ PhNO₂ solvent | FeCl₃ cat./ ArH excess | AlCl₃–MeNO₂ cat. ArH excess |
|---|---|---|---|---|
| Bromobenzene | | 0.15 | | 0.15 |
| Chlorobenzene | | 0.17 | | 0.19 |
| Fluorobenzene | | 0.29 | | 0.34 |
| Benzene | 1 | 1 | 1 | 1 |
| Ethylbenzene | | 11.9 | 13.0 | 17.4 |
| Toluene | 2,445 | 13.5 | 14.8 | 18.3 |
| o-Xylene | | 38.0 | 62.4 | 69.6 |
| p-Xylene | 14,200 | 43.9 | 74.0 | 73.2 |
| m-Xylene | 247,000 | 110 | 177.3 | 201.3 |
| Mesitylene | $\sim 5 \times 10^6$ | 632 | 920 | 1,875 |

which refer to the relative rates obtained in the chlorination of alkyl- and halogeno-benzenes under various conditions of catalysis by $FeCl_3$ or $AlCl_3$–$MeNO_2$, in some cases with nitrobenzene as solvent[275]. The ferric chloride-catalysed reactions were first-order in aromatic and the high positional selectivity which accompany these intermediate substrate selectivities were considered to be another example of reactions in which the substrate selectivity arises from a rate-determining formation of a $\pi$-complex. A small inverse kinetic isotope effect ($k_H/k_D = 0.87 \pm 0.05$) was obtained for the ferric chloride-catalysed chlorination of benzene and its deuterated isomer in nitromethane and was attributed to the cause described previously (p. 107). The significance of the results of Olah et al.[275] has, however, been thrown open to serious doubt by Caille and Corriu[276], who measured the relative rates of chlorination of toluene and benzene, by means of a direct rather than a competitive kinetic technique. They find that with nitrobenzene or nitromethane as solvent and with aluminium chloride as catalyst, the kinetic equation is

$$\text{Rate} = k_3[Cl_2][AlCl_3][ArH] \tag{131}$$

The relative rate of chlorination of toluene and benzene is 247 (at 0 °C) and 186 (at 15 °C), both with nitrobenzene, and is 215 (at 0 °C ) in nitromethane[276]; repeating the experiments using the competitive techniques gave values (at 15 °C, nitrobenzene as solvent) of 150 and 160, and these were dependent on the initial relative concentrations of toluene and benzene, which should not be so. Generally it was shown that relative reactivities obtained by the competition techniques were lower than obtained by the direct method, and the implied conclusion is that the value of 13–18 obtained by Olah *et al.*[275] (using different catalysts) should be considerably higher.

### (d) Chlorination by molecular sulphuryl chloride

Bolton and de la Mare[277] have examined the kinetics of chlorination of anisoles and methylanisoles by sulphuryl chloride in chlorobenzene and these followed the rate equation

$$\text{Rate} = k_2[\text{ArH}][\text{SO}_2\text{Cl}_2] \tag{132}$$

They argued that pre-equilibria to form $Cl^+$ or $SO_2Cl^+$ may be ruled out, since these equilibria would be reversed by an increase in the chloride ion concentration of the system whereas rates remained constant to at least 70 % conversion during which time a considerable increase in the chloride ion concentration (the by-product of reaction) would have occurred. Likewise, a pre-equilibrium to form $Cl_2$ may be ruled out since no change in rate resulted from addition of $SO_2$ (which would reverse the equilibrium if it is reversible). If this equilibrium is not reversible, then since chlorine reacts very rapidly with anisole under the reaction condition, kinetics zeroth-order in aromatic and first-order in sulphur chloride should result contrary to observation. The electrophile must, therefore, be $Cl^{\delta+} \ldots SO_2Cl^{\delta-}$ and the polar and non-homolytic character of the transition state is indicated by the data in Table 68; a cyclic structure (VII) for the transition state was considered as fairly probable.

(VII)

Second-order rate coefficients ($10^7 k_2$ at 25.0 °C) for a number of aromatics were: benzene, toluene, *m*-xylene, and mesitylene, all $\gg$ 1.0; anisole, 6.0; 3-methylanisole, 200; 2,3-dimethylanisole, 240; 3,5-dimethylanisole, 3,800, and 1,3-dimethoxybenzene containing the following substituents: H, 5,800; 4-Br, 180; 4-Cl, 160; 4-I, 2,900; 4-NO$_2$, 14; 4-MeO, 5,500; 4-CO$_2$Me, 420; 5-CO$_2$Me, 62; 5-Br, 2,600; 5-Cl, 2,700; 5-CN, 260[277, 277a]. A good correlation of the substituted dimethoxybenzene data with the Hammett equation was obtained with $\rho = -4.0$; this value was taken to further confirm the heterolytic nature of the rate-determining step. The incursion of an apparently homolytic reaction was detected if the aromatics were distilled from diethyl ether, anomalously high but quite reasonably

TABLE 68

RATE COEFFICIENTS FOR REACTION OF $SO_2Cl_2$ WITH 3,5-DIMETHYLANISOLE AT 25 °C[277]

| 10 [ArOMe] (M) | 10 [SO₂Cl₂] (M) | Solvent | $10^4k_2$ |
|---|---|---|---|
| 1.361 | 1.00 | PhCl | 3.96 |
| 2.506 | 1.00 | PhCl | 3.82 |
| 2.523 | 0.490 | PhCl | 3.78 |
| 2.528 | 1.286 | PhCl | 3.92 |
| 2.525 | 1.030 | PhCl+glass | 3.94 |
| 2.760 | 1.103 | PhCl+0.02 $M$ $I_2$ | 3.74 |
| 2.050 | 1.103 | PhCl+0.04 $M$ $Bz_2O_2$ | 3.76 |
| 2.221 | 0.991 | PhCl+0.182 $M$ $SO_2$ | 3.60 |
| 2.326 | 2.000 | PhH | 0.40 |
| 1.635 | 1.678 | $o$-$C_6H_4Cl_2$ | 13.0 |
| 1.553 | 0.962 | PhNO₂ | >3,800 |

second-order rate coefficients being obtained; steam distillation of the aromatics from ferrous sulphate caused these second-order rate coefficients to drop sharply and the radical initiator was believed to be diethyl peroxide.

The kinetics of chlorination by sulphuryl chloride in nitromethane have been found to follow equation (132), though in this solvent a significant proportion of reaction also occurred *via* molecular chlorine formed in a rapid pre-equilibrium[277a]. Second-order rate coefficients, $10^6k_2$ (which contain a contribution from the latter reaction), were obtained for some aromatics at 25 °C as follows: benzene, 9.6; toluene, 2,000; ethylbenzene, 1,300; *t*-butylbenzene 710; *o*-xylene, 15,000; *p*-xylene, 7,300; *m*-xylene, 380,000; fluorobenzene, 20; *m*-fluorotoluene 3,600. This data gave a Hammett correlation with $\rho = -7.2$, again inconsistent with a homolytic mechanism. The reactions gave no kinetic isotope effect so the first step of the reaction was assumed to be rate-determining.

### 5.2.2 Molecular bromination

#### (a) Molecular bromine as electrophile

Bradfield *et al.*[278] first studied the kinetics of molecular bromination using aromatic ethers in 50 % aqueous acetic acid at 18 °C. They showed that the kinetics are complicated by the hydrogen bromide produced in the reaction which reacts with free bromine to give the tribromide in $Br_3^-$, a very unreactive electrophile. To avoid this complication, reactions were carried out in the presence of 5–10 molar excess of hydrogen bromide, and under these conditions second-order rate coefficients (believed to be $10^2k_2$ by comparison with later data) were obtained as follows after making allowance for the equilibrium $Br_2 + Br_3^- \rightleftharpoons Br_3^-$, for which $K = 50$ at 18 °C: 4-chloroanisole (1.12), 4-bromoanisole (1.20), 4-

chlorophenetole (2.55), 4-bromophenetole (2.80), 4-chlorophenyl isopropyl ether (8.00), 4-bromophenyl isopropyl ether (8.50). This was, therefore, one of the earliest demonstrations that *para* chlorine deactivities rather more strongly than bromine and that *para* alkyl groups activated in the inductive order in this reaction for which the kinetics followed the equation

$$\text{Rate} = k_2[\text{ArH}][\text{Br}_2] \tag{133}$$

Subsequent kinetic studies of the bromination of 2- and 4-chlorophenyl ethers in 75 % aqueous acetic acid at 20 °C showed mixed second- and third-order kinetics, *viz.*

$$\text{Rate} = k_2[\text{ArH}][\text{Br}_2] + k_3[\text{ArH}][\text{Br}_2]^2 \tag{134}$$

and the ratio of second- to third-order rates was constant, being five times smaller for the 2-chloro relative to the 4-chloro series[193]. This is entirely consistent with the large steric hindrance expected on having four bromine atoms close to the reaction site as required by the proposed termolecular mechanism. Initial reaction rates were measured with the concentration of added hydrogen bromide being nil or up to ten times that of the bromine concentration and the rate coefficients $(10^4 k_2)$ for 4-chlorophenyl isopropyl ether correspondingly increased from 19.0 to 26.0, the third-order coefficients being 319 times as great. Other second-order rate coefficients were: 4-chloroanisole (2.30), 4-chlorophenetole (5.70), 4-chloro-phenyl-*n*-propyl ether (7.40) and 4-nitroanisole[194] (0.20) at 19.8 °C; these results again show the marked inductive activation by the alkyl groups, and also the wide spread of rates expected for a neutral and, therefore, relatively unreactive electrophile. If the assumption regarding the units of the rate coefficients of the earlier work is correct, then comparison of the two sets of data indicates that reaction is more rapid in the more aqueous media. Bromination of 2-chloro-anisole gave second-order rate coefficients which decreased from 550 to 450 as the hydrobromic acid concentration was increased and there seems to be no explanation for this, but the much greater reactivity of the *ortho* relative to the *para* compound (substitution occurring *para* and *ortho* to the methoxy group, respectively) reflects the high steric hindrance to molecular bromination, again consistent with the large neutral molecule being the electrophile.

The most extensive study of the effect of conditions upon the kinetics of bro-mination was made by Robertson *et al.*[231, 279], who measured the rates of bro-mination of alkylbenzenes, acetanilide, aceto-*p*-toluidide, mesitylene, anisole and *p*-tolyl methyl ether in acetic acid at 24 °C. They found that at relatively high con-centrations of bromine ($M/40$–$M/100$) the reaction is second-order in bromine, *i.e.* the rate equation is

$$\text{Rate} = k_3[\text{ArH}][\text{Br}_2]^2 \tag{135}$$

but at low concentrations of bromine ($M/1000$ or less) the rate expression changes to (133) and this change in order with bromine concentration is now recognised to be a general observation for molecular bromination under fairly non-polar conditions. The reaction was shown to be acid-catalysed (sulphuric acid increased the rate) and hydrogen bromide produced in the reaction exhibited a catalytic effect (as an acid) and an anti-catalytic effect (because of the bromide ion). The effect of the latter was shown for the reaction with acetanilide, where the $k_3$ values (at 24 °C) decreased from 14.8 at 10 % reaction to 5.7 at 50 % reaction; in this example the acid catalysis would be masked by protonation of the nitrogen substituent. With mesitylene, however, (no protonation) the appropriate coefficients were 14.5 and 13.7 respectively, and the different effect in these two examples shows the weak catalytic effect. Measurement of the rates of bromination of acetanilide between 24 and 49 °C showed that the second-order reaction had a much higher activation energy ($\sim 11$) than the third-order reaction ($\sim 3.3$) and consequently the overall kinetic order is reduced by raising the temperature, e.g. it is 3.0 at 24 °C and 2.5 at 50 °C. The addition of water was found to confirm the increase in reaction rate noted above and also to decrease the reaction order, whilst non-polar solvents such as chloroform and carbon tetrachloride decreased the rate and increased the order as shown by the relative rates and kinetic orders (in parentheses) in Table 69. Less reactive compounds such as m-xylene and

TABLE 69

RELATIVE RATES (AND KINETIC ORDERS) FOR BROMINATION IN HOAc–CHCl$_3$ (OR CCl$_4$) MIXTURES AT 24 °C[231,279]

| | | | | |
|---|---|---|---|---|
| CHCl$_3$ in AcOH(%) | 0 | 50 | 75 | 90 |
| Anisole ($M/80$) | 1(3.0) | 0.30(2.8) | 0.12(3.1) | 0.08(3.6) |
| CCl$_4$ in AcOH (%) | 0 | 60 | | |
| Acetanilide ($M/40$) | 1(2.9) | 0.20(3.5) | | |

naphthalene gave up to fourth-order kinetics depending on the concentrations. For the latter compound, rates were obtained different from those of Lauer and Oda[280], the divergencies being greater at lower initial reactant concentrations, shorter reaction times, and for more slowly reacting compounds, showing Lauer and Oda's work to have been vitiated by unsaturated impurities in the solvent.

The above data was interpreted as showing that the second-order reaction arises from equilibria (136), (137), and (138)

$$ArH + Br_2 \rightleftharpoons ArHBr_2 \quad (fast) \tag{136}$$

$$ArHBr_2 \underset{k'_{-1}}{\overset{k'_1}{\rightleftharpoons}} ArHBr^+ + Br^- \quad (slow) \tag{137}$$

$$ArHBr^+ \overset{k'_2}{\rightleftharpoons} ArBr + H^+ \quad (fast) \tag{138}$$

and the third- and higher-order reactions arise from interaction of one or more bromine molecules with the first formed intermediate to aid breaking of the bromine–bromine bond, as in the equilibrium

$$ArHBr_2 + Br_2 \rightleftharpoons ArHBr^+ + Br_3^- \tag{139}$$

It then follows that at low bromine concentrations this latter process is less likely, consequently the kinetic order is reduced. More ionic media will facilitate equilibrium (137) without the need for intervention of equilibrium (139) and *vice versa*, so that the observed variation in the kinetic order with this condition then follows. The absence of high kinetic orders in molecular chlorination also becomes rationalised since the $Cl_3^-$ ion is not as stable as $Br_3^-$.

In this work the relative reactivities of the alkylbenzenes (PhR) varied as follows: (R = ) Me, 100; Et, 76; *i*-Pr, 44; *t*-Bu, 23. This relates to 85 % aqueous acetic acid at 24 °C and with 88 % acid at 25 °C the same relative rate of toluene to *t*-butylbenzene was obtained[261], showing the hyperconjugative order to prevail for this reaction.

Robertson *et al.*[233] examined the effects of catalysts and measured relative catalysed to uncatalysed reaction rates for bromination of *m*-xylene in acetic acid at 24 °C, with bromine (*M*/80) and catalyst (*M*/20), as follows: HClO₄ (3.1), LiCl (1.0), H₂SO₄ (0.3), and NaOAc (0.2). The very weak catalysis by sodium acetate, a strong proton acceptor, indicated that equilibrium (138) was not significantly rate-determining (see also below, p. 125). In general the catalytic activity paralleled the molecular conductivity and this was consistent with assistance of breaking the bromine–bromine bond in the intermediate, either by direct interaction with one end of the dipole, or though increase in the ionising power of the solution. The observed change in the kinetic order from third to second on increasing the concentration of hydrogen bromide was considered to arise from the consequential reduction in free bromine concentration, and from the establishment of a second-order reaction catalysed by hydrogen bromide. The fact that the reaction is neither strongly acid-catalysed, nor retarded by sodium acetate, was considered to respectively rule out $Br^+$ and BrOAc as electrophiles in pure acetic acid[234].

Robertson *et al.*[261] measured rates of bromination of some aromatic hydrocarbons in acetic acid containing sodium acetate (to eliminate protonation of the aromatic by liberated hydrogen bromide) and lithium bromide (to reduce the rate to a measurable velocity ) at 25 °C, the second-order rate coefficients for 3-nitro-N,N-dimethylaniline and anisole being 14.2 and 0.016 respectively; the former compound was thus stated to be about $10^{12}$ times as reactive as benzene (though no measurement of the latter rate coefficient, inferred to be $1.33 \times 10^{-11}$, could be found in the literature) and this large rate spread gives one further indication of the unreactive nature of the electrophile. Other rates relative to benzene were:

phenol ($1.1 \times 10^{11}$), reaction occurring on the neutral molecule since addition of sulphuric acid did not decrease the rate; diphenylmethane (51); diphenyl ether ($1.4 \times 10^7$). Just how these relative rates were obtained is not clear, however, since bromination of the individual compounds was carried out under different conditions.

A re-examination of some of the kinetic work described above has confirmed the general conclusions. Keefer et al.[281] found that the kinetics of bromination of mesitylene in acetic acid followed equation (134) the second- and third-order coefficients at 25.4 °C (16 °C) being 0.00043 (0.0002) and 2.63 (1.9), respectively, the corresponding activation energies being 15 and 7. Addition of carbon tetrachloride or benzene (10 % by volume) suppressed the second-order reaction and reduced the third-order rate by 50 %, and water (13 % by volume) increased the second- and third-order coefficients by factors of 1,400 and 50, respectively, the greater effect on the second-order rate being consistent with the proposed mechanism; addition of salts increased the rate (except for addition of sodium acetate) this acceleration being likewise greater for the second-order reaction. Similarly, for the bromination of 4-bromophenol in acetic acid, Rajaram and Kuriacose[281a] obtained second- and third-order coefficients of $6.65 \times 10^{-3}$ and 1.14 respectively, at 20 °C, the corresponding activation energies (20 °C range only) being 11.3 and 1.24. The same kinetic features were likewise observed for bromination of mesitylene, 1,2,3,5-tetramethyl- and pentamethyl-benzenes in 90 vol. % aqueous acetic acid, the relative rates being 1 : 2.3 : 4.8, and the ratio of the third- to second-order rate coefficients was found to be greater for the more reactive aromatics[282]. This is unexpected, since assistance of Br–Br bond breaking should be less necessary for the more reactive (and therefore, electron-supplying) aromatics. The relative rate pentamethylbenzene to mesitylene agrees well with that (4.25) obtained by de la Mare and Robertson[231], who found the relative rates with o-xylene, m-xylene, mesitylene and pentamethylbenzene as 1 : 93 : 39,500 : 168,000, using 99.9 % acetic acid at 24.0 °C. Use of deuterated acetic acid caused a small decrease in both rate coefficients, and in the presence of deuterated water the second-order coefficient decreased whereas the change in the third-order coefficient was indeterminate; the balance of evidence was that the O–H bond of the acid was weakened, but not broken, in the rate-determining step.

The kinetics of bromination in 85 % aqueous acetic acid of the polynuclear hydrocarbons biphenyl, naphthalene, and phenanthrene, as well as benzene, were examined by Mason[283]. Third-order kinetics were obtained except for phenanthrene which gave second-order kinetics, this being (incorrectly) attributed to the removal of hydrogen bromide through addition to the 9,10 bond, so that $Br_2$ would not be removed as $Br_3^-$. This would, of course, increase the kinetic order and not decrease it. The reduced kinetic order can quite simply be explained by the greater reactivity of the phenanthrene molecule relative to the other compounds. In the less polar glacial acetic acid, naphthalene was found to give almost fourth-order kinetics as observed by Robertson et al.[279], and relative reactivities in the aqueous acid were found to be benzene (1.0), biphenyl ($1.1 \times 10^3$), and naphthalene ($1.4 \times 10^5$).

Berliner et al.[284-8] have examined kinetics of bromination in aqueous acetic acid in an attempt to find the acid concentration at which the change in kinetic order principally occurs, though it follows from the earlier work that this will depend upon the aromatic reactivity. In 50 % acid the bromination of naphthalene was second-order overall[284], and at constant ionic strength the rate coefficient showed a dependence on $[Br^-]$ according to equation (140)

$$k_{obs.} = k_2 K/(K + [Br^-]) \tag{140}$$

where $K$ is the dissociation constant for $Br_3^-$. Hence a plot of $k_{obs.}$, versus $K/(K + [Br^-])$ gave a straight line of slope $k_2 = 0.341$ at 24.9 °C, and since this line did not pass through the origin the discrepancy was attributed to bromination by $Br_3^-$ accounting for 0.5 % of the total reaction, though in subsequent papers the non-coincidence of origin and line was not considered to be meaningful (since there was sometimes a negative intercept)[285-8]. The activation energy (10–35 °C), after correction for the bromine–tribromide ion equilibrium, was found to be 15.3, log $A$ = 10.2 and $\Delta S^{\ddagger} = -14.0$. In considering the role of water in lowering the kinetic order it was argued that equilibrium (136) would be fast (since complex formation between halogen and aromatics appears to be instantaneous) and that the subsequent equilibrium

$$ArHBr_2 + H_2O \rightleftharpoons ArHBr^+ + H_2OBr^- \tag{141}$$

analogous to (139) would be rate-determining.

Bromination of biphenyl in 50 % aqueous acetic acid at 35 °C also yielded a second-order rate coefficient of 1.17; $E_a$ (25–45 °C) = 17.1, log $A$ = 9.5 and $\Delta S^{\ddagger} = -17$[285]. Because reaction rates are much faster in this aqueous medium, it was possible to measure the reactivity of benzene (under first-order conditions) and the rate coefficient was $2.17 \times 10^{-5}$ at 45 °C, $E_a$ (25–45 °C) = 18.1, log $A$ = 7.1 and $\Delta S^{\ddagger} = -28.0$. In a solution 0.1 $M$ in NaBr and 0.4 $M$ in NaClO$_4$ relative reactivities were found to be: benzene (1.0); biphenyl ($1.4 \times 10^3$): naphthalene ($1.19 \times 10^5$), and the agreement with the data of Mason[283] is good; partial rate factors of $f_2^{Ph} = 3.94 \times 10^3$, $f_{3,4}^{C_4H_4} = 1.79 \times 10^3$, and $f_{2,3}^{C_4H_4} = 1.77 \times 10^5$ were obtained.

In 60 % aqueous acetic acid at 25 °C naphthalene brominated according to second-order kinetics[286] with a rate coefficient of 0.108, i.e. 3.16 times slower than in 50 % aqueous acid, so that the change in rate with change of water content in a medium of high water content is very much smaller than in a medium of little or no water content (cf. ref. 281) as might be expected; $E_a$ (20–40 °C) = 16.6, log $A$ = 9.9 and $\Delta S^{\ddagger} = -15.3$. Clearly the reduction in rate with decrease in water concentration arises from a decrease in the entropy which is not unreasonable for a reaction in which ions are formed from neutral molecules.

In 75 % aqueous acetic acid, the bromination of fluorene at 25 °C obeys second-order kinetics in the presence of bromide ion and higher orders in its absence[287], with $E_a$ (17.85–44.85 °C) = 17.4, log $A$ = 10.5 and $\Delta S^{\ddagger}$ = −12.4; however, these values were not corrected for the bromine–tribromide ion equilibrium, the constant for which is not known in this medium, and so they are not directly comparable with the preceeding values. In the absence of bromide ion the order with respect to bromine was 2.7–2.0, being lowest when $[Br_2]_{initial}$ was least. Second- and third-order rate coefficients were determined for reaction in 90 and 75 wt. % aqueous acetic acid as 0.0026 and 1.61 ($k_3/k_2$ = 619), 0.115 and 12.2 ($k_3/k_2$ = 106) respectively, confirming the earlier observation that the second-order reaction becomes more important as the water content is increased. A value of $7.25 \times 10^6$ was determined for $f_{3,4}^{C_7H_6}$ (*i.e.* the 2 position of fluorene).

Second-order rate coefficients have been determined for the bromination of 2,6-dimethylnaphthalene at 25 °C in 75 and 90 wt. % aqueous acetic acid as 9.12 and 0.733 respectively[288], and the difference in the rate coefficients here (12.4) is noticeably different from those for fluorene (44) and presumably arises from the different reactivities of the two compounds. It follows then that a very reactive compound would give much the same rate coefficient regardless of the water content of the medium, which seems mechanistically reasonable. For 1,8-dimethylnaphthalene in 90 % aqueous acetic acid at 25 °C, $k_2$ = 1.75, whereas the very similar acenaphthene gives $k_2$ = 450 under the same conditions[288]; these rates were each determined by the method indicated above[284]. The cause of the enhanced reactivity of acenaphthene is not known but it has been confirmed by others[289], though the reactivity with respect to 1,8-dimethylnaphthalene was found to be a factor of only 40 in acetic acid and not 257 as above[753].

A re-examination[290] of the bromination of phenol and anisole in acetic acid (m.p. 15.8 °C) gave second-order kinetics with $E_a$ (20–40 °C) = 4.8 for phenol, whereas anisole gave kinetic orders which decreased from approximately third to approximately second with decreasing concentration of reactants and increasing temperature, as expected; the decrease in kinetic order was accompanied by an increase in activation energy, again as expected. In 75 % aqueous acetic acid the reaction with anisole was second-order, the rate being 1,210 times as fast as in glacial acetic acid.

Mixed second- and third-order kinetics were found for thiophen in acetic acid at 25 °C and a rate relative to benzene of $1.7 \times 10^9$, again showing the large rate spread consistent with an unreactive electrophile[265]. For other heterocyclic derivatives, rate coefficients were determined with either 50 % aqueous or 100 % acetic acid as solvent at a constant ionic strength but a variable bromide ion concentration, the slope of the $[Br^-]$ *versus* $k_{obs}[1+K[Br^-]]$ plots being linear and giving the second-order rate coefficients at 25 °C as follows[291]; 2-methoxycarbonylthiophen (50 % aqueous, $1.69 \times 10^{-3}$); 2-methoxycarbonylfuran (50 % aqueous, $1.97 \times 10^{-1}$); thiophen (50 % aqueous, $4.35 \times 10^{3}$; 100 %, $6.27 \times 10^{-1}$); 2-methoxycarbonylpyrrole (100 %, $6.20 \times 10^{2}$); again the large difference in the rate of bromination in anhydrous and in aqueous acetic acid is apparent.

Kinetic studies of molecular bromination have been carried out using a variety of solvents other than acetic acid. The bromination of 2-nitroanisole by bromine in water revealed that molecular bromine is the reactive species and that the tribromide ion is very unreactive[191]. By making allowance for the concentration of free bromine (which differs from the stoichiometric concentration through reaction with bromine ion), good second-order rate coefficients were obtained by application of equation (133) with $k_2 = 0.062$ at 25 °C; the dominance of the bimolecular mechanism is to be expected here in view of the trend observed on making acetic acid media more aqueous.

Bell et al.[292, 294, 295] have examined the kinetics of bromination of a number of compounds in aqueous solution. The rates of bromination of N,N-dimethyl- and N,N-diethylaniline, and their o-, m-, and p-derivatives in strongly acid solution were measured at 25 °C, with the bromine concentration in the range $10^{-3}$–$10^{-7}M$. and second-order rate coefficients in the range $10^{6}$–$10^{10}$ were found[292]. Rates decreased with increasing acidity of the media, showing that the free amine was the only aromatic reactant, hence quoted relative rates were determined from a linear plot of log rate versus acidity function $H_0$ which was extrapolated for each compound to $H_0 = -1.43$ (3.5 $M$ H$_2$SO$_4$), so that $\log k_2 = \log k \ (H_0 = 1.43) + pK_a + 1.43$. Reaction rates were lower at a higher bromide ion concentration (in this polar medium there would be no catalytic effect of hydrogen bromide) and the values of $\log k_2$ at [Br$^-$] = 0.025 $M$ and 0.05 $M$ respectively were: N,N-Me$_2$- (8.84, 8.50); 2-Me-N,N-Me$_2$- (6.87, 6.39); 3-Me-N,N,-Me$_2$- (9.21, 9.02); 4-Me-N,N-Me$_2$- (8.55, 8.30); 2-Me-N,N-Et$_2$- (6.13, 5.98); 3-Me-N,N-Et$_2$- (9.80, 9.63); 4-Me-N,N-Et$_2$-aniline (9.12, 8.80). Significantly, whilst 3- and 4-methyl substitution increases the reaction rate, 2 substitution decreases it and this was reasonably attributed to inhibition of conjugation between the substituent and the aromatic ring; consistent with this, the 2-ethyl compounds were even less reactive whereas the 3 and 4 compounds were more reactive, an effect analogous to the inductive activation order by substituent alkyl groups observed with aromatic ethers. Rates with some of these compounds have been re-measured and the second-order kinetics confirmed[293].

The rates of bromination of 3-nitrophenol in aqueous solution at 25 °C have been measured at various concentrations of perchloric acid and sodium bromide[294]. An increase in both caused a decrease in rate; the latter again shows that Br$_3^-$ is much less reactive than Br$_2$, whilst the former shows that reaction occurs principally on the 3-nitrophenoxide ion and the difference from the observation for phenol in acetic acid (above p. 117) is undoubtedly partly due to the greater stability of the 3-nitrophenoxide relative to the phenoxide ion. The kinetics were analysed in terms of reaction between 3-nitrophenol and $Br_2$ $(k_2^a)$, 3-nitrophenoxide ion and $Br_2(k_2^b)$, and 3-nitrophenoxide ion and $Br_3^-(k_2^c)^{294}$; the corresponding second-order rate coefficients were 101, $1.32 \times 10^9$, and $2.8 \times 10^7$ these values being subject to errors due to the assumption that the activity coefficients for ions of 3-nitrophenol were the same as those for perchloric acid, since the former values were not available[295]. It follows from the above rates that only 2 % of the overall reaction occurs *via* the tribromide ion.

The reaction rates for phenoxide ions are thus similar to those observed for dialkylanilines (and also enolate ions) and seem to represent an upper limit for brominating rate in aqueous solution. Consequently, the reactions have an almost zero activation energy and there is an apparent lack of deactivation by the nitro group. That bromination by $Br_3^-$ occurs in this reaction is not surprising, since the high reactivity of the phenoxide ion means that it will not discriminate very much between electrophiles of differing reactivity.

Investigation of rates of bromination of anisoles and phenols in aqueous solution at 25 °C showed that, for 2-bromoanisole, the rate was little affected by changes in concentrations of hydrogen and bromide ions as expected, since there is no equilibrium to form phenoxide ion, and consequently reaction with tribromide ion is negligible[295]. Using solutions of constant ionic strength it was shown that kinetic salt effects were operating (sodium nitrate and perchlorate had much larger effects than sodium bromide). By extrapolating plots of rate *versus* ionic strength to zero ionic strength, the second-order rate coefficients at the latter at 25 °C were as follows: 4-bromoanisole (5.0), 2-chloroanisole $(2 \times 10^2)$, 2-bromoanisole $(2 \times 10^2)$, 3-fluoroanisole $(1 \times 10^4)$, anisole $(4 \times 10^4)$; at 0 °C, the rate coefficient for anisole was $6.1 \times 10^3$, hence the activation energy is low.

For substituted phenols, rates were analysed as before for reaction between the various species possible and the rate coefficients obtained (at 25 °C) are gathered in Table 70 from which it can be seen that bromination by $Br_3^-$ occurs to the extent of 1–3 % only.

More recently, the kinetics of bromination of benzene in water have been examined[296]. The reaction is second-order overall and the slope of the plot $k_{obs}$.

### TABLE 70

SECOND-ORDER RATE COEFFICIENTS FOR REACTION OF ArH WITH BROMINE IN AQUEOUS SOLUTION AT 25 °C[295]

| ArH | $k_2{}^a$ | $k_2{}^b$ | $k_2{}^c$ | $k_2{}^c/k_2{}^b$ |
|---|---|---|---|---|
| Phenol | $1.8 \times 10^5$ | | | |
| 4-Bromophenol | $3.2 \times 10^9$ | $7.8 \times 10^9$ | | |
| 2,4-Dibromophenol | $5.5 \times 10^2$ | $1.5 \times 10^9$ | $5.0 \times 10^7$ | 0.033 |
| 3-Nitrophenol | $1.0 \times 10^2$ | $1.3 \times 10^9$ | $2.8 \times 10^7$ | 0.022 |
| 2,6-Dinitrophenol | | $5.4 \times 10^6$ | $1.0 \times 10^5$ | 0.018 |
| 2,4-Dinitrophenol | | $1.0 \times 10^6$ | $1.3 \times 10^4$ | 0.013 |

*versus* $K/(K+[Br^-])$ gave a second-order rate coefficient of $1.18 \times 10^{-3}$ at 50 °C. Measurement of the coefficients between 40 and 60 °C gave $E_a = 21.0$, log $A =$ 10.9, and $\Delta S^{\ddagger} = -10.8$ (at 25 °C). Extrapolation of the Arrhenius plot to 25 °C gave $k_2 = 8.34 \times 10^{-5}$ so that it is now possible to calculate partial rate factors under these conditions for some of the compounds described above. The rate coefficient may be compared with the value of $2.7 \times 10^{-6}$ obtained with 50 % aqueous acetic acid as solvent ($E_a = 19.6$, log $A = 7.97$, $\Delta S^{\ddagger} = 24.1$), and it is apparent that the increase in rate in the more aqueous medium arises principally from a change in the entropy; this is not uncommon for reactions carried out in more polar media.

Nitromethane has been used as a solvent for molecular bromination[297]. The bromination of polymethylbenzenes in nitromethane, acetic acid, and 1 : 1 mixtures of these solvents at 30 °C, showed that rates were much faster (about 330-fold) in nitromethane than in acetic acid. With nitromethane, in the bromine concentration range 0.01–0.02 $M$, the reaction was third-order in bromine. The relative deactivating effects of *m*-halogen substituents were measured in terms of the time taken for 10 % reaction to occur, and these values are given in Table 71 from which the relative reactivities in the different solvents are apparent: the deactivating effects of the *m*-nitro substituent were obtained by comparison with the reactivity of chloromesitylene at different concentrations (0.035, 0.055 $M$) of reactants. The results for the nitro compounds were interpreted in the same way

TABLE 71

RELATIVE RATES FOR REACTION OF ArH WITH $Br_2$ (BOTH 0.01 $M$) IN AcOH–MeNO$_2$ AT 30 °C[297]

| ArH | Substituent | Time (min) for 10 % reaction in | | | Relative rates |
| --- | --- | --- | --- | --- | --- |
| | | MeNO$_2$ | 1 : 1 Mixture | AcOH | |
| 1,3,5-Trimethylbenzenes | H | | | 5.40 | 62,800 |
| | F | 28.3 | 150 | | 35.9 |
| | Cl | 48.8 | 263 | | 20.6 |
| | Br | 54.9 | | | 18.2 |
| | I | 21.2 | | | 47.2 |
| | NO$_2$ | | | | $2.97 \times 10^{-3}$ |
| 1,2,3,5-Tetramethylbenzenes | H | | | 2.99 | 113,000 |
| | F | 14.5 | | | 69 |
| | Cl | 27.6 | | | 36.3 |
| | Br | 28.2 | | | 35.5 |
| | I | 5.34 | | | 187 |
| | NO$_2$ | | | | $6.67 \times 10^{-3}$ |
| 1,2,4,5-Tetramethylbenzenes | H | | 5.47 | 339 | 1,000 |
| | F | | | 147 | 2,310 |
| | Cl | 13.8 | 75.4 | | 72.6 |
| | Br | 33.1 | 173 | | 30.9 |
| | I | 25.0 | | | 40.0 |
| | NO$_2$ | | | | $7.34 \times 10^{-4}$ |

as those for their molecular chlorination (see p. 104), and the data for the halogeno compounds in each series correlated satisfactorily with each other showing that steric effects were constant in each series. Correlation of the data with $\sigma^+$-values yielded a $\rho$-value from the reaction of about $-8.6$. This is much lower than for molecular bromination of unsubstituted aromatics and shows the transition state to be shifted nearer to the ground state for reaction of these very reactive substrates.

Kinetic studies of molecular bromination have been carried out using non-polar solvents. Carbon tetrachloride was used as a solvent for studies on mesitylene, which gave data for the reaction catalysed by water and hydrogen bromide[298]. The reactions were first-order in mesitylene at low concentration, but this order apparently decreased at high concentration. In the presence of water the rate was proportional to the concentration of bromine and the square root of the concentration of hydrogen bromide. Erroneous kinetics were observed if the mesitylene was not sufficiently pure, and by measuring reaction rates with an excess of mesitylene, rate coefficients at 25.2 °C of order 2.5 were obtained and these were roughly independent of an eight-fold change in bromine concentration. From rate measurements at 17.6 and 33.6 °C, a *negative* activation energy of $-10.9$ was obtained, and this was attributed to it being a composite value which included the negative energy from the step in which $H_3O^+$ is formed from hydrogen bromide and water, this ion being subsequently employed in breaking the Br–Br bond in the intermediate $ArHBr_2$, *i.e.* $H_3O^+$ replaces $Br_2$ in equilibrium (139) the dependence of rate upon the square root of the hydrogen bromide concentration following from this. However, a subsequent determination of the kinetic isotope effect in the presence of $H_2O$ and $D_2O$ gave a value of $k_{H_2O}/k_{D_2O} = 1.25$ which showed that this mechanism cannot be correct since a higher value should have been obtained[282]. Nevertheless, negative activation energies have been observed in a number of studies of bromination in non-polar solvents, especially in those catalysed by iodine (see below). The bromination of phenol in carbon tetrachloride gave third-order kinetics with rate coefficients at a maximum at 19 °C. This was attributed to an appreciable extent to the stability of the first formed complex, $PhOHBr_2$, decreasing with increasing temperature (it was estimated to have a heat of formation of $-27.6$ kcal. mole$^{-1}$) so that a balance of this stability *versus* reactivity could give the observed overall temperature dependence[290].

Finally, kinetic studies of molecular bromination have been carried out using trifluoroacetic acid as a solvent[299]. The reaction is pure second-order as might be expected with a polar hydroxylic solvent, and the rate coefficients for toluene were $2.88 \times 10^{-4}$ (0 °C) and $3.81 \times 10^{-3}$ (35 °C) with $E_a = 11.9$ and $\Delta S^{\ddagger} = -33.2$. At 25 °C, rate coefficients were: benzene, $7.62 \times 10^{-7}$; toluene, $19.7 \times 10^{-4}$; $p$-xylene, $69.1 \times 10^{-4}$, so that the electrophile in this reaction is extremely selective. The absence of higher orders in bromine suggests that trifluoroacetic acid, with its known facility for hydrogen bonding is very effective at breaking the bromine–

bromine bond in the rate-determining step of the reaction. Consistently, one would expect the addition of water to have a much smaller effect than with acetic acid, and this is observed, for addition of 15 % of water produced a rate increase of 18.5, whereas with acetic acid the rate increases by a factor of 150[299]. The very high selectivity of the electrophile apparent from the above data, makes this solvent ideal for specific monobromination[300], and reaction tends to cease once one mole of bromine has been consumed.

There have been a number of measurements of kinetic isotope effects in molecular bromination. Farrell and Mason[301] found that the bromination of dimethylaniline and its 2,4,6-trideuterated isomer by bromine in the presence of excess bromide ion in 6.4 $M$ sulphuric acid at 25 °C gave a fairly small isotope effect, $k_H/k_D = 1.8$. Subsequently, it was found that this value was independent of the amine concentration but varied according to the bromide ion concentration[302], being 1.9 at 0.0125 $M$ Br⁻ and 1.1 at 0.25 $M$ Br⁻. The isotope effect was also position-dependent, being absent at the 3 and 4 positions, whereas the 2,6-dideuterated compound gave a higher isotope effect of 2.6. This is entirely consistent with C–H bond breaking becoming partially rate-determining in sterically hindered situations, i.e. $k'_{-1}$ becomes comparable to $k'_2$ (see p. 8). Since the kinetic order in bromine was found to be 1.3, the reduced isotope effect at higher bromide ion concentration is consistent with a lowering of the kinetic order through bromide ion removing $Br_2$, thereby reducing the possibility of $Br_2$ reacting with first formed complex $ArHBr_2$ in a bulky transition state. It is not consistent, however, with the observation that the $o:p$ ratio for the bromination of dimethylaniline in aqueous acid decreases from 2.3 to 0.15 as the bromide ion concentration is increased from 0.025 $M$ to 0.5 $M$. Under the latter conditions second-order kinetics were obtained, and the greater steric hindrance was attributed to increasing substitution by $Br_3^-$ at higher bromide ion concentration[293].

Christen and Zollinger[303] have made an extensive study of kinetic isotope effects in bromination of the disodium salt of 2-naphthol-6,8-disulphonic acid with hypobromous acid and with bromine in aqueous buffers at 20 °C. Both brominating agents give the same rate (within 20 %) and the reactions are first-order

TABLE 72

FIRST-ORDER RATE COEFFICIENTS ($10^4k_1$) FOR REACTION OF 2-NAPHTHOL-6,8-DI-SULPHONATE WITH $Br_2$ IN AQUEOUS BUFFERS AT 20 °C[303]

| | | pH | | |
|---|---|---|---|---|
| $10^3$ M [HOBr] | $10^3$ M [ArH] | 3.85 | 4.50 | 5.25 |
| 1.28 | 5.00 | 2.77 | 2.78 | 3.00 |
| 3.22 | 2.50 | 3.17 | 3.17 | 3.20 |
| 5.30 | 2.50 | 6.19 | 4.06 | 4.06 |

TABLE 73

| $10^3$[NaOAc] (M) | pH | $10^4 k_1$ |
|---|---|---|
| 50 | 4.54 | 2.82 |
| 50 | 5.09 | 2.89 |
| 150 | 5.24 | 5.94 |

in each reactant, the rates being dependent upon the concentration of the buffer. The rates were almost independent of pH as shown by the first-order rate coefficients in Table 72, but the bromination was catalysed by sodium acetate (Table 73) and by pyridine, the rate being almost linearly dependent on the pyridine concentration. Further a kinetic isotope effect $k_H : k_D$ of $2.08 \pm 0.01$ was obtained, independent of a four-fold change in the concentration of the aromatic and a two-fold change in the concentration of the hypobromous acid. With molecular bromine the isotope effect varied from 1.48 to 2.34 as the ratio of aromatic to bromine was increased, and this variation was attributed to the catalytic effect of bromine in breaking the Br–Br bond in the intermediate and which operates to lower the isotope effect, hence this is lower when the bromine concentration is larger. In the presence of excess bromine, the isotope effect was found to increase with increasing concentration of bromide ion and this effect (which is the exact opposite of that observed by Farrell and Mason[302]) was attributed to bromination by unreactive Br$_3^-$ resulting in an increase in the isotope effect.

Not only were the reaction rates for bromination by bromine and by hypobromous acid very similar, but the corresponding activation energies (determined over a 20 °C range) were between 11.8 and 12.6 (for Br$_2$) and 12.5 and 12.7 (for HOBr). Thus all this kinetic data is consistent with the rapid formation of an intermediate which is identical for both brominating reagents, and from which the slow loss of a proton subsequently occurs.

A kinetic isotope effect, $k_H/k_D = 1.4$, has been observed in the bromination of 3-bromo-1,2,4,5-tetramethylbenzene and its 6-deuterated isomer by bromine in nitromethane at 30 °C, and this has been attributed to steric hindrance to the electrophile causing $k'_{-1}$ to become significant relative to $k'_2$ (see p. 8)[268]. A more extensive subsequent investigation[304] of the isotope effects obtained for reaction in acetic acid and in nitromethane (in parentheses) revealed the following values: mesitylene, 1.1; pentamethylbenzene 1.2; 3-methoxy-1,2,4,5-tetramethylbenzene 1.5; 5-t-butyl-1,2,3-trimethylbenzene 1.6 (2.7); 3-bromo-1,2,4,5-tetramethylbenzene 1.4; and for 1,3,5-tri-t-butylbenzene in acetic acid–dioxan, with silver ion catalyst, $k_H/k_D = 3.6$. All of these isotope effects are obtained with hindered compounds, and the larger the steric hindrance, the greater the isotope effect. The larger effect with nitromethane as solvent was attributed to its poorer ability to facilitate ionisation of the transition state, as also indicated by the slower reaction rates in nitromethane than in acetic acid (see Table 71, p. 122).

Berliner and Schueller[305] have put a different interpretation upon the effect of bromide ion concentration on the isotope effect. It follows from equilibrium (137), p. 115, that an increase in bromide ion concentration will cause $k'_{-1}$ to become larger relative to $k'_2$ and hence the isotope effect should become larger as the bromide ion concentration increases, i.e. the same conclusion arrived at by Christen and Zollinger[303], and the magnitude of the change in the isotope effect with bromide ion concentration was calculated. For the bromination of biphenyl and its 4,4'-dideutero derivative in 50 % aqueous acetic acid in the presence of 0.1 $M$ sodium bromide and 0.4 $M$ sodium perchlorate at 25 °C, the appropriate second-order rate coefficients were 5.44 (H) and 4.75 (D)$\times 10^{-4}$ and 15.8 (H) and 13.8 (D)$\times 10^{-4}$ at 25 °C and 35 °C, respectively leading to $k_H/k_D = 1.14$ which was not sensitive to changes in [Br$^-$] as calculated. Hence it was assumed that this effect arises not from partial breaking of the C–H bond in the rate-determining step, but rather from hyperconjugation in the transition state with overcompensates the effect which would arise from change in hybridisation from $sp^2$ to $sp^3$ on going from ground to transition state. It would be useful to find other examples of a kinetic isotope effect which is insensitive to changes in bromide ion concentration.

The bromination of 1,3,5-trimethoxybenzene in dimethylformamide exhibits no isotope effect, but introduction of one and then two bromine substituents gives isotope effects of 3.6 (at 25 °C) and 4.8 (at 65 °C), respectively[306], the latter value being equivalent to about 5.9 at 25 °C. The increased effect was attributed to the reactivity being reduced through inhibition of conjugation with the ring rather than the inductive effect of the bromo substituents, though both processes are obviously important; the transition state for the reaction becomes more like the products, or, in other words, $k'_{-1}$ becomes larger relative to $k'_2$ and an isotope effect results. That the inductive effect is not insignificant can be gauged from the isotope effect obtained with the mono- and di-methyl compounds[307], which gave values of 2.05 and 2.95 respectively at −20 °C and these values would be smaller still at 25 °C; the difference between these values and those for the bromo compounds obviously arises to a large extent from the difference in the inductive effects of substituents. Contrary to the observation above, in this case the isotope effect was dependent upon the bromide ion concentration.

Variation of the isotope effect with bromide ion concentration has also been observed for the bromination of 4-methoxybenzenesulphonic acid and its *ortho* dideuterated derivative at 0 °C, the value of $k_H/k_D$ changing from 1.0 with no Br$^-$ to 1.31 at 2.0 $M$ Br$^-$ [308].

The bromination of phenol in acetic acid, containing lithium bromide and perchlorate at a constant total concentration of 0.2 $M$, gave kinetic isotope effects

TABLE 74

SECOND-ORDER RATE COEFFICIENTS ($k_2$) AND KINETIC ISOTOPE EFFECTS FOR REACTION OF $Br_2$ WITH PhOH IN AcOH AT 25 °C[309]

| Reactants | [LiBr] (M) | 0.20 | 0.15 | 0.10 | 0.05 |
|---|---|---|---|---|---|
| | [LiClO$_4$] (M) | 0.00 | 0.05 | 0.10 | 0.15 |
| PhOH in AcOH | | 0.533 | 0.717 | 1.08 | 2.03 |
| PhOD in AcOD | | 0.317 | 0.384 | 0.584 | 1.05 |
| $k_H/k_D$ | | 1.69 | 1.87 | 1.86 | 1.94 |

as shown in Table 74[309]; these show that by the time the transition state has been reached, the O–H bond in the substituent has broken appreciably. One would expect, however, that the $k_H/k_D$ value should increase with increasing bromide ion concentration, as in the case of ring C–H breakage.

The theoretical difficulties encountered in attempting to relate the isotope effect in iodination with the occurrence of base catalysis are also apparent in bromination, for Vainshtein and Shilov[224] have shown that the isotope effects $k_H/k_D$ are 1.0 ($Br^+$), 2.6 ($Br_2$) and 1.0 (HOBr) whereas the corresponding reactivities of the electrophiles in the systems used are $> 1400 : 1.0 : 0.0015$.

(b) *Molecular hypobromous acid as electrophile*

The bromination of sodium-*p*-anisate in dilute solution of phosphate buffers (pH 7–8) by molecular hypobromous acid has been investigated at 21.5 °C. At pH 7.6 decomposition of the acid is appreciable and gives bromine so that the rate increases with time, but it is possible to obtain second-order rate coefficients for 75 % of the reaction in the presence of suspended silver phosphate (which regenerates hypobromous acid from bromine and water). The kinetics follow equation (142)

$$\text{Rate} = k_2[\text{HOBr}][\text{ArCOO}^-] \tag{142}$$

which is equivalent to

$$\text{Rate} = k_2^{\text{HOBr}}[\text{HOBr}][\text{ArCOO}^-] + k_2^{\text{Br}_2}[\text{Br}_2][\text{ArCOO}^-] \tag{143}$$

since molecular bromine is produced from the hypobromous acid. Now, if $K_{\text{hydrolysis}} = [\text{Br}^-][\text{H}^+][\text{HOBr}]/[\text{Br}_2]$, then equation (143) can be replaced by equation (144)

$$\text{Rate} = [\text{HOBr}][\text{Ar}^-] \left\{ k_2^{\text{HOBr}} + k_2^{\text{Br}_2} \frac{[\text{Br}^-][\text{H}^+]}{K_{\text{hydrolysis}}} \right\} \tag{144}$$

However, in bromination by molecular bromine, $Br^-$ and $H^+$ are regenerated, and since these ions are both used initially in making bromine from hypobromous acid according to the equilibrium

$$Br^- + H^+ + HOBr \rightleftharpoons Br_2 + H_2O \qquad (145)$$

then their concentration remains constant. Thus a plot of $k_2$ versus $[Br^-][H^+]/K_{hydrolysis}$ was linear; hence $k_2^{Br_2}$ and $k_2^{HOBr}$ were evaluated as 119 and 0.05 respectively, i.e. hypobromous acid is approximately 2000 times less reactive than molecular bromine[192].

(c) Molecular bromine phosphate as electrophile

In the above work the velocity of bromination was found to increase along with the concentration of phosphate buffer, whereas when the acid component of the buffer is varied, no such rate change occurred. The rate change was, therefore, attributed to the equilibrium

$$HOBr + H_2PO_4^- \rightleftharpoons BrHPO_4^- + H_2O \qquad (146)$$

in which bromine phosphate is the more reactive electrophile, and this would be consistent with the second dissociation constant for phosphoric acid being much greater than the dissociation constant for water.

(d) Molecular bromine sulphate as electrophile

Bromine sulphate $BrHSO_4$ has been proposed as a possible molecular brominating species, since the catalysis by sulphuric acid of the bromination of benzoic acid by hypobromous acid was much greater than by perchloric acid of the same acidity[198]. Its reactivity was considerably less than that of $H_2OBr^+$ so that an enhanced rate spread is observed and its reactions only become noticeable with the least deactivated (i.e. most reactive) compounds employed in this particular study.

(e) Molecular bromine acetate as electrophile

Whilst molecular hypobromous acid can be a brominating species, it is not believed to be the active species in acetic acid solution. The bromination of 4-nitroanisole by hypobromous acid in 75 % aqueous acetic acid at 19.8 °C gave a second-order rate coefficient of 0.162, so that the brominating species here appears to be more reactive than molecular bromine[194]. In addition, the presence of 0.05 M sodium acetate caused the rate coefficient to fall to only 0.040, and both these observations were contrary to expectation if hypobromous acid was the brominating species, but are quite consistent with it being bromine acetate, BrOAc. Also, the addition of chloride ion caused the reaction to become immeasurably slow, due to the formation of the much less reactive bromine chloride.

Bromine acetate has also been proposed[310] as an intermediate species in the bromination of anisole, phenetole, and methyl p-tolyl ether by 2,4,6-tribromo-N-bromoacetanilide in acetic acid at 25 °C, since this latter compound was stable to both the ethers and to acetic acid, but in the presence of both, bromination of the ethers occurred, presumably *via* bromine acetate formed as in the equilibrium

$$ArNBrAc + HOAc \rightleftharpoons ArNHAc + BrOAc \tag{147}$$

The kinetics of the bromination of biphenyl by hypobromous acid in 75 % aqueous acetic acid have been examined at temperatures between −3.8 and +20.1 °C[199]. The reaction rate obeys equation (92) (p. 86) and the second-order rate coefficient was 4.5 at 0 °C, compared with $5.4 \times 10^{-4}$ for bromination by molecular bromine in 50 % aqueous acetic acid at 25 °C, *i.e.* the latter reaction is slower in a more polar medium and at a higher temperature. Since extrapolation of the former rate coefficient to 25 °C gives a value of 15, the brominating species giving rise to it must be at least 20,000 times as reactive as molecular bromine. The kinetic data was analogous to that for reactions involving the formation of chlorine acetate as a chlorinating species, hence it was concluded that bromine acetate was the species in this case. The activation energy for the reaction was 7.9 *cf.* 17.1 found for molecular bromine in 50 % aqueous acetic acid[285], and this further indicates the greater reactivity of bromine acetate. However, the partial rate factors derived for this species, *viz.* $f_4^{Ph} = 3,000$, *cf.* 4,000 for molecular bromine, seem quite at variance with the kinetic data since the former value should be much smaller for a species so much more reactive than molecular bromine. It should be noted that partial rate factors for chlorination by chlorine acetate are considerably smaller than those obtained with molecular chlorine.

### (f) Bromination by a molecular species with a catalyst

The kinetics of bromination with the complex formed between bromine and dioxan have been examined using benzene (which is unattacked) as solvent[311], and it is probably appropriate to regard this as a catalysed bromination in view of the effect of dioxan upon the polarity of the bromine–bromine bond. With anisole, phenetole, and isopropoxybenzene, third-order kinetics are obtained, *viz.*

$$Rate = k_2[ArH][dioxan\ dibromide]^2 \tag{148}$$

and the rate is increased and the kinetic order reduced, by hydrogen bromide. Because the complex is slightly more polar than molecular bromine, reaction rates are slightly higher, *e.g.* phenol at 20 °C gives third-order rate coefficients of 0.85 with molecular bromine and 0.65 with dioxan dibromide. Dioxan can also increase the rate of bromination of phenols (but not ethers) due, it is believed, to hydrogen bonding of the side-chain hydrogen to the oxygen of the dioxan; conversely, addition of a strong organic acid (trichloroacetic acid) to the medium causes a reduction of reactivity through hydrogen bonding of the carboxylic hydrogen to the phenolic oxygen, and a similar effect has been observed in hydrogen exchange of anisole in trifluoroacetic acid (see p. 252). Reaction rates at 25 and 50 °C, and activation energies were determined for a range of aromatic ethers, PhOR where R = Me, Et, $n$-Pr, $i$-Pr, Ph, $CH_3CO$, and PhCO, and for 2-, 3-, 4-methoxy- and 3,5-dimethoxy-anisoles as well as $p$-diethoxybenzene. For 2-, 3-, and 4-methyl- and bromo-phenols, the activation energies were negative, and the decrease in rate with increasing temperature was regular over the range 10–50 °C; the anomaly was attributed to the preformation of a complex analogous to that in equilibrium (136), the stability of which decreases with increasing temperature. The relative reactivities of ethers, PhOR, was R = Me (100); Et (274); $i$-Pr (665), $i.e.$ very close to those found with bromine in 50 % aqueous acetic acid (see p. 113).

The catalysis of molecular bromination by iodine has received considerable attention, and the first study was carried out by Bruner[312], who found that the reaction with benzene (no solvent) gave high kinetic orders with respect to halogen. Price, and Price and Arntzen[313] examined the effects of catalysts on the bromination of phenanthrene in carbon tetrachloride at 25 °C, and found that of $I_2$, $AlCl_3$, $SbCl_5$, $PCl_3$, $PCl_5$ and $SnCl_4$, the first was the most effective catalyst, the kinetics being consistent with those of Bruner which he analysed as three-halves in bromine and five-halves in iodine. The interpretation was that $BrI_3$ was the electrophile with $IBr_2^-$ aiding bond-breaking in the intermediate, but in deriving this it was assumed that $[IBr]^2$ was proportional to $[I_2][Br_2]$ and this is only so if it is considerably dissociated, which is not the case in carbon tetrachloride. Also, as Robertson $et$ $al.$[314] pointed out, the mechanism did not involve free bromine (except to produce IBr) and hence the occurrence of a maximum in a plot of reaction rate $versus$ iodine concentration could not be explained. These workers re-analysed Bruner's data in terms of the equation

$$\text{Rate} = k_4[Br_2]^2[IBr]^2 + k_4'[Br_2][IBr]^3 \tag{149}$$

and for the iodine catalysed bromination of mesitylene in carbon tetrachloride and chloroform at 24 °C, they found the kinetics to be most compatible with the rate equation

$$\text{Rate} = k_4[ArH][Br_2]^2[IBr] + k_4'[ArH][Br_2][IBr]^2 \tag{150}$$

which was somewhat vaguely interpreted as involving a mechanism in which $IBr_3(Br^+IBr_2^-)$ would replace $Br_2$ as electrophile and IBr would replace $Br_2$ in assisting removal of $Br^-$ from the intermediate. The rate maximum in this reaction occurred at an $I_2/Br_2$ ratio of 0.5 compared with 0.61 for Bruner's work with benzene, and $IBr_5$ and $I_2Br_4$ were suggested as possible electrophiles in the latter work. Now it has been argued that if $x = [Br_2]_{initial}$ and $y = [I_2]_{initial}$ then $[IBr] = 2y$ and $[Br_2] = x - y$. Hence $m = [IBr]/[free\ bromine] = 2y/x - y$ so that $y/x$, the initial ratio of concentrations $= m/(m+2)$. This value is, therefore, 0.33, 0.50 or 0.60 according to whether the order in IBr is one, two, or three relative to $Br_2$[315], and this is consistent with the experimental evidence for the last terms in equations (149) and (150).

Further interpretation of Bruner's results were that the loss of hydrogen bromide from the first formed $ArHBr_2$ complex was proportional to $[IBr]^3$ and the bromination of toluene in carbon tetrachloride at 25 °C was found to give a maximum rate at $[I_2][Br_2] = 0.8$, and said to follow the rate equation

$$\text{Rate} = k_4[ArHBr_2][IBr]^3 \tag{151}$$

but this would seem, however, to require a maximum at 0.6. This reaction also showed an unexplained induction period and was believed to be catalysed by hydrogen bromide[316].

No induction period, however, was found by Blake and Keefer[317] in a re-investigation of the iodine-catalysed bromination of mesitylene in carbon tetrachloride, and like the original workers[314] they found the rate to be proportional to $[ArH][Br_2][IBr]^2$, but they envisaged dimers of IBr aiding removal of the bromide ion from the first formed intermediate $ArHBr_2$. Consistent with the above calculation the rate maximum occurred at an $[I_2]/[Br_2]$ ratio of 0.51. Fourth-order rate coefficients obtained at 14.1, 25.2, and 36.0 °C were each approximately $10^{-6}$, indicating a zero activation energy which seems to be common with this solvent, being attributed to decreasing stability of $ArHBr_2$ with increasing temperature.

Kinetics of the iodine-catalysed bromination of phenol[290] and anisole[315] in acetic acid, at 30 °C, in the concentration ranges 0.1–0.001 $M$ and 0.16–0.01 $M$, respectively, followed equation (152)

$$\text{Rate} = k_3[PhOR][IBr]^2 \tag{152}$$

Separate experiments on the iodine-catalysed bromination of these compounds revealed a rate maximum at $[I_2]/[Br_2] = 0.35$, from which it follows that the concentrations of molecular bromine and iodine monobromide are equal, i.e. the latter catalyses bond-breaking in the former in the intermediate. Since iodine monobromide is dissociated into iodine and bromine, dissociation constant $K$, $[Br_2]\sqrt{K}$ is proportional to $[IBr]$ and hence equation (152) may be rewritten in the form

$$\text{Rate} = k_3'[PhOR][Br_2][IBr] \tag{153}$$

which predicts the maximum rate when $[Br_2] = [IBr]$. At higher iodine concentrations, the reaction rate is depressed below that which is obtained without iodine (as much as 300-fold) and it was proposed that the dissociation of iodine monobromide to iodine and bromine must be slower than the rate of subsequent bromination by the bromine produced. The relative rates of bromination of phenol and anisole was found to be 74, the respective activation energies being 9.7 and 11.5. With carbon tetrachloride as solvent, for phenol in the concentration range 0.02–0.002 $M$ the kinetic form was again that of equation (151) and, as with acetic acid, a maximum in rate occurred at an iodine : bromine concentration ratio of $ca.$ 0.35. With the chlorinated solvent, however, a rate maximum occurred at 19 °C, as was observed in the molecular bromination of phenol in this medium, and the instability of the $PhOHBr_2$ intermediate with increasing temperature was considered to be the cause. In the case of anisole, the appropriate kinetic equation was

$$\text{Rate} = k_4[\text{ArH}][Br_2][IBr]^2 \tag{154}$$

and here the maximum occurred at an iodine : bromine concentration ratio of 0.49 in keeping with this; in the temperature range $-15.5$ to $+40$ °C a negative activation energy of $-3.1$ was obtained. The higher kinetic order for anisole was attributed to the need for greater assistance, with this less reactive compound, for breaking of the Br–Br bond in the intermediate. The observation of even higher orders in iodine monobromide in the bromination of toluene is likewise consistent[315].

Finally, the brominations of mesitylene, 1,2,4,5-tetramethyl- and pentamethyl-benzene in chloroform (which is more polar than carbon tetrachloride) are first-order in bromine and iodine monobromide[318], so that this is entirely consistent with the pattern developed above, $i.e.$ the more polar the solvent and the more reactive the compound, the fewer the number of molecules of iodine monobromide that are involved in the rate-determining step. Measurements of rates between 25 and 42 °C revealed no significant trend owing to the variability of the rate coefficients determined at any temperature, but even so it is clear that there is no appreciable activation energy for these compounds, and there may have been temperature inversion for some of them.

The kinetics of bromination catalysed by metal halides have been examined. The relative activities of such catalysts towards bromination of benzene was found to be: $MgBr_2 < ZnBr_2 \approx HgBr_2 < CdBr_2 < BeBr_2$, the halides of the Group 2 metals, $viz.$ magnesium, barium, strontium and calcium, having very little catalytic effect[319]. A more detailed study[281] of the catalysis of bromination of mesitylene at 25.4 °C in acetic acid (a reaction the kinetics of which follow equation (134)) showed that zinc chloride catalysed the reaction in direct proportion to its concentration, and reduced the activation energy (determined, however, only between

16.4 and 25.4 °C) for the second-order reaction from 15 to 7, but had no effect on the third-order reaction for which $E_a = 7$. This is consistent with Br–Br bond-breaking in the $ArHBr_2$ intermediate being aided by the catalysts for the second-order reaction, whereas this function is performed by another bromine molecule in the third-order reaction, consequently zinc chloride has no effect, though, of course, it follows from the reduction in activation energy that the proportion of molecules reacting by the "bimolecular" reaction will increase under the influence of the catalyst. Other salts, viz. $LiClO_4$, $NaClO_4$ and $LiCl$, also increased the rate but less effectively, this being interpreted as normal salt effects.

Subsequently, rate coefficients were determined for the zinc chloride-catalysed bromination of benzene, toluene, i-propyl-benzene, t-butylbenzene, xylenes, p-di-t-butylbenzene, mesitylene, 1,2,4-trimethyl-, sym-triethyl-, sym-tri-t-butyl-, 1,2,3,5- and 1,2,4,5-tetramethyl- and pentamethylbenzenes, all at 25.4 °C and in acetic acid, and it was shown that the reaction was inhibited by $HBr.ZnCl_2$ which accumulates during the bromination and was considered to cause the first step of the reaction (formation of $ArHBr_2$) to reverse[320]. The second-order coefficients for bromination of o-xylene at 25.0 °C were shown to be inversely dependent upon the hydrogen bromide concentration and the reversal of equilibrium (155)

$$ArHBr_2 \cdots\cdots ZnCl_2 + HOAc \underset{}{\overset{fast}{\rightleftharpoons}} HBr \cdot ZnCl_2 + ArHBr^+OAc^-$$

$$(155)$$

was thought to be responsible[321] though this view was subsequently abandoned in favour of the zinc chloride-catalysed formation of bromine acetate from bromine and acetic acid, equilibrium (156),

$$HOAc + Br_2 + ZnCl_2 \rightleftharpoons BrOAc + HZnCl_2Br \qquad (156)$$

this reagent being more reactive than molecular bromine.

An investigation of the relative rates of bromination of benzene, toluene, m- and p-xylene by bromine in acetic acid, catalysed by mercuric acetate, revealed relative rates almost identical with those obtained with molecular bromine[322], though as in the bromination of biphenyl by bromine acetate (p. 129) it is quite inconsistent for a much more reactive electrophile to have the same selectivity. Relative rates were (molecular bromination values in parenthesis) benzene 1.0; toluene, 480 (610); p-xylene, $2.1 \times 10^3$ ($2.2 \times 10^3$); m-xylene $2.0 \times 10^5$ ($2.1 \times 10^5$).

Studies of the relative rates of the zinc chloride-catalysed bromination of alkyl- and halogeno-benzenes in nitromethane at 25 °C have lead to the suggestion that the rate-determining step of the reaction is formation of $\pi$-complex, since low substrate selectivity was found to be coupled with high (i.e. normal) positional selectivity[323]. Under some conditions (column 1 in Table 75) the low selectivity

TABLE 75

RELATIVE RATES OF $ZnCl_2$-CATALYSED REACTION OF $Br_2$ WITH ArH AT 25 °C[323]

| ArH | $Br_2$ added to ArH/$FeCl_3$/MeNO$_2$ | $Br_2$/MeNO$_2$ added to ArH/$FeCl_3$/MeNO$_2$ | $Br_2$/$FeCl_3$/MeNO$_2$ added to ArH/$FeCl_3$ |
|---|---|---|---|
| Benzene | 1.0 | 1.0 | 1.0 |
| Toluene | 2.3 | 7.4 | 3.6 |
| Ethylbenzene | 2.4 | | |
| o-Xylene | 2.4 | 34.3 | 3.9 |
| m-Xylene | 2.7 | 534 | 5.6 |
| p-Xylene | 1.7 | 31.5 | 4.3 |
| Mesitylene | 10.2 | >1,000 | 15.9 |
| Fluorobenzene | 0.69 | 0.27 | 0.48 |
| Chlorobenzene | 0.35 | 0.12 | 0.20 |
| Bromobenzene | 0.30 | 0.10 | 0.16 |

was known to be due to diffusion control since dibromination occurred, but the absence of this under other conditions (columns 2 and 3) coupled with the relative constancy of the $k_{Ph \cdot Ch_3}$ or $k_{Ph \cdot Cl} : k_{PhH}$ ratio for a variation (up to 36-fold) in the ratio of reagent concentrations was considered as evidence for adequate mixing prior to reaction, the difference in the selectivities apparent in columns (2) and (3) being attributed to preformation of a more reactive electrophile, e.g. $Br^+ FeCl_3Br^-$ under condition (3). However, as the concentrations of ferric chloride and bromine were reduced under condition 3 by a factor of thirty (the ratio remaining the same) the selectivity increased tenfold and the ortho : para ratio decreased. This was attributed to a solvation effect, the solvation of the electrophile becoming greater at higher dilution, resulting in a bulkier, less reactive, electrophile. However, Caille and Corriu[276] measured the relative rates of aluminium chloride-catalysed bromination of benzene and toluene in nitrobenzene and nitromethane at 0 and 30 °C by direct and competitive methods. The ratio of the third-order rate coefficient was found to be approximately double that found from the competitive measurements, and also very temperature-dependent and the values from the competitive measurements showed greater variation with change in the rates of substrate concentration, than found by Olah et al.[323]; the rate determining formation of a π-complex, therefore, seems less likely (see also p. 45). Olah et al.[323] found that a small isotope effect $k_H/k_D$ of $1.08 \pm 0.03$ was obtained in the silver perchlorate-catalysed bromination of benzene and its hexadeutero isomer in nitromethane; ferric chloride could not be used for this measurement since it promotes significant hydrogen–deuterium exchange during the time of the halogenation.

Finally, peroxyacetic acid had been used as a catalyst in the bromination of benzene in acetic acid at 60 °C[324]. The rate of consumption of peroxyacid appeared to be independent of the concentration of benzene and the kinetics followed equation (157), viz.

$$\text{Rate} = k_2[CH_3CO_3H][Br_2] \tag{157}$$

The rate-determining step was, therefore considered to be reaction of bromine with peroxyacetic acid to give a species (suggested as bromine acetate)which subsequently and rapidly, brominates. Formation of bromine acetate was believed to take place according to the reaction scheme represented by equilibrium (158) (which is analogous to the mercuric oxide oxidation of bromine) followed by either equilibrium (159), (160) or (161), *viz.*

$$CH_3COOOH + Br_2 \overset{\text{slow}}{\rightleftharpoons} CH_3COOH + Br_2O \tag{158}$$

$$Br_2O + H_2O \overset{\text{fast}}{\rightleftharpoons} 2\ HOBr \tag{159}$$

$$CH_3COOH + HOBr \overset{\text{fast}}{\rightleftharpoons} CH_3COOBr + H_2O \tag{160}$$

$$2\ CH_3COOH + Br_2O \overset{\text{fast}}{\rightleftharpoons} 2\ CH_3COOBr + H_2O \tag{161}$$

This is not unreasonable since it was found that the second-order rate coefficient for bromination of benzene by bromine acetate at 60 °C was $3.20 \times 10^{-2}$, *i.e.* twenty times faster than under the above conditions for which $10^3 k_2$ ranged from 1.43 to 1.66 for a threefold variation in benzene concentration and a twofold variation in bromine concentration.

### 5.2.3 *Molecular iodination*

#### (*a*) *Molecular iodine as electrophile*

The possibility that a number of iodinations with iodine and iodine monochloride may involve the molecular species rather than the positive species are formerly believed has been considered under positive iodination (see pp. 92–97). Briefly, iodination rates have been found to be proportional to $[I^-]^{-2}$ and $[I^-]^{-1}$, generally with a mixed dependence, and the former term has been interpreted as involving iodination by hypoiodous acid (or more likely, positive iodine) and the latter term attributed to iodination by molecular iodine.

Additional kinetic evidence supporting molecular iodine as an iodinating species is sparse. Li[325] found that the iodination of tyrosine in acetate buffers at 25 °C showed the mixed inverse dependence on iodide ion concentration noted above, so that part of the reaction appeared to involve the molecular species. Subsequently, Doak and Corwin[326] found that the kinetics of the iodination of (N-Me)-4-carboethoxy-2,5-dimethyl- and (N-Me)-5-carboethoxy-2,4-dimethyl-pyrroles in phosphate buffers in aqueous dioxane at 26.5 °C obeyed equation (162), *viz.*

$$\text{Rate} = k_2[\text{pyrrole}][\text{free } I_2] \tag{162}$$

The observed rate coefficient will be less than the true rate coefficient since some iodine is converted to triiodide ion through the equilibrium for which $K[I_3^-] = [I_2][I^-]$, i.e. $k_2 = k_{\text{obs}}([I_2]+[I_3^-])/[I_2]$ which reduced to $k_{\text{obs}} = k_2K/(K+[I^-])$ and which again shows the inverse dependence on iodide ion concentration. This relationship was found to be approximately true at high iodide ion concentration, but at low concentration $k_{\text{obs}}$ became relatively greater and this was attributed to concurrent iodination by hypoiodous acid (or rather, by positive iodine) since this obviously becomes more important as $[I^-]$ decreases, due to the dependence of rate upon $[I^-]^{-2}$ rather than upon $[I^-]^{-1}$.

The kinetics of iodination of azulene with iodine in dilute aqueous sodium iodide at 25 °C followed the equation

$$\text{Rate} = \frac{\sum k_i[B_i][\text{ArH}][I_2]}{[I^-]} \tag{163}$$

where B is a general base[327]. The rate equation implies iodination by molecular iodine and the kinetics here are less complicated since the neutral azulene molecule is not liable to appreciable protonation at the low acidities employed (pH 5.2–7.5) nor does azulene function as an effective general base. The mechanism deduced, equilibria (164) and (165)

$$\text{ArH} + I_2 \underset{k'_{-1}}{\overset{k'_1}{\rightleftharpoons}} \text{ArHI}^+ + I^- \tag{164}$$

$$\text{ArHI}^+ + B \overset{k'_2}{\rightleftharpoons} \text{ArI} + \text{BH}^+$$

satisfies the observed kinetics provided the steady-state treatment is applicable to the intermediate and $k'_{-1}[I^-] \gg k'_2[B]$, i.e. $k = k'_1k'_2/k'_{-1}$. A kinetic isotope effect $k_H/k_D$ of from 3.2 to 6.5 was obtained and a plot of log $k$ versus $pK_a$ for the base was linear except for the bases water and 2,4,6-trimethylpyridine; the deviation in the latter case was attributed to a steric effect, as also indicated by the large primary isotope effect of 6.5. This was supported by the fact that data for the non-hindered pyridine fitted the correlation line and gave a much smaller isotope effect.

(b) *Molecular iodine monochloride as electrophile*
   Approximately third-order kinetics, *viz.*

$$\text{Rate} = k_3[\text{ArH}][\text{ICl}]^2 \tag{166}$$

were obtained in iodination by iodine monochloride in acetic acid as solvent, though the kinetic order was temperature-dependent, acetanilide giving a value of 3.0 at 24 °C but 2.5 at 50 °C[328]. Steric hindrance complicated the kinetics by causing a change-over to iodine chloride-catalysed chlorination, with modification of the kinetic order. The amount of chlorination obtained for the following substrates were: acetanilide (10 %), anisole (20 %), pentamethylbenzene (30 %), and p-tolyl methyl ether (50 %). The kinetic observations were all based on the initial few percent of reaction since the reactions are inhibited by the formation of hydrogen chloride which removes iodine monochloride as $HICl_2$.

The zinc chloride-catalysed iodination of aromatics by iodine monochloride in acetic acid at 25.2 °C is first-order in each reagent[329]. The relative rates are given in Table 76 and the reactivities of the higher substituted aromatics are less

TABLE 76

RELATIVE RATES FOR REACTION OF ArH WITH ICl–ZnCl$_2$ IN HOAc AT 25.2 °C[329]

| ArH | $k_{rel}$ |
|---|---|
| Benzene | <0.001 |
| Toluene | 0.14 |
| o-Xylene | 1.3 |
| m-Xylene | 23 |
| p-Xylene | 1 ($k_3 = 0.0001$) |
| 1,2,4-Trimethylbenzene | 75 |
| Mesitylene | 1,800 |
| 1,2,4,5-Tetramethylbenzene | 23 |
| 1,2,3,5-Tetramethylbenzene | 1,600 |
| Pentamethylbenzene | 740 |
| 1,3,5-Trimethylbenzene | 224 |

than predicted from the reactivities of lower members, indicating substantial steric hindrance consistent with the electrophile being effectively $I^+ \ldots ClZnCl_2^-$. From a comparison of the third-order rate coefficient for the catalysed and uncatalysed iodinations of mesitylene it was deduced that iodine monochloride is a more effective catalyst than zinc chloride for this reaction, which may be compared with the greater effectiveness of zinc chloride than bromine in catalysing molecular bromination[281, 320].

Third-order kinetics, equation (166), have also been obtained[330] for the iodination of mesitylene and pentamethylbenzene by iodine monochloride in carbon tetrachloride, the negative activation energies of −4.6 and −1.6 (from measurements at 25.2 and 45.7 °C) obtained being attributed to a mildly exothermic preformation of ArHICl complexes (cf. molecular bromination, p. 123) which subsequently react with two further molecules of iodine monochloride to give the products, viz. equilibria (167) and (168)

$$ArH + ICl \rightleftharpoons ArHICl \qquad (167)$$

$$ArHICl + 2\ ICl \rightleftharpoons ArI + HCl + 2\ ICl \tag{168}$$

With trifluoroacetic acid as solvent, toluene and o-xylene gave second-order kinetics and for the activation energy for toluene was 12.7 (from data at 1.6 and 25.2 °C), i.e. considerably less than for the zinc chloride-catalysed reaction in acetic acid[330].

### (c) Molecular iodine acetate as electrophile

The kinetics of iodination by a species believed to be iodine acetate have been examined. In all those reactions described under positive iodination, for which molecular iodine has been proposed as an alternative electrophile, iodine acetate is also a possible electrophile. For example, in the iodination of phenol at constant pH, it was observed that the rate of iodination was greatly affected by the concentration of the buffer, being approximately a linear function of the buffer acid and this was not a salt effect since addition of potassium nitrate did not affect the rate[214]. This catalysed reaction was shown to arise from the iodination of phenoxide ion by iodine acetate, for if the rate of this reaction is given by equation (169) then this is equivalent to equation (170), viz.

$$Rate = k_{cat}[AcOI][PhO^-] \tag{169}$$

$$Rate = k_{cat}\,KK_{I_3^-}\,K_a\frac{[PhOH][I_3^-][AcOH]}{[H^+]^2[I^-]^2} \tag{170}$$

where $K_{I_3^-} = [I_2][I^-]/[I_3^-]$ and $K_a = [PhO^-][H^+]/[PhOH]$. This equation shows the experimentally observed dependence on hydrogen and iodide ion concentrations, supporting the proposal of iodine acetate as the electrophile. With phosphate buffers the same catalysis effect was observed (again not a salt effect) and the reactive species here was, therefore, considered to be $HIPO_4^-$ formed via equilibrium (171)

$$I_2 + H_2PO_4^- \rightleftharpoons IHPO_4^- + H^+ + I^- \tag{171}$$

similarly to bromine phosphate, equilibrium (146), p. 128.

Iodine acetate also has been proposed[331] as the electrophile in the peroxyacetic catalysed-iodination of benzene in acetic acid at 50 °C, which obeys the kinetic equation

$$Rate = k_2[CH_3CO_3H][I_2] \tag{172}$$

analogous to that observed for bromination under the same condition (see p. 135), i.e. in both cases the rate of halogenation (measured by rate of halogen consumption) is independent of the benzene concentration. For iodination, $10^3k_2 = 1.26$, which is slightly less than that for bromination (1.54) carried out at 60 °C. The reaction scheme proposed was equilibrium (173), followed by (175) or more probably by (174) and then (175a), *viz.*

$$CH_3CO_3H + H_2O + I_2 \overset{slow}{\rightleftharpoons} CH_3COOH + 2\ HOI \tag{173}$$

$$CH_3COOH + HOI \overset{fast}{\rightleftharpoons} CH_3COOI + H_2O \tag{174}$$

$$HOI + PhH \overset{fast}{\rightleftharpoons} PhI + H_2O \tag{175}$$

$$CH_3COOI + PhH \overset{fast}{\rightleftharpoons} PhI + CH_3COOH \tag{175a}$$

Iodine acetate would seem to be unambiguously present in the iodination of pentamethylbenzene in acetic acid by iodine and mercuric acetate, since the latter components form an equilibrium mixture of iodine acetate and acetoxy-mercuric iodide and mercuric acetate speeds up the iodination[332]. Second-order rate coefficients of 0.078 (25 °C) and 0.299 (45 °C) were obtained, and these values are intermediate between those obtained for the reaction of bromine acetate with benzene ($2.5 \times 10^{-3}$) and toluene (1.2) at 25 °C, indicating that bromine acetate is the stronger electrophile.


## 6. Reactions with electrophilic carbon

### 6.1 FRIEDEL–CRAFTS ALKYLATION

Reliable kinetic studies of the Friedel–Crafts alkylation reaction

$$ArH + RX \xrightarrow[\text{acid}]{\text{Lewis}} ArR + HX \tag{176}$$

are comparatively recent and this reflects the experimental difficulties accompanying quantitative studies of the reaction. Principal among these are the need for absence of water, homogeneity of reaction mixture, non-isomerisation of reaction products (and reagents) and a sufficiently low speed of reaction for convenient measurement; all of these factors are difficult to attain. Thus some early kinetic studies[333] failed to use homogeneous conditions, and another[334] measured the reaction rate *via* rate of hydrogen bromide evolution but the complexing of this product with the aromatic reagent (which was not accounted for) nullifies the results that were obtained.


*6.1.1 Alkylation with alkyl halides in organic solvents other than nitromethane*

The first reliable kinetic study of alkylation appears to have been that of

Olivier and Berger[335], who measured the first-order rate coefficients for the aluminium chloride-catalysed reaction of 4-nitrobenzyl chloride with excess aromatic (solvent) at 30 °C and obtained the rate coefficients ($10^5 k_1$): PhCl, 1.40; PhH, 7.50; PhMe, 17.5. These results demonstrated the electrophilic nature of the reaction and also the unselective nature of the electrophile which has been confirmed many times since. That the electrophile in these reactions is not the simple and intuitively expected free carbonium ion was indicated by the observation by Calloway that the reactivity of alkyl halides was in the order RF > RCl > RBr > RI, which is the reverse of that for acylation by acyl halides[336]. The low selectivity (and high steric hindrance) of the reaction was further demonstrated by Condon[337] who measured the relative rates at 40 °C, by the competition method, of isopropylation of toluene and isopropylbenzene with propene catalyzed by boron trifluoride etherate (or aluminium chloride); these were as follows: PhMe, 2.09 (1.10); PhEt, 1.73 (1.81); Ph-iPr, (1.69); Ph-tBu, 1.23 (1.40). The isomer distribution in the reactions[337,338] yielded partial rate factors of $f_o^{Me}$, 2.37; $f_m^{Me}$, 1.80; $f_p^{Me}$, 4.72; $f_o^{i\text{-}Pr}$, 0.35; $f_m^{i\text{-}Pr}$, 2.2; $f_p^{i\text{-}Pr}$, 2.55[337,339].

The most valuable and comprehensive kinetic studies of alkylation have been carried out by Brown *et al.* The first of these studies concerned benzylation of aromatics with 3,4-dichloro- and 4-nitro-benzyl chlorides (these being chosen to give convenient reaction rates) with catalysis by aluminium chloride in nitrobenzene solvent[340]. Reactions were complicated by dialkylation which was especially troublesome at low aromatic concentrations, but it proved possible to obtain approximately third-order kinetics, the process being first-order in halide and catalyst and roughly first-order in aromatic; this is shown by the data relating to alkylation of benzene given in Table 77, where the first-order rate coefficients $k_1$ are calculated with respect to the concentration of alkyl chloride and the second-order coefficients $k_2$ are calculated with respect to the products of the

TABLE 77

RATE COEFFICIENTS FOR REACTION OF $3,4\text{-}Cl_2C_6H_3CH_2Cl$ WITH PhH IN $PhNO_2$,
CATALYSED BY $AlCl_3$, AT 25 °C[340]

| [PhH] (M) | [RCl] (M) | [AlCl$_3$] (M) | $10^4 k_1$ | $10^4 k_2$ |
|---|---|---|---|---|
| 0.335 | 0.333 | 0.333 | 2.67 | |
| 0.335 | 1.00 | 0.333 | 2.61 | |
| 0.335 | 0.111 | 0.333 | 2.89 | |
| 0.335 | 0.333 | 0.111 | 0.83 | |
| 0.335 | 0.333 | 0.667 | 4.55$^a$ | |
| 1.00 | 0.333 | 0.333 | 5.42 | 5.92 |
| 0.50 | 0.333 | 0.333 | 2.97 | 5.75 |
| 1.67 | 0.333 | 0.333 | 7.20 | 4.67 |

$^a$ This lower value than expected may be the result of experimental difficulties which arose from the high viscosity of the solution.

benzene and alkyl chloride concentrations. Values of $k_1$ are independent of the concentration of alkyl chloride and are approximately linear in the concentration of aluminium chloride showing the reaction to be first-order in each reagent. Values of $k_2$, obtained by dividing the $k_1$ values at constant [RCl] and [AlCl$_3$], by [PhH] are not constant but decrease at high benzene concentration; this departure from the expected third-order behaviour was attributed to a solvent effect arising from the dilution of nitrobenzene with benzene (see also below).

Analysis of the first-order rate coefficient in terms of the two consecutive reactions which were occurring, yielded values of $5.3 \times 10^{-4}$ and $2.64 \times 10^{-4}$; the latter value was confirmed as arising from reaction on the first reaction product, 3,4-dichlorodiphenylmethane, because separate 3,4-dichlorobenzylation of this gave a rate coefficient of $2.98 \times 10^{-4}$. The first-order (overall) rate coefficients obtained at 15 °C $(0.665 \times 10^{-4})$ and 35 °C $(6.1 \times 10^{-4})$ yielded $E_a = 19.6$, and $\log A = 14.3$, the rate ratio for the consecutive reactions being the same (0.5) at both temperatures; later studies have tended to confirm this order of activation energy.

Third-order rate coefficients were determined for reaction of some aromatics with 3,4-dichlorobenzyl chloride at 25 °C as follows: chlorobenzene, $0.745 \times 10^{-3}$; benzene, $1.58 \times 10^{-3}$; toluene $2.60 \times 10^{-3}$; $m$-xylene, $3.30 \times 10^{-3}$ and these show the unselective nature of the reaction. With 4-nitrobenzyl chloride, benzene gives a third-order rate coefficient of $4.78 \times 10^{-6}$, which diminishes to 2.4 and $1.9 \times 10^{-6}$ as the benzene composition of the solvent was increased to 50 and 83 vol. %, respectively.

The interpretation of these observations was as follows:

(a) Rate-determining ionisation of the benzyl halide followed by rapid attack of the aromatic is eliminated by the differing substrate reactivities.

(b) Rapid ionisation of the halide followed by rate-determining attack on the aromatic is eliminated by virtue of the slower reaction rate observed with 4-nitrobenzyl chloride than with 3,4-dichlorobenzyl chloride.

(c) The mechanism must, therefore, involve rate-determining nucleophilic attack by the aromatic on the polar halide aluminium halide complex equilibrium (178) the complex being formed as in equilibrium (177), viz.

$$\text{RCl} + \text{AlCl}_3 \underset{k'_{-1}}{\overset{k'_1}{\rightleftharpoons}} \text{RCl.AlCl}_3 \tag{177}$$

$$\text{ArH} + \text{RCl.AlCl}_3 \overset{k'_2}{\rightleftharpoons} [\text{Ar}\underset{R}{\overset{H}{<}}]^+ \text{AlCl}_4^- \tag{178}$$

$$[\text{Ar}\underset{R}{\overset{H}{<}}]^+ \text{AlCl}_4^- \rightarrow \text{ArR} + \text{HCl} + \text{AlCl}_3 \tag{179}$$

leading to the rate equation

$$\text{Rate} = k_1' \, k_2' \, \frac{[\text{AlCl}_3][\text{RCl}][\text{ArH}]}{k_{-1}' + k_2'[\text{ArH}]} \tag{180}$$

Since nitrobenzene is a much stronger base than alkyl halides, the concentration of RCl.AlCl$_3$ will be small and hence $k_{-1}'$ will be large and, therefore, much greater than $k_2'$. Equation (180), therefore, reduces to a third-order expression which includes the equilibrium constant $(k_1'/k_{-1}')$ of the first step and this accounts for the lower rates with 4-nitrobenzyl chloride since it is a poorer base than the 3,4-dichloro compound.

In an attempt to extend this study to other alkylating species, the kinetics of methylation of toluene and benzene by methyl bromide, were investigated by Brown and Jungk[341]; at 0 °C first-order rate coefficients (in alkyl halide) were > 0.1 for reaction of both substrates with methyl bromide, but for the reaction of toluene with methyl iodide a rate coefficient of *ca.* $5.5 \times 10^{-4}$ was obtained. This confirms the qualitative observation of Calloway[336], and has been interpreted by Brown and Jungk in terms of methyl bromide forming a stronger bond with the metal halide, so that the carbon–bromine bond will be more highly polarised than the corresponding carbon–iodine bond resulting in a more rapid reaction with the former; this is consistent with the proposed mechanism. The following partial rate factors were determined by the competition method: $f_o^{\text{Me}}$, 6.07; $f_m^{\text{Me}}$, 1.96; $f_p^{\text{Me}}$, 6.56 (MeBr, 0–5 °C); $f_o^{\text{Me}}$, 7.02; $f_m^{\text{Me}}$, 1.73; $f_p^{\text{Me}}$, 11.05 (MeI, $-1$ to $+2$ °C). These values indicate the greater reactivity of methyl bromide since greater reactivity is usually indicated by lower selectivity (but see pp. 129, 173, 181, 191), and the lower *ortho : para* ratio for the reaction with methyl iodide confirms the local presence of the halide ion in the transition state complex.

A similar attempt to measure the relative rates of ethylation, i-propylation, and *t*-butylation of benzene and toluene by direct rate measurements failed because the reactions were too fast. However, competition studies, using aluminium bromide as catalyst yielded the relative reactivities of 2.95 (methylation), 2.4 (ethylation), and 1.65 (*i*-propylation)[342]. Now if the reaction involved free carbonium ion, the greater stability of the secondary ions would mean a lower reactivity and higher selectivity. The reverse is, however, observed, so that the aromatic must be involved in the rate-determining step of the alkylation, with a gradual increase in the degree of ionic dissociation of the carbon–halogen bond in the transition state with increasing branching of the alkyl group R.

In order to reduce the rate of this reaction to a measurable value, a further study was carried out using 1,2,4-trichlorobenzene as solvent and aluminium tribromide as catalyst[343]. Under these conditions, isopropylation was still too fast to measure directly, and the relative rates of methylation, ethylation and isopropylation were 1 : 57 : > 2,500. For methylation with methyl bromide, fairly good third-order rate coefficients were obtained ranging from 2.78 to $4.35 \times 10^{-3}$, (average $3.66 \times 10^{-3}$) for benzene, and $18.25 \times 10^{-3}$ for toluene, giving $k_{rel.}$ as 5.1 at 25 °C. Rate coefficients for benzene at 35 and 45 °C were 8.92 and $13.5 \times 10^{-3}$ respectively, so that $E_a = 14.6$ and $\Delta S^{\ddagger} = -20.4$; the latter value is consistent with the substantial extent of ordering required in the transition state for the nucleophilic displacement mechanism. For ethylation, rate coefficients were 0.23 (benzene) and 0.655 (toluene) giving $k_{rel.}$ at 2.8 at 25 °C; rate coefficients at 35 ° and 45 °C were 0.448 and 0.609 so that $E_a = 10.7$ and $\Delta S^{\ddagger} = -25.1$. Again the results show that as the alkyl group is better able to accommodate the positive charge and the carbon–halogen bond polarity increases, nucleophilic participation by the aromatic decreases.

These alkylation studies were extended to the use of gallium tribromide as catalyst in an attempt to get measurable rates of reaction throughout the range of primary, secondary, and tertiary alkyl halides[344]. Using the aromatic as solvent and the alkyl halide in excess with respect to catalyst, Smoot and Brown[344] found that the reaction was apparently zeroth-order with respect to ethyl bromide and second-order in gallium tribromide as indicated by some representative data in Table 78. Although the order in aromatic could obviously not be determined, the

TABLE 78

RATE COEFFICIENTS FOR REACTION OF PhH WITH EtBr CATALYSED BY $GaBr_3$, AT 25 °C[344]

| [EtBr] (M) | [GaBr₃] (M) | $10^2 k_2 (k_0 [GaBr_3]^{-2})$ |
|---|---|---|
| 0.410 | 0.0121 | 14.8 |
| 0.410 | 0.0243 | 15.9 |
| 0.423 | 0.0301 | 15.8 |
| 0.846 | 0.0301 | 12.9 |
| 0.411 | 0.0423 | 16.2 |
| 0.411 | 0.0463 | 15.5 |

Average rate coefficients ($10^2 k_2$) at 15, 25 and 40 °C were 7.55, 15.9 and 42.5, hence $E_a = 12.4$ and $\Delta S^{\ddagger} = -22.6$; $10^2 k_2$ for toluene (25 °C) was 39.2.

fact that ethylation of toluene (at 25 °C) proceeded 2.47 times as fast as for benzene indicated that the aromatic was involved in the rate-determining step. Consequently, the transition state of the reaction was envisaged as involving nucleophilic attack of the aromatic upon a polarised 1 : 1 ethyl bromide–gallium bromide complex, the substitution being aided by further polarisation through the presence of the additional molecule of gallium bromide.

The lower reaction rates obtained with this catalyst permitted measurements of the reaction rates of benzene and toluene with a range of alkyl halides including $i$-propyl and $t$-butyl bromides, the rate being followed in some cases by the

TABLE 79

RATE COEFFICIENTS AND KINETIC PARAMETERS FOR REACTION OF PhH
AND $PhCH_3$ WITH RBr, CATALYSED BY $GaBr_3$[344, 345]

| $R$ | Temp. (°C) | $10^2 k_2$ (benzene) | $k_{rel}$ | $E_a$ | $\Delta S^{\ddagger}$ | $10^2 k_2$ (toluene) | $k_{rel}$ | $E_a$ | $\Delta S^{\ddagger}$ |
|---|---|---|---|---|---|---|---|---|---|
| Me | 0 | | 1.0 | 12.5 | −29.3 | 0.278 | 1.0 | 14.0 | −20.0 |
| | 15 | 0.228 | | | | 1.43 | | | |
| | 25 | 0.500 | | | | 2.85 | | | |
| | 40 | 1.31 | | | | 7.67 | | | |
| Et | −23.2 | | 32 | 12.4 | −22.6 | 0.747 | 13.7 | 12.1 | −21.5 |
| | 0 | | | | | 5.70 | | | |
| | 15 | 7.53 | | | | 19.1 | | | |
| | 25 | 15.9 | | | | 39.2 | | | |
| | 40 | 42.5 | | | | 102 | | | |
| $n$-Pr | 15 | 15.65 | 67 | 12.9 | −19.2 | 23.6 | 16.5 | 11.8 | −22.7 |
| | 25 | 33.3 | | | | 44.9 | | | |
| | 40 | | | | | 121.5 | | | |
| $i$-Pr | −78.5 | | 64,000 | | | 35 | 20,000 | 8.3 | −19.3 |
| | −63.7 | | | | | 146 | | | |
| | −45.3 | | | | | 800 | | | |
| | 25.0 | 32,000[a] | | | | 58,000[a] | | | |
| $t$-Bu | 25.0 | | | | | $23 \times 10^{5b}$ | $8 \times 10^{5b}$ | | |

[a] Values calculated from data and $k_{rel}$ values at other temperatures.
[b] These values may be in error by an order of magnitude.

measurement of the pressure of hydrogen bromide evolved; the data are given in Table 79[344, 345]. The low negative entropies of activation are consistent with the proposed mechanism, and significantly there is no change in this parameter for the $i$-propyl compound, which indicates that a free carbonium ion is not present in the transition state of reaction of this compound. The relative small increase in rate from $i$-propyl to $t$-butyl may, therefore, mean that a free carbonium ion is not present in the reaction of the latter either. The smaller spread of rates for reaction with toluene shows that the ability of the alkyl group to tolerate a positive charge is less important than in the corresponding reaction with benzene, *i.e.* toluene is more nucleophilic as expected for the proposed mechanism.

Gallium bromide was used as the catalyst in a determination[346] of the partial rate factors, by the competition method, for ethylation of some substituted benzenes in 1,2-dichloroethane at 25 °C. No direct rate measurements were made and the results, summarised in Table 80, show the low selectivity and high steric hindrance in the reaction.

The conclusion of Brown *et al.*[346, 347] *viz.* that the increased reactivity along a series of alkyl halides is due to the increased polarisation of the carbon–halogen bond has been challenged by Allen and Yats[348], who found constant *meta* : *para* isomer ratios for methylation, ethylation, and isopropylation of toluene, and since this ratio reflects the selectivity and hence reactivity of the electrophile they con-

TABLE 80

PARTIAL RATE FACTORS FOR REACTION OF PhX WITH EtBr, CATALYSED BY $GaBr_3$ IN $1,2\text{-}Cl_2C_2H_4$ AT 25 °C

| X | $f_o$ | $f_m$ | $f_p$ |
|---|---|---|---|
| Me | $2.69(4.9)^a$ | $1.47(2.3)^a$ | $5.70(9.4)^a$ |
| Ph | 0.905 | 0.695 | 2.23 |
| F | 0.364 | 0.116 | 0.738 |
| Cl | 0.271 | 0.102 | 0.538 |
| Br | 0.096 | 0.087 | 0.433 |

[a] Values for benzylation in the absence of solvent[347].

cluded that the reactivity of each halide is much the same, the difference in rate arising from a difference in the *concentration* of the *polarised alkyl species*. The conflict would seem to be not so much one of interpretation, but of experimental results, since the data of Brown indicate that the selectivity (measured in terms of $k_{rel.}$ values) does change with changing alkyl group; this may, however, be misleading because the overall $k_{rel.}$ value depends upon substitution at all positions and the more highly branched the alkyl group the less will be the *ortho* substitution in this highly sterically hindered reaction. Clearly it would be valuable to have partial rate factors determined under the conditions in which the rate coefficients were measured.

The conclusions of Brown and Grayson[340] relating to the reaction order in aluminium chloride has been challenged by Lebedev[349], who measured the rates of aluminium chloride-catalysed cyclohexylation of benzene by cyclohexyl chloride in nitrobenzene and other solvents. The reaction order was 1.0 with respect to alkyl chloride and benzene but varied from 0.35 to 1.5 in catalyst, the order increasing as the polarity of the medium decreased. Some representative data are given in Table 81, the quoted rate coefficients being calculated as apparent first-order in cyclohexyl chloride. Similar results were obtained in the reaction of benzene with *t*-butyl and *t*-amyl chlorides, the order in catalyst rising as high as 2.1 for the reaction of *t*-butyl chloride with benzene in carbon disulphide[350]. The rates of this reaction in a range of solvents were also determined, and the relative values are as follows, the dielectric constants of the solvents being given in parentheses[350]: carbon tetrachloride 1.0 (2.24); benzene 2.5 (2.28); carbon disulphide 4.0 (2.64); 1,1,2,2,-tetrachloroethane 30 (8.0); 1,2-dichloroethane 110 (10.5); nitrobenzene 360 (36.1), so that there is clearly a parallel between the polarity of the medium and the reaction rate. Lebedev has interpreted the results to mean that in non-polar media the catalyst in these reactions would be two molecules of aluminium chloride (which would give the reaction order of *ca.* 2 observed) whereas in polar media it would be the ion $(PhNO_2)_2AlCl_2^+$ formed as in the equilibrium

TABLE 81

FIRST-ORDER RATE COEFFICIENTS AND ORDER IN $AlCl_3$ FOR REACTION OF $C_6H_{11}Cl$ WITH PhH IN $PhNO_2$[349]

| [PhH] (vol. %) | [PhNO_2] (vol. %) | Temp. (°C) | [AlCl_3] (M) | $10^4 k_1$ | Order |
|---|---|---|---|---|---|
| 12.5 | 85.0 | 30.0 | 0.025 | 3.34 ⎫ | |
| | | | | | 0.64 |
| 12.5 | 85.0 | 30.0 | 0.049 | 5.13 ⎬ | |
| | | | | | 0.59 |
| 12.5 | 85.0 | 30.0 | 0.100 | 7.84 ⎭ | |
| | | | | | 0.44 |
| 12.5 | 85.0 | 30.0 | 0.203 | 10.7 ⎰ | |
| 12.5 | 85.0 | 30.0 | 0.410 | 13.7 ⎱ | 0.35 |
| 50 | 47.5 | 40.0 | 0.049 | 5.45 ⎫ | |
| | | | | | 1.05 |
| 50 | 47.5 | 40.0 | 0.100 | 11.5 ⎬ | |
| | | | | | 1.05 |
| 50 | 47.5 | 40.0 | 0.203 | 24.3 ⎭ | |
| 75 | 22.5 | 40.0 | 0.100 | 3.27 ⎫ | |
| 75 | 22.5 | 40.0 | 0.203 | 9.00 ⎭ | 1.44 |
| 75 | 22.5 | 50.0 | 0.100 | 7.50 ⎫ | |
| 75 | 22.5 | 50.0 | 0.203 | 2.08 ⎭ | 1.46 |

$$2\,PhNO_2 + 2\,AlCl_3 \rightleftharpoons [(PhNO_2)_2AlCl_2]^+ + AlCl_4^- \qquad (181)$$

The concentration of this ion was stated in an earlier paper as being proportional to $[AlCl_3]^{\frac{1}{2}}$, and this would account for the experimental results though it was not clear how this proportionality was arrived at[349]. In a later paper it was written as proportional to $[Al_2Cl_6]^{\frac{1}{2}}$, which seems correct but does not then account for the experimental results.

TABLE 82

RATE COEFFICIENTS AND ACTIVATION ENERGIES FOR REACTION OF ArH WITH $C_6H_{11}Cl$ IN $PhNO_2$[351]

| ArH | $10^4 k_2$ | Temp. (°C) | $k_{rel}$ | $E_a$ | ArH | $10^4 k_2$ | Temp. (°C) | $k_{rel}$ | $E_a$ |
|---|---|---|---|---|---|---|---|---|---|
| Benzene | 7.58 | 30 | 1.0 | 18.7 | 1,4-Diphenylbutane | 21.4 | 30 | 2.81 | |
| Toluene | 17.0 | 30 | 2.24 | 18.2 | Biphenyl | 12.3 | 30 | 1.63 | 18.4 |
| Ethylbenzene | 15.8 | 30 | 2.09 | 18.1 | Naphthalene | 23.9 | 30 | 3.14 | 17.8 |
| i-Propylbenzene | 14.0 | 30 | 1.85 | 18.4 | Fluorobenzene | 2.15 | 30 | 0.283 | 19.9 |
| s-Butylbenzene | 12.8 | 30 | 1.69 | | Fluorobenzene | 6.25 | 40 | 0.283 | 19.9 |
| t-Butylbenzene | 11.0 | 30 | 1.45 | | Chlorobenzene | 2.52 | 40 | 0.114 | 19.5 |
| Cyclohexylbenzene | 13.8 | 30 | 1.82 | | Bromobenzene | 2.14 | 40 | 0.099 | 20.1 |
| m-Xylene | 24.5 | 30 | 3.23 | 17.3 | 2-Chlorotoluene | 4.47 | 40 | 0.202 | 19.7 |
| p-Xylene | 20.8 | 30 | 2.74 | | 3-Chlorotoluene | 4.53 | 40 | 0.205 | 19.6 |
| Diphenylmethane | 15.7 | 30 | 2.02 | 17.8 | 4-Chlorotoluene | 2.95 | 40 | 0.134 | 19.5 |
| 1,2-Diphenylethane | 20.8 | 30 | 2.75 | 17.7 | Anisole | 28.1 | 40 | 1.27 | 17.0 |

TABLE 83

FIRST-ORDER RATE COEFFICIENTS AND ACTIVATION ENERGIES FOR REACTION OF RX WITH BENZENE IN $PhNO_2$ CATALYSED BY $AlCl_3$[352]

| RX | $[AlCl_3]$ (M) | $10^4 k_1$ | Temp. (°C) | $E_a$ | RX | $[AlCl_3]$ (M) | $10^4 k_1$ | Temp. (°C) | $E_a$ |
|---|---|---|---|---|---|---|---|---|---|
| Cyclohexyl chloride | 0.203 | 10.7 | 30 | 18.7 | Benzyl chloride | 0.049 | 14.2 | 20 | 13.6 |
| i-Propyl chloride | 0.203 | 12.3 | 30 | | 3-Methylbenzyl chloride | 0.049 | 53.3 | 20 | 12.8 |
| s-Butyl chloride | 0.203 | 13.3 | 30 | 18.4 | 4-Fluorobenzyl chloride | 0.049 | 22.5 | 20 | 13.4 |
| 2-Chloropentane | 0.203 | 13.8 | 30 | | 4-Chlorobenzyl chloride | 0.049 | 6.20 | 20 | 14.6 |
| 3-Chloropentane | 0.203 | 12.8 | 30 | 18.0 | 4-Bromobenzyl chloride | 0.049 | 2.97 | 20 | |
| 2-Chlorohexane | 0.203 | 14.2 | 30 | | t-Butyl chloride | 0.049 | 1.75 | 20 | 14.9 |
| 3-Chlorohexane | 0.203 | 13.5 | 30 | | Benzyl chloride | 0.025 | 4.43 | 20 | |
| 3-Chlorobenzyl chloride | 0.203 | 13.7 | 30 | 18.7 | 2-Methylbenzyl chloride | 0.025 | 33.4 | 20 | |
| 2-Chlorobenzyl chloride | 0.203 | 55.7 | 30 | 19.9 | 4-Methylbenzyl chloride | 0.025 | 113 | 15 | 11.0 |
| Cyclohexyl chloride | 0.41 | 13.6 | 30 | | Cyclohexyl chloride[a] | 0.049 | 1.52 | 20 | |
| Cyclohexyl bromide | 0.41 | 2.25 | 30 | 19.0 | t-Amyl chloride[a] | 0.049 | 76.7 | 20 | |
| Cyclohexyl bromide | 0.41 | 6.16 | 40 | | t-Amyl chloride[b] | 0.049 | 10.5 | 20 | |
| i-Propyl bromide | 0.41 | 6.09 | 40 | | t-Butyl chloride[b] | 0.049 | 33.4 | 20 | |

[a] Medium composed of 85.1 vol. % $PhNO_2$, 12.5 vol. % PhH.
[b] Medium composed of 50 vol. % $PhNO_2$, 31.1 vol. % $CCl_4$ and 12.5 vol. %.

The rates of the aluminium chloride-catalysed reaction of cyclohexyl chloride with a wide range of aromatics in nitrobenzene were determined between 20 and 50 °C. Second-order rate coefficients (first-order in aromatic and cyclohexyl chloride) are given together with the activation energies in Table 82[351]. The low response of the alkylation reaction to substituent effects is once again apparent. A plot of the logarithm of rate coefficients against the activation energies gave a straight line within a quoted experimental error of 0.4 kcal.mole$^{-1}$. However, as some of the activation energies were determined only over a 10° range, this claim is of dubious validity.

The rates of reaction for a wide range of alkyl halides with benzene (as solvent) catalysed by aluminium chloride and at various temperatures have been determined by Lebedev[352], the first-order rate coefficients (first-order in alkyl halide) being in Table 83. A number of points emerge from this work *viz.* that the reactivity of the halides is chlorides > bromides as previously observed[336], and that the reactivity of a given halide is decreased by electron withdrawal in the alkyl moiety, hence the free carbonium ion cannot be involved in the rate-determining step of the reaction; a dependence of rate upon the catalyst concentration is also apparent.

The kinetics of alkylation by triphenylmethyl compounds have been studied. Hart and Cassis[353] found that the alkylation of phenol and *o*-cresol by triphenylmethyl chloride in *o*-dichlorobenzene gave non-linear kinetic plots which were, however, rendered linear by presaturation of the reaction mixture with hydrogen chloride, precise third-order kinetics, equation (182)

$$\text{Rate} = k_3[\text{phenol}][\text{Ph}_3\text{CCl}][\text{HCl}] \tag{182}$$

then being obtained. This autocatalysis by hydrogen chloride was regarded as arising from hydrogen chloride-assisted polarisation of the alkyl chloride. In the absence of added hydrogen chloride it was possible to analyse the kinetics in terms of a mixture of those given by equations (182) and (183), *viz.*

$$\text{Rate} = k_2[\text{phenol}][\text{Ph}_3\text{CCl}] \tag{183}$$

The reaction was also found to be inhibited by addition of dioxan and tetrahydropyran, the rate decrease being proportional to the ether concentration. The results were rationalised by the assumption that 2 : 1 and 1 : 1 phenol : ether complexes were formed, respectively. The inhibition was attributed to participation of the hydroxyl group in solvation of the halogen atom of the alkyl halide, though this seems much less likely than a straightforward modification of the electron-supplying effect of the substituent[354].

All of the foregoing work indicates incomplete ionisation of the carbon–halogen bond in the rate-determining transition state of the reaction. This does not mean, however, that a free carbonium ion may never be found. Thus no autocatalysis was observed in the reaction of triphenylmethyl perchlorate with phenol and phenolic ethers in nitrobenzene, the rate coefficients given in Table 84 referring to the *para* position, since exclusive *para* substitution occurred as a result of the large steric hindrance for the reaction[355]. The lack of autocatalysis was believed to result from the fact that the perchlorate is highly ionised without the aid of added acids and may give the free carbonium ion as electrophile. The activation energies were all with in experimental error of 14.0 but the unusual activation order of OMe < OH < OEt < O-*i*-Pr was attributed to hydrogen bonding of the phenolic hydrogen with the nitro group of the solvent, thereby

## TABLE 84

RATE COEFFICIENTS ($10^5 k_1$) FOR REACTION OF $Ph_3C.ClO_4$ WITH ArOR[355]

| | Temp. (°C) | | | |
|---|---|---|---|---|
| R | 30 | 40 | 50 | 60 |
| H | 1.16 | 2.58 | 5.02 | 11.0 |
| Me | 0.615 | 1.42 | 2.58 | 5.12 |
| Et | 1.53 | 3.82 | 7.26 | |
| i-Pr | 1.71 | 3.92 | | |

facilitating electron release by the substituent. In other words this amounts to solvent assistance of hyperconjugation. Modification of the electronic property of the hydroxy substituent by the solvent in this manner would seem, however, to conflict with the previous observations of Hart et al.[353,354].

### 6.1.2 Alkylation by alkyl halides in nitromethane

The general applicability of the Brown mechanism for alkylation reactions has been challenged by Olah et al.[356], who propose that the rate-determining step is formation of a π- rather than a σ-complex. Measurement of the relative rates of the aluminium chloride: nitromethane-catalysed benzylation of a range of aromatics in homogeneous nitromethane solution at 25 °C gave the relative rates given in Table 85, and these correlate well with π-complex stabilities. However, the low substrate selectivities were not matched by low positional selectivities as found by Brown et al. in their work and it was, therefore, proposed that the substrate selectivity is determined in the rate-determining reaction of the aromatic with the polarised alkyl halide-catalyst complex to give a π-complex, and that the positional selectivity is determined in the subsequent rearrangement to the σ-

## TABLE 85

RELATIVE RATES FOR THE REACTION OF $PhCH_2Cl$ WITH ArH, CATALYSED BY $AlCl_3$-$MeNO_2$ (REACTANT RATIO $1:10:1$) AT 25 °C[356]

| ArH | $k_{rel}$ | ArH | $k_{rel}$ |
|---|---|---|---|
| Benzene | 1.0 | p-Xylene | 4.35 |
| Toluene | 3.20 | Mesitylene | 5.20 |
| Ethylbenzene | 2.45 | Fluorobenzene | 0.46 |
| n-Propylbenzene | 2.22 | Chlorobenzene | 0.24 |
| n-Butylbenzene | 2.08 | Bromobenzene | 0.18 |
| o-Xylene | 4.25 | Iodobenzene | 0.28 |
| m-Xylene | 4.64 | | |

complex. The possibility that these results arise through non-competitive conditions (which one would expect with the benzylation reaction since it is very rapid) seemed to be ruled out by the fact that raising the toluene : benzene ratio from 1 : 4 to 4 : 1 produced a fairly constant relative rate (14 % variation) and also by the fact that no disubstitution was observed; the fluorobenzene : benzene rate ratio, however, varied by 30 % for a 16-fold change in reactant concentration ratio. This evidence is not conclusive, however, for if the reaction conditions are a long way from being truly competitive, alteration in the reagent concentrations will have only a small effect on the reactivity ratio. Similar reactivity ratios for toluene relative to benzene were obtained with other catalysts, $viz.$ $AlCl_3.MeNO_2$ (3.20), $FeCl_3.MeNO_2$ (3.24) and $AgBF_4$ (2.95), all at 25 °C. Hydrogen exchange did not occur under the latter conditions and the kinetic isotope effect ($k_H/k_D$) could therefore be determined and was found to be 1.12. This was regarded as a secondary isotope effect showing that in the rate-determining transition state, some progress towards $sp^3$ bonding from the initial $sp^2$ bonding must have been made.

Low substrate selectivity accompanying high positional selectivity was also found in isopropylation of a range of alkyl and halogenobenzenes by $i$-propyl bromide or propene in nitromethane, tetramethylene sulphone, sulphur dioxide, or carbon disulphide, as indicated by the relative rates in Table 86. The toluene : benzene reactivity ratio was measured under a wide range of conditions, and varied with $i$-propyl bromide (at 25 °C) from 1.41 (aluminium chloride–sulphur

TABLE 86

RELATIVE RATES FOR THE $AlCl_3$–$MeNO_2$-CATALYSED ISOPROPYLATION OF AROMATICS AT 25 °C[357]

| ArH | $k_{rel}$ with isopropylbromide | $k_{rel}$ with propene |
|---|---|---|
| Benzene | 1 | 1.0 |
| Toluene | 2.03 | 1.95 |
| $o$-Xylene | 2.21 | 1.73 |
| $m$-Xylene | 2.80 | 2.49 |
| $p$-Xylene | 2.70 | 2.29 |
| 1,2,3-Trimethylbenzene | 4.31 | 3.98 |
| 1,2,4-Trimethylbenzene | 3.25 | 2.75 |
| 1,3,5-Trimethylbenzene | 0.35[a] | 3.31 |
| 2-Methyl-$i$-propylbenzene | 3.42 | |
| 3-Methyl-$i$-propylbenzene | 3.50 | |
| 4-Methyl-$i$-propylbenzene | 2.82 | |
| Fluorobenzene | 0.23 (0.28)[b] | 0.23 |
| Chlorobenzene | 0.10 (0.13)[b] | 0.11 |
| Iodobenzene | 0.08 (0.11)[b] | 0.07 |

[a] This appears to be a typographical error in the original paper.
[b] Data for $FeCl_3$ as catalyst.

dioxide) to 2.88 (stannic chloride–tetramethylene sulphone) and with propene from 1.45 (100 % sulphuric acid, $-10\,°C$) to 3.27 (100 % sulphuric acid–tetramethylene sulphone, 25 °C). The first-order dependence of the rate upon the aromatic concentration was considered proven by the observation that a change in the toluene : benzene concentration ratio from 1 : 9 to 9 : 1 gave an approximately constant (25 % variation) relative rate. The relative rate was independent of the catalyst concentration (for reactions with aluminium chloride) and a small kinetic isotope effect $k_H/k_D$ of 1.17 was again observed[357].

Extension of the studies to *t*-butylation in nitromethane showed that the reaction of *t*-butylbromide with toluene and benzene gave substrate and positional selectivities which were highly dependent upon the nature and concentration of the catalyst, being in the range 1.2–1.9 for ferric chloride, silver tetrafluoroborate, and aluminium chloride with high *meta* orientation (low selectivity), and in the range 14.5–16.6 for zinc iodide, silver perchlorate, and stannic chloride, with low *meta* orientation (high selectivity)[358]. For ferric chloride, however, reduction in the catalyst concentration from 0.5 $M$ to 0.005 $M$ increased the $k_{rel.}$ value to 12 and increased the selectivity; the different results obtained with the former group of catalysts was, therefore, attributed to thermodynamic control of the rates and products. With *i*-butene, no isomerisation of products occurs and $k_{rel.}$ values of 15.2–16.3 were obtained with aluminium chloride and sulphuric acid. The relative reactivities of a range of alkyl- and halogeno-benzenes are given in Table 87; the relative reactivity of toluene to benzene showed not more than 20 % variation for an 81-fold change in the aromatic concentration ratio again suggesting the reaction to be first-order in aromatic. No substitution was observed for *p*-xylene, 1,2,4 and 1,3,5-trimethylbenzenes, presumably because of the very high steric hindrance in the reaction. Again also, a small secondary kinetic isotope effect $k_H/k_D = 1.16$–1.21 was observed in the reaction. From this work it was

TABLE 87

RELATIVE RATES FOR THE $SnCl_4$-CATALYSED *t*-BUTYLATION OF ArH IN $MeNO_2$ AT 25 °C[358]

| ArH | $k_{rel}$ with t-butylbromide | $k_{rel}$ with isobutene |
|---|---|---|
| Benzene | 1 | 1 |
| Toluene | 16.6 | 15.2 |
| o-Xylene | 44.3 | 47.8 |
| m-Xylene | 2.5 | 3.82 |
| 1,2,3-Trimethylbenzene | 170 | 110 |
| Fluorobenzene | 0.12 (0.16)[a] | 0.19 |
| Chlorobenzene | 0.07 (0.03)[a] | 0.06 |
| Iodobenzene | 0.02 (0.2)[a] | 0.03 |

[a] Data for $AlCl_3 . MeNO_2$ as catalyst.

concluded that the previously observed[346,347] low substrate and positional selectivities observed in *tert.*-butylation must be in error through intervention of thermodynamic control. The greater selectivity in *tert.*-butylation was attributed to the greater stability and hence lower reactivity of the (partially formed) *t*-butyl carbonium ion, as previously proposed by Brown *et al.* (above, *e.g.* ref. 342).

Since there is inherent in reactions which give low selectivities, the possibility that non-competitive conditions are responsible, Olah and Overchuck[359] have measured directly the rates of benzylation, isopropylation, and *tert.*-butylation of benzene and toluene with aluminium and stannic chlorides in nitromethane at 25 °C. Apparent second-order rate coefficients were obtained (assuming that the concentration of catalyst remains constant), but it must be admitted that the kinetic plots showed considerable departure from second-order behaviour. The observed rate coefficients and $k_{rel}$. values determined by the competition method are given in Table 88, which seems to clearly indicate that the competitive ex-

TABLE 88

RATE COEFFICIENTS AND $k_{rel}$ VALUES FOR ALKYLATION OF BENZENE AND TOLUENE IN MeNO$_2$ AT 25 °C[359]

| RCl | Catalyst | $10^4k_2$ (benzene) | $10^4k_2$ (toluene) | $k_{rel}$ | $k_{rel}$ (competition) |
|---|---|---|---|---|---|
| PhCH$_2$Cl | AlCl$_3$–MeNO$_2$ | 2.2 | 7.7 | 3.45 | 3.2 |
| i-PrCl | AlCl$_3$–MeNO$_2$ | 1.7 | 2.8 | 1.65 | 2.03 |
| t-BuCl | SnCl$_4$ | 0.13 | 1.7 | 13.1 | 16.6 |

periments were indeed competitive.    Furthermore, the isomer distributions obtained under direct kinetic conditions were similar to those obtained under competitive conditions.

### 6.1.3 Alkylation with alkyl amines

Relative rates of alkylation of toluene and benzene using a mixture of nitrosonium hexafluorophosphate, nitromethane (or acetonitrile) and aliphatic amine as the alkylations agent have been determined at 25 °C as follows[360]: 1.5 (ethylamine), 2.5 (*i*-propylamine) and 3.5 (benzylamine); nothing more as yet is known about the kinetics of alkylation with these new alkylating reagents.

### 6.1.4 Alkylation with alkyl sulphonates in organic solvents

The kinetics of alkylation of aromatics with sulphonic acid esters have been investigated using both dichloroethane and the aromatic in excess, as solvent, and in the presence of benzenesulphonic acid[361]. For reaction of a series of sub-stituted benzyl benzenesulphonates with benzene (in excess), the kinetics were third-order, being first-order in ester and second-order in added sulphonic acid, *viz.*

$$\text{Rate} = k_3[\text{PhSO}_2\text{OR}][\text{PhSO}_2\text{OH}]^2 \tag{184}$$

Since the product of the alkylation reaction is the sulphonic acid, the reactions showed autocatalysis. The reaction mechanism was envisaged, therefore, as in-volving an intimate ion pair formed *via* equilibrium (185)

$$\text{PhSO}_2\text{OR} + 2\ \text{PhSO}_2\text{OH} \rightleftharpoons \left[ \begin{array}{c} \overset{+}{R} \text{---} O \overset{\diagup O_2SPh}{\diagdown H} \\ \\ Ph \cdot S\overset{\diagdown}{O}_2 \overset{\diagup O}{\diagdown} H \cdots \overset{-}{O} \cdot O_2SPh \end{array} \right] \tag{185}$$

in which each ion is specifically solvated by a molecule of sulphonic acid and that this ion pair subsequently reacts with benzene. This mechanism was considered to be applicable to other benzylations in hydrocarbon solvents since benzene-sulphonic acid had the same kinetic effect on benzylation by benzyl halides. Further evidence to support the mechanism was the fact that the reaction was completely inhibited by bases. Rate coefficients and kinetic parameters were determined as given in Table 89; in view of the small temperature range over which rate measurements were made, little reliance can be placed upon these activation

TABLE 89

RATE COEFFICIENTS AND KINETIC PARAMETERS FOR REACTION OF Ph.SO$_2$OCH$_2$Ar WITH PhH AT VARIOUS TEMPERATURES[361]

Temperatures in parentheses

| Substituent in Ar | $10^3k_3$ | $E_a$ | $\Delta S^{\ddagger}$ |
|---|---|---|---|
| 3-NO$_2$ | 0.161(50) | | |
| 3-NO$_2$ | 0.279(55) | 24.9 | −0.78 |
| 3-NO$_2$ | 0.516(60) | | |
| 4-Cl | 34.5  (40) | | |
| 4-Cl | 69.9  (40) | 14.9 | −19.9 |
| 4-Cl | 145   (60) | | |
| H | 70.2  (30) | | |
| H | 97.4  (40) | 6.8 | −43.6 |
| H | 141   (50) | | |
| H | 192   (60) | | |
| 4-CH$_3$ | 741   (36) | | |
| 4-CH$_3$ | 767   (42) | 3.9 | −48.7 |
| 4-CH$_3$ | 967   (50) | | |

energies which indicate that the relative reactivities change at a temperature as low as 80 °C. The enormous variation in the entropies of activation, if correct, suggest that the reaction mechanism is extremely substrate- (as well as temperature-) dependent, and the attempted analysis in terms of a linear free energy equation is hardly meaningful ($\rho$ was quoted as $-4.77$ at 50 °C). In addition, the reduction in rate arising from electron-withdrawing substituents was stated to be consistent with the proposed mechanism since the carbonium ion would be destabilised by such electron withdrawal. But this is precisely the reason that carbonium ions with electron-withdrawing substituents are more reactive, and the rejection by Nenit-zescu et al.[361] of the bimolecular nucleophilic substitution mechanism arises from an incorrect interpretation of the facts. It was stated that $E_a$ versus $\Delta S^{\ddagger}$ plots were linear, though the worth of such plots is dubious (see p. 9). In an earlier publication[362] the reactivities of p- and m-xylene relative to benzene were reported as 2.6 and 6.7 respectively, so that substituents appear to have a larger effect in the electrophile than in the aromatic substrate which indicates that the charge is predominantly on the former in the rate-determining step of the reaction; the numerical values underestimate the reality of the situation because the charge is in the side chain of the electrophile and consequently the substituents produce a smaller effect upon the charge in the side chain than in the aromatic ring.

With dichloroethane as solvent, approximately fourth-order kinetics were obtained, the order in aromatic being approximately one. The rates of reaction of 4-chlorobenzyl benzenesulphonate with mesitylene and pentamethylbenzene at 50 °C were 0.12 and 1.1 respectively, and for reaction of the 3-nitrobenzyl ester with pentamethylbenzene, 0.009 (at 70 °C) and 0.012 (at 80 °C); these results correspond to the previously observed orders of reactivity.

### 6.1.5 Alkylation with alkenes

The reaction rates for the alkylation of benzene by cyclohexene in the presence of hydrogen chloride gave rate coefficients which depended only upon the con-centration of added hydrogen chloride (Table 90); this was stated as showing that HCl reacts with cyclohexene to give cyclohexyl chloride, which then reacts with benzene in the rate-determining step of the reaction[363]; there is, however, too little experimental evidence to decide which reaction step is unambiguously rate-determining.

Further evidence against the formation of a free carbonium ion in the alkylation reaction is obtained from the fact that in the presence of boron trifluoride–phos-phoric acid catalyst, but-1-ene, but-2-ene, and i-butene react at different rates with alkylbenzenes, yet they would each give the same carbonium ion. In addition, only the latter alkene gave the usual activation order (in this case the hyper-

TABLE 90

FIRST-ORDER RATE COEFFICIENTS FOR REACTION OF CYCLOHEXENE WITH BENZENE AT 30 °C[363]

| $[AlCl_3]$ $(M)$ | $[HCl]$ $(M)$ | $10^3k_1$ |
|---|---|---|
| 0.049 | 0.012 | 0.60 |
| 0.100 | 0.015 | 0.86 |
| 0.203 | 0.020 | 1.0 |

conjugative order) by alkyl substituents, deactivation by alkyl groups being otherwise observed, and it was concluded that only those alkenes which are strongly polarised by the catalyst will give the normal substituent effect[364].

### 6.1.6 *Alkylation with alcohols and derivatives*

Some fairly extensive kinetic studies of alkylation using alcohols and derivatives as alkylating agents have been carried out by Gold *et al.*[365-367]. Bethell and Gold measured the rate of reaction of diphenylmethanol with mesitylene, and anisole, and 4,4-dichlorodiphenylmethanol with mesitylene at 25 °C in a range of acetic acid–aqueous sulphuric acid mixtures[365]. With aromatic in 10-fold excess, first-order kinetics were obtained, from which second-order rate coefficients were derived, a representative sample being given in Table 91. Reaction was envisaged

TABLE 91

RATE COEFFICIENTS FOR REACTION OF $Ph_2CHOH$ WITH ArH IN SULPHURIC ACID–ACETIC ACID CONTAINING 0.55 $M$ $H_2O$ AT 25 °C[365]

| *Mesitylene* | | *Anisole* | | |
|---|---|---|---|---|
| $[H_2SO_4]$ $(M)$ | $10^4k_2$ | $[H_2SO_4]$ $(M)$ | $10^4k_2$ | $10^4k_2$ (*no water*) |
| 0.729 | 2.59 | 0.306 | | 1.53 |
| 0.871 | 4.26 | 0.612 | | 11.2(11.6)[b] |
| 0.886 | 1.87[a] | 0.919 | | 32.7(34)[c] |
| 1.008 | 7.08 | 1.225 | | 82.7(87)[c] |
| 1.19 | 4.87[a] | 1.478 | | 140.7(141)[c] |
| 1.238 | 13.3 | 0.371 | 0.90 | |
| 1.358 | 18.3 | 0.772 | 7.10 | |
| 1.660 | 40.3 | 1.149 | 28.0 | |
| 1.760 | 49.1 | 1.349 | 45.2 | |
| 1.914 | 27.25[a] | 1.410 | 65.8 | |

[a] Rate coefficients for reaction with 4,4-dichlorodiphenylmethanol.
[b] Rate coefficients for reaction with diphenylmethyl acetate.
[c] Rate coefficients for reaction with [4-$^3$H]-anisole as aromatic.

as occurring with *free* carbonium ions under these conditions, and the fact that electron-withdrawing substituents in the alcohol gave a reduced reaction rate was regarded as arising from a reduced rate of protonation (to give $ROH_2^+$), this reduction outweighing the increased reactivity of the resultant carbonium ion. The increase in rate coefficient with increasing acid concentration paralleled the rates of ionisation of compounds similar to the alcohol employed, suggesting that a carbonium ion is involved in the rate-determining step of the reaction; the similar rates obtained with diphenylmethyl acetate shows that this is either very rapidly hydrolysed to the alcohol or forms the carbonium ion directly. Rate coefficients in the non-aqueous media were higher than in the aqueous media, which accords with the absence of competition for protonation by the sulphuric acid in the former medium. Rate coefficients for the tritiated and untritiated anisoles were negligibly different so that loss of hydrogen is here not rate-determining as would be expected for a reaction with an electrophile bearing a full positive charge.

The effects of some methoxy substituents were determined for the reaction with diphenylmethanol at a range of sulphuric acid concentrations in acetic acid containing 0.55 *M* water at 25 °C (Table 92)[366]. The logarithms of the rate coeffi-

TABLE 92

RATE COEFFICIENTS FOR REACTION OF $Ph_2CHOH$ WITH ArH IN $H_2SO_4$–HOAc–$H_2O$ AT 25 °C[366]

| ArH | $[H_2SO_4]$ (M) | $10^4 k_2$ |
|---|---|---|
| Anisole | 0.633 | 5.52 |
| | 1.774 | 1,560 |
| 1,4-Dimethoxybenzene | 1.241 | 29.0 |
| | 2.475 | 286 |
| 1,2-Dimethoxybenzene | 0.993 | 102 |
| | 1.659 | 404 |
| 1,3-Dimethoxybenzene | 0.102 | 28.1 |
| | 0.407 | 451 |
| Benzene | 3.75 | 0.0359 |
| | 4.89 | 0.263 |

cients plotted linearly against the acidity function with near unit slope which was considered indicative of an $S_N2 C^+$ mechanism *i.e.* rate-determining attack of the preformed carbonium ion upon the aromatic though this evidence would currently be regarded as an insufficient diagnosis of the mechanism (see p. 5). However, there were significant differences in slope, which did not vary in a regular manner with the reactivity; consequently, the relative substituents effects varied in an irregular manner with acidity. Some unidentified medium effect was proposed as being responsible, and possibly this was hydrogen bonding of the methoxyl oxygen with solvent hydrogen.

This work was extended to the reaction of diphenylmethyl chloride and

TABLE 93

RATE COEFFICIENTS FOR REACTION OF MeOPh WITH Ar$_2$CHCl OR Ar$_2$CHOAc AT 25 °C, CATALYZED BY ZnCl$_2$[367]

| [ZnCl$_2$] (M) | $10^5k_2$ (Ph$_2$CHCl) | $10^5k_2$ (Ph$_2$CHOAc) | $10^5k_2$ [(4-MeOPh) Ph.CHCl] |
|---|---|---|---|
| 0.011 |       |       | 13.6 |
| 0.022 |       |       | 26.1 |
| 0.046 | 4.13  |       |      |
| 0.092 | 8.90  |       |      |
| 0.108 |       |       | 103  |
| 0.162 |       |       | 50.7 |
| 0.184 | 22.5  |       |      |
| 0.351 | 26.4  | 0.354 |      |
| 0.369 | 33.5  |       |      |
| 0.398 |       |       | 19.0 |
| 0.461 | 25.8  |       |      |
| 0.597 |       |       | 57.5 |
| 0.702 | 13.8  | 4.21  |      |
| 0.796 |       |       | 110  |
| 0.922 | 27.8  |       |      |
| 1.053 | 31.4  | 17.1  |      |
| 1.404 | 72.5  | 49.7  |      |

acetate, and of 4-methoxydiphenylmethyl chloride, with anisole, catalysed both by zinc chloride, and zinc chloride–hydrogen chloride in acetic acid solvent[367]. The increase of rate with increasing catalyst concentration (Table 93) parallels the effect of the catalyst upon the ionisation rates of indicators of similar structure. It can be seen from the data in Table 93 that whereas rate coefficients for reaction with acetate increase uniformly with increasing catalyst concentrations, for both halides the rates pass through a maximum, In addition, at low zinc chloride concentrations the halide reacts very much faster than the corresponding acetate, but this difference diminishes with the increase of the zinc chloride concentration. This result was attributed to the alkyl chloride being hydrolysed to the acetate, the hydrogen chloride liberated acting as a co-catalyst so that the reaction rates with the chloride always exceeded those of the acetate. If the co-catalytic effect of hydrogen chloride diminishes as the concentration of zinc chloride increases, then the diminishing rate difference (and possibly the observed maxima) as the latter increased can be accounted for. The effect of liberated hydrogen chloride was proved by measuring reaction rates in solutions containing either acetate or an equimolar concentration of the chloride, and then in solutions of the former containing the same quantity of hydrogen chloride that would have been liberated by the solutions of the latter; the solutions with added hydrogen chloride then gave similar rate coefficients to those containing the alkyl chloride (Table 94); the catalytic effect of hydrogen chloride is also apparent from the rate coefficient

TABLE 94

RATE COEFFICIENTS FOR REACTION OF $Ph_2CHX(0.027\ M)$ WITH ANISOLE AT 25 °C[367]

| $X$ | $[ZnCl_2]$ $(M)$ | $[Added\ HCl]$ $(M)$ | $10^5 k_2$ |
|---|---|---|---|
| Cl | 0.070 | | 6.67 |
| OAc | 0.070 | | very slow |
| OAc | 0.070 | 0.027 | 6.80 |
| Cl | 0.351 | | 26.4 |
| Cl | 0.351 | 0.022 | 76.9 |
| Cl | 1.053 | | 31.4 |
| OAc | 1.053 | 0.027 | 32.8 |

obtained with diphenylmethyl chloride both in the absence and presence of added hydrogen chloride.

The reaction of triphenylmethanol with cresol has also been studied in aqueous sulphuric acid–acetic acid mixtures and in media containing 0–22.6 wt. % sulphuric acid, kinetics being obtained which were first-order in both alcohol and phenol for a ten-fold change in the concentration of each[368]. Rate coefficients $(10^4 k_2,$ at 55 °C) varied from 0 in 95 wt. % acetic acid, through 48.6 (5.33 at 25 °C) in 20 wt. % sulphuric acid–5 wt. % water–75 wt. % acetic acid, to 17 in 40 wt. % sulphuric acid–5 wt. % water–55 wt. % acetic acid. The increase in rate with acidity was attributed to the formation of the triphenylmethyl cation from the alcohol, and the maximum in the rate to formation of the conjugate acid of the cresol. By correcting for the rate of the cation-forming equilibrium, the second-order rate coefficients (which are thus medium independent) for reaction of the cation with cresol were determined as $7.35 \times 10^{-4}$ (25 °C) and $61.6 \times 10^{-4}$ (55 °C) giving $E_a = 13.5$ and log $A = 7.02$; these values were attributed respectively to high reactivity of the cation and high steric hindrance to its substitution.

### 6.1.7 Cyclialkylation

An interesting alkylation reaction which has received a considerable amount of kinetic study is cyclialkylation, *viz.* reaction (186)

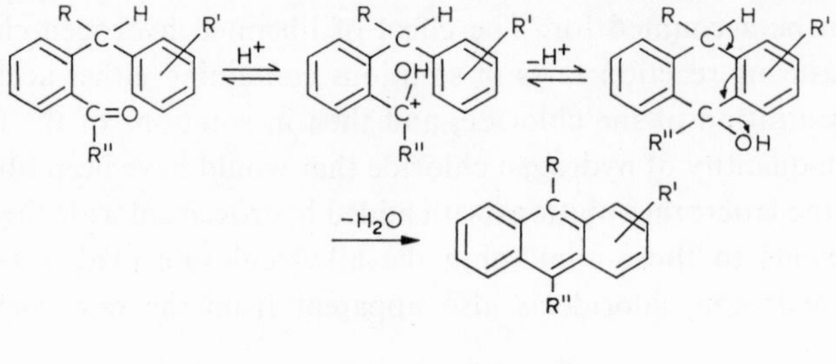

(186)

referred to also as cyclodehydration, though it could reasonably be considered as an acylation reaction.

Early methods of following the reaction relied upon quantitative recovery of the anthracene derivative from the reaction mixture and in view of the extreme insolubility of these derivatives, this is one of the few reactions that can be accurately studied by product recovery methods. More recently, of course, the UV spectroscopic method has been used, the formation of the anthracene spectrum with time being measured.

The initial study by Berliner[369] used R = Me, R' = H, and R'' variable, the reaction being carried out in a refluxing mixture of acetic acid (10 ml), hydrobromic acid (4 mole, 48 %) and water (1 ml). The rate coefficients (the reaction was first-order in aromatic) are given in Table 95 and it appeared that the reaction

TABLE 95

FIRST-ORDER RATE COEFFICIENTS FOR CYCLIALKYLATION OF 2-R''COC$_6$H$_4$.CHMe.Ph BY AQUEOUS HBr–HOAc AT 133–4 °C[369]

| $R''$ | Me | Et | $n$-Pr | $n$-Bu | $n$-Pentyl | $n$-Hexyl | Ph | Benzyl |
|---|---|---|---|---|---|---|---|---|
| $10^4 k_1$ | 7.67 | 3.0 | 1.65 | 0.58 | 0.60 | 0.60 | 0.267 | 1.52 |

was facilitated by electron supply in R'' (the alkyl groups activating in the hyperconjugative order) though the alternative possibility, that a steric effect was responsible, could not be ruled out. The rates were decreased by a decrease in acid concentration and there was some evidence that the rate is dependent upon a power of hydrogen bromide concentration greater than one[370]. In addition, at a constant hydrogen bromide concentration, the rate decreased very rapidly as the mole fraction of water in the solvent was increased. Thus apparent first-order rate coefficients ($10^6 k_1$ at 100 °C) in a medium of 99.6 mole % acetic acid were 33.6 (0.987 $M$ HBr) and 51.3 (1.223 $M$ HBr), whilst in a medium of ca. 56.5 mole % acetic acid values were 0.636 (0.987 $M$ HBr) and 1.10 (1.223 $M$ HBr); the effect of water is obvious, and the dependence on concentration of hydrogen bromide was thought to indicate a dependence upon the (unknown) acidity function of the media.

Further studies[371–373] with the compounds R = R' = H, and R'' variable but to include this time a range of meta- and para-substituted benzyl compounds (for which the steric effect might be supposed to be constant) gave a much smaller resultant effect for the substituent in the benzyl group indicating that the overall substituent effect is principally steric in origin (Table 96). This was further indicated by the results for ortho substituents, all of which caused deactivation regardless of their usual electronic effect. It is rather difficult to determine what the overall electronic effect in R'' actually is (and indeed the meta-trifluoromethyl substituent produces opposite results at the two different temperatures employed)

TABLE 96

FIRST-ORDER RATE COEFFICIENTS FOR CYCLIALKYLATION OF 2-R″COC$_6$H$_4$.CHR.Ph BY AQUEOUS HBr–HOAc AT 117.5 °C[371–373]

| R″ | $10^5k_1$ | R″ | $10^5k_1$ | R″ | $10^5k_1$ | R″ | $10^5k_1$ at 150 °C [c] |
|---|---|---|---|---|---|---|---|
| H | 150 | Ph | 1.22 | 3-MeC$_6$H$_4$ | 1.22 | Ph | 15.3 |
| Me | 19.4 | 4-MeC$_6$H$_4$ | 1.17 | 3-FC$_6$H$_4$ | 1.47 | 2-MeC$_6$H$_4$ | 2.64 |
| Et | 8.35 | 4-FC$_6$H$_4$ | 0.76 | 3-ClC$_6$H$_4$ | 1.47 | 2-FC$_6$H$_4$ | 10.7 |
| Benzyl | 6.4 | 4-ClC$_6$H$_4$ | 1.14 | 3-BrC$_6$H$_4$ | 1.39 | 2-ClC$_6$H$_4$ | 1.92 |
| Ph[a] | 3.62 | 4-BrC$_6$H$_4$ | 1.17 | 3-CF$_3$C$_6$H$_4$ | 1.78 | 2-BrC$_6$H$_4$ | 0.91 |
| Ph[b] | 3.62 | 4-CF$_3$C$_6$H$_4$ | 2.59 | | | 3-CF$_3$C$_6$H$_4$ | 8.8 |

[a] For R = Me.
[b] For R = Ph; in all other cases R = H.
[c] Incorrectly quoted as reciprocal rates in ref. 373.

but on balance, electron withdrawal appears to aid reaction as might be expected since this increases the positive charge on the reacting carbonium ion, thereby accelerating the substitution step. However, the same electron withdrawal will retard protonation of the carbonyl carbon in the first place so it is not surprising that the overall electronic effect observed is a small one. The very strong effect of *ortho*-fluorine compared to the other substituents is compatible with a strong −I effect increasing the reactivity of the carbonium ion, but it was suggested that hydrogen bonding to the hydroxyl hydrogen attached to the carbonium ion would bring about the same result. Infrared analysis failed to show evidence of hydrogen bonding, though it is not surprising in view of the very small rate factors involved (see also p. 130).

The last two entries in the first column of Table 96 show that the reaction is speeded up by electron supply in the alkyl bridge as expected for an electrophilic substitution; the seeming unexpected similarity of the rates for methyl and phenyl substitution in this bridge arises from a statistical effect of the latter compound having two rings available for substitution. The expected rate acceleration by substituents in the ring undergoing substitution was observed with the compounds R = H, R″ = Ph, R variable, the rate coefficients being given in Table 97[374]. Note that a 3 substituent is *para* to the substitution site, whereas the 2 and 4 substituents are *meta* to this site. The difference in the values for the 2- and 4-methyl

TABLE 97

RATE COEFFICIENTS FOR CYCLIALKYLATION OF 2-PhCOC$_6$H$_4$.CH$_2$.C$_6$H$_4$R″ BY AQUEOUS HBr–HOAc AT 117.5 °C[371]

| R″ | H | 2-Me | 3-Me | 4-Me | 3-CF$_3$ | 4-F |
|---|---|---|---|---|---|---|
| $10^5k_1$ | 1.22 | 4.28 | 55.5 | 3.85 | no reaction in 10 days | no reaction in 3 days |

substituents is interesting (if real) since the methyl group adjacent to the bridge (and thus expected to be less electron-supplying as a result of buttressing) is in fact the more electron supplying. Possibly this reaction would reward further accurate kinetic studies.

An interesting application of this work has been to measure rate of reaction of the compounds in which $R = R' = H$, $R'' = $ pyridyl or phenyl, in acetic acid at 100 °C, which are $(10^6 k_1)$ as follows[375]: $(R =)$ Ph, 2.02; 2-Py, 7.65; 3-Py, 7.47; 4-Py, 23.2. As expected, the reaction rates are increased through electron withdrawal by the pyridyl ring, though the effects at the various positions do not correspond with those expected for the neutral pyridine molecule, nor for the protonated molecule which should be present here. This discrepancy has been attributed to the possibility that the 2-pyridyl compound is only monoprotonated, as otherwise there would be two adjacent protonated sites. Unfortunately, it is not possible to analyse these results in terms of the known $\sigma^+$-values[376] for the neutral pyridine molecule, since the effect of other substituents in the phenyl ring (and hence the $\rho$-factor) are not known under the conditions of this study; again this might be an area for further investigation.

The effects of substituents in the ring undergoing substitution have also been measured in the cyclisation of 4-anilino-pent-3-en-2-ones to give 2,4-dimethyl-quinoline and its derivatives[377–380], *viz.*

$$(187)$$

The mechanism (strong $H_2SO_4$ solvent) was originally thought to involve the protonated species (VIII) rather than the alternative (IX) since logarithms of rates

(VIII)                    (IX)

(Table 98) correlated with the Hammett acidity function $H_0$ rather than with log $[ROH]/[R^+]$[377]. However, subsequent spectroscopic analysis and colligative property measurements showed that anils are present in strong sulphuric acid entirely as the monoprotonated species and since reaction is not instantaneous, it must be occurring on some other species, assumed to be diprotonated. For this to be true a plot of log rate *versus* the acidity function $H_+$ would have to be linear; the values of $H_+$ were not known in this medium but were considered to parallel

TABLE 98

RATE COEFFICIENTS AND ARRHENIUS PARAMETERS FOR CYCLIALKYLATION OF 4-
(X-ANILINO)-PENT-3-EN-2-ONES IN SULPHURIC ACID AT 25 °C[377]

| X = H | | | | X = 4-Me | | | |
|---|---|---|---|---|---|---|---|
| $H_2SO_4$ (wt. %) | $10^6 k_1$ | $E_a$ | $\log A$ | $H_2SO_4$ (wt %) | $10^5 k_1$ | $E_a$ | $\log A$ |
| 88.6 | 2.85 | | | 84.5 | 3.98 | | |
| 89.2 | 3.51 | | | 85.5 | 5.50 | | |
| 91.2 | | 15.6 | 9.1 | 86.8 | | 17.3 | 10.5 |
| 93.3 | 10.3 | | | 87.6 | 12.4 | | |
| 94.9 | | 17.4 | 10.9 | 89.2 | 20.5 | | |
| 95.2 | 19.0 | | | 91.2 | | 16.5 | 10.4 |
| 95.5 | 20.1 | | | 94.9 | | 15.9 | 10.4 |
| 97.3 | 35.0 | | | 95.5 | 141 | | |

the $H_0$ values in their dependence upon the medium composition, and the linear correlation of log rate with the latter was thus thought to be in fact a correlation against $H_+$; the diprotonated species would presumably be protonated at the carbonyl oxygen *and* at nitrogen.

The greater reactivity of the compound with a methyl group *para* to the nitrogen in the ring undergoing substitution, was considered to arise from the greater basicity of the nitrogen here so that protonation to the reactive species would occur more readily. It is doubtful, however, if this explanation is necessary since the factor observed (6) is close to that (*ca.* 4) obtained in the earlier work described above for compounds which also have a methyl group *meta* to the substitution site. The activation energy obtained for the unsubstituted compound in these studies appears to decrease with increasing acidity, whereas for the 4-methyl compound the reverse is observed; it seems likely that the value for the unsubstituted compound in 91.2 wt. % acid is in error.

The reaction rates for the 2,3-dimethyl compound in 70.5–80.8 wt. % sulphuric acid (and also in 63.6–71.4 wt. % perchloric acid) showed this to react 430 times faster than the unsubstituted compound[378]. Since substituents in the 2 and 4 positions may be regarded as approximately equivalent in this reaction (and a methyl group in the latter position increases the rate by a factor of about 6), the effect of a methyl group *para* to the reaction site can be evaluated as $430 \times 2/6 \sim 140$. (The factor of 2 is a statistical correction for the fact that in the 2,3-dimethyl compound there is only one position available for substitution compared to two in the 4-methyl and unsubstituted compounds.) The 3,4-dimethyl compound reacted 1,180 times faster than the unsubstituted compound so that from this the effect of a methyl group *para* to the reaction site can be evaluated as $1,180/6 \approx 195$; there is thus reasonable agreement considering the assumptions made and clearly the $\rho$-factor under these conditions is greater than for the reactions of the benzophenones described above. This enforces, therefore, the argument in favour of the 4-methyl substituent effect arising from direct modification of the reactivity of the site of substitution. Further rate coefficients ($10^5 k_1$) were determined for substituents *para* to the reaction site in 93.8 wt. % sulphuric acid at 25 °C as follows[378]: (X = ) H, 6.2; 3-F, 69.3; 3-Cl, 13.8; 3-Br, 11.6; 3-I, 31.7. These results are exceptional in that all the *p*-halogens appear to activate, no reason being at present available. No cyclisation could be affected for the 4-chloro compound and this is consistent with the strong deactivation expected for a halogen substituent *meta* to the reaction site.

A determination[380] of the kinetic isotope effect for the reaction (using the 2,4,6-trideuterated compound) gave the following rate coefficients: 95.7 wt. % sulphuric acid, $10^4 k_1(H) = 2.05$, $10^4 k_1(D) = 1.35$; 89.2 wt. % sulphuric acid, $10^5 k_1(H) = 3.52$, $10^5 k_1(D) = 2.48$, so that $k_H/k_D = 1.5$. This does not necessarily mean, however, that the loss of water is rate-determining since the change in hydridisation from $sp^2$ to $sp^3$ that occurs at the carbon atom undergoing substitution can account for this factor if the substitution step is rate-determining.

Rates of cyclisation in mixture of ammonium sulphate and sulphuric acid were also measured[377]. Although acidity functions were not known for such media, the logarithms of the rate coefficients (Table 99) correlated with $[H_2SO_4]/[HSO_4^-]$

TABLE 99

RATE COEFFICIENTS FOR CYCLISATION OF 4-ANILINE-PENT-3-EN-2-ONE IN $(NH_4)_2SO_4$–$H_2SO_4$ AT 25 °C[377]

| $[(NH_4)_2SO_4]$ (M) | 1 | 2 | 2.5 | 3 | 4 |
|---|---|---|---|---|---|
| $10^4 k_1$ | 21.8 | 9.25 | 5.23 | 2.87 | 0.488 |

which is equivalent to a linear plot against $H_0$. However, the conclusions of this study would probably need to be modified in view of the later belief that reaction occurs on the diprotonated rather than the monoprotonated species.

## 6.2 HALOMETHYLATION

### 6.2.1 Chloromethylation

The first kinetic study used chloromethyl methyl ether as chlorinating reagent and acetic acid as solvent, *viz.* reaction (188)[381]

$$ArH + CH_2ClOCH_3 = ArCH_2Cl + CH_3OH \qquad (188)$$

Rates were measured at different temperatures between 15 and 100 °C, and relative rates were then quoted at 65 °C as follows: benzene, 1.0; toluene, 3.0; $m$-xylene, 24.0; mesitylene, 600; anisole, 1,300; 2,5-dimethylanisole, 100,000; chloro-mesitylene 2.0; nitrobenzene and nitromesitylene gave rates too slow to measure. However, it is not apparent how the reaction rates at 65 °C were obtained since the activation energies were not determined.

A later study[382] using paraformaldehyde and hydrogen chloride as the source of the chloromethyl moiety, reaction (189)

$$ArH + CH_2O + HCl = ArCH_2Cl + H_2O \tag{189}$$

and acetic acid as solvent gave very different results to those quoted above, and it was implied that the former results were in error though the different system used made this assumption unjustifiable. The need to follow rates only over the first 10 % of reaction was stressed in view of the consecutive reactions which can occur, $e.g.$ alkylation by the benzyl halide product of the initial reaction. The reaction was followed by measuring the uptake of hydrogen chloride (as loss of chloride ion) and since the consecutive reactions can produce hydrogen chloride, the need for caution follows. Relative rates at 85 °C were: benzene 1.0; toluene, 3.1; n-butylbenzene, 2.9; $t$-butylbenzene, 2.8; $p$-xylene, 1.6; mesitylene, 13, diphenylmethane, 0.77; bromobenzene, 0.48; diphenylsulphide, 0.88; diphenyl ether, 1.5; anisole, 23; 4-methylcresyl ether, 7.4; most of these values disagree with the earlier work but are themselves suspect as consideration of some of the individual results, $e.g.$ that for diphenylmethane, indicates.

The reliability of this work has also been questioned by Brown and Nelson[339], who could not get any reaction at all with benzene under the conditions quoted by Szmant and Dudek[382]. If the benzene value is in error it could account for the low selectivities that were observed. By using the competition technique with a deficiency of paraformaldehyde (but an excess of hydrogen chloride) they obtained a toluene : benzene reactivity ratio of 112 and partial rate factors of $f_o^{Me} = 117$, $f_m^{Me} = 4.37, f_p^{Me} = 430$.

Different results yet again have been obtained by Mironov et al.[383], who used a heterogeneous mixture which gave no reaction without shaking, and a constant reactivity ratio in the range of shaking speed employed so that diffusion control was not considered important. Reaction rates were obtained over the first 10 % of reaction at 85 °C (or at 50 ° for the most reactive compounds) and the relative rates of chloromethylation of the chloromethyl products were also obtained (Table 100). The scale of reactivities seems to be of the order of that obtained by Brown and Nelson, but there are significant differences as shown by the partial rate factors of $f_p^{Me} = 45.6, f_m^{Me} = 2.97$, and $f_p^{Me} = 84.6$ giving $\rho = -4.9$, and these differences may arise from the different solvents employed. Nevertheless, considerable doubt must remain and clearly the reaction merits further investigation.

Only one detailed kinetic study of chloromethylation involving measurement

TABLE 100

RELATIVE RATES OF CHLOROMETHYLYLATION OF ArH BY $CH_2O$ AND HCl AT 85 OR 50 °C[383]

| ArH | $k_{rel}$ | $k_{rel}$ (di-chloromethylation) |
|---|---|---|
| Benzene | 1 | 0.447 |
| Toluene | 30 | |
| o-Xylene | 46.5 | 2.56 |
| p-Xylene | 33.9 | 1.50 |
| m-Xylene | 1,220 | 11.7 |
| 1,2,4-Trimethylbenzene | 8,010 | 12.8 |
| Mesitylene | 4,800 | 320 |
| 1,2,3,4-Tetramethylbenzene | 1,500 | 1.0 |
| Ethylbenzene | 7.81 | 0.333 |
| Diphenylether | 70.3 | 14.4 |

of rate coefficients has been reported[384]. The chloromethylation of mesitylene in acetic acid (containing 10 % water) at 60 °C was followed by iodimetrically measuring the loss of formaldehyde and chloride ion. The reaction rate was given by

$$\text{Rate} = k_2[\text{ArH}][\text{HCHO}] \tag{190}$$

the first-order dependence on the concentration of aromatic aldehyde holding for a twofold variation in each, and the rate was independent of the chloride ion concentration. The logarithm of the rate coefficients (Table 101) plotted linearly

TABLE 101

RATE COEFFICIENTS FOR CHLOROMETHYLATION OF MESITYLENE BY FORMALDEHYDE-HYDROGEN CHLORIDE IN 90 VOL % ACETIC ACID AT 60 °C[384]

| $[HCl]$ (M) | 0.2 | 0.4 | 0.6 | 0.8 | 1.0 |
|---|---|---|---|---|---|
| $10^3 k_2$ | 0.17 | 0.56 | 1.28 | 1.88 | 3.21 |

against the acidity functions of the aqueous acetic acid–hydrochloric acid media with a slope of $-0.96$. In a medium containing 0.60 $M$ sulphuric acid, the rate coefficient ($10^3 k_2$) was 2.23, and 2.17 in the presence of 0.40 $M$ lithium chloride so that rate-determining attack by species such as $CH_2^+Cl$ is ruled out. The mechanism was evaluated from these data as

$$\text{HCHO} \overset{H^+}{\rightleftharpoons} CH_2^+OH \quad \text{(fast)} \tag{191}$$

$$\text{ArH} + CH_2^+OH \rightleftharpoons ArCH_2OH + H^+ \quad \text{(slow)} \tag{192}$$

$$ArCH_2OH \overset{HCl}{\longrightarrow} ArCH_2Cl \quad \text{(fast)} \tag{193}$$

Further evidence to support this mechanism was the fact that solutions of alcohol and hydrogen chloride of varying initial relative concentrations, gave, after five minutes, equilibrium mixtures of identical relative concentrations.

Some results which are consistent with this mechanism have been obtained by Ishii and Yamashita[385], who found that the kinetics of the reaction of *m*-xylene with formaldehyde and hydrogen chloride (to give the 4-substituted product) were third-order overall. However, this was followed by a slow di-chloromethylation which was of zeroth-order, but no interpretation or further mechanistic details are available.

### 6.2.2 *Bromomethylation*

The above mechanism for chloromethylation seems to be general for halo-methylation since bromomethylation gives the same *ortho* : *para* ratio for toluene, ethylbenzene, and *i*-propylbenzene, which is entirely in accord with the halogen being substituted in a non rate-determining step of the reaction[386].

### 6.3 ACYLATION

Kinetic studies of acylation, which are limited almost exclusively to the Lewis acid-catalysed reaction represented by

$$ArH + RCOX \xrightarrow{MX_n} ArCOR + HX \qquad (193a)$$

do not have the difficulties experienced in studies of alkylation with regard to isomerisation of reagent and products, and occurrence of consecutive reactions, but are complicated by uncertainty in the nature of the electrophile. This may be the free acylium ion $RCO^+$, or a polarised entity such as $(RCOCl)^{\delta+} MX_n^{\delta-}$, different structures for which have been proposed (for leading references see ref. 387); in addition, recent work has shown that isolable 1 : 1 complexes of the type $RCOF : MF_{3,5}^{\delta-}$ are reactive acetylating agents[388], and may themselves be the electrophile in these cases. Two further complications are the occurrence of complexing of the ketonic product with the catalyst, and the fact that acetylation gives greater steric hindrance than benzoylation contrary to expectations.

### 6.3.1 *Relative efficiency of catalysts and the kinetic order in catalyst concentration*

The first significant kinetic study of acylation was that of Steele[333a], who measured rates of benzoylation of toluene (excess) by benzoyl chloride catalysed by aluminium chloride at 25 °C, the reaction being followed by measuring the production of hydrogen chloride which was swept from the reaction mixture by a stream of hydrogen, and this technique has been found advantageous by later workers. One equivalent of aluminium chloride was found to catalyse the reaction of only one equivalent of benzoyl chloride, so that the ketone product must enter into a very stable combination with the catalyst, and this observation has been confirmed by many other workers, numerous examples being in the literature[389]. Using aluminium chloride: benzoyl chloride ratios of 0.65–2.3, at low ratios the reaction was found to be initially zeroth-order, this being attributed to formation of a 1 : 1 complex of the two reagents which are in equilibrium with the complex so that the latter is in constant concentration provided free reagent is available; when this ceases to be so, kinetics first-order in acyl chloride : aluminium chloride should occur as observed. At high ratios, the reaction became second-order in catalyst ($k_2 = 45 \times 10^{-5}$), and this was thought to be due to reaction between the above complex and one formed from toluene and aluminium chloride. When ferric chloride was used, kinetics second-order in catalyst ($k_2 = 17 \times 10^{-6}$) were obtained under all the conditions examined and this showed the greater effectiveness of aluminium chloride relative to ferric chloride. This work was repeated by Martin et al.[333b], who made a correction for the fact that the removal of the hydrogen chloride from the solution by hydrogen is a first-order process, and thus the rate is slower than assumed by Steele; these new results obtained, however, substantially confirmed the earlier observations. The catalytic effect of mixture of aluminium and ferric chlorides were found to reach a maximum at a concentration of the latter of > 50 %, which is surprising in view of the generally observed greater reactivity of the former when used alone.

The dependence of rate on an order of catalyst greater than one was also found by Menschutkin[390], who measured the rates of reaction of a range of aromatics with benzoyl chloride, catalysed by antimony trichloride or antimony tribromide at 155 °C, the former catalyst being less effective than aluminium chloride. The reaction was second-order in catalyst and the mechanism was incorrectly assumed to involve reaction of benzoyl chloride with a rapidly formed complex of aromatic and catalyst dimer. Rate coefficients ($10^4 k_2^*$ where $k_2^*$ is believed to be $k_4 [SbCl_3$ or $SbBr_3]^2$) were obtained as follows (data for the bromide in parenthesis), but the units are not known: benzene, 2.24 (7.10); toluene, 27.2 (104); ethylbenzene, 46.6 (58.9); i-propylbenzene, 51.0; i-amyl benzene, 52.0; p-xylene, 47.6 (140); o-xylene, 72.5 (272); m-xylene, 178 (94.0); p-methylpropylbenzene 89.0; mesitylene, 1,070, 1,2,4-trimethylbenzene, 3,160; chlorobenzene 0.22; bromobenzene 0.049; nitrobenzene, 0.045; data for di- and tri-phenylmethanes and also diphenyl was obtained but appeared to be affected by side reactions. This was the first demonstration of the electrophilic nature of the reaction and also the characteristic high steric hindrance as shown by the results for the halogenobenzenes; the selectivity of the reaction for some of the compounds also seemed to be markedly dependent upon the catalyst which may be an additional manifestation of steric hindrance. These results also showed the greater efficiency of antimony tribromide relative to the trichloride.

Dermer et al.[391] found the order of decreasing catalyst efficiency in the reaction of acetyl chloride with toluene was as follows: $AlCl_3 > SbCl_5 > FeCl_3 > TeCl_2 > SnCl_4 > TiCl_4 > TeCl_4 > BiCl_3 > ZnCl_2$. Many of the catalysts gave the maximum product yield at a catalyst : acetyl chloride ratio greater than 1 : 1, and the yield with some catalysts (especially Al, Mo, Fe and Sb halides) decreased with reaction time which again suggests complexing of product with catalyst. Premixing the catalyst and acid halide gave the same result[392], and the relative efficiency of bromides was: $AlBr_3 > (AlCl_3) > FeBr_3 > SbBr_3 > ZnBr_2 > TiBr_4 > TeBr_4 > MoBr_4 > WBr_4 > CdBr_2 > SnBr_4 > HgBr_2$; the reaction orders in catalyst were much the same as for the chlorides[393]; the greater efficiency of bromides noted by Menschutkin is thus confirmed. Calloway[336] found that aluminium fluoride was inert as a catalyst and this again implies that the reactivity of the aluminium halides increases down Group VII.

Some of the above qualitative orders do not correspond with the quantitative values observed by Jensen and Brown[394], who found relative rates at 25 °C for benzoylation of toluene, benzene, or chlorobenzene with benzoyl chloride excess to be: $SbCl_5$, 1,300; $FeCl_3$, 570; $GaCl_3$, 500; $AlCl_3$, 1.0; $SnCl_4$, 0.003; $BCl_3$, 0.0006; $SbCl_3$, "very small". These relative rates should, however, only be regarded as approximate in view of the difficulties associated with the kinetics as described below.

The aluminium chloride-catalysed reaction of benzene with excess benzoyl chloride gave simple second-order kinetics, viz.

$$\text{Rate} = k_{(1+n)}[\text{ArH}][\text{PhCO.Cl.AlCl}_3]^n \tag{194}$$

where $n = 1$, and the derived rate coefficients were independent of the initial concentration of reagents; some representative results are given in Table 102[395]. The kinetics departed from second-order when one mole of benzophenone had been obtained per mole of aluminium chloride, and this was ascribed to the greater basicity of benzophenone relative to benzoyl chloride such that the former preferentially complexes with the catalyst. The reaction was also highly sensitive to changes in the polarity of the solvent as shown by the rate coefficient obtained on adding 74 vol. % of cyclohexane (Table 102); intermediate volumes produced intermediate rate reductions.

With gallium chloride, ferric chloride and antimony pentachloride the rate coefficients were dependent upon the concentration of chlorobenzene and the square of the concentration of the catalyst, but the third-order coefficients varied with the initial concentration of the catalyst (Table 103)[394]. The overall kinetic equation was, therefore,

TABLE 102

RATE COEFFICIENTS FOR REACTION OF PhCOCl (EXCESS) WITH PhH, CATALYSED BY AlCl₃, AT 49.9 °C[395]

| $[AlCl_3]$ (M) | $[PhH]$ (M) | $10^4 k_2$ |
|---|---|---|
| 0.228 | 0.470 | 3.85(0.175)[a] |
| 0.223 | 0.699 | 3.77 |
| 0.220 | 0.904 | 3.70 |
| 0.470 | 0.235 | 3.85 |
| 0.452 | 0.452 | 3.77 |
| 0.915 | 0.475 | 3.77 |

[a] In the presence of 74 vol % of cyclohexane.

$$\text{Rate} = k_3 [MCl_n]^2 [ArH][MCl_n]_0^{-1} \tag{195}$$

(where $[MCl_n]_0$ is the initial concentration of catalyst) which should strictly be written as

$$\text{Rate} = k_3 [MCl_n.PhCOCl]^2 [ArH][MCl_n.PhCOCl]_0^{-1} \tag{196}$$

since the catalyst is complexed with benzoyl chloride. The squared terms in the kinetic equation probably arise from a tendency for this catalyst to form dimeric metal halide molecules since they are all known to do this; the dependence upon the initial concentration of catalyst probably derived from a tendency for the product to complex with the catalyst.

TABLE 103

RATE COEFFICIENTS FOR THE CATALYSED REACTION OF PhCOCl (EXCESS) WITH PhCl AT 25 °C[394]

| Catalyst | $[MCl_n]$ (M) | $[PhCl]$ (M) | $10^4 k_3$ | $10^4 k_5 [MCl_n]_0$ |
|---|---|---|---|---|
| GaCl₃ | 0.1014 | 0.305 | 26.8 | 2.72 |
|  | 0.202 | 0.304 | 12.0 | 2.42 |
|  | 0.302 | 0.303 | 8.13 | 2.45 |
|  | 0.510 | 0.305 | 4.89 | 2.48 |
|  | 0.301 | 0.152 | 8.21 | 2.47 |
| FeCl₃ | 0.278 | 0.278 | 10.8 | 3.00 |
|  | 0.454 | 0.454 | 5.83 | 2.67 |
| SbCl₅ | 0.181 | 0.181 | 36.5 | 6.57 |
|  | 0.208 | 0.208 | 30.4 | 6.31 |

Values of $10^4 k_3$ $[GaCl_3]_0$ at 35.0 and 52.4 °C were 8.12 and 21.6 respectively giving $E_a = 15.4$, $\log A = 7.68$, $\Delta H^\ddagger = 14.8$ and $\Delta S^\ddagger = -25.4$.

*References pp. 388–406*

For the reaction of stannic chloride with toluene (this aromatic being used here because of the lower effectiveness of the catalyst), different kinetics were obtained the rate expression being

$$\text{Rate} = k_2[\text{SnCl}_4]_0[\text{PhCH}_3] \tag{197}$$

which indicates that stannic chloride is not tied up by the ketone formed in the initial stages of reaction as is the case with other powerful catalysts. The lack of a squared term in catalyst concentration is consistent with the known lack of dimerisation of stannic chloride. At later stages in the reaction, as ketones are formed in higher concentrations, evidence suggested a change in the kinetic expression

$$\text{Rate} = k_2[\text{SnCl}_4][\text{PhCH}_3] \tag{198}$$

Since stannic chloride is known not to complex with benzoyl chloride, the true kinetic expression should, therefore, be

$$\text{Rate} = k_3[\text{SnCl}_4][\text{PhCH}_3][\text{PhCOCl}] \tag{199}$$

Values of $k_1$ (the first-order rate coefficient, *viz.* rate = $k_1$[aromatic]), $k_2$ and $k_3$ are given in Table 104, and the third-order coefficients show slightly greater consistency than the second-order coefficients.

The boron trichloride-catalysed benzoylation of toluene gave a similar kinetic form, the average value of $10^6 k_2$ being 3.17. Again $k_3$ values should strictly be calculated since there is no association between catalyst and benzoyl chloride.

The above results show consistency with the known properties of the catalysts except for aluminium chloride, the tendency of which to dimerize would lead one

TABLE 104

RATE COEFFICIENTS FOR REACTION OF PhCOCl (EXCESS) WITH PhCH₃, CATALYSED BY SnCl₄ AT 25 °C[394]

| [SnCl₄] (M) | [PhCH₃] (M) | [PhCOCl] (M) | $10^6 k_1$ | $10^6 k_2$ | $10^6 k_3$ |
|---|---|---|---|---|---|
| 0.199 | 0.405 | 8.03 | 3.28 | 16.5 | 2.05 |
| 0.396 | 0.396 | 7.85 | 6.12 | 15.5 | 1.97 |
| 0.882 | 0.391 | 7.36 | 11.1 | 12.6 | 1.72 |
| 0.337 | 0.591 | 7.73 | 5.23 | 15.6 | 2.02 |
| 0.399 | 0.207 | 8.02 | 6.76 | 17.0 | 2.12 |
| 0.398 | 0.935 | 7.36 | 5.57 | 14.0 | 1.90 |

At 39.9 and 49.9 °C, values of $10^6 k_3$ were 4.92 and 8.03 respectively giving $E_a = 10.9$, log $A = 2.3$, $\Delta H^{\ddagger} = 10.3$ and $\Delta S^{\ddagger} = -48.0$; no reason was given for this extremely large negative entropy of activation.

to expect that with this catalyst, the rate equation should involve the square of the catalyst concentration. The failure to observe this has led Brown to speculate that the extremely strong affinity of aluminium chloride for oxygen, successfully competes with its affinity for chlorine such that formation of the dimeric species fails to compete with complexing of catalyst with benzoyl chloride.

Though the kinetic results above can be rationalised by reasonable premises, one experimental observation is markedly at variance and this is that the percentage of *ortho*-benzoylation of toluene is constant under all conditions; clearly there is still much to be understood about the role of the catalyst in these reactions.

### 6.3.2 The mechanism of acylation and the nature of the acylating species

A large number of workers have made kinetic studies aimed at elucidating the mechanism of the acylation reactions. Olivier[396] studied the benzoylation of benzene (excess) by benzoyl chloride (or bromide) catalysed by aluminium chloride (or bromide) at 30 °C. The velocity of reaction increased with the concentration of acyl halide–aluminium halide complex but greater than the first power of this. Part of this increase in rate seemed to be due to an increase in solvent polarity, since addition of an equivalent amount of benzophenone–aluminium halide complex caused a similar increase in rate. Addition of excess aluminium halide gave a very large increase in the rate which was not a catalytic effect, whereas an excess of acyl halide did not; this may also be due to increase in medium polarity. In carbon disulphide as solvent, the reaction ceased after one mole of acyl halide per mole of catalyst had been used up, due to complexing of the catalyst with the ketone product. Ulich and Frogstein[397] repeated some of the experiments of Steele[333a] and Olivier[396] and confirmed the essentials of their findings. Thus when equivalent amounts of aluminium chloride and benzoyl chloride were used, the reaction with benzene (excess) was first-order in the complex, and a rate coefficient, $k_1$, of $2.8 \times 10^{-5}$ was obtained at 30 °C.

Neither of the above groups of workers made any detailed proposals concerning the nature of the acylating species. Lebedev[398], however, proposed that on the basis of UV spectra and heats of complex formation, the reactive species were $(2\,R_2CO.AlCl_2^+)(AlCl_4^-)$ and $(2\,RCOCl.AlCl_2^+)(AlCl_4^-)$ for the reactions of ketones and acyl halides respectively. Greenwood and Wade[399], however, concluded from a study of the molecular addition complexes formed from gallium chloride and acetyl or benzoyl chlorides, that, although these are powerful Friedel–Crafts catalysts and must be ionic since they are conducting in solution, there is no real supporting evidence for the precise structures proposed by Lebedev.

Tedder[400] proposed that dual mechanisms, analogous to those proposed for alkylation, may apply to acylation also. Thus substitution may occur *via* a free acylium ion, referred to as the ionic mechanism, or *via* nucleophilic displacement of the aromatic upon a polarised acyl halide–catalyst complex. Tedder suggested that only the less reactive aromatics would react *via* the latter mechanism, though the converse seems to be much more likely. A consequence of this duality is that by the former mechanism, electron supplying groups will decrease the reactivity of the electrophile through delocalisation of the positive charge, so that the reactivity order will be $PhCO^+ > MeCO^+$; however, by the latter mechanism the converse order would prevail since the polarisation of the carbon–halogen bond will be aided by electron supply.

There seems to be no doubt that in certain circumstances the free acylium ion is formed. For example, Baddeley and Voss[401] found that 2,4,6-tribromobenzoyl halides underwent halogen exchange in the acyl moiety, indicating the formation of the free acylium ion which was also inferred by the lack of steric hindrance when this ion benzoylated. They suggested that the variation in steric hindrance which accompanies change of solvent (for example in the $\alpha : \beta$ ratio for naphthalene)[402] arises from the degree of solvation of the acylium ion. Some relative rates of aluminium bromide-catalysed 2,4,6-tribromobenzoylation were determined as follows: naphthalene, 0.66; benzene, 1.0; *p*-xylene, 1.92; *m*-xylene, 2.56; mesitylene, 5.75. The spread of rates here is surprisingly small, indicating a very reactive electrophile, which is consistent with the free acylium ion, and the results for mesitylene and *m*-xylene show the relative lack of steric hindrance in the reaction.

Baddeley and Voss also showed that the aluminium halide-catalysed acylation by acid anhydrides proceeds *via* the initial formation of the acyl halide, equilibrium (200)

$$(RCO)_2O + AlX_3 \rightleftharpoons RCOX + RCOAlX_2 \tag{200}$$

since hydrolysis of the reaction mixture gave acyl halide together with carboxylic acid. It is difficult to reconcile this conclusion, however, with the results of Man and Hauser[403], who found that in the reaction with bromobenzene, acetic anhydride was a *more* reactive acylating agent than acetyl chloride, whilst with toluene the reverse was true. This implies that reaction may occur before the establishment of equilibrium (200) and the extent to which this takes place will depend on the reactivity of the substrate. However, since partial rate factors for the unhindered *para* position were not obtained, it is difficult to ascertain the extent to which steric hindrance affects these results; if there is substantial steric hindrance then the free acylium ion is unlikely to be the electrophile here. The relative reactivities of acyl chlorides with toluene were found to be acetyl > benzoyl > 2-ethylbutyryl, and the order of the first two acyl groups was the same for the corresponding derivatives of 4-nitrophenol, *i.e.* 4-nitrophenyl acetate > 4-nitrophenyl benzoate. These results (which have been confirmed with respect to the relative ease of benzoylation and acetylation) are at variance with the expected reverse order in view of the greater selectivity of acetylation (see below).

Gore[404] suggested that the ionic mechanism was only important when the aromatic or the acyl halide has a high steric requirement, and this argument was based upon the results of Baddeley and Voss[401] in which naphthalene was less reactive than benzene in acylation it being assumed that the low reactivity was due to the occurrence of $\beta$ substitution only. This argument is unsoundly based, however, because the $\beta$ position should still be more reactive than benzene, and, if the $\alpha$ position was hindered, the reactivity of mesitylene should have been markedly reduced which was not observed; the abnormal reactivity of naphthalene is, there-fore, likely to arise from some cause other than a change in mechanism.

A further kinetic study of the effect upon the reactivity of acyl halides of varying the acyl group has been carried out by Yamase[405]. For the aluminium chloride-catalysed reaction of acetyl halides with mesitylene in carbon disulphide the order was the same as that first observed by Calloway[336], viz. MeCOI > MeCOBr > MeCOCl > MeCOF. Replacement of the methyl group by ethyl, i-propyl and t-butyl gave the respective orders: I < Br > Cl > F, I < Br $\approx$ Cl > F and I < Br < Cl < F. For the corresponding cyclohexoyl and benzoyl halides the order was I < Br $\approx$ Cl > F, and these results have been interpreted as showing effect of steric hindrance in causing a change from the nucleophilic displacement mechanism to the ionic mechanism as formation of the ionised complex becomes more difficult, assuming, as is most probable (see p. 171) that the catalyst bonds to oxygen rather than halogen. Yamase argued that, in contrast to the accepted mechanism for alkylation, both the ionic and substitution mech-anisms for acylation give the reactivity order I > Br > Cl > F, and this may be true since the catalyst bonds to oxygen and not to halogen as is the case of the alkyl halides. The change in reactivity sequence can then be rationalised only if the reaction via the ionic mechanism proceeds at a slower rate than reaction by the displacement mechanism. Steric hindrance will arise not only from the size of the halogen, but also from the size of the alkyl component of the acyl group and, of course, from the aromatic. Steric hindrance would not be such an important factor for alkylation in view of the different point of attachment of the catalyst, the carbon–oxygen double bond being shorter than the carbon–halogen bond.

Yamase and Goto[406] determined first- and second-order rate coefficients for the aluminium chloride-catalysed reaction of halide derivatives of benzoic acid ($10^5 k_1$ = F, 1.73; Cl, 4.49; Br, 4.35; I, 0.81) and phenylacetic acid ($10^5 k_2$ = F, 12; Cl, 21; Br, 9; I, 6) with benzene. The maxima in the rates for the acid chloride are best accommodated by the assumption that a highly (but not completely) polarised complex takes part in the transition state. Polarisation of such a complex would be aided by electron supply, and consistently, the acetyl halides are about a hundred times as reactive as the benzoyl compounds (see p. 180, also Tables 105 and 108).

Brown and Jensen[395] suggested that the rate equation (194) for the reaction of benzene with excess benzoyl chloride could be interpreted according to the mechanisms given by the reactions (201) and (202), (203) and (204) and (205) and (206) which refer to nucleophilic attack of the aromatic upon the polarised acyl halide-catalyst complex, upon the free acylium ion, and upon an ion pair derived from the acyl halide-catalyst complex, viz.

$$
\overset{+}{ArH} + \underset{\underset{Cl}{|}}{RC\!=\!\overset{+}{O}\!-\!AlCl_3^-} \underset{}{\overset{k_1'}{\rightleftharpoons}} \underset{\underset{Cl}{|}}{\overset{+}{ArH}\!-\!\underset{\overset{|}{R}}{C}\!-\!OAlCl_3^-} \tag{201}
$$

$$
\underset{\underset{Cl}{|}}{\overset{+}{ArH}\!-\!\underset{\overset{|}{R}}{C}\!-\!OAlCl_3^-} \overset{k_2'}{\rightarrow} \underset{\overset{|}{R}}{ArC}\!=\!\overset{+}{O}\!-\!AlCl_3^- + HCl \tag{202}
$$

$$
\underset{\underset{Cl}{|}}{\overset{+}{RC}\!=\!\overset{+}{O}\!-\!AlCl_3} \underset{k_{-1}''}{\overset{k_1''}{\rightleftharpoons}} \overset{+}{RCO} + AlCl_4^- \tag{203}
$$

$$
\overset{+}{RCO} + ArH \underset{k_{-2}''}{\overset{k_2''}{\rightleftharpoons}} \overset{+}{ArHCOR} \underset{AlCl_4^-}{\overset{k_3''}{\longrightarrow}} \underset{\overset{|}{R}}{ArC}\!=\!\overset{+}{O}\!-\!AlCl_3^- + HCl \tag{204}
$$

$$
\overset{+}{RCO}\,AlCl_4^- \ (\text{ion pair}) + ArH \underset{k_{-2}'''}{\overset{k_2'''}{\rightleftharpoons}} \overset{+}{ArHCOR} + AlCl_4^- \tag{205}
$$

$$
\overset{+}{ArHCOR} \underset{AlCl_4^-}{\overset{k_3'''}{\longrightarrow}} \underset{\overset{|}{R}}{ArC}\!=\!\overset{+}{O}\!-\!AlCl_3^- + HCl \tag{206}
$$

Thus, reactions (201) and (202) give the required kinetics if either of the steps with rate coefficients $k_1'$ or $k_2'$ are rate-determining, reactions (203) and (204) would do so if the final step ($k_3''$) were rate-determining (but would give Rate = $[PhCOCl.AlCl_3]^{\frac{1}{2}}[ArH]$ if step $k_2''$ were rate-determining) and reactions (205) and (206) would do so if steps $k_2'''$ or $k_3'''$ were rate-determining. Some of these possibilities can be ruled out by the result that saturation of the reaction mixture with hydrogen chloride did not affect the rate so that the last steps cannot be rate-determining.

Subsequently rates of benzoylation of a range of aromatics were determined under the same conditions (Table 105)[407]. The high negative entropy of activation is consistent with the high degree of ordering required for the polarised acyl chloride–aluminium chloride complex to be the electrophile.

TABLE 105

RATE COEFFICIENTS ($10^4 k_2$) AND KINETIC PARAMETERS FOR REACTION OF PhCOCl (EXCESS) WITH ArH CATALYSED BY $AlCl_3$[407]

| ArH | Temp. ($^\circ C$) | | | | | $E_a$ | $log A$ | $\Delta H^\ddagger$ | $\Delta S^\ddagger$ |
| | 0 | 15 | 25 | 39.9 | 49.9 | | | | |
|---|---|---|---|---|---|---|---|---|---|
| Benzene | | | 0.495 | 1.67 | 3.78 | 15.8 | 7.28 | 15.1 | −27.2 |
| Toluene | 7.16 | 23.8 | 54.3 | | | 13.3 | 7.50 | 12.6 | −26.7 |
| t-Butylbenzene | | | 35.8 | | | | | | |
| Chlorobenzene | | | $5.66 \times 10^{-3a}$ | | | | | | |
| o-Xylene | | | 555 | | | | | | |
| m-Xylene | | | 1,950 | | | | | | |
| p-Xylene | | | 69.2 | | | | | | |

[a] Calculated from data at 70 °C, assuming $\Delta S^\ddagger = -27$; this gives $k_{rel} = 0.0114$, cf. 0.0260 obtained by direct competition in nitrobenzene at 25 °C, the latter yielding $f_o^{Cl} = 0.00328$, $f_m^{Cl} = 0.00016$, and $f_p^{Cl} = 0.150$[408a].

In exceptional circumstances the acylium ion (or the polarised complex) can decompose to give an alkyl cation so that alkylation accompanies acylation. This occurs in the aluminium chloride-catalysed reaction of pivaloyl chloride which gives acylation with reactive aromatics such as anisole, but with less reactive aromatics such as benzene, the acylium ion has time to decompose, viz.

$$MeCCO^+ \rightleftharpoons MeC^+ + CO \qquad (207)$$

before reacting, and does so because of the high stability of the t-butyl cation; this stability was increased by an increasing number of alkyl groups in the tertiary cation so that the ratio of alkylation to acylation increased dramatically here. No simple kinetic order was apparent in the reaction of pivaloyl chloride with anisole; the usual electrophilic substitution reactivity order was obtained however, i.e. anisole > toluene > benzene[409].

### 6.3.3 The kinetic effect of solvents

Brown and Young[410,411] studied the benzoylation of benzene and toluene by benzoyl chloride catalysed by aluminium chloride at 25 °C in nitrobenzene as solvent. For a given concentration of reagents, the rate equation was

$$Rate = k_{(2+n)}[ArH][PhCOCl][AlCl_3]^n \qquad (208)$$

but the rates decreased in magnitude with increasing concentration of aluminium chloride. For benzene, $n = 0.9$, and for toluene $n = 1.4$, giving overall orders of approximately 3 and 3.5. These precise orders were obtained by assuming that if all the reactants have the same concentration, then $-dc/dt = kc^n$, whence $\log(-dc/dt) = \log k + n \log c$. The rate was measured in terms of loss of benzoyl chloride with time and the slope of the resultant plot gave values of 2.9 and 3.4 for reaction with benzene and toluene respectively. No reason for the variation in kinetic order could be found and in order to determine relative rates of benzoylation of benzene and toluene, rates were calculated for the former compound as for a reaction of order 3.5; this gave kinetic plots linear to about 50 % of reaction and a relative reactivity of *ca.* 150 from which partial rate factors were determined (refs. [410, 411]) as follows: $f_o^{Me}$, 32.6; $f_m^{Me}$, 5.0; $f_p^{Me}$, 83.1. Some typical kinetic data obtained in this study are given in Table 106 and the rather unsatisfactory kinetic order for toluene is clearly apparent. These studies in nitrobenzene were extended to measurements of a wide range of alkyl and polyalkylbenzenes, the rate coefficients with initial (and equal) concentrations of reagents being given in Table 107[342, 412].

Smeets and Verhulst[413] obtained cleanly second-order kinetics, equation (194) where $n = 1$ in the acylation of aromatics using bromobenzene as solvent

TABLE 106

RATE COEFFICIENTS FOR THE REACTION OF PhCOCl WITH PhH AND PhMe IN PhNO$_2$ CATALYSED BY AlCl$_3$, AT 25 °C[410, 411]

| [PhH] (M) | [PhCOCl] (M) | [AlCl$_3$] (M) | $10^5 k_3$ | $10^5 k_{3.5}$ |
|---|---|---|---|---|
| 0.199 | 0.398 | 0.397 | 5.43 | |
| 0.398 | 0.400 | 0.401 | 5.64 | 10.4 |
| 0.409 | 0.402 | 0.401 | 5.67 | 10.1 |
| 0.395 | 0.400 | 0.401 | 5.64 | |
| 0.795 | 0.403 | 0.402 | 5.38 | |
| 0.406 | 0.203 | 0.402 | 6.38 | |
| 0.394 | 0.594 | 0.397 | 5.93 | |
| 0.397 | 0.399 | 0.200 | 8.37 | |
| 0.400 | 0.441 | 0.800 | 3.50 | |
| 1.162 | 0.395 | 0.777 | 3.43 | |
| 0.608 | 0.208 | 0.403 | 6.51 | |
| 0.598 | 0.606 | 0.598 | 4.13 | |

| [PhMe] (M) | [PhCOCl] (M) | [AlCl$_3$] (M) | $10^5 k_3$ | $10^5 k_{3.5}$ |
|---|---|---|---|---|
| 0.400 | 0.401 | 0.400 | 588 | 1,550 |
| 0.400 | 0.401 | 0.401 | 642 | 1,680 |
| 0.402 | 0.400 | 0.401 | 593 | 1,480 |
| 0.404 | 0.203 | 0.398 | 825 | |
| 0.610 | 0.204 | 0.399 | 784 | |
| 0.398 | 0.400 | 0.801 | 496 | |
| 1.213 | 0.406 | 0.810 | 480 | |
| 0.200 | 0.200 | 0.198 | 875 | 3,100 |

TABLE 107

AVERAGE RATE COEFFICIENTS ($10^2 k_{3.5}$) FOR REACTION OF ArH WITH PhCOCl IN PhNO$_2$
CATALYSED BY AlCl$_3$ AT 25 °C[342,412]

| ArH | Initial concentrations (M) of each reagent | | |
| --- | --- | --- | --- |
| | 0.400 | 0.200 | 0.100 |
| Benzene | 0.0103 | | |
| Toluene | 1.595 | 3.67 | 8.20 |
| Ethylbenzene | | 3.08 | |
| i-Propylbenzene | | 2.98 | |
| t-Butylbenzene | | 2.58 | |
| o-Xylene | | 32.8 | |
| m-Xylene | | 92.6 | 210 |
| p-Xylene | | 3.38 | |
| 1,2,3-Trimethylbenzene | | | 710 |
| 1,2,4-Trimethylbenzene | | | 405 |
| 1,3,5-Trimethylbenzene | | | 6,650 |
| 1,2,3,4-Tetramethylbenzene | | | 1,885 |
| 1,2,3,5-Tetramethylbenzene | | | 11,100 |
| 1,2,4,5-Tetramethylbenzene | | | 585 |
| Pentamethylbenzene | | | 7,400 |

(Table 108). Their data gave a toluene : benzene reactivity ratio of 53 at 40 °C. Equation (194), $n = 1$, was also found by Jensen[414] to apply to the benzoylation of toluene and p-xylene by benzoyl chloride catalysed by aluminium chloride in ethylene dichloride as solvent, though for the reaction with naphthalene, third-

TABLE 108

RATE COEFFICIENTS AT 40 °C AND ACTIVATION ENERGIES FOR THE AlCl$_3$-CATALYSED
REACTION OF RCOCl WITH ArH[413].

| R | ArH | $10^4 k_2$ | $E_a^a$ |
| --- | --- | --- | --- |
| PhCH$_2$ | PhBr | 1.83 | 13.5 |
| PhCH$_2$ | PhCl | 2.41 | 12.8 |
| Me | PhBr | 0.916 | 14.3 |
| Me | PhCl | 1.29 | 13.5 |
| Et | PhBr | 0.565 | 14.8 |
| Et | PhCl | 0.797 | 12.9 |
| n-Pr | PhBr | 0.449 | 14.0 |
| n-Pr | PhCl | 0.797 | 12.9 |
| Ph | PhBr | 0.00296 | 19.4 |
| Ph | PhCl | 0.00258 | 20.4 |
| Ph | PhMe | 19.6 | 15.0 |
| Ph | PhH | 0.373 | 13.4 |
| PhCH=CH | PhMe | 4.92 | 15.4 |
| PhCH=CH | PhH | 0.155 | 12.7 |

<sup>a</sup> Neither the temperature ranges employed for these determinations, nor the corresponding individual rate coefficients were quoted.

*References pp. 388–406*

order kinetics, equation (194) where $n = 2$, were obtained. However, both the second- and third-order rate coefficients were dependent upon the initial concentration of the benzoyl chloride–aluminium chloride complex. Also, addition of excess benzoyl chloride did not affect the rate or isomer distribution for toluene or $p$-xylene, whereas it reduced both the rate and the $\alpha : \beta$ ratio for naphthalene. Thus with the initial concentration of reagents = 0.222 $M$, the rate coefficient ($10^3 k_2$) was constant at 1.12 for 0–0.45 $M$ added benzoyl chloride in the reaction with toluene, and constant at $ca$. 2.1 for 0–0.273 $M$ added benzoyl chloride in the reaction with $p$-xylene. For naphthalene, however, the third-order rate coefficient ($10^2 k_3$) for the $\alpha$ position decreased from 2.0 to 1.0 and for the $\beta$ position de-

TABLE 109

RATE COEFFICIENTS FOR REACTION OF PhCOCl WITH ArH, CATALYSED BY AlCl$_3$, IN VARIOUS SOLVENTS AT 25 °C[415,417]

| Solvent | [PhCOCl.AlCl$_3$] (M) | [excess PhCOCl] (M) | [ArH] (M) | $10^3 k_2$ |
|---|---|---|---|---|
| 1,2,4-Trichlorobenzene | 0.0946 | 0 | 0.0942(toluene) | 0.18 |
| | 0.0906 | 0 | 0.276 | 0.18 |
| | 0.275 | 0 | 0.275 | 0.30 |
| | 0.276 | 0.203 | 0.277 | 0.28 |
| | 0.274 | 0 | 0.0992 | 0.30 |
| | 0.642 | 0 | 0.277 | 0.50 |
| | 0.312 | 0 | 0.156 | 0.420[a] |
| | 0.310 | 0 | 0.617 | 0.422[a] |
| | 0.0242 | 0 | 0.311 | 0.200[a] |
| | 0.105 | 0 | 0.311 | 0.268[a] |
| | 0.500 | 0 | 0.311 | 0.522[a] |
| | 1.024 | 0 | 0.312 | 0.925[a] |
| 1,2-Dichlorobenzene | 0.0949 | 0 | 0.289 | 0.43 |
| | 0.293 | 0 | 0.293 | 0.49 |
| | 0.677 | 0 | 0.289 | 0.78 |
| Dichloromethane | 0.0973 | 0 | 0.289 | 0.63 |
| | 0.307 | 0 | 0.307 | 0.78 |
| | 0.312 | 0.204 | 0.312 | 0.85 |
| | 0.307 | 0 | 0.682 | 0.77 |
| 1,2-Dichloroethane | 0.100 | 0 | 0.1 | 0.763 |
| | 0.100 | 0 | 0.3 | 0.771 |
| | 0.100 | 0.200 | 0.1 | 0.776 |
| | 0.200 | 0 | 0.1 | 0.960 |
| | 0.300 | 0 | 0.1 | 1.08 |
| | 0.300 | 0 | 0.3 | 1.09 |
| | 0.100 | 0 | 0.1 | 0.0874(0 °C)[b] |
| | 0.100 | 0 | 0.1 | 1.59(35 °C)[b] |
| | 0.300 | 0 | 0.3(benzene) | 0.00928 |
| | 0.100 | 0 | 0.1($o$-xylene) | 9.09 |
| | 0.100 | 0 | 0.1($m$-xylene) | 25.8 |
| | 0.100 | 0 | 0.1($p$-xylene) | 1.59 |

[a] Data refers to use of AlBr$_3$ and PhCOBr.
[b] Yields $E_a = 13.9$, $\log A = 7.08$, $\Delta H^\ddagger = 13.35$, $\Delta S^\ddagger = -28.1$.

creased from 0.72 to 0.61 on adding 0.4 $M$ benzoyl chloride, the implication here being that the size of the attacking reactant increases considerably in the presence of excess benzoyl chloride, though the reason for this is obscure.

A quantitative study of the kinetic effect of different solvents has been carried out by Brown et al.[415]. Rate coefficients were found to increase along the series: 1,2,4-trichlorobenzene, 1,2-dichlorobenzene, methylene dichloride, ethylene dichloride, for the aluminium chloride-catalysed reaction of benzoyl chloride with toluene, the relation reactivities being the ratio 1 : 2.4 : 3.5 : 4.1 (: 30 for excess benzoyl chloride as solvent). The reaction was first-order in toluene, first-order in the benzoyl chloride–aluminium chloride complex and the rate was independent of the concentration of excess benzoyl chloride. However, rate coefficients increased linearly with the initial concentration of the complex, and this was presumed to be a salt effect arising from the polar nature of the complex. Some representative rate coefficients are given in Table 109. For the aluminium bromide-catalysed reaction of benzoyl bromide in 1,1,4-trichlorobenzene the same kinetic form was observed and rates (Table 109) were ca. 20 % higher than for reaction involving the chloro compounds. With aluminium bromide in excess over benzoyl bromide, the rate increased dramatically as previously observed by Olivier[397] using the chloro compounds (p. 171) and the kinetics approached third-order, being second-order in catalyst though the reason for this was not given. The data in Table 110 show that the rate increases ca. 300-fold and 800-fold for the addition of 1- and 2-molar excess of aluminium bromide.

TABLE 110

EFFECT OF EXCESS AlBr₃ ON THE AlBr₃-CATALYSED REACTION OF PhCOBr WITH TOLUENE IN 1,2,4-TRICHLOROBENZENE AT 25 °C[415]

| [PhCOBr: AlBr₃] (M) | [PhCH₃] (M) | [excess AlBr₃] (M) | $10^3k_2$ | $10^3k_3$ |
|---|---|---|---|---|
| 0.309 | 0.309 | 0 | 0.405 | |
| 0.300 | 0.300 | 0.154 | 6.05 | |
| 0.309 | 0.309 | 0.309 | | 128 |
| 0.311 | 0.310 | 0.465 | | 225 |
| 0.309 | 0.309 | 0.617 | | 327 |

An explanation for the effect of excess catalyst has been offered by Corriu et al.[416], who measured the rates of the aluminium chloride-catalysed reaction of benzoyl chloride with benzene, toluene, and o-xylene. The observed rate coefficients were analysed in terms of a mixture of second- and third-order reactions (the latter being second-order in the halide–catalyst complex), the following results being obtained: benzene (40 °C), $k_2 = 2.5 \times 10^{-5}$, $k_3 = 3.3 \times 10^{-5}$; toluene (2.5 °C), $k_2 = 0.75 \times 10^{-4}$, $k_3 = 3.83 \times 10^{-4}$; o-xylene (0 °C), $k_2 = 1.83 \times 10^{-3}$, $k_3 = 4.50 \times 10^{-3}$. They suggest the equilibrium

$$2 \; \overset{+}{\underset{\underset{\text{Cl}}{|}}{\text{Ph.C=O}}}\text{–AlCl}_3^- \; \rightleftharpoons \; \overset{+}{\underset{\underset{+ \, \text{Cl–AlCl}_3^-}{|}}{\text{Ph.C=O}}}\text{–AlCl}_3^- + \text{PhCOCl}$$

$$\text{(X)} \hspace{4cm} \text{(XI)}$$

(209)

is set up, and that the species (X) and (XI) give rise to the kinetic orders of 2 and 3 respectively, in other words the concentration of (XI) is increased by the addition of excess aluminium chloride.

The use of ethylene dichloride as solvent was extended by Brown et al.[417] to the determination of the kinetics of benzoylation of other aromatics, using benzoyl chloride catalysed by aluminium chloride, and the data are included in Table 109; the relative reactivities are thus: benzene, 1.0; toluene, 117; o-xylene, 1,393; m-xylene, 3,960; and p-xylene, 243 and these values are closely similar to those obtained with nitrobenzene as solvent. No exact comparison of the coefficients with those of Corriu et al.[416] is possible because of the different temperatures employed, but the rates appear to be comparable for the two sets of data after allowing for reasonable temperature dependencies.

Using 1,2-dichloroethane as solvent, Brown et al.[418] have also studied the acetylation reaction, with acetyl chloride and aluminium chloride as reagents at 25 °C. The appropriate data for benzene are given in Table 111 and by comparison with Table 109 it appears that acetylation occurs some 300 times as fast as benzoylation.

### TABLE 111

RATE COEFFICIENTS FOR REACTION OF PhH WITH MeCOCl, CATALYSED BY $\text{AlCl}_3$, IN $\text{C}_2\text{H}_4\text{Cl}_2$ AT 25 °C[418]

| $[PhH]$ $(M)$ | $[MeCOCl.AlCl_3]$ $(M)$ | $[excess \; MeCOCl]$ $(M)$ | $[C_2H_4Cl_2]$ $(M)$ | $10^3 k_2$ |
|---|---|---|---|---|
| 0.120 | 0.200 | 0.020 | 0.060 | 2.47 |
| 0.200 | 0.200 | 0.020 | 0.100 | 2.65(av.)0.339 (0 ° C)[a] |
| 0.400 | 0.200 | 0.020 | 0.200 | 2.54 |
| 0.200 | 0.200 | 0.040 | 0.100 | 3.31[b] |
| 0.200 | 0.200 | 0.100 | 0.100 | 2.84 |
| 0.200 | 0.200 | 0.200 | 0.100 | 2.80 (av.) |

[a] $E_a = 13.3$, $\log A = 7.33$, $\Delta H^{\ddagger} = 12.7$, $\Delta S^{\ddagger} = -27.6$.
[b] High value attributed to decomposition of the complex.

Relative rates of aluminium chloride-catalysed acetylation of a range of aromatics with acetyl chloride at 0 °C had earlier been measured by McDuffie and Dougherty[419] as follows: 4-chlorotoluene, 0.0137; bromobenzene, 0.0242; chlorobenzene, 0.0314; 2-chlorotoluene, 0.271; benzene, 1.0; toluene, 13.3; mesitylene, 27.1; *m*-xylene, 100. Comparison of these results with the benzoylation data of Menschutkin[390] (p. 167) shows very similar relative reactivities for toluene and *m*-xylene, whereas those for mesitylene indicated that acetylation is far more sterically hindered than benzoylation and this has been noted by later workers. A possible explanation is that the more electron-withdrawing phenyl group (relative to methyl) will result in the carbonyl carbon atom possessing a higher positive charge in the benzoyl halide, hence there will be less need for participation of the catalyst in the acyl halide–metal halide complex in the rate-determining transition state, *i.e.* the reacting species is more nearly the benzoyl cation in benzoylation than the acetyl cation in acetylation.

The work of Brown *et al.*[418] gives the relative reactivity of toluene to benzene as 125–144, which increased as the concentration of reactants increased. This value is much higher than that obtained by McDuffie and Dougherty[419] above, and by Ogata and Oda[420] who found a value of 8.35. The differences were attributed[418] to the reaction in the earlier work (which was in heterogeneous solution) taking place in the highly polar catalyst phase. Partial rate factors were determined as $f_o^{Me} = 4.5$, $f_m^{Me} = 4.8$, and $f_p^{Me} = 749$, which may be compared with the corresponding values for benzoylation of 32.6, 5.0, 831 (in nitrobenzene) 30.7, 4.8, 589 (in excess benzoyl chloride) and 32.6, 4.9, 626 (in dichloroethane). This leads to an immediate contradiction that the faster reaction is also the most selective, no reason for this being apparent, and also confirms the greater steric hindrance to acetylation. This difference in steric requirement was rationalised[419] by making the unreasonable assumption that the phenyl ring is constrained in a conformation which gives less hindrance than the methyl group, *i.e.* it is not rotating, and the van der Waals radius of 1.7 A would be less than that for the methyl group which is 2.0 A when freely rotating; a large difference in entropy of activation should then be apparent and this was not so. As noted above, the length of the oxygen–catalyst bond is probably a much more important factor, and in view of the elec-

TABLE 112

PARTIAL RATE FACTORS FOR ACETYLATION AND BENZOYLATION[421]

| | Acetylation | | | Benzoylation | | |
|---|---|---|---|---|---|---|
| ArH | $f_o$ | $f_m$ | $f_p$ | $f_o$ | $f_m$ | $f_p$ |
| PhMe | 29.5 | 4.9 | 633 | 4.5 | 4.8 | 749 |
| PhEt | 10.9 | 10.3 | 563 | 1.0 | 10.4 | 753 |
| Ph-*i*-Pr | 8.6 | 11.1 | 519 | | 11.5 | 745 |
| Ph-*t*-Bu | 0 | 11.4 | 398 | | 13.0 | 658 |
| Ph.Ph | | | 245[a] | | | 248 |
| Fluorene (2 position) | | | 38,000[b] | | | |

[a] $10^3 k_2$ for biphenyl = 0.568.
[b] Determined by a competition method.

*References pp. 388–406*

TABLE 113

RELATIVE RATES FOR ACETYLATION OF ArH BY MeCOCl, CATALYSED BY AlCl$_3$, IN C$_6$H$_4$Cl$_2$ AT 25 °C[422]

| ArH | $k_{rel}$ | ArH | $k_{rel}$ | ArH | $k_{rel}$ |
|---|---|---|---|---|---|
| Benzene | 1 | Mesitylene | 2,920 | 1,2,3,4-Tetramethyl-benzene | 7,300 |
| Toluene | 128 | 1,2,3-Trimethylbenzene (4 position) | 1,740 | Pentamethylbenzene | 13,200 |
| o-Xylene | 2,130 | 1,2,3-Trimethylbenzene (5 position) | 6,530 | Fluorobenzene[a] | 0.252 |
| m-Xylene (4 position) | 335 | 1,2,4-Trimethylbenzene (5 position) | 1,760 | Chlorobenzene[b] | 0.0209 |
| m-Xylene (5 position) | 8.7 | 1,2,4,5-Tetramethyl-benzene | 102 | Bromobenzene[c] | 0.0140 |
| p-Xylene | 23.5 | 1,2,3,5-Tetramethyl-benzene | 7,430 | | |

[a] Giving $f_p^F = 1.51$.
[b] Giving $f_m^{Cl} = 0.0003$, $f_p^{Cl} = 0.125$.
[c] Giving $f_p^{Br} = 0.084$.

tron-supplying properties of methyl relative to phenyl this would be expected to be shorter in the acetyl intermediate compared to the benzoyl intermediate.

The greater steric hindrance to acetylation was also shown by a comparison of the rate of $(10^3 k_2)$ of acetylation of toluene (0.763), ethylbenzene (0.660), i-propylbenzene (0.606) and t-butylbenzene (0.462) with those (determined by the competition method) for benzoylation; both sets of data (Table 112) were obtained with dichloroethane as solvent at 25 °C, all reagent concentrations being 0.1 $M$[421]. Relative rates of acylation other aromatics under the same conditions have also been obtained and are given in Table 113[422]. The different steric requirements for acetylation and benzoylation are further shown by the following respective relative rates for acylation of naphthalene derivatives in chloroform at 0 °C: naphthalene (1 position) 1.00, 1.00, (2 position) 0.31, 0.04; 2,3-dimethylnaphthalene (1 position) 1.59, 172, (5 position) 7.14, 38.2, (6 position) 3.68, 7.7[422a].

### 6.3.4 Use of kinetic studies of acylation as a measure of electrophilic reactivities

The acetylation reaction has been used by a number of workers for this purpose. Using acetyl chloride with aluminium chloride as catalyst in carbon disulphide or 1,2-dichloroethane, Cram et al.[423] measured the relative rates of acetylation of 2,2- (> 29), 3,4- (11.2), 4,4- (1.6), and 6,6-paracyclophanes (1.0), and the enhanced reactivity of the 2,2 compound was attributed to transannular stabilisation of the electron-deficient transition state. Goldfarb et al.[424] obtained the relative acetylation rates of aromatics using acetyl chloride catalysed by stannic chloride at 0.3 °C in chlorobenzene as follows: 1,3-dimethoxybenzene > 2,5-dimethylthiophene > 2-methylthiophene > 2,5-diethylthiophene > thiophene > anisole > 2-bromothiophene > *m*-xylene > *o*-xylene > *p*-xylene > toluene > benzene. Novikova[425] measured rates of benzoylation using benzoyl chloride (and *o*, *m*, and *p*-iodobenzoyl chlorides) catalysed by aluminium chloride and obtained the reactivity sequence at 50 °C: naphthalene > toluene > benzene > fluoro- > iodo- > bromo- > chloro- > diiodo- > dibromo-benzene; the order for the halogenobenzenes differs from that for acetylation given in Table 113 presumably due to the different steric requirement. Goodman and Gore measured by the competition method the relative rates of benzoylation of dichlorobenzenes in nitrobenzene at 25 °C which were: *ortho*, 3 position, $1.48 \times 10^{-5}$, 4 position, $1.59 \times 10^{-4}$; *meta*, 2 position, $1.23 \times 10^{-4}$, 4 position, $4.95 \times 10^{-4}$, 5 position $< 2 \times 10^{-6}$; *para*, $2.8 \times 10^{-8}$; with the exception of the 4 position of *m*-xylene, agreement with values calculated from the data in the footnote of Table 111 was poor, with no very obvious reasons[425a].

Gore *et al.*[426] have used chloroform as a solvent for acetylation catalysed by aluminium chloride and at 45–55 °C find that a 2-methoxy substituent in naphthalene increases the reactivity of the 1 position 1.72 times, of the 6 position 3.8 times, and of the 8 position, 0.9 times; the former and latter of these results indicate a considerable steric effect. Likewise, a 2-bromo substituent caused the reactivity of the 6 and 8 positions to be 0.63 and 0.58 times that of the corresponding positions in the unsubstituted compound. At 20–25 °C the relative reactivities of some polycyclics were as follows[427]: 1-naphthyl, 1.0; 3-phenanthryl 0.64; 9-phenanthryl, 0.02; 1-phenanthryl, 0.29; 2-naphthyl, 0.28; 2-phenanthryl, 0.12; 4-phenanthryl, 0.0085. Some of these results seem to be due to steric hindrance, and the large difference in reactivity of naphthalene and biphenyl seems erroneous.

TABLE 114

RELATIVE REACTIVITIES OF REACTION OF FERROCENE DERIVATIVES WITH MeCOCl AND AlCl$_3$ AT 0 °C[428]

| Compound | Position of substitution | | | |
| | *p* | *1'* | *2* | *3* |
|---|---|---|---|---|
| 1,1'-Diphenylferrocene | 0.22 | | 0.71 | 0.43 ⎫ |
| 1,1'-Bis-*p*-methoxyphenylferrocene | | | 1.95 | 1.0  ⎬ a |
| 1,1'-Bis-*p*-bromophenylferrocene | | | 0.33 | 0.30 ⎭ |
| Phenylferrocene | 0.23 | 1.00 | 0.77 | 0.47 ⎫ |
| *p*-Methoxyphenylferrocene | | 1.00 | 1.16 | 0.60 ⎬ b |
| *p*-Ethylphenylferrocene | | 1.00 | 1.40 | 4.2  ⎭ |

Sets a and b are not related to each other; for set a the reactivity of a single position in ferrocene = 1.0, and for set b the reactivity of a single position in the unsubstituted ring was taken as 1.0 so that the values for each compound in set b are not strictly comparable with each other.

The acetylation reaction has also been employed to evaluate electronic effects in the ferrocene molecule. Rosenblum and Howells[428] measured the relative reactivities using acetyl chloride with aluminium chloride as catalyst at 0 °C as given in Table 114. The relative reactivities of benzene, ferrocene, and acetylferrocene under the same conditions were found to be $1.0 : 3.3 \times 10^6 : 1.9 \times 10^2$ and the large difference in these last two results (which relate to the electron withdrawal in one ring by a substituent in the other) were interpreted to mean that electrophilic substitution of ferrocene occurs *via* initial attack at iron[429], though later work by the same group failed to find evidence to confirm that such metal participation was essential in electrophilic substitution[430].

Benkeser *et al.*[431] measured the relative rates of acetylation of alkylferrocenes using acetic anhydride catalysed by boron trifluoride etherate, in dichloromethane at 0 and 25 °C. The relative reactivities are given in Table 115, and are interesting

TABLE 115

RELATIVE RATES OF REACTION OF $Ac_2O$ WITH $R \cdot C_5H_4FeC_5H_5$ IN $CH_2Cl_2$ CATALYSED BY $BF_3Et_2O$ AT 0 AND 25 °C (HALF TIME AT EACH TEMPERATURE)[431]

| | Position | | |
|---|---|---|---|
| R | 1′ | 2 | 3 |
| Me | 2.2 | 2.5 | 3.4 |
| Et | 2.0 | 1.7 | 3.8 |
| *i*-Pr | 1.9 | 1.1 | 4.7 |
| *t*-Bu | 1.7 | 0.44 | 5.5 |

for they show that while substitution follows the inductive order at the unhindered 3 position, the steric hindrance manifest at the 2 position also shows up for substitution in the other ring although the alternative possibility that the unsubstituted ring is activated in the conjugative order cannot be discounted; presumably this would not be the case for benzoylation and results for this would be interesting.

Acetylation with acetic anhydride was used in a determination of the relative reactivities of five-membered heterocycles by a competition method using iodine or stannic chloride as catalysts and 1,2-dichloroethane as solvent at 25 °C[432]. Acetic anhydride was preferred to acetyl chloride since the acid by-product is weaker and did not cause decomposition of furan, neither did the mild catalysts used. The relative rates are given in Table 116 along with those determined for the trifluoroacetylation with trifluoroacetic anhydride in the same solvent at 75 °C[433]; no details for the experimental method were given and it should be noted that trifluoroacetic anhydride boils at about 36 °C. This reagent is sufficiently reactive that no catalyst is required as expected for the strong electron withdrawal by the trifluoromethyl group. As can be seen from the table, trifluoroacetylation

TABLE 116

RELATIVE RATES FOR REACTION OF ArH WITH ACID ANHYDRIDES IN $C_2H_4Cl_2$[433]

| ArH | $(CF_3CO)_2O$ (75 °C) | $(CH_3CO)_2O, I_2$ (25 °C) | $(CH_3CO)_2O, SnCl_4$ (25 °C) |
|---|---|---|---|
| Anisole | | 0.377 | 0.320 |
| 2-Chlorothiophen | | 0.080 | 0.714 |
| Thiophen | 1 | 1 | 1 |
| Furan | $1.5 \times 10^2$ | 9.33 | 11.9 |
| 2-Methylthiophen | $3 \times 10^2$ | 15.6 | 17.6 |
| 2-Methylfuran | $3.5 \times 10^4$ | | |
| 2-Methoxythiophen | $1 \times 10^6$ | | |
| N-Methylpyrrole | $2 \times 10^8$ | | |

is very much more selective than the other reactions and this would be unlikely if a free ion was reacting, so presumably here reaction occurs *via* nucleophilic displacement of aromatic upon the polarised anhydride.

Finally, Kimoto[434] has measured the relative reactivities of a series of substituted diphenyl ethers (4-$XC_6H_4OPh$) in acetylation as follows: (X = ) MeO, 1,166; Me, 687; Cl, 121; H, 90.5; Ac, 35; $NO_2$, 1.3. The greater reactivity of the chloro- relative to the unsubstituted compound seems erroneous.

### 6.3.5 Cycliacylation

The acyl analogue of the cyclialkylation reaction is of course known, but until very recently there have been no kinetic studies of the reaction. As might be expected in view of the higher probability of collision of reagents in the reaction, milder conditions are sufficient and cycliacylation of 2-benzoylbenzoic acid (to give anthraquinone)[434a] and of 2-phenylbenzoic acid (to give fluorenone)[434b] may be brought about by concentrated sulphuric acid. The greater ease of reaction would lead one to suppose that a kinetic isotope effect might be observed and this is so, for the cycliacylation of 2-(2′-deuterophenyl)benzoic acid shows a product isotope effect as does 2-(2′-deuterobenzoyl)benzoic acid, the product retaining 56–62 % of one deuterium. For the latter compound, second-order rate coefficients ($10^4k_2$) were determined as ranging from 1.27 (H) and 0.994 (D) in 97.13 wt. % sulphuric acid through to 2.91 (H) and 2.42 (D) in 102.9 % acid, the average isotope effect $k_H/k_D$ in this range being 1.20. As in the case of cyclialkylation this is not, however, conclusive evidence for rate-determining breaking of the carbon–hydrogen bond since it is possible to apply the arguments about hybridisation changes at the carbon undergoing substitution (see p. 163).

## 7. Mercuration

Kinetic studies of mercuration have been carried out under heterogeneous and homogeneous conditions, and the first study by Westheimer *et al.*[435] using the former, showed that phenylmercuric nitrate was formed from benzene and mercuric nitrate in a reaction first-order in each reagent, and that the reaction was catalysed by nitric and sulphuric acids. Rate coefficients were determined under first-order conditions, and increasing the nitric acid concentration from 2 to 60 % increased the coefficients from $5 \times 10^{-7}$ to $3 \times 10^{-4}$; this increase did not arise from the increased solubility of benzene in the more acidic medium since this changed by a factor of only five. Salts appeared not to accelerate the reaction but this was subsequently shown to be erroneous because the change in solubility of benzene arising from the addition of salt was not allowed for[436].

A more extensive study showed the kinetics to follow the equation

$$\text{Rate} = k_{1 \text{ obs.}} [\text{PhH}][\text{Hg}^{2+}] = k_2^0 [\text{PhH}][\text{Hg}^{2+}] + k_3^A [\text{PhH}][\text{Hg}^{2+}][\text{A}^-]$$

(210)

where $k_2^0$ and $k_3^A$ are the rate coefficients for uncatalysed and catalysed reactions, respectively, and the accelerating effects of added acids and salts was found to be $\text{HClO}_4 > \text{H}_2\text{SO}_4 > \text{NaClO}_4 > \text{HNO}_3 = \text{NaNO}_3 \gg \text{Cl}^-$ (retards reaction). The catalytic effect was attributed to the anions, and increased with increasing anion concentration[436]. The experiments with added salts had to be carried out in the presence of a small amount of acid to suppress hydrolysis of the mercuric salt, and urea was added to solutions containing nitric acid or nitrates to prevent decomposition of the mercurated product by oxides of nitrogen.

Further investigation of the kinetics under heterogeneous conditions showed that the reaction is not acid-catalysed, since no correlation existed between the logarithm of the second-order rate coefficient and either of the acidity functions $H_0$ or $H_R$. However, a very satisfactory correlation existed between the rate and the vapour pressure of water above the solution, *i.e.* log $k$ correlated very satisfactorily with log $a_{\text{H}_2\text{O}}$. The acceleration by anions was, therefore, attributed to progressive removal of water molecules from the solvated mercury cation to give a more reactive species[437]. The reaction was also found to give a very large isotope effect, $k_\text{H}/k_\text{D} = 4.68-6.75$, being greater the lower the concentration of catalysing anions in the medium; this is one of the three reactions for which an isotope effect has been found for benzene (see also p. 59, 80) and is attributed to the weakness of the carbon–mercury bond which makes $k'_{-1}$ large relative to $k'_2$ in the equilibria

$$\text{Hg}^+\text{X} + \text{PhH} \underset{k'_{-1}}{\overset{k'_1}{\rightleftharpoons}} \text{PhHgH}^+\text{X} \quad \text{(fast)}$$

(211)

$$PhHgH^+X + H_2O \overset{k'_2}{\rightleftharpoons} PhHgX + H_3O^+ \quad \text{(slow)} \tag{212}$$

Obviously in the more strongly catalysed reaction, $k'_{-1}$ decreases and hence so does the isotope effect. Mercuration was also investigated under homogeneous conditions using 97 % acetic acid as solvent, and mercuric acetate as the source of mercuric ions. Here the addition of perchloric acid produced a very large rate increase *e.g.* 0.05 *M* acid increased the rate of benzene mercuration 1,840-fold whereas sodium perchlorate caused only a slight rate increase, the former effect was considered to arise largely from liberation of ionized mercury compound (and consequently a more reactive electrophile) according to equilibrium (213)

$$Hg(OAc)_2 + HClO_4 \rightleftharpoons Hg(OAc)^+ ClO_4^- + AcOH \tag{213}$$

and the effect of perchlorate ion is then only secondary[436].

The acceleration by anions under both conditions was attributed to displacement of one of the water molecules presumed to be tetrahedrally coordinated with the mercuric ion, the subsequent reaction being then envisaged as displacement of the anion or water molecule by the aromatic; the anions which cause reaction to take place more slowly were presumed to be more tightly bound to the mercuric ion than water. It has, however, been pointed out that less tightly bound anions would be unlikely to displace the more tightly bound water molecules in the first place[438].

An extensive investigation has been recently made of the kinetics of mercuration of benzene by mercuric acetate and mercuric perchlorate in acetic acid solvent, and the kinetic details are somewhat complicated[439].

(*a*) The reaction was first-order in benzene, and approximately first-order in mercuric salt.

(*b*) The reaction was catalysed strongly by perchloric acid, as observed above and in other investigations[440, 441] and the catalysis diminished the higher the acidity of the medium.

(*c*) In the region of weaker catalysis the catalytic effect was independent of the concentration of the metal salt, but in the region of strong catalysis, the catalytic effect was greater the lower the concentration of the metal salt.

(*d*) An increase in the initial concentration of mercuric acetate produced a decrease in the rate coefficient, whereas with mercuric perchlorate the reverse occurred; both these effects were more pronounced at lower concentrations of perchloric acid, and the effect remained for mercuric perchlorate (but not for the acetate) even in the absence of added acid.

(*e*) Observations were based on initial portions of reaction because departure from true second-order behaviour was observed, partly due to polymercuration (though this was reduced to a minimum by using a large excess of benzene) and the most significant departures from second-order kinetics were observed in the media of lowest acidities.

(*f*) The reaction was retarded by water, and this retardation was greater for mercuric perchlorate in the absence of added perchloric acid than for mercuric acetate in the presence of perchloric acid.

(*g*) Addition of sodium perchlorate produced a small retardation in the rate of mercuration by mercuric perchlorate in acetic acid containing 0.2 $M$ water, and this may be compared with earlier results in which this salt produced a small acceleration in 97 % aqueous acetic[436], and a large acceleration in the rate of mercuration by mercuric acetate in anhydrous acetic acid[440].

Although mercuration by the two salts above appears at first sight to be fundamentally different, analysis of the above data shows that this is not the case[439]. Firstly, mercuric perchlorate in acetic acid is partially converted to mercuric acetate and perchloric acid; thus, the first step is given by equilibrium (214)

$$Hg^{2+}(ClO_4^-)_2 + HOAc \rightleftharpoons HgOAc^+ClO_4^- + HClO_4 \qquad (214)$$

and the second step is the reverse of equilibrium (213), the equilibrium constant for which is 70, which means that an acetic acid solution of mercuric perchlorate (0.1 $M$) contains very little of the latter, about 90 % of acetoxymercury perchlorate ion pairs and 10 % of mercuric acetate, and is about 0.11 $M$ in perchloric acid. Since the latter is a catalyst for mercuration, it follows that an acetic acid solution of mercuric perchlorate will mercurate more rapidly than an equimolar solution of mercuric acetate. Since also the amount of perchloric acid produced will depend upon the amount of salt used initially, the observation that the mercuration rate increases with the latter then follows as does the observation that mercuration by mercuric acetate is independent of the initial concentration of the salt. Confirmation of the relevance of equilibria (213) and (214) was given by the fact that an acetic acid solution of a given amount of mercuric perchlorate gives mercuration rates closely similar to those for a solution of an equivalent amount of mercuric acetate together with twice the molar amount of perchloric acid[439].

The variation in the second-order rate coefficients with time and with change in initial concentration of mercuric salt can also be explained on the basis of equilibria (213) and (214). At low acidities, conversion of mercuric acetate to acetoxymercury perchlorate is incomplete, and, therefore, decreasing the concentration of the acetate *increases* the concentration of free perchloric acid which thus increases the conversion of the acetate into the more reactive perchlorate, hence the second-order rate coefficients increase. Decreasing the concentration of mercuric perchlorate will, however, *decrease* the concentration of free perchloric acid and this effect will be particularly marked since solvation of the perchlorate produces two equivalents of perchloric acid; the second-order rate coefficients will, therefore, decrease. In both cases, substitution changes the concentration of unreacted mercury salt, and hence the position of the equilibria during a kinetic run, the departure from second-order kinetics during a run then follows. It follows also that for mercuration with mercury acetate in the absence of perchloric acid, neither of the equilibria (213) and (214) apply, so that the second-order rate coefficients should be, and are, independent of the initial concentration of acetate. The equilibria still apply, however, for mercuration with mercury perchlorate in the absence of perchloric acid; the rate coefficients should therefore be, and are, dependent upon the initial concentration of perchlorate. At high acidities all of the mercury acetate will be converted to the acetoxymercury perchlorate ion pair, and reduction in the concentration of the acetate will *increase* the concentration of free perchloric acid and aid conversion of the ion pair to the mercuric perchlorate ion triplet: the rate coefficients will therefore increase. For the perchlorate, reduction in the concentration will *reduce* the concentration of free perchloric acid and hence reduce formation of the ion triplet; the rate coefficients will therefore decrease. However, in both cases the equilibrium will be less disturbed during a kinetic run than at low acidities, consequently there is little departure from second-order kinetics during a run.

The catalytic effect of perchloric acid has already been explained[436], and the anticatalytic effect of water follows from the equilibrium constant of 34 for equilibrium (215)

$$H_2O + HClO_4 \rightleftharpoons H_3O^+ClO_4^- \tag{215}$$

which means that water competes very significantly with unchanged mercury(II) for the available perchloric acid, thereby lowering the reaction rate. The smaller anticatalytic effect in mercuration by 0.5 $M$ mercuric acetate in the presence of 0.5 $M$ added acid than for 0.1 $M$ or 0.5 $M$ mercuric perchlorate in the absence of added acid then simply arises from the fact that in the former case almost all of the mercury is positively charged and the catalytic effect by this mechanism is saturated, whereas in the latter case only half is in the positive form, and consequently the retarding effect of water is greater.

The accelerating effect of sodium perchlorate on the rate of mercuration by mercuric acetate in acetic acid arises out of the equilibrium

$$Na^+ClO_4^- + HOAc \rightleftharpoons Na^+OAc^- + HClO_4 \tag{216}$$

From this equilibrium it follows that 0.1 $M$ sodium perchlorate will produce $6 \times 10^{-5}$ $M$ perchloric acid, *i.e.* sufficient to produce a significant acceleration for mercuration with mercuric acetate in the absence of added perchloric acid, but not otherwise. The acceleration of the rate of mercuration with mercuric perchlorate in 97 % aqueous acetic acid (but not with acetic acid containing a concentration of water of 0.2 $M$) was attributed to the fact that in the former medium for which the mole fraction of water is 0.1, the dissolved salts will appreciably lower the activity of free water since they will be solvated and this will cause equilibrium (215) to reverse thereby increasing the acidity of the solution, and hence the rate of mercuration; this effect of added salts is thus considered to be rather different from the effect in aqueous media.

If mercuration under the homogeneous conditions described above is brought about essentially by acetoxymercury perchlorate ion pairs then the ratio $k_{obs}(Hg^{II}]/[HgOAc^+ClO_4^-]$ should be constant but this was found not to be the case, a variation by a factor of over two being observed in the acidity range studied[439]. The discrepancy was attributed to mercuration by mercury perchlorate ion triplets produced *via* the reverse of equilibrium (214) for which the equilibrium constant is $K^{++}$. Now if

$$\text{Rate}/[\text{PhH}][\text{Hg(II)}] = k_{2\,obs} = k_2^+[HgOAc^+ClO_4^-]/[\text{Hg(II)}] + k_2^{++}[Hg^{++}(ClO_4^-)_2]/[\text{Hg(II)}]$$

then

$$k_{2\,obs}.[\text{Hg(II)}]/[HgOAc^+ClO_4^-] = k_2^+ + k_2^{++}K^{++}[\text{HClO}_4]$$

so that a linear relationship should hold between $k_{2\,obs}[\text{Hg(II)}]/[HgOAc^+ClO_4^-]$ and $[\text{HClO}_4]$, which was observed, and the values of slope and intercept indicated that $k_2^{++}$ was probably large as would be expected for a reaction with a dipositive ion.

Unionized mercuric acetate is also a mercurating species, for the second-order rate coefficient for mercuration of benzene by mercuric acetate in acetic acid at 25 °C is $0.41 \times 10^{-7}$. If mercuration took place *via* ionized acetate ion pairs $HgOAc^+OAc^-$ for which $K$, the equilibrium constant can be estimated at $2 \times 10^{-8}$, then since the rate of mercuration by this ion pair will be approximately the same as by the acetoxymercury perchlorate ion pair for which $k_2^+$ the second-order rate coefficient has been determined (above) as $0.37 \times 10^{-3}$ at 25 °C, the observed second-order rate should be $2 \times 10^{-8} \times 0.37 \times 10^{-3} = 0.74 \times 10^{-11}$. This is so different from the rate actually observed that mercuration by the ion pair can be eliminated which leaves ionized mercurcy acetate as the only possible mercurating species[439].

As in the case of heterogeneous mercuration, a large kinetic isotope effect $k_H/k_D$ of $6.0 \pm 0.1$ at 25 °C has been observed in the mercuration of benzene with mercuric acetate $(0.051\ M)$ in acetic acid containing 0.32 $M$ perchloric acid and 0.23 $M$ water[442]. Under these conditions the equilibrium to form acetoxymercury perchlorate ion pairs is nearly complete and consequently good second-order kinetics were obtained throughout a kinetic run. The possibility that the first step of the reaction could be the formation of a semi-stable intermediate of high concentration rather than a high-energy intermediate of low concentration was considered to be eliminated by the observation that kinetics carried out with one reagent in excess were first-order in that reagent whereas the former possibility would lead to a zeroth-order dependence on the excess reagent. The mechanism given by equilibria (211) and (212) therefore, applies. It has also been pointed out that a reaction with a slow proton loss should be base-catalysed; the most basic substances present in the reaction media are water, acetic acid, and mercuric acetate. For mercuration of benzene in the absence of perchloric acid isotope effects of 3.20 (25 °C), 2.58 (50 °C), and 1.85 (90 °C) were obtained in media which were 0.3 $M$ and 1.7 $M$ in water and the rate coefficients were similar for each medium[443]; hence water can be excluded, as can mercuric acetate since the reactions were only first order in this compound. This leaves acetic acid as the only alternative base, though an intramolecularly assisted proton removal, *viz.*

$$HPhHgOAc^+ClO_4^- \longrightarrow \left[ Ph \underset{H====O}{\overset{Hg---O}{\diagup}} C-CH_3 \right]^+ ClO_4^- \longrightarrow PhHg^+ClO_4^- + HOAc \qquad (217)$$

has been considered as a possibility.

The rates of mercuration of a number of aromatic hydrocarbons have been determined using mercuric acetate in acetic acid, in some cases using perchloric acid as catalyst. Rate coefficients for benzene and alkylbenzene have been determined under both conditions at a range of temperatures and the data are gathered in Tables 117 and 118[441-448]; in the absence of perchloric acid, mercuric acetate reacted with the solvent at a rate that was not insignificant compared with the rate of aromatic mercuration, and a correction was made for this side reaction.

It is apparent that, although the reaction rates of the uncatalysed reaction are less at any particular temperature, the selectivity of the electrophile under uncatalysed conditions (for which $\rho = -4.0$) is also less than under catalysed conditions (for which $\rho = -5.1$), indicating a more reactive electrophile and these facts are difficult to reconcile. Now if the mechanism represented by equilibria (211) and (212) is correct, then different mercurating electrophiles should give much the same reaction rate; since they do not, there must be an appreciable energy barrier in the first equilibrium. The isotope effect in catalysed mercuration (*ca.* 6.0) is considerably greater than in uncatalysed mercuration (*ca.* 3.0; the energy barrier in going from the Wheland intermediate to the rate-determining transition state in the former reaction must, therefore, be *higher* than in the latter reaction. Since the overall reaction rate is the product of the equilibrium constant for the first step multiplied by the rate of the second step, it follows that the energy barrier for the equilibrium is lower for catalysed mercuration than for uncatalysed mercuration, and this is entirely in accord with the electrophile being more positive in the former case. The larger isotope effect under catalysed conditions must also mean that the rate-determining transition state must occur later, *i.e.* be further

TABLE 117

RATE COEFFICIENTS ($10^6 k_2$) AND KINETIC PARAMETERS FOR REACTION OF ArH WITH Hg(OAc)$_2$ IN HOAc[441-448]

| ArH | Temp. (°C) | | | | $\Delta H^{\ddagger}$ | $\Delta S^{\ddagger}$ |
|---|---|---|---|---|---|---|
| | 25 | 50 | 70 | 90 | | |
| Benzene | 0.041(67)[a] | 0.75 | 5.65(5.72, 70.2 °C) | 34.2(35.1, 90.5 °C) | 21.6 | −19.6 |
| Toluene | 0.265(530)[a] | 3.77 | (24.7, 70.2 °C) | (128, 90.5 °C) | 19.7 | −22.6 |
| t-Butylbenzene | | 2.45 | 15.2 | 84 | 19.9 | −22.8 |
| Biphenyl | | 1.74 | 11.9 | 66.5 | 21.6 | −23.6 |
| Fluorene | 2.11[b] | | | | | |
| Fluorobenzene | | 0.47 | 3.03 | 18.0 | | |
| Bromobenzene | | | | 3.69, 2.74[c] | | |
| p-Methylanisole | 9.3 | | | | | |
| p-t-Butylanisole | 10.6 | | | | | |
| Diphenylether | 2.67 | | | | | |
| Acetanilide | 1.91 | | | | | |
| Anisole | 18.5 | | | | | |
| $k_{PhMe}/k_{PhH}$ | 6.4(7.9)[a] | 5.0(7.0)[a] | 4.3(5.9, 75 °C)[a] | 3.6 | | |

[a] Value appropriate to a solution containing 0.29 M perchloric acid.
[b] Measured for a solution in benzene, a correction being applied for benzene mercuration.
[c] Estimated rate for competitive mercuration with benzene.

from the Wheland intermediate than under uncatalysed conditions. Now it has been argued that the normal pattern of substituent effects is observed in these reactions because in the first equilibrium, the substituents determine the relative stabilities of the resulting Wheland intermediates in the usual way, and that the nearer the subsequent rate-determining transition state is to this Wheland intermediate, the larger should be the $\rho$ factor[448]. But this leads one to the conclusion that uncatalysed mercuration should have the larger $\rho$ factor which is not observed. In attempting to explain these results it was also considered that the Wheland intermediate for catalysed mercuration might be doubly positively charged and that this extra positive charge would facilitate subsequent ejection of the proton leading to the faster overall reaction rate[448]. How this can be reconciled with the *larger* isotope effect found under these conditions was not, however, indicated.

Kinetic studies of mercuration have also been used as a test for hyperconjugation. Toluene and toluene-$\alpha\alpha\alpha$-$d_3$ were mercurated at 25 °C by mercuric acetate (0.5 M) in acetic acid containing water (0.25 M) and perchloric acid (0.050 M) and an isotope effect, $k_H/k_D = 1.00\pm0.03$, obtained. This insignificant effect was considered as evidence against the participation of hyperconjugation in electron supply by a methyl group[449].

The mercuration of 2-substituted thiophenes by mercuric acetate at temperatures between 16.9 and 50.0 °C has been shown to be second-order over the first 20 % of reaction[449a]. The activation energies for thiophen and its 2-acetyl, 2-methyl,

TABLE 118

RATE COEFFICIENTS ($10^5 k_2$) FOR REACTION OF ArH WITH $Hg(OAc)_2$ IN HOAc (OR HOAc/$HClO_4$)[441–448]

| ArH | Uncatalysed at 50 °C | Perchloric acid (0.50 M) catalysed at 25 °C |
|---|---|---|
| Benzene | 0.075 | 43 |
| Toluene | 0.382 | 387 |
| Ethylbenzene | 0.325 | 330 |
| $i$-Propylbenzene | 0.303 | 303 |
| $t$-Butylbenzene | 0.248 | 271 |
| $n$-Propylbenzene | | 310 |
| $i$-Butylbenzene | | 292 |
| $s$-Butylbenzene | | 334 |
| $o$-Xylene | 1.26 | 1,370 |
| $m$-Xylene | 2.70 | 38,400 |
| $p$-Xylene | 0.640 | 538 |
| 1,2,3-Trimethylbenzene | 5.37 | |
| 1,2,4-Trimethylbenzene | 3.84 | |
| 1,3,5-Trimethylbenzene | 16.5 | |
| 1,2,3,4-Tetramethylbenzene | 10.0 | |
| 1,2,4,5-Tetramethylbenzene | 20.3 | |
| 1,2,3,5-Tetramethylbenzene | 2.37 | |
| Pentamethylbenzene | 17.5 | |
| Fluorobenzene | | 16.0 |
| Chlorobenzene | | 2.27 |
| Bromobenzene | | 2.00 |
| Iodobenzene | | 1.82 |
| Nitrobenzene | | 0.0177 |
| Diphenyl | | 179 |
| Naphthalene | | 523 |
| Phenylmercuric acetate | | 29.5 |

and 2-carboethoxy derivatives were 13.8, 12.0, 13.2 and 17.9, respectively, which do not correspond with the observed reactivities in the 5-positions which correlated with $\sigma_p$ values ($\rho = -3.47$ at 50 °C). Furthermore, the thiophene : benzene reactivity ratio was $9.7 \times 10^4$ at 25 °C and $3.3 \times 10^5$ at 50 °C which is greater than for nitration (850), a more selective reaction.

Finally, rates of mercuration have been measured using mercuric trifluoroacetate in trifluoroacetic acid at 25 °C[450]. The kinetics were pure second-order, with no reaction of the salt with the solvent and no isomerisation of the reaction products; rate coefficients ($10\,k_2$) are as follows: benzene, 2.85; toluene, 28.2; ethylbenzene, 24.4; $i$-propylbenzene, 21.1; $t$-butylbenzene, 17.2; fluorobenzene, 0.818; chlorobenzene, 0.134; bromobenzene, 0.113. The results follow the pattern noted above in that the reaction rates are much higher (*e.g.* for benzene, 690,000 times faster than for mercuration with mercuric acetate in acetic acid) yet the $\rho$ factor is larger ($-5.7$); if the pattern is followed fully, one could expect a larger isotope effect for this reaction as well. A further pattern which emerges is that in the reactivity sequence uncatalysed mercuration ($CH_3COOH$), catalysed mercuration ($CH_3COOH$), uncatalysed mercuration ($CF_3COOH$), the $\log f_o : \log f_p$ rates decrease from 0.54 to 0.43 to 0.335, showing clearly the increasing size of the electrophile along the series.

## 8. Hydrogen exchange

Electrophilic aromatic hydrogen exchange reactions fall into two classes, namely those reactions catalysed by acid and those reactions catalysed by base. Of these the former are by far the most common and have been subjected to the most extensive and intensive kinetic studies.

### 8.1 ACID-CATALYSED HYDROGEN EXCHANGE

Kinetic studies of hydrogen exchange under acid-catalysed conditions have been twofold. Firstly, there have been those studies aimed primarily at determining the mechanism of the reaction, both with regard to the nature of the transition state and the species undergoing reaction; hence much work has been carried out to determine if pre-equilibrium proton transfers to carbon occur, and also to determine if reaction occurs with a free base or corresponding conjugate acid (protonation at the substituent). Secondly, there have been those studies aimed primarily at evaluating substituent effects in electrophilic aromatic substitution. Some of these studies necessarily overlap, but in order to try and simplify the presentation of a very large amount of data, it is divided into these two categories.

The reaction of an acid with an aromatic leading to hydrogen exchange, *viz.*

$$ArH + H*A \rightleftharpoons ArH* + HA \tag{218}$$

is as might be expected a straightforward second-order reaction which in the majority of cases has been studied under first-order conditions using an excess of acid. This is particularly appropriate when the label is initially in the aromatic since it is then possible to work with very small amounts of compound, and kinetic studies have been successfully carried out with only a few milligrams of compound obtained from chromatographic procedures. In this way also, the hazards of working with tritium-labelled compounds have been minimized.

*8.1.1 Kinetic studies of the mechanism of acid-catalysed hydrogen exchange*

The first of these studies was carried out by Ingold *et al.*[451], who showed qualitatively that the efficiencies of acids in promoting hydrogen–deuterium exchange in benzene was: $D_2SO_4 > D_2SeO_4 > D_3O^+ > DOPh > D_2O$, that the rates for substituted benzenes PhX increased along the series: (X = ) $SO_3H <$ H < OMe < $NMe_2 < O^-$ and that electron-supplying substituents caused more rapid exchange at the *ortho* and *para* than at the *meta* positions. The reaction was, therefore, clearly established as an electrophilic substitution. Koizumi *et al.*[452] made a number of studies of the rates of deuteration of phenol, nitrophenols, aniline hydrochloride, furan, thiophen, pyrrole, and indole derivatives in basic and weakly acidic (hydrochloric acid) media. The principle conclusions were: (*a*) that phenol would undergo exchange in both acidic and basic media with activation energies of 27.3 and 24.8, respectively, the lower value arising from reaction upon the anion (*b*) the activation energies for *m*- and *p*-nitrophenols in alkali were greater, being 29.0 and 28.4, respectively[453], consistent with an electrophilic substitution (*c*) that aniline hydrochloride exchanges *via* the free base[454] since the rate is not appreciably increased in more acidic media, and consistently the activation energies were 20.8 and 29.5 for acidic and basic media respectively which would only follow if the same species was undergoing reaction in both media (*d*) that furan and thiophen are less reactive than pyrrole[455] and that (*e*) for indoles, exchange occurred first on nitrogen, then at the $\beta$ carbon and finally at the $\alpha$ carbon as the acidity was increased up to 0.8 *M* hydrochloric acid[456].

Fifteen years elapsed before the publication of further work, this pause arising from the military significance and unavailability of deuterium. One of the first papers to appear was that of Gold and Long[457], who measured rates of de-deuteration of 9-deuteroanthracene (in carbon tetrachloride) by sulphuric acid at 25 °C. Rate coefficients (some of which can only be approximate in view of the very high quoted values) increased with increasing acidity (Table 119) and this was assumed (in view of the distribution of anthracene between carbon tetrachloride and sulphuric acid) to arise from the increased solubility of anthracene in the stronger acid, since the rate of proton loss from the protonated intermediate appeared to be relatively insensitive to acid concentration; this was explained by the assumption that the increase in the concentration of the acidic species causing proton attachment to the intermediate with increasing acid concentration was compensated

TABLE 119

FIRST-ORDER RATE COEFFICIENTS FOR REACTION OF [9-$^2$H]-ANTHRACENE IN $CCl_4$ WITH $H_2SO_4$ AT 25 °C[457]

| $H_2SO_4$(wt. %) | 84.0 | 85.0 | 85.4 | 87.8 | 90.0 |
|---|---|---|---|---|---|
| $10^5 k_1$ | 52 | 100 | 170 | 430 | 2,300 |

by a decrease in the concentration of the basic species promoting proton loss from the intermediate. However, since these two species must be related as conjugate acid and base by virtue of the symmetry of the reaction and an increase in protonation of the aromatic by the former results in an increase in concentration of the latter, this argument is untenable.

A subsequent study, using clearly homogeneous conditions, of the dedeuteration of [2-$^2$H]-4-nitro-, -chloro-, and -methyl-phenols by aqueous sulphuric acid showed that the rate coefficients clearly increased with increasing acidity (Table 120) and plots of log rate coefficient *versus* the acidity function $-H_0$ were linear,

TABLE 120

FIRST-ORDER RATE COEFFICIENTS ($10^7 k_1$) FOR REACTION OF [2-$^2$H]-4-X-PHENOL WITH HA AT 25 °C[458]

| $H_2SO_4$(wt. %) | 9.8 | 14.5 | 19.3 | 27.0 | 34.5 | 40.5 | 45.3 | 47.0 | 53.0 | 59.0 | 63.5 |
|---|---|---|---|---|---|---|---|---|---|---|---|
| $-H_0$ | 0.15 | 0.50 | 0.85 | 1.35 | 1.85 | 2.36 | 2.75 | 2.90 | 3.56 | 4.18 | 4.69 |
| $X = 4\text{-}Me$ | 0.61 | 1.73 | 4.00 | 13.3 | 47.0 | | 43.8 | | | | |
| $X = 4\text{-}Cl$ | | | | | | 1.57 | | 4.73 | 15.9 | 33.2 | 159 |

| $H_2SO_4$(wt. %) | 64.3 | 70.3 | 71.1 | 78.8 | 81.3 | 82.9 | 87.2 | 92.2 | 97.6 |
|---|---|---|---|---|---|---|---|---|---|
| $-H_0$ | 4.80 | 5.58 | 5.68 | 6.65 | 7.03 | 7.31 | 7.85 | 8.40 | 9.17 |
| $X = 4\text{-}NO_2$ | 1.16 | 4.40 | 5.33 | 48.5 | 123 | 270 | 508 | 1,010 | 1,930 |

| $HCl$(wt. %) | 7.10 | 11.7 | 12.1 | 13.65 | 15.2 | 16.05 | 19.0 | 21.55 | 22.0 |
|---|---|---|---|---|---|---|---|---|---|
| $-H_0$ | 0.71[a] | 1.21[a] | 1.23 | 1.40 | 1.58[a] | 1.68 | 2.03 | 2.38[a] | 2.43 |
| $X = 4\text{-}Me$ | 3.95 | 13.3 | 4.88[b] | 9.44[b] | 38.9 | 18.7[b] | 54.6[b] | 306 | 164[b] |

| $H_3PO_4$(wt. %) | 63.6 | 68.1 | 72.7 | 76.4 |
|---|---|---|---|---|
| $-H_0$ | 1.90 | 2.25 | 2.65 | 3.00 |
| $X = 4\text{-}Cl$ | 1.17 | 2.67 | 5.60 | 14.8 |

[a] Different values were given in ref. 459 and were amended in ref. 460.
[b] Data refers to reaction of [2-$^3$H]-4-MeC$_6$H$_4$OH[460].

the respective slopes being 0.94, 0.90, and 1.08[458]. (Below $-H_0 = 8.0$, however, the slopes were less steep so that *over the whole acid range, the dependence was non-linear*.) A further study of the dedeuteration of the latter two compounds in aqueous phosphoric and hydrochloric acids respectively (Table 120) also showed linear plots of log rate coefficient *versus* $-H_0$ with slopes of 1.00 and 1.14[459, 460].

A linear correlation of log rate coefficient with acidity function $H_0$ was considered to have been obtained in the dedeuteration of deuterobenzene in aqueous sulphuric acid (Table 121) the slope of the plot being given as 1.36[461]. However, rigorous examination of the data shows the plot to be a curve which is amplified

## TABLE 121

FIRST-ORDER RATE COEFFICIENTS ($10^7 k_1$) FOR REACTION OF ArH* WITH AQUEOUS $H_2SO_4$ AT 25 °C

| $H_2SO_4$ (wt. %) | $C_6H_5$ | | $4\text{-}MeC_6H_4$ | | | $3\text{-}MeC_6H_4$ | | $2\text{-}MeC_6H_4$ | | $4\text{-}Bu^tC_6H_4$ | $2\text{-}Bu^tC_6H_4$ |
|---|---|---|---|---|---|---|---|---|---|---|---|
| | $^2H$ (Ref. 461) | $^3H$ (Refs. 166, 464) | $^2H$ (Ref. 463) | $^3H$ (Refs. 166, 464) | $^3H$ (Ref. 465) | $^2H$ (Ref. 463) | $^3H$ (Refs. 166, 464) | $^2H$ (Ref. 463) | $^3H$ (Ref. 465) | $^3H$ (Refs. 465, 466) | $^3H$ (Ref. 465) |
| 53.0 | | | 1.36 | | | | | | | | |
| 58.5 | | | 9.40 | | | | | | | | |
| 63.9 | | | 56.6 | | | | | | | | |
| 64.5 | | | | | 50.7 | | | | 49.6 | | |
| 66.5 | 2.68 | | | | | 8.4 | | | | | |
| 68.0 | | | | | | | | 356 | | | |
| 68.23 | | | | 224 | | | | | | | |
| 68.7 | 5.75 | | | | | | | | | | |
| 68.9 | | | 498 | | | | | | | | |
| 69.0 | | | | | 320 | | | | 316 | | |
| 70.08 | | | | 637 | | | | | | | |
| 70.70 | | | | | 753 | | | | 753 | | |
| 71.34 | | | | 1,040 | | | | | | 748 | |
| 71.41 | | | | | | | | | | 805 | |
| 73.25 | | 11.8 | | 3,000 | 2,710 | | 58.9 | | 2,800 | | 768 |
| 73.5 | 35.2 | | | | | | | | | | |
| 74.85 | | 28.7 | | | 4,915 | | | | | | |
| 75.3 | | | | 6,980 | | | 144 | | 5,185 | | |
| 75.8 | 115 | | | | | | | | | | |
| 77.67 | | 102 | | | | | 553 | | | | |
| 79.2 | 453 | | | | | | | | | | |
| 79.8 | | 341 | | | | | 1,950 | | | | |
| 81.14[a] | | 873 | | | | | 4,970 | | | | |
| 82.82 | 3,450 | | | | | | | | | | |
| 83.2 | | 3,290 | | | | | | | | | |
| 84.03 | | 6,300 | | | | | | | | | |

[a] In 88.66 wt. % acid, Olsson and Melander[467] obtained values of 5.22 (3.36), 13.5 (8.73), 630 (325), and 702 (392) for the dedeuteration (detritiation) of benzene and the 3, 4, and 2 positions of toluene respectively, under heterogeneous conditions.

TABLE 122

FIRST-ORDER RATE COEFFICIENTS ($10^7 k_1$) AND ARRHENIUS PARAMETERS FOR REACTION
OF [$^3$H]-$C_6H_4X$ WITH 66.4 WT. % $H_2SO_4$[468]

| Temp. (°C) | X = H | X = 4-Me | X = 4-Bu$^t$ |
|---|---|---|---|
| 24.22 | | 70.5 | 49.7 |
| 39.25 | | 459 | 337 |
| 50.30 | | 1,700 | 1,320 |
| 65.75 | 42.3 | 7,240 | 6,500 |
| 75.70 | | 18,700 | 15,300 |
| 90.8 | 645 | | |
| $E_a$ | 26.8 | 22.3 | 23.1 |
| log $A$[a] | 11.9 | 11.2 | 11.7 |

[a] Pre-exponential factor calculated from $k_1$ and $E_a$.

if the Long and Paul acidity function values[462] for the acid are used, the slope
of the curve varying from 1.2 to 1.7 with increasing acidity. Rate coefficients for
dedeuteration of toluene were also measured (Table 122) and a correlation of log
rate coefficient *versus* acidity function with a slope of 1.40 appeared to exist[463].

On the basis of these correlations, Gold and Satchell[463] argued that the A-1
mechanism must apply (see p. 4). However, a difficulty arises for the hydrogen
exchange reaction because of the symmetrical reaction path which would mean
that the slow step of the forward reaction [equilibrium (2) with E and X = H]
would have to be a fast step [equivalent to equilibrium (1) with E and X = H]
for the reverse reaction, and hence an impossible contradiction. Consequently,
additional steps in the mechanism were proposed such that the initial fast equilibrium formed a π-complex*, and that the hydrogen and deuterium atoms exchange positions in this π-complex in two slow steps *via* the formation of a σ-
complex; finally, in another fast equilibrium the deuterium atom is lost, *viz.*

$$\tag{219}$$

Since much work has gone into eventually disproving this mechanism for hydrogen exchange it is worth examining the soundness of the experimental data
upon which the theory was based. Firstly, the log rate coefficient *versus* acidity
function plots should have had unit slopes and the fact that they did not was attributed to salt effects. Secondly, different rate coefficients were obtained at the

* Called an "outer" complex by the original workers, the meaning of the term "outer" being
unspecified. Subsequent workers in this field have all argued that "outer" complexes are
merely π-complexes. Likewise, the σ-complexes were referred to originally as "inner" complexes.

same $H_0$ values for different acids (see Table 120) and this was again attributed to salt effects. Thirdly, the plots were, in any case, not linear and this is clearly proven by the data for detritiation of benzene and the *meta* and *para* positions of toluene given in Table 121, and which were corrected for the rate of the sulphonation side reaction, the slopes of the curves ranging from 1.5–2.2. It may be noted here that the dedeuteration and detritiation data for toluene are nicely consistent in that the latter reaction takes place approximately 1.6 times slower than the former, in excellent agreement with the value of 1.8 obtained by Olsson[467], the increase of rate with increasing acidity being very slightly greater for detritiation. The data for benzene, however, is in serious disagreement, which can be traced to the results for dedeuteration of benzene which appears to be in error by a large factor. In detritiation, partial rate factors determined from rates for benzene and toluene obtained in the same acid varied from 245–254 for $f_p^{Me}$ and 5.0–5.7 for $f_m^{Me}$ depending upon acidity, and the former factor is $>$ 170 even at 66 °C (Table 122), whereas the dedeuteration data gives values of 88 and 1.9 respectively, and it should be noted that the dedeuteration kinetics were carried out with different acid mixtures for toluene and benzene. Olsson[469] has found that if allowance is made for the differing solubilities of benzene and toluene in sulphuric acid, partial rate factors for dedeuteration (detritiation) in 80.6 wt. % acid under heterogeneous conditions, are obtained as follows: $f_m^{Me}$, 6.4 (6.4); $f_p^{Me}$, 298 (240); $f_o^{Me}$, 332 (289); which is in excellent agreement with the data* of Eaborn and Taylor[464]. Since there is also excellent agreement between all three sets of data for values of $f_p^{Me} : f_m^{Me}$ which are 44–46 (dedeuteration) and 38–44 (detritiation) there is no doubt whatever that the benzene dedeuteration data are in error and should be disregarded. It is probably significant that data obtained by Gold and Satchell for sulphonation of benzene by sulphuric acid are also in disagreement with other work (the rate coefficients for a stated acid concentration being higher than those obtained by other workers, see p. 69) and it seems quite possible that the method for determining the sulphuric acid concentration was at fault.

The greater dependence of rate coefficient upon acidity for detritiation compared to dedeuteration is apparent for benzene as it was for toluene and is more marked, but in view of the errors in the benzene work (which appear to arise only from measuring the acid concentration but could possibly arise from some feature of the kinetic method) and element of doubt must remain here. Nevertheless, this phenomenon (which is understandable on the basis that when the reactions are infinitely fast they will then both take place at the same rate, and the more reactive the compound and the stronger the acid, the more closely this situation is approached) seems to be general, for Gold *et al.*[460] found that the log rate coefficient

---

* It is interesting to note that Olsson's data, taken along with that of Eaborn and Taylor, confirms the findings of Eaborn *et al.*[465] that the $f_o^{Me} : f_p^{Me}$ values increase with increasing sulphuric acid concentration.

for detritiation of [2-$^3$H]-$p$-cresol by hydrochloric acid correlated with $-H_0$ with a slope of 1.23 which was greater than for dedeuteration (1.14). With 13–25 wt. % hydrochloric acid (Table 120) detritiation rates were 2–3 times slower than dedeuteration rates (in reasonable agreement with the results obtained above in sulphuric acid) and this difference was attributed to the differing masses of the isotopes in the rearrangement step of the A-1 mechanism, reaction (219). However, it is equally consistent with the A-$S_E$2 mechanism which can be written as

$$\text{ArH}^* + \text{H}^+ \underset{k'_{-1}}{\overset{k'}{\rightleftharpoons}} [\text{ArH}^*\text{H}]^+ \underset{k'_{-2}}{\overset{k'_2}{\rightleftharpoons}} \text{ArH} + {}^*\text{H}^+ \tag{220}$$

The forward isotope effect for this reaction is given by

$$\text{Forward isotope effect} = \frac{k'_2}{k'_{-1} + k'_2} = \frac{1}{(k'_{-1}/k'_2) + 1} \tag{221}$$

from which equation (222)

$$\frac{k_D}{k_T} = \frac{(k'_{-1}/k'_2)_T + 1}{(k'_{-1}/k'_2)_D + 1} \tag{222}$$

follows. By taking reasonable theoretical values of $k'_{-1}/k'_2$ as 4–6 for dedeuteration and 7–10 for detritiation, the observed isotope effect $k_D/k_T$ should be *ca.* 1.6*. Olsson and Melander[467, 470] found that the kinetic isotope effect decreased from 1.92 for exchange at the *ortho* position in toluene to 1.56 for exchange in benzene, both in aqueous sulphuric acid. Values of $k_H/k_D$ increased from 3.4 (benzene) to 5.5 (*ortho* position in toluene) and the corresponding values of $k_H/k_T$ were 5.9 and 11.5. It follows from the symmetry of the reaction that $k_H/k_{D(T)} = (k'_1/k'_{-2})_{D(T)}$ and hence methyl substitution lowers the first energy barrier more than the second (presumably through aiding proton attachment), so it also follows that the partial rate factor for the first step of the reaction (exchange of hydrogen for hydrogen) would be larger and values of $f_m^{Me} = 6.4, f_p^{Me} = 422$ and $f_o^{Me} = 435$ were calculated.

The substrate isotope effect then is no evidence for the A-1 mechanism. Gold et al.[460] also found a solvent isotope effect, the detritiation of (2-$^3$H)-$p$-cresol in aqueous 13.65 wt. % hydrochloric acid occurring 1.6 times faster in $D_2O$ than in $H_2O$ (Table 123) this being attributed to the product of two opposing effects in the A-1 mechanism. The first equilibrium to form the $\pi$-complex would be more rapidly established in $H_2O$ leading to a higher concentration of the complex, but rearrangement of the deuterated complex in the rate-determining step was postulated as being faster than for the protonated complex, the observed solvent

---

* A more rigorous treatment of isotope effects is given later (p. 211).

TABLE 123

VARIATION OF FIRST-ORDER RATE COEFFICIENTS WITH ATOM FRACTION OF
DEUTERIUM IN REACTION OF [2-$^3$H]-$p$-CRESOL WITH 13.65 WT % AQUEOUS HCl AT 25 °C[460]

| *Atom fraction of D, in HCl* | 0 | 0.15 | 0.3 | 0.45 | 0.60 | 0.75 | 0.90 | 0.96 | 1.00 |
|---|---|---|---|---|---|---|---|---|---|
| $10^7 k_1$ | 9.17 | 10.6 | 11.4 | 12.2 | 13.0 | 14.0 | 15.1 | 15.6 | 15.8 |

isotope effect thereby confirming the proposed reaction mechanism. However, the effect of the last equilibrium upon the rate was not considered and in any case the detritiation of [4-$^3$H]-$m$-xylene in aqueous trifluoroacetic acid (Table 124) was faster in $H_2O$ than in $D_2O$[471] so that the effect is not general, and any mechanism which requires only the former order of solvent isotope effect is necessarily invalid. Gold *et al.* also found that the dependence of rate coefficient upon the atom fraction of deuterium (Table 123) was non-linear in accordance with the prediction of the Gross–Butler theory[23] for a reaction involving a fast proton transfer. The difference between linearity and non-linearity was very small and in any event the theory is no longer regarded as having any diagnostic value (p. 6).

TABLE 124

VARIATION OF SOLVENT ISOTOPE EFFECT WITH MEDIUM COMPOSITION IN REACTION
OF [4-$^3$H]-$m$-XYLENE WITH $(CF_3CO)_2O$–$H_2O(D_2O)$ AT 25.0 °C[471]

| $(CF_3CO)_2O$ (*mole %*) | $10^7 k_1$ (*H*) | $10^7 k_1$ (*D*) | $(CF_3CO)_2O$ (*mole %*) | $10^7 k_1$ (*H*) | $10^7 k_1$ (*D*) |
|---|---|---|---|---|---|
| 24.95 |       | 16.2  | 43.82  | 237   |       |
| 25.00 | 29.95 |       | 44.45  |       | 123   |
| 27.50 | 59.4  |       | 46.80  | 198   |       |
| 28.55 |       | 39.6  | 46.90  |       | 112.5 |
| 28.85 |       | 42.0  | 48.75  | 159   |       |
| 31.02 | 117   |       | 48.95  |       | 80.5  |
| 32.95 |       | 78.0  | 50.00[a]| 128  |       |
| 34.30 | 179   |       | 56.90  |       | 69.0  |
| 36.00 | 213   |       | 51.3   | 112   |       |
| 37.60 | 240   |       | 51.92  |       | 63.4  |
| 37.70 |       | 118   | 53.7   | 82.0  |       |
| 39.65 |       | 127.5 | 55.0   |       | 56.2  |
| 40.25 | 251   |       | 56.4   | 52.4  |       |
| 41.05 |       | 135   | 58.15  |       | 31.5  |
| 42.05 | 246   |       | 60.45  | 32.6  |       |

[a] 50.00 mole % anhydride is 100 mole % acid. Media of lower and higher concentrations contain free water and free anhydride respectively. In this (protium-containing) medium, rate coefficients ($10^7 k_1$) for [2-$^3$H]-$m$-xylene in the presence of 0.10 $M$ added salts were: 56 (none); 69 (LiO$_2$CCF$_3$); 71 (NaO$_2$CCF$_3$); 910 (LiCl); 13,600 (LiBr); 11,600 (LiClO$_4$), and 245 (LiClO$_4$ 0.10 $M$+LiO$_2$CCF$_3$ 0.10 $M$) 400 (LiClO$_4$ 0.10 $M$+LiO$_2$CCF$_3$ 0.05 $M$); 1,510 (LiClO$_4$ 0.10 $M$+LiO$_2$CCF$_3$ 0.01 $M$).

*References pp. 388–406*

In the work of Eaborn et al.[471], the solvent isotope effect $k_{H_2O}/k_{D_2O}$ varies from ca. 1.93 in 60 mole % acid, passes through a maximum of 2.00 at ca. 75 mole % acid and then gradually decreases towards a value of 1.3 in media containing free anhydride. Likewise, the maximum in the rate coefficient occurs at ca. 80 mole % acid. This latter was attributed to two opposing factors, namely the increase in the solvating power of the medium as more water is added (trifluoroacetic acid being a very poor solvating medium) and the opposed decrease in strength of the acidic species responsible for the exchange, probably through replacement of the dimeric $(CF_3COOH)_2$ by the more weakly acidic monohydrate $CF_3COOH.H_2O$. The smaller change in rate coefficient with change in water concentration of the medium for added $D_2O$ compared to added $H_2O$ was attributed to the poorer solvating ability of the former, and the occurrence of the solvent isotope effect maximum and the rate maximum at the same concentration is a direct consequence of this. Qualitatively identical results were obtained in protodesilylation (p. 325) except that the magnitude of the solvent isotope effect was larger, that in detritiation being reduced through the near symmetry of the reaction path. The large rate acceleration by lithium chloride, bromide and perchlorate, was proved to be due to the corresponding free acids formed via the equilibrium

$$CF_3COOH + X^- \rightleftharpoons CF_3COO^- + HX \tag{223}$$

these acids being more effective than trifluoroacetic acid in promoting exchange. Consistently, the lithium perchlorate solutions containing lithium trifluoroacetate (which would force this equilibrium to the left) gave smaller rate increases. The poorer rate acceleration by lithium chloride was due to the (visible) dissolution of the formed hydrogen chloride. The lack of a significant effect of added lithium and sodium trifluoroacetates was due either to the effective acidic species in the pure acid being the undissociated acid or (less likely) to an ionic salt effect compensated by reduction in the concentration of the species $CF_3COOH^+$ formed by autoprotolysis; 0.1 $M$ solutions of lithium and sodium trifluoroacetate produced similarly small rate increases (10 %) in detritiation of [9-$^3$H]-phenanthrene at 100 °C[540].

Further substrate and solvent isotope effects were measured by Batts and Gold[472] for the dedeuteration and detritiation of labelled 1,3,5-trimethoxybenzene in aqueous protium- and deuterium-containing perchloric acid. Contrary to the observations above, they found the rate coefficients for dedeuteration to detritiation to be independent of the concentration of the catalysing acid (Table 125). Detritiation in the deuterium-containing aqueous perchloric acid media occurred 1.68 times faster than in the protium-containing media.

However, using acetate buffers, the observations of Eaborn et al.[471] were confirmed in that rates were faster in the protium- than in the deuterium-containing media. Furthermore, an earlier observation of Kresge and Chiang[473] (see below)

## TABLE 125

RATE COEFFICIENTS, SOLVENT AND SUBSTRATE ISOTOPE EFFECTS FOR REACTION OF
$[1\text{-}^{2(3)}H]\text{-}2,4,6\text{-}(MeO)_3C_6H_2$ WITH AQUEOUS $HClO_4$ AT 25 °C[472]

| $10[HClO_4]$ (M) | $10^5 k_1{}_H^D$ | $10^3 k_2{}_H^D$ | $10^5 k_1{}_H^T$ | $10^3 k_2{}_H^T$ | $10^5 k_1{}_D^T$ | $10^3 k_2{}_D^T$ | $k_2{}_H^D/k_2{}_H^T$ | $k_2{}_D^T/k_2{}_D^T$ |
|---|---|---|---|---|---|---|---|---|
| 0.0609 | 8.95 | 14.7 | 4.11 | 6.75 | | | 2.18 | |
| 0.122 | 17.2 | 14.1 | 7.85 | 6.43 | | | 2.19 | |
| 0.240 | | | 15.9 | 6.63 | | | | ⎫ |
| 0.243 | | | | | 27.2 | 11.2 | | 1.67 |
| 0.366 | | | 23.9 | 6.52 | | | | ⎫ |
| 0.379 | | | | | 41.7 | 11.0 | | 1.67 |
| 0.609[a] | 88.9 | 14.6 | 40.0 | 6.57 | | | 2.22 | |
| 0.610 | | | 39.8 | 6.52 | | | | ⎫ |
| 0.631 | | | | | 68.8 | 10.9 | | 1.69 |
| 1.22 | 196 | 16.1 | 92.7 | 7.60 | | | 2.11 | |
| 10.6 | 2,640 | 24.9 | 1,180 | 11.1 | | | 2.24 | |
| 14.2 | 4,850 | 34.1 | 2,370 | 16.7 | | | 2.05 | |
| 20.4 | 13,400 | 65.7 | 5,990 | 29.4 | | | 2.24 | |

[a] With a medium 0.05 $M$ in perchloric acid, Kresge and Chiang[473] obtained a value of 13.3 for $10^3 k_2{}_H^D$, 6.20 for $10^3 k_2{}_H^T$, and $k_2{}_H^D/k_2{}_H^T = 2.15$.

was confirmed in that the reaction was *general acid-catalysed* so that if this was accepted as a diagnosis of mechanism, then the mechanism could not be A-1. The existence of general acid catalysis was shown by the fact that subtraction of the rate at a specific value of $[H^+]$ or $[D^+]$ observed with aqueous perchloric acid (and assuming the entire value here to be specific acid-catalysed) from that at the same hydrogen ion concentration in acetate buffers left a substantial residual rate $k_2{}_{HA}^D$ etc. (Table 126) which must arise from catalysis by some other acid species. The derived substrate isotope effect $k_D/k_T$ of 1.89 is close to that found using perchloric acid and in other work noted above and the solvent isotope effect is within the range found by Eaborn *et al.*

Other examples of linear correlations of log rate coefficients with $-H_0$ have been reported though most of them are deficient in some respect. Satchell[474] observed such a correlation in the dedeuteration of anisole and benzene in various aqueous or acetic acid solutions of sulphuric acid, and aqueous perchloric acid, media at 25 °C. First-order rate coefficients are given in Table 127 along with those for dedeuteration of $[4,5\text{-}^2H_2]\text{-}1,3,5$-trimethoxybenzene and -2-hydroxy-1,3-dimethoxybenzene[475]. For the *ortho*- and *para*-monodeuterated anisoles the slopes of the log rate *versus* $-H_0$ plots were the same for aqueous sulphuric acid and for acetic acid–sulphuric acid so that it was concluded that the A-1 mechanism therefore applied to non-aqueous media as well as to aqueous media. The fact that the slope of the log rate coefficient *versus* $-H_0$ plot was less (1.12) for benzene in aqueous perchloric acid than that (1.36) previously found in aqueous

## TABLE 126

RATE COEFFICIENTS, SOLVENT AND SUBSTRATE ISOTOPE EFFECTS FOR REACTION OF
$[1\text{-}{}^2{}^{(3)}\text{H}]\text{-}2,4,6\text{-}(\text{MeO})_3\text{C}_6\text{H}_2$ WITH ACETATE BUFFERS AT 25 °C[472]

| $\dfrac{[H(D)A]}{[NaA]}$ | Ionic strength | $10^6[H^+]$ | $10^6[D^+]$ | $10^7 k_1{}_H^D$ | $10^7 k_1{}_H^T$ | $10^7 k_1{}_D^T$ |
|---|---|---|---|---|---|---|
| 0.10 | 0.47 | 3.31 |      | 6.21 | 3.18 |      |
| 0.15 | 0.20 | 4.64 |      | 4.10 | 2.09 |      |
| 0.25 |      | 7.73 | 2.44 | 6.92 | 2.43 | 1.12 |
| 0.50 |      | 15.5 | 4.89 | 13.8 | 7.18 | 4.67 |
| 0.75 |      | 23.2 | 7.33 |      | 11.0 | 7.42 |
| 1.00 |      | 30.9 | 9.77 | 28.2 | 14.5 | 9.87 |

| $10^7 k_2{}_H^D[H^+]^*$ | $10^7 k_2{}_H^T[H^+]^*$ | $10^7 k_2{}_D^T[D^+]^*$ | $10^7 k_1{}_{HA}^D$ | $10^7 k_1{}_{HA}^T$ | $10^7 k_1{}_{DA}^T$ | $k_1{}_{HA}^D{:}k_1{}_{HA}^T$ | $k_1{}_{DA}^T{:}k_1{}_{HA}^T$ |
|---|---|---|---|---|---|---|---|
| 0.480 | 0.218 |       | 5.73 | 2.95 |      | 1.94 |       |
| 0.673 | 0.305 |       | 3.43 | 1.78 |      | 1.93 |       |
| 1.12  | 0.509 | 0.268 | 5.80 | 3.12 | 2.16 | 1.87 | 0.688 |
| 2.25  | 1.02  | 0.538 | 11.5 | 6.16 | 4.13 | 1.82 | 0.690 |
|       | 1.52  | 0.806 |      | 9.50 | 6.61 |      | 0.696 |
| 4.48  | 2.03  | 1.08  | 23.7 | 12.5 | 8.79 | 1.19 | 0.698 |

\* Values interpolated from the data in Table 125 and ascribed to reaction *via* specific hydronium ion only.

sulphuric acid was attributed either to salt effects or to non-coincidence of the available $H_0$ scales.

From this work partial rate factors of $f_2^{\text{MeO}} = 2.3 \times 10^4$ and $f_4^{\text{MeO}} = 5.5 \times 10^4$ were quoted but since they were obtained *via* a large extrapolation using the erroneous data for benzene in Table 121 they may be in error by a factor of about 5. There may be an additional complication, for it is apparent from the data for [3-²H]-anisole and [²H]-benzene, that there is a markedly different dependence of rate upon acidity in the perchloric acid media (the slope of the log rate coefficient *versus* $-H_0$ plot for the former compound being 0.98), the differences in rates therefore, increasing with increasing acidity and rate whereas the reverse is expected from reactivity–selectivity principle. These anomalies very probably arise through increasing hydrogen bonding of the methoxy substituent with increasing acidity, thereby increasing electron withdrawal from the aromatic ring; the rates for the *ortho*- and *para*-deuterated anisoles may be similarly affected, though, of course, the rate coefficients were measured for weaker acid media in which this bonding would be less effective. In the acetic acid–sulphuric acid media (which give higher rates than for aqueous media of the same sulphuric acid composition as also found in other work, see p. 261, and indicating that the increase in acidity more than compensates for the poorer solvation of the transition state)

## TABLE 127

RATE COEFFICIENTS ($10^7 k_1$) FOR REACTION OF [$^2$H]-Ar WITH AQUEOUS $H_2SO_4$ OR $HClO_4$ AND HOAc–$H_2SO_4$ AT 25 °C[474,475]

**[$H_2SO_4$] (M) in $H_2O$**

| Substituent in Ar | Position of [$^2$H] | 3.51 | 4.70 | 4.86 | 5.18 | 5.22 | 5.90 | 5.95 | 6.77 | 6.82 | 7.34 | 7.59 | 7.88 |
|---|---|---|---|---|---|---|---|---|---|---|---|---|---|
| | $-H_0 =$ | 1.44 | 2.15* | 2.05 | 2.22 | 2.37* | 2.55 | 2.59 | 3.00 | 3.06 | 3.34 | 3.44 | 3.60 |
| MeO | 2 | 2.93 | | 18.5 | 14.9 | | 36.3 | 88.0 | 255 | 121 | 291 | 839 | 521 |
| MeO | 4 | 520 | | | | | | | | | | | |
| 1,2,3-(MeO)$_3$ | 4,6 | | 1,750 | | | 2,910 | | | | | | | |
| 2-HO-1,3-(MeO)$_2$ | 4,6 | | 3,000 | | | 5,060 | | | | | | | |

**[$H_2SO_4$] (M) in HOAc**

| Substituent in Ar | Position of [$^2$H] | 0.232 | 0.550 | 0.632 | 0.916 | 1.23 | 1.72 | 2.64 | 2.68 | 3.52 |
|---|---|---|---|---|---|---|---|---|---|---|
| | $-H_0 =$ | 1.93 | 2.49 | 2.56 | 2.80 | 3.00 | 3.28 | 3.73 | 3.76 | 4.18 |
| MeO | 2 | 5.0 | 14.0 | 30.0 | 50.0 | 51.0 | 230 | 820 | 2,100 | |
| MeO | 4 | | | | | | 450 | | | |

**[$HClO_4$] (M) in $H_2O$**

| Substituent in Ar | Position of [$^2$H] | 3.00 | 10.04 | 10.43 | 11.01 | 11.71 |
|---|---|---|---|---|---|---|
| | $-H_0 =$ | | 5.45 | 5.85 | 6.55 | 7.35 |
| MeO | 3 | | 7.20 | 28.2 | 32.4 | 175 |
| H | 1 | 2.00 | | | | 1,350 |

\* These $H_0$ values were taken from the compilation of Long and Paul[462], the others were not.

TABLE 128

RATE COEFFICIENTS FOR REACTION OF [4-$^2$H]-ANISOLE WITH $SnCl_4$ OR $ZnCl_2$ IN HOAc OR HCl–HOAc AT 25 °C[476]

| | | | | | | | | | | |
|---|---|---|---|---|---|---|---|---|---|---|
| $10[SnCl_4]$ (M) | 0.15 | 0.59 | 1.18 | 1.47 | 2.20 | 8.96 | 7.32 | 8.79 | 11.0 | 29.0 |
| $10^7 k_1$ | 3.05[a] | 3.01 | 7.10 | 28.8[a] | 19.5 | 63.1 | 225[a] | 211 | 617[a] | 15,900 |
| $10[ZnCl_2]$ (M) | 0.15 | 0.24 | 0.40 | 0.48 | 1.01 | 1.44 | 2.20 | 2.90 | 3.50 | 4.80 |
| $10^7 k_1$ | 3.24[b] | 13.6[c] | 5.81[b] | 21.9[c] | 7.90[b] | 57.0[c] | 7.22[b] | 105[c] | 6.04[b] | 139[c] |
| $10[ZnCl_2]$ (M) | 20.7 | 26.8 | 31.9 | | | | 6.50 | 11.2 | 14.8 | 19.0 |
| $10^7 k_1$ | 4.05 | 8.15 | 16.6 | | | | 3.71[b] | 177[c] | 201[c] | 3.10 |

[a] In presence of 0.001 $M$ HCl.
[b] In presence of 0.05 $M$ HCl.
[c] In presence of 0.42 $M$HCl.

a side reaction occurred causing loss of anisole at approximately one-twentieth of the rate of exchange and this reaction was almost certainly sulphonation.

The results for the trimethoxy compound show the reactivity to be about 60 times lower than that expected from the additivity principle which points to a steric effect which may be hindrance to the incoming reagent, mutual inhibition of resonance by the methoxy groups through inability to attain a planar configuration, or mutual enhancement of hydrogen bonding. The greater reactivity of the hydroxy relative to the methoxy compound was attributed to hyperconjugation, though the possibility of an unresolved steric effect must make this conclusion a tentative one.

Satchell[476] also measured the first-order rate coefficients for dedeuteration of [4-$^3$H]-anisole by acetic acid or acetic acid–hydrochloric acid media containing zinc and stannic chlorides (Table 128). The rates here paralleled the indicator ratio of 4-nitrodiphenylamine and 4-chloro-2-nitroaniline, so that the implication is that a linear relationship exists between log $k$ and the unknown $H_0$ values. The results also show the rate-enhancing effect of these Friedel–Crafts catalysts, presumably through additional polarisation of the catalysing acid, for in the absence of them, exchange between acetic acid and anisole would be very slow. Other studies relating to the effect of these catalysts are reported below (p. 238).

An exceptionally badly reported kinetic study in which a linear correlation of rate coefficient with acidity function was claimed was that of Mackor et al.[477], who studied the dedeuteration of benzene and some alkylbenzenes in sulphuric acid–trifluoroacetic acid at 25 °C. Rates were given only in the form of a log rate coefficient versus $-H_0$ plot and rate coefficients and entropies of activation (measured relative to p-xylene) together with heats of activation (determined over a temperature range which was not quoted) were also given (Table 129). However,

TABLE 129

KINETIC DATA FOR REACTION OF [$^2$H]-Ar WITH $H_2SO_4$–$CF_3COOH$ OR $H_2SO_4$–$CH_3COOH$ AT 25 °C[477]

| ArH | log $k_{rel}$[a] | $\Delta S^{\ddagger}$ | $\Delta H^{\ddagger}$ | Solvent | $[H_2SO_4]$ (moles/1000g) |
|---|---|---|---|---|---|
| [$^2$H]-Benzene | −3.1 | −9.0 | 16.6 | $CF_3CO_2H$ | 2.0 |
| [3-$^2$H]-Toluene | −2.6 | | Not given | | |
| [2-$^2$H]-Toluene | −0.73 | +1.0 | 16.3 | $CF_3CO_2H$ | 0.25 |
| [4-$^2$H]-Toluene | −0.56 | −0.8 | 16.0 | $CF_3CO_2H$ | 0.25 |
| [4-$^2$H]-Ethylbenzene | −0.57 | −1.5 | 15.3 | $CF_3CO_2H$ | 0.25 |
| [4-$^2$H]-t-Butylbenzene | −0.54 | −0.1 | 15.8 | $CF_3CO_2H$ | 0.25 |
| [2-$^2$H]-p-Xylene | 0 | 0 | 15.0 | $CF_3CO_2H$ | 0.10 |
| [2-$^2$H]-p-Xylene | | | 16.0 | $CF_3CO_2H$ | 0.015 |
| [4-$^2$H]-m-Xylene | 2.3 | +16.3 | 17.8 | $CF_3CO_2H$ | 0.015 |
| [4-$^2$H]-m-Xylene | | | 23.5 | $CH_3CO_2H$ | 6.0 |
| [$^2$H]-Mesitylene | 4.4 | +24 | 23.0 | $CH_3CO_2H$ | 3.5 |

[a] Values relative to the data for p-xylene.

the quoted log $k_{rel}$ values do not correspond to those which may be interpolated from the graph, nor is the relevance of the acid concentration given in Table 129 apparent: one possibility seems to be that it is the concentration required to give an exchange rate for the relevant compound, identical to that for p-xylene. In view of these deficiencies, the data in this paper must be treated with considerable caution. The correlation of log rate coefficient with $-H_0$ is seen also to show departure from linearity even over the small $-H_0$ range ($< 1.8$ units) studied, and one straight line plot was drawn through two points only.

Thus it can be seen that evidence for the A-1 mechanism, even if one accepted that this followed from a linear rate coefficient–acidity function correlation, was scant. On the other hand, there have been a very large number of carefully documented studies in which general acid catalysis has been observed leading to the A-$S_E2$ mechanism for the reaction, or it has been shown that the conclusions from an acidity function dependence are not rigorous. One such study has already been described above, and Satchell[478] also found that in the detritiation of [4,6-$^3H_2$]-1,2,3-trimethoxybenzene by potassium bisulphate, dichloro- and tri-fluoroacetic acids, plots of log $k_1$ versus $-H_0$ were linear with a slope of ca. 1.0

TABLE 130

FIRST-ORDER RATE COEFFICIENTS FOR REACTION OF [4,6-$^3H_2$]-1,2,3-(MeO)$_3$C$_6$H WITH HX AT 25 °C[478]

| [KHSO$_4$] (M) | $10^7k_1$ | [CHCl$_2$COOH] (M) | $10^7k_1$ | [CF$_3$COOH] (M) | $10^7k_1$ |
|---|---|---|---|---|---|
| 0.5 | 1.84 | 0.12 | 0.34 | 0.99 | 4.80 |
| 1.0 | 4.41 | 0.24 | 0.48 | 2.16 | 12.2 |
| 2.0 | 10.5 | 1.21 | 1.51 | 4.04 | 19.8 |
| 2.5 | 14.4 | 2.42 | 2.40 | 6.73 | 137 |
| | | 4.84 | 6.00 | | |
| | | 7.24 | 11.0 | | |
| | | 12.13 | 1,230 | | |

for the former two compounds, whereas for the latter acid, rates increased faster with increasing acid concentration than was required for an acidity function dependence (Table 130) and it was concluded that for reaction with this acid at least, general acid catalysis occurs.

Kresge and Chiang[15] showed that the dependence of rate coefficient upon the Hammett acidity function was not a safe criterion of mechanism and also explained how different dependencies of rate upon acidity function for different compounds could arise. Rate coefficients for detritiation of [$^3H$]-2,4,6-trimethoxybenzene followed the acidity function $h_0$ in perchloric acid up to 3 $M$, i.e. $k = 0.484 (h_0)^{1.07}$, so that a plot of log $k$ versus $-H_0$ would have a slope of 1.07, which implies the A-1 mechanism. (New acidity function values[479] would make this slope 1.14.) However, the diagnostic value of such a correlation was revealed as meaningless by the fact that trimethoxybenzene turns out not to be a Hammett base. The degree of protonation of this compound is proportional to the acidity function $h'_R$ (where $h'_R = h_R.a_{H_2O}$ and $H_R(-\log h_R) = pk_R + \log [ROH]/[R^+]$ rather than to $h_0$ (where $H_0(-\log h_0) = pk_{BH} + \log [B]/[BH^+]$). Since in the acid range studied $h'_R \approx (h_0)^2$, for the A-1 mechanism to apply, a correlation of log rate *versus* $-H_0$ should have been obtained with a slope of 2.0. If by contrast, a proton transfer has not yet occurred in the transition state, then this will be composed of the aromatic and a nearly intact hydronium ion, so that the rate will depend on $[H_3O^+]$ and the linear log $k$ *versus* $-H_0$ plot should have a slope of $< 1.0$. Thus with reactive compounds which will have transition states nearer to the ground state, proton transfer will be less complete and a slope of *ca.* 1.0 will be obtained and this will increase to *ca.* 2.0, the less reactive the substrate. This is entirely confirmed by the data above which give slopes of *ca.* 1.0 for phenols, through 1.5 for toluene to 2.0 for benzene. Also the transition state for detritiation will be nearer to the products than that for dedeuteration so that a greater slope will be obtained, as was noted for benzene and toluene. Further confirmation that compounds of different reactivity have different acidity dependencies for this protonation was provided by the fact that protonation of *m*-dimethoxybenzene increases even more rapidly with acid strength than is required by a dependence upon $h'_R$.

Kresge and Chiang[480] measured the rate coefficients for detritiation of [1-$^3$H]-2,4,6-trimethoxybenzene in acetate buffers and found the first-order rate coefficient $(10^7k_1)$ to increase from 2.5 at 0.01 $M$ acetic acid to 8.3 at 0.1 $M$ acetic acid, whereas if the reaction was specific acid-catalysed no change in rate should have been observed. A similar technique to that described above for separation of the rate coefficients due to hydronium ions and other acids was used, the values for the former being obtained using dilute hydrochloric acid at which acidities no undissociated acid was present (Table 131). Rate coefficients were then measured

TABLE 131

RATE COEFFICIENTS FOR REACTION OF [$^3$H]-2,4,6-(MeO)$_3$C$_6$H$_2$ WITH HCl
(IONIC STRENGTH = 0.1) AT 25 °C[480]

| $10^2[HCl](M)$ | 5.0 | 4.0 | 3.0 | 2.0 | 1.0 |
|---|---|---|---|---|---|
| $10^5k_1$ | 32.4 | 25.0 | 20.8 | 12.8 | 8.0 |
| $10^3k_2$ | 6.5 | 6.35 | 7.0 | 6.35 | 8.0 |

with solution of different buffer concentrations, but constant buffer ratios, at a constant ionic strength of 0.1 at 25 °C (Table 132). Linear plots of rate coefficient *versus* the buffer acid concentration gave positive values when extrapolated to zero acid concentration and these values, given in Table 132, were taken to be due to reaction with the small concentration of hydronium ion in the particular medium. Subtraction of these first-order rate coefficients from the observed rate

TABLE 132

RATE COEFFICIENTS FOR REACTION OF $[^3H]$-2,4,6-$(MeO)_3C_6H_2$ WITH HX (IONIC STRENGTH = 0.1) AT 25 °C[480]

| $10^2[HX]$ (M) | $10^7k_1$ | $10^7k_1$ $(H_3O^+)$ | $10^6k_2$ (HA) | $10^2[HX]$ (M) | $10^7k_1$ | $10^7k_1$ $(H_3O^+)$ | $10^6k_2$ (HA) |
|---|---|---|---|---|---|---|---|
| $X = OAc$ | [HA]/[NaA] = 1.0 | | | $X = OOCH$ | [HA]/[NaA] = 2.0 | | |
| 10.0 | 8.42 | 2.04 | 6.33 | 20 | 56.0 | 33.0 | 11.5 |
| 5.0 | 4.63 | 2.04 | 5.16 | 14 | 49.2 | 33.0 | 11.5 |
| 2.5 | 3.83 | 2.04 | 7.16 | 10 | 44.2 | 33.0 | 11.15 |
| 1.0 | 2.72 | 2.04 | 6.13 | 60 | 40.2 | 33.0 | 12.0 |
| $X = OAc$ | [HA]/[NaA] = 0.2 | | | $X = OOCCH_2F$ | [HA]/[NaA] = 1.0 | | |
| 2.0 | 1.66 | 0.376 | 7.00 | 10 | 275 | 186.5 | 88.4 |
| 1.0 | 1.06 | 0.376 | 6.83 | 5.0 | 209 | 173.5 | 70.0 |
| 0.2 | 0.515 | 0.376 | 7.00 | 2.9 | 192 | 167 | 86.7 |
| | | | | 2.5 | 188 | 160 | 113 |
| | | | | 1.0 | 130 | 123 | 66.7 |

coefficients permitted calculation of the second-order rate coefficients for reaction of the general acid as given in Table 132. In fluoroacetate buffers, hydronium ion concentrations were not constant so that a linear correlation of rate *versus* the concentration of the buffer acid was not obtained, and a method was devised for calculating these at each buffer concentration; the less satisfactory method is reflected in the greater variation in the calculated second-order rate coefficients for reaction with fluoroacetic acid.

For water, the second-order rate coefficient was determined as $9.5 \times 10^{-12}$ by extrapolation from data at higher temperatures and using the presence of hydroxide ion to suppress any reaction with hydronium ion. For reaction with solutions of biphosphate and ammonium ions, since reaction *via* hydronium ions in these media is negligible (*ca.* 1 % of the total rate), the second-order rate coefficients were evaluated from exchange data at a single acid concentration as $k_2$ $(H_2PO_4^-) = 3.89 \times 10^{-7}$ and $k_2(NH_4^+) = 5.0 \times 10^{-9}$, the latter value being corrected for the water-catalysed reaction.

Base catalysis was shown not to be significant on two grounds. Firstly, the second-order rate coefficients for the two sets of acetate buffer data are the same within experimental error, and secondly, the addition of base of concentrations 0.05 and 0.2 M to the reaction with water caused a negligible change in the rate coefficient.

In summary, this work which covered seventeen pK units and nine powers of ten in rate, gave an excellent Brönsted correlation

$$k_{HA} = G(K_{HA})^\alpha \tag{224}$$

(where $K_{HA}$ is the dissociation constant of the acid HA, and G is a constant, and with $\alpha = 0.518$). Although the aromatic here is very reactive, the very weak acids employed mean that a range of reactivity was covered equivalent to the difference in reactivity between trimethoxybenzene and benzene. The value of $\alpha$ shows that in the rate-determining transition state, the exchanging proton is half-transferred from catalysing acid to the aromatic and again this is inconsistent with the A-1 mechanism for which this transfer should be complete.

Kresge and Chiang[473, 481] have also presented a clear analysis of the problem associated with measurement of kinetic isotope effects in hydrogen exchange. Firstly, there are two steps involving hydrogen transfer, both of which are rate-determining to a similar extent, and, secondly, two isotopes must be used to measure a single exchange rate and a third is required in order to obtain an isotope effect. The rate of the forward reaction is given by the equation

$$\text{Rate} = \frac{k_1' k_2'}{k_{-1}' + k_2'} = \frac{k_1'}{(k_{-1}'/k_2') + 1} \tag{225}$$

from which is obtained, for example, equation (226)

$$k_{obs}(\text{protodetritiation}) = \frac{k_1'(H)}{[k_2'(H)/k_2'(T)] + 1} \tag{226}$$

for protodetritiation where the letter in parentheses indicates the form of hydrogen being transferred to and from the aromatic. Likewise, the appropriate equations for protodedeuteration, deuterodetritiation, etc. can be set up and solved for the isotope effects for the individual steps $k_1'(H)/k_1'(D)$, $k_2'(H)/k_2'(D)$ etc. However, this neglects secondary isotope effects, principal among which is the fact that a change in hybridisation occurs at the position undergoing substitution in each step of the reaction, and these changes do not cancel each other out because the hydrogen which becomes attached in the first step is not the one that becomes detached in the second. Thus $k_1'(H)$ for protodedeuteration is not the same as $k_1'(H)$ for protodetritiation and the equation of the form of (226) must be modified to the form

$$k_{obs}(\text{protodetritiation}) = \frac{k_1'(H-T)}{[k_2'(H-T)/k_2'(T-H)] + 1} \tag{227}$$

where $k_1'(H-T)$ means that hydrogen is being transferred in the first step *to* a carbon bearing a tritium atom, $k_2'(H-T)$ means that hydrogen is being transferred in the second step *from* a carbon atom already bearing a hydrogen and a tritium atom, etc. This increases the number of unknowns, but use of the Swain relationships[482], *viz.*

$$[k(H)/k(D)^{1.442} = k(H)/k(T)$$

and

$$[k(T)/k(D)]^{1.442/0.442} = k(T)/k(H)$$

together with the method of successive approximations, leads to a solution.

For the hydrogen exchange of 1,3,5-trimethoxybenzene in aqueous perchloric acid at 24.6 °C, second-order rate coefficients ($10^2 k_2$) were obtained as follows: 0.620 (protodetritiation); 1.048 (deuterodetritiation); 1.384 (deuterodeprotonation); 2.79 (protodedeuteration) and analysis of these results into their components by the method outlined above yielded the *second-order* rate coefficients for the individual steps as follows: $k'_1(H–T) = 13.2 \times 10^{-2}$; $k'_1(D–T) = 3.69 \times 10^{-2}$; $k'_1(H–D) = 12.6 \times 10^{-2}$; $k'_1(D–H) = 3.14 \times 10^{-2}$. Hence the transfer of hydrogen in the first step takes place $13.2/3.69 = 3.58$ times faster than deuterium, the transfer of deuterium to carbon bearing hydrogen takes place $3.14/3.69 = 0.85$ times slower than transfer to a carbon bearing tritium, and hence by the Swain relationship, 0.90 times slower than transfer to a carbon bearing deuterium, both of these being secondary isotope effects, and $k'_2(H–D)/k'_2(D–H)$ and $k'_2(H–T)/k'_2(T–H)$ are 8.1 and 20.3 respectively. The latter two values, of course, are products of primary and secondary isotope effects, and assuming the hybridisation factors in the second step are the reciprocals of those in the first step (this is a reasonable assumption for trimethoxybenzene since the value of $\alpha$ is approximately 0.5 so that the proton is half transferred in the transition state which is therefore symmetrical) then $k'_2(H–D)/k'_2(D–H)$ (primary effect) $= 8.1 \times 1.11 = 9.0$. This is a very large value but is to be expected since the value of 0.5 for the Brönsted coefficient $\alpha$ (and hence the maximum symmetry of the transition state with respect to proton transfer) leads one to expect a maximum isotope effect.

The secondary isotope effects come out to be inverse, as they should be[483], and likewise have a value of about 12 % per atom[484]. If these effects are neglected, *i.e.* the experimental data are used to solve the equations of the form of (226), then the following values are obtained: $k'_1(H)/k'_1(D) = 1.25$; $k'_2(H)/k'_2(D) = 2.51$; $k'_2(H)/k'_2(T) = 6.84$, and these are much too low. Alternatively, if the simplified equation for protodedeuteration and protodetritiation are used and the Swain equation used to eliminate $k'_2(H)k'_2(T)$, this being the common method of dealing with these systems, a value of 7.2 for $k'_2(H)/k'_2(D)$ is obtained, or 8.0 after correction for secondary isotope effects, and this is significantly lower than that obtained by the more rigorous method.

General acid catalysis has been observed by Kresge *et al.*[485] in the detritiation of [³H]-2,4-dimethoxybenzene over a range of temperatures. The rate of reaction in formic acid-formate buffers at ionic strength $\mu = 0.1$ (by addition of sodium chloride) was given approximately by $7.95 \times 10^{-6}[H^+] + 1.79 \times 10^{-8}[HCOOH]$ at 25 °C and $10.8 \times 10^{-3}[H^+] + 2.14 \times 10^{-5}[HCOOH]$ at 96° C, the dependence

on hydronium ion concentration being determined from studies in dilute perchloric acid media. At 25 °C reaction rates were so slow that about 3 % reaction
only was followed, the zeroth-order values thus obtained being converted to
first-order by dividing by the total initial radioactivity. The fact that general acid
catalysis was observed for this compound means that this mechanism is now
proved for compounds with a span of reactivity covering seven powers of ten.
Furthermore, linear plots of log rate *versus* $-H_0$ values for 10–40 wt. % perchloric
acid were obtained at 15.0, 25.7, 35.9 and 45.0 °C (Table 133), the slopes being

TABLE 133

FIRST-ORDER RATE COEFFICIENTS FOR REACTION OF [³H]-2,4-(MeO)₂C₆H₃ WITH
AQUEOUS HClO₄[485]

| $HClO_4$ (wt. %) | $-H_0$ | $10^7 k_1$ | $HClO_4$ (wt. %) | $-H_0$ | $10^7 k_1$ |
|---|---|---|---|---|---|
| *Temp. = 15.0 °C* | | | *Temp. = 24.7 °C* | | |
| 10.04 | 0.36 | 6.865 | 9.94 | 0.35 | 22.3 |
| 18.75 | 0.90 | 29.95 | 18.82 | 0.91 | 97.0 |
| 22.84 | 1.15 | 57.8 | 26.82 | 1.39 | 335 |
| 26.70 | 1.38 | 107 | 33.95 | 1.86 | 1,180 |
| 33.84 | 1.86 | 375 | 37.10 | 2.11 | 2,260 |
| 37.01 | 2.10 | 726 | 40.21 | 2.37 | 4,440 |
| 40.34 | 2.39 | 1,480 | *Temp. = 45.0 °C* | | |
| *Temp. = 35.0 °C* | | | 10.08 | 0.36 | 224 |
| 9.995 | 0.35 | 76.65 | 14.61 | 0.64 | 487 |
| 18.96 | 0.91 | 319.5 | 18.97 | 0.91 | 933 |
| 26.93 | 1.40 | 1,115 | 23.11 | 1.16 | 1,750 |
| 33.96 | 1.86 | 3,740 | 26.94 | 1.40 | 3,110 |
| 37.24 | 2.12 | 7,020 | 30.65 | 1.63 | 5,630 |

1.15, 1.14, 1.11, and 1.085 respectively. Here again, application of the Zucker–
Hammett hypothesis leads one to the entirely wrong conclusion, and it seems
certain that general acid catalysis is observed for all compounds in hydrogen
exchange; significantly, the slopes increase with decreasing temperature, corresponding to less proton transfer in the transition state the lower the temperature,
whereas according to the A-1 mechanism the slopes should be temperature-
independent.

The detritiation of [³H]-2,4,6-trimethoxybenzene by aqueous perchloric acid was
also studied, the second-order rate coefficients ($10^7 k_2$) being determined as 5.44,
62.0, and 190 at 0, 24.6, and 36.8 °C, respectively, whilst with phosphate buffers,
values were 3.75, 13.8, and 42.1 at 24.6, 39.9, and 55.4 °C, respectively. The summarised kinetic parameters for these studies are given in Table 134, and notable
among the values are the more negative entropies of activation obtained in catalysis
by the more negative acids. This has been rationalised in terms of proton transfer

TABLE 134

KINETIC PARAMETERS FOR REACTION OF [³H]Ar WITH HX[485]

| $ArH$ | $HX$ | $\Delta H^{\ddagger}$ | $\Delta S^{\ddagger}$ |
|---|---|---|---|
| [³H]-2,4-dimethoxybenzene[a] | $H_3O^+$ | 21.5 | −9.7 |
| [³H]-2,4-dimethoxybenzene[b] | $H_3O^+$ | 22.1 | −7.9 |
| [³H]-2,4-dimethoxybenzene | HCOOH | 21.1 | −23.1 |
| [³H]-2,4,6-trimethoxybenzene | $H_3O^+$ | 15.6 | −16.3 |
| [³H]-2,4,6-trimethoxybenzene | $H_2PO_4^-$ | 14.7 | −38.4 |

[a] Dilute solutions.
[b] Concentrated solutions.

from hydronium ion being accompanied by liberation of some solvating water molecules which will raise the entropy of the system, making a positive contribution to $\Delta S^{\ddagger}$. Proton transfer from a neutral or negatively charged acid will result in an increase in solvation, since here new ionic species are being formed and there will, therefore, be a negative contribution to $\Delta S^{\ddagger}$.

Long *et al.* have carried out many kinetic investigations similar to the above and have arrived at like conclusions. In their work, azulene and its derivatives were used since monoprotonation at carbon is assured and there is no possibility of ambiguity arising from protonation at oxygen or diprotonation such as could arise from the use of bases containing oxygen; however, the similarity of the conclusions to those of Kesge *et al.* suggests that the compounds used by the latter are perfectly satisfactory. Long and Schulze[16,486] confirmed the findings of Kresge and Chiang[15] in that log [BH$^+$]/[B] for the protonation of azulene and 1-methylazulene was more nearly proportional to $H_R$ than to $H_0$. However, for 1-nitroazulene the correlation was with $H_0$ even though protonation was shown to occur at the 3 carbon. Rate coefficients for detritiation of azulene were given by $k_1 = 0.19$ [H$^+$] (see below), but, in the acid range studied (pH 2–4), $h_0$ and $h_R$ are almost identical so that no mechanistic conclusions were possible. By contrast, for nitroazulene, $k_1 = 5.3 \times 10^{-6} . h_0^{1.05}$ so that there is clearly considerable specificity in the interaction of the system with acids; this was rationalised in terms of the high electrolytic concentrations of the media involved with consequent large and probably specific medium effects on the activity coefficients of the various species present. A further complication is that although the slopes of the correlations of log [BH$^+$]/[B] with $-H_0$ were 1.9 (azulene), 1.6 (methylazulene), 1.7 (*p*-dimethylaminobenzylazulene), 1.8 (chloroazulene) all of these compounds protonating at carbon, 1.1 (formylazulene, protonation at oxygen) and 0.9 (carboxylazulene, protonation site uncertain), the correlation of log rate with acidity function $H_0$ had similar slopes in each case as described below.

Thomas and Long[487] measured second-order rate coefficients for detritiation of [1-³H]-azulene and [3-³H]-guaiazulene in a range of acids at 25 °C (Table

TABLE 135

SECOND-ORDER RATE COEFFICIENTS FOR REACTION OF [1-$^3$H]-AZULENE (A) AND [3-$^3$H]-GUAIAZULENE (G) WITH HA AT 25 °C[487]

| HA | $10^2 k_2(G)$ | $10^4 k_2(A)$ | HA | $10^2 k_2(G)$ | $10^4 k_2(A)$ |
|---|---|---|---|---|---|
| Dimethylacetic acid | | 3.14 | p-Anisidinium ion | | 8.0 |
| Acetic acid[a] | 1.38 | 2.50 | p-Toluidinium ion | | 11.0 |
| Glycollic acid | 3.60 | 9.58 | Anilinium ion | | 13.4 |
| Formic acid | 5.10 | 12.1 | o-Anisidinium ion | | 38.0 |
| Iodoacetic acid | | 41.6 | m-Anisidinium ion | | 44.0 |
| Chloroacetic acid | | 45.7 | p-Bromoanilinium ion | | 72 |
| Oxalate ion | | 17.6 | Hydronium ion[b] | 610 | 1,810 |
| Malonate ion | | 98.0 | Water | $7 \times 10^{-7}$ | $2.1 \times 10^{-6}$ |
| Maleate ion | | 33.0 | Biphosphate ion | 0.174 | |
| | | | Ammonium ion | $7.2 \times 10^{-4}$ | |

[a, b] Rates ($10^4 k_2$) of detritiation of [1-$^3$H]-cyclo[3,2,2]azine were 0.848 and 480 in these media, respectively.

135). For azulene, good Brönsted plots of log rate *versus* p$K_a$ were obtained, the value of $\alpha$, the slope, being 0.61 (anilinium ions), 0.88 (dicarboxylic acid mono-anions), and 0.67 (carboxylic acids), and the points for water and hydronium ion came at the extreme ends of a plot on which the separate groups of points approximately fell. Significantly, the more basic acids gave a higher value of $\alpha$, which is entirely in accord with theoretical expectation since the transition state for reaction with these acids will occur later along the reaction coordinate, and proton transfer will be more complete in the rate-determining transition state. Likewise, if the aromatic is more basic, less transfer of the proton will have occurred in the transition state and a lower value of $\alpha$ should be expected. Accordingly, the data for the detritiation of [3-$^3$H]-guaiazulene (Table 135) gave a Brönsted correlation with $\alpha = 0.54$.

Thomas and Long[488] also measured the rate coefficients for detritiation of [1-$^3$H]-cycl[3,2,2]azine in acetic acid and in water and since the rates relative to detritiation of azulene were similar in each case, a Brönsted correlation must similarly hold. The activation energy for the reaction with hydronium ion (dilute aqueous hydrochloric acid, $\mu = 0.1$) was determined as 16.5 with $\Delta S^{\ddagger} = -11.3$ (from second-order rate coefficients ($10^2 k_2$) of 0.66, 1.81, 4.80, and 11.8 at 5.02, 14.98, 24.97, and 34.76 °C, respectively). This is very close to the values of 16.0 and $-10.1$ obtained for detritiation of azulene under the same condition[499] (below) and suggests the same reaction mechanism, general acid catalysis, for each.

The detritiation of [1-$^3$H]-azulene in aqueous hydrochloric acid has been examined more extensively by Schulze and Long[489], using media of constant ionic strength (0.1) at 25 °C. Average first-order rate coefficients for detritiation by 0.001, 0.005, and 0.01 M hydrogen ion were $1.75 \times 10^{-4}$, $9.35 \times 10^{-4}$, and

$1.77 \times 10^{-3}$ giving an average value of $k_2$ of 0.180. With formic acid buffers the linear correlation of rate coefficients with buffer acid concentration was given by $k_1 = 0.183[H^+] + 1.16 \times 10^{-3}[HCOOH]$, and the similarity of the second-order coefficients for reaction with hydronium ion under both conditions confirmed the correctness of the approach. In mixtures of $H_2O$ and $D_2O$ containing 0.005 $M$ hydrogen ion, the rates were correlated by means of the "cubic" Gross–Butler equation, but since the reaction is general acid-catalysed this equation was again shown to have no diagnostic value.

The rate coefficients for detritiation of some 3-substituted-1-tritiated azulenes were also measured in a range of aqueous perchloric acid media at 25 °C (Table 136). The log rate coefficients correlated with $-H_0$ with slopes of 1.05, 1.08,

TABLE 136

FIRST-ORDER RATE COEFFICIENTS FOR REACTION OF [1-$^3$H]-3-X-AZULENES WITH
AQUEOUS $HClO_4$ AT 25 °C[480]

| $[HClO_4]$ (M) | $10^5 k_1$ (X = NO$_2$) | $[HClO_4]$ (M) | $10^5 k_1$ (X = CN) | $[HClO_4]$ (M) | $10^5 k_1$ (X = CHO) |
|---|---|---|---|---|---|
| 1.09 | 0.90 | 0.35 | 2.39 | 0.24 | 2.75 |
| 1.56 | 1.70 | 0.58 | 4.89 | 0.50 | 8.31 |
| 1.93 | 2.49 | 1.13 | 15.1 | 1.12 | 21.3 |
| 2.80 | 9.27 | 1.56 | 28.2 | 1.55 | 37.2 |
| 3.89 | 17.1 | 1.98 | 54.7 | 1.96 | 70.6 |
| 3.98 | 26.8 | 2.42 | 100 | | |
| 4.46 | 76.8 | 2.74 | 146 | | |
| 4.88 | 117 | 3.12 | 252 | | |
| 5.36 | 206 | 3.68 | 454 | | |

and 1.20 for the formyl, nitro, and cyano compounds, respectively so that an increase in slope accompanies decreasing basicity as noted above (p. 208) and argues as it did there, against the A-1 mechanism.

The dedeuteration of [1-$^2$H]-azulene has also been studied. In a preliminary study, Colapietro and Long[490] found that rate coefficients correlated with the concentration of buffer acid in solutions of pH 3–5 and data obtained in formate, acetate, and chloroacetate buffers gave a value of $\alpha$ of 0.5. Gruen and Long[491] subsequently found that the observed first-order rate coefficients for dedeuteration in formate buffers at 25 °C was given by $0.45[H^+] + 2.5 \times 10^{-3}[HCOOH]$ and in acetate buffers by $0.45[H^+] + 5.1 \times 10^{-4}[CH_3COOH]$. The two values for the second-order rate coefficient for reaction with hydronium ion agree exactly with that obtained from a third study of the rates of dedeuteration in aqueous hydrochloric acid media ($\mu = 0.1$), given in Table 137; in the absence of added sodium chloride an average value of 0.41 was obtained, hence it was concluded that there was a small ionic strength effect.

TABLE 137

FIRST-ORDER RATE COEFFICIENTS FOR REACTION OF [²H]-AZULENE WITH AQUEOUS
HCl AT 25 °C[491]

| Added NaCl, $\mu = 0.1$ | | | No added NaCl | | |
|---|---|---|---|---|---|
| $10^4[H^+]$ (M) | $10^4 k_1$ | $k_2$ | $10^4[H^+]$ (M) | $10^4 k_1$ | $k_2$ |
| 9.89 | 4.35 | 0.449[a] | 5.95 | 2.41 | 0.405 |
| 9.93 | 4.44 | 0.450[a] | 9.87 | 4.40 | 0.446 |
| 20.7 | 10.4 | 0.503 | 9.87 | 3.75 | 0.379 |
| 33.1 | 14.4 | 0.436 | 100.4 | 4.35 | 0.433 |
| 43.2 | 18.2 | 0.422 | 137.7 | 5.72 | 0.416 |
| 52.2 | 23.3 | 0.446 | 157.0 | 6.07 | 0.387 |
| 62.3 | 28.5 | 0.457 | 163.4 | 6.45 | 0.395 |
| 85.0 | 37.3 | 0.444 | | | |
| 108 | 46.0 | 0.426 | | | |

[a] In error in original paper.

Comparison of these results with those obtained for detritiation, gives values of $k_H/k_D$ of 2.5($H_3O^+$), 2.1(HCOOH) and 2.0($CH_3COOH$) which, taken in conjunction with the solvent isotope effect data obtained by Schulze and Long and solution of the simplified equations of the form of (226), gives values of $k_1'(H)/k_1'(D)$ and $k_2'(H)/k_2'(D)$ of 4.3 and 9.2 respectively. These are greater than those obtained from the work of Kresge and Chiang[473, 481] (using the same method, see p. 212) and the value of 9.2 is greater than the theoretical maximum of 7. The discrepancy was attributed to the possibility of tunneling, the failure to take into account with the possible differences in vibrational energies in the transition state and the effect of bending modes in the initial and transition states. These calculated isotope effects would, of course, be even greater if the effect of secondary isotope effects was accounted for (see p. 212).

Comparison of the values of $k_2'(H)/k_2'(D)$ with $\Delta pK$ values between acid and protonated aromatic base gives the following: hydronium ion 9.2 ($-0.02$); formic acid 6.3($-5.51$); acetic acid 6.1($-6.52$). This is in agreement with the postulate that the kinetic isotope effect should be a function of the position in the transition state of the hydrogen atom which is being transferred[492]. It follows that there should be a maximum effect when the protonated substrate and the catalysing acid have similar acidities, i.e. the transition state will be symmetrical and $\alpha$ equal to 0.5. This has been further amplified by the data of Longridge and Long[493], who have measured rates of proton transfers to and from aromatic bases using a fast flow apparatus. The first-order rate coefficients for the approach to equilibrium are given by

$$k_1 = \frac{2.303}{t} \log\left(\frac{[AzH_2^+]_{eq} - [AzH_2^+]_0}{[AzH_2^+]_{eq} - [AzH_2^+]_t}\right) \qquad (228)$$

where AzH is azulene. Now since the overall rate coefficient $k_1 = k_1$ (forward)$+k_1$ (reverse) it can be shown that $k_1$ (forward) is given by

$$k_1(\text{forward}) = \frac{k_1}{1+([\text{AzH}]_{\text{eq}}/[\text{AzH}_2{}^+]_{\text{eq}})} \tag{229}$$

from which the second-order coefficient for the first step of the reaction $k_1(\text{H})$ can be obtained through dividing by the hydrogen ion concentration. From these calculations, the values of $k_1(\text{H})$ were evaluated as 244, 92.0, and 16.3 for 4,6,8-trimethylazulene (26.4 °C), guaiazulene (25.0 °C) and guaiazulene-2-sulphonate (25 °C) respectively, at the indicated temperatures. Rate coefficients for hydrogen exchange of guaiazulene with hydronium ion were already available (Table 135) and were obtained for the other compounds by the previously described technique of using dilute aqueous hydrochloric acid and buffer solutions (Table 138);

TABLE 138

RATE COEFFICIENTS FOR REACTION OF ArH WITH HCl OR HOAc–NaOAc AT 25 °C[493]

HCl

| ArH | [1-³H]-4,6,8-trimethylazulene | | | | | | [1-³H]-guaiazulene-2-sulphonate | |
|---|---|---|---|---|---|---|---|---|
| $10^5[\text{H}^+](M)$ | 89 | 95 | 114 | 169 | 195 | 196 | 499 | 786 |
| $10^5 k_1$ | 400 | 375 | 454 | 635 | 752 | 780 | 422 | 677 |
| $10^2 k_2$ | 450 | 397 | 398 | 382 | 384 | 400 | 84.6 | 86.1 |

HOAc–NaOAc

| ArH | [1-³H]-4,6,8-trimethylazulene | | | | | | | | [1-³H]-guaiazulene-2-sulphonate | | | | |
|---|---|---|---|---|---|---|---|---|---|---|---|---|---|
| $10^2[\text{HOAc}](M)$ | 0 | 0.5 | 1 | 2 | 4 | 6 | 8 | 10 | 0 | 1 | 3 | 6 | 10 |
| $10^6 k_1$ | 114[a] | 158 | 161 | 249 | 351 | 522 | 598 | 691 | 243[a] | 990 | 430 | 580 | 830 |

[a] Calculated from the hydrochloric acid data.

hence for trimethylazulene $k_{\text{obs}} = 4.0[\text{H}^+]+6.4\times10^{-3}[\text{HOAc}]$ and for guaia-zulene-2-sulphonate $k_{\text{obs}} = 0.853[\text{H}^+]+5.9\times10^{-4}[\text{HOAc}]$. By neglecting secondary isotope effects, the values of $k_2'(\text{H})/k_2'(\text{T})$ were then evaluated as 14.1, 26.1 and 18.1 for guaiazulene, trimethylazulene, and guaiazulene-2-sulphonate, respectively, from which corresponding $k_2'(\text{H}) : k_2'(\text{D})$ values of 6.0, 9.6, and 7.4 were calculated by the Swain relationship. The summarised results are given in Table 139, from which it can be seen that there is a rough correlation between $\Delta \text{p}K$ and $k_2'(\text{H})/k_2'(\text{D})$ the latter passing through a maximum when $\Delta \text{p}K$ is approximately zero. This relationship appears to hold over a wider range than indicated by the data in Table 139, for if the isotope effects for dedeuteration of a range of aromatics are plotted against their reactivities (which will be a function

TABLE 139

CORRELATION OF ISOTOPE EFFECT FOR DEPROTONATION WITH $\Delta pK$ (SUBSTRATE-CATALYST)[493]

| Substrate | Catalyst | $k'_2(H)/k'_2(D)$ | $\Delta pK$ |
|---|---|---|---|
| Guaiazulene | $H_2O$ | 6.6 | +2.9 |
| Trimethylazulene | $H_2O$ | 9.6 | +1.8 |
| Guaiazulene-2-sulphonate | $H_2O$ | 7.4 | +0.9 |
| Azulene | $H_2O$ | 9.2 | −0.50 |
| Trimethoxybenzene | $H_2O$ | 7.2 | −3.6 |
| Azulene | $HCOO^-$ | 6.3 | −5.0 |
| Azulene | $CH_3COO^-$ | 6.1 | −7.8 |

of the unknown $\Delta pK$ values) there appears to be a correlation of the same kind (Table 140)[493,494].

Kresge et al.[494a] have measured the secondary isotope effect for the tritiation of 1,3-dimethoxybenzene and 1,3,5-trimethoxybenzene in protium oxide and in deuterium oxide containing 1–6 $M$ perchloric acid. Since the tritium incorporation into the aromatic is, by nature of the experimental method, the only reaction under observation, one thereby obtains a measure of the effect of the adjacent protium–oxygen and deuterium–oxygen bonds upon the breaking of the tritium–oxygen bond in the hydronium ion. Since in deuterium oxide the aromatic substrate would gradually become deuterated and hence complicate the kinetics, the incorporation of the tritium tracer was followed by only 2 % of reaction. For both substrates, rates were faster in deuterium oxide and ca. $10^3$ times faster overall for the trisubstituted compound. However, the secondary isotope effect turned out to be identical ($k_H/k_D = 0.59 \pm 0.01$) in each case. This was unexpected in view of the variation of the primary isotope effect noted above and possible explanations are that the reactivity difference is too small for any effect to show

TABLE 140

CORRELATION OF $k_H/k_D$ WITH REACTIVITY FOR DEDEUTERATION OF $[^2H]$-Ar[493,494]

| Aromatic | Position of $^2H$ | log $k_{rel}$ | $k_H/k_D$ |
|---|---|---|---|
| Benzene | 1 | 0 | 3.4 |
| Toluene | 3 | 0.8 | 3.4 |
| Toluene | 2 | 2.6 | 4.6 |
| Toluene | 4 | 2.6 | 5.5 |
| Anisole | 2 | 4.3 | 7.2 |
| Anisole | 4 | 4.8 | 6.7 |
| 2,4,6-Trimethoxybenzene | 1 | 10.0 | 6.7 |
| Azulene | 1 | 11.5 | 9.2 |
| Guaiazulene-2-sulphonate | 3 | 12.2 | 7.4 |
| 4,6,8-Trimethylazulene | 1 | 12.8 | 9.6 |
| Guaiazulene | 3 | 13.0 | 6.0 |

up, or that secondary effects reach a limiting value earlier in terms of position of the transition state along the reaction coordinate.

General acid catalysis has also been found by Challis and Long[495] in detritiation of methyl-substituted indoles. Earlier work by Koizumi and Titani[456] and by Hinman and Whipple[496] established the reactivities of various nuclear positions relative to each other and to the hydrogen on nitrogen by deuteration. For [3-$^3$H]-2-methylindole in acetate buffers at pH 5.01 and $\mu = 0.1$ at 25 °C, $k_{obs} = 42[H_3O^+]+0.071[HOAc]$ and the data for other buffer acids gave a value for $\alpha$ of 0.58. For the more reactive [3-$^3$H]-1,2-dimethylindole, $k_{obs} = 72.4$ $[H_3O^+]+0.176[HOAc]$, so exchange of both compounds follows the established pattern of general acid catalysis. Exchange of [3-$^3$H]-2-methylindole also took place under basic conditions but much more rapidly than the 1,2-dimethyl compound and it was therefore concluded that the base functions by forming the anion of [3-$^3$H]-2-methylindole, which is then sufficiently reactive to undergo exchange with the very weak general acid water; by contrast the 1,2-dimethyl compound cannot form the reactive anion and can only react with water in its neutral form.

Challis and Long[497] have used the fast flow technique described above (p. 217) to measure the equilibrium protonation of azulene in a range of aqueous perchloric acid media at 7.5 °C and hence the rates of the forward protonation and reverse deprotonation, the overall exchange rate being the sum of these. Some representative values are given in Table 141. Coupled with data obtained at other temperatures

TABLE 141

RATE COEFFICIENTS FOR PROTONATION ($k_1$ FORWARD) AND DEPROTONATION ($k_1$ REVERSE) OF AZULENE IN AQUEOUS HClO$_4$ AT 7.5 °C[497]

| [HClO$_4$] | $k_1$ (forward) | $k_1$ (reverse) | $k_1$ (exchange) |
|---|---|---|---|
| 1.46 | 6.3 | 44.9 | 51.3 |
| 1.61 | 8.7 | 39.6 | 48.3 |
| 1.92 | 14.6 | 34.8 | 52.4 |
| 2.50 | 28.1 | 21.9 | 50.0 |
| 2.69 | 30.6 | 20.5 | 51.1 |
| 3.00 | 57.6 | 16.1 | 73.7 |
| 3.91 | 166 | 7.7 | 174 |

and kinetic data for detritiation of azulene, the free energies of activation for each stage of the two-step $S_E2$ process were: initial protonation, 17.5; loss of a proton from the intermediate, 15.1; loss of a triton from the intermediate, 16.6; addition of a triton to the protium containing aromatic, (i.e. the first step in tritiation) 19.0. Thus for detritiation under these conditions, the second energy barrier is 1.5 kcal.mole$^{-1}$ higher than the first and the Wheland intermediates is less stable than the ground state by 24 kcal.mole$^{-1}$. Extrapolation of the protonation rate to the low acidities under which hydrogen exchange rates were measured gave a value which agreed well with that which could be calculated from the exchange data and this consistency was argued as further evidence in favour of the two-step A-$S_E2$ mechanism.

Kresge et al.[498] have drawn attention to the fact that detritiation of [$^3$H]-2,4,6-trihydroxy- and [$^3$H]-2,4,6-trimethoxy-benzenes by concentrated aqueous perchloric acid gives correlations of log rate coefficient with $-H_0$ with slopes of 0.80 and 1.14 respectively. Protonation to give the carbon conjugate acids is, however, governed by $h_0^{1.10}$ and $h_0^{1.95}$, respectively, which suggests that the difference in kinetic acidity dependence is a property of the substrate and should not be interpreted as a major difference in mechanism. The kinetic difference can be eliminated by an appropriate comparison of kinetic and equilibrium acidity dependencies. In equation (230)

$$k_1 = k[\text{H}^+](f_{\text{ArH}}f_{\text{H}^+}/f_{\text{ArH.H}^+})^\alpha \tag{230}$$

where $k$ is the medium-independent rate coefficient, the activity coefficient for the transition state is intermediate between the coefficients of the limiting structures, i.e. $f^{\ddagger} = (f_{\text{ArH}} \cdot f_{\text{H}^+})^{1-\alpha} \cdot (f_{\text{ArH.H}^+})^\alpha$ and $\alpha$ is the measure of the degree of proton transfer in the transition state ($0 < \alpha < 1$). Since the indicator ratio $I = [\text{ArH.H}^+]$ [ArH] and the acidity constant of ArH.H$^+$ is $K$ then equation (230) becomes

$$k_1/[\text{H}^+] = k \, K^\alpha (I/[\text{H}^+])^\alpha \tag{231}$$

A plot of log $(k_1/[\text{H}^+])$ versus log $(I/[\text{H}^+])$ gave the values of $\alpha$ as 0.48 and 0.44 for the hydroxy and methoxy compounds, respectively. It therefore appears that a single reaction mechanism applies even though there is a different kinetic acidity dependence. Kresge et al. stressed that changes in reaction rate with acidity should not be related to a single acidity function since the dependence of the equilibrium protonation upon the medium acidity may differ for each compound. Therefore, since the dependence of rate upon acidity function is related to the structure, use of the dependence as a diagnosis of mechanism is valueless.

Two papers have appeared in which it has been suggested that the mechanism may be dependent upon the reactivity of the aromatic substrate. Melander[499] drew attention to the similarity of the dependence of rate coefficient upon the acidity function for the detritiation of benzene[464], d log $k_1$/d$(-H_0) \approx 2.2$, to that for the protonation of diarylethylenes[500], d log $([\text{BH}^+]/[\text{B}])$/d$(-H_0) = 2.2$ with 84–89 wt. % sulphuric acid and 2.7 in 95.8 wt. % acid. Consequently, these reactions have a dependence upon acidity even greater than that given by use of the acidity function $h_R$. The comparable acidity dependence in both systems was interpreted to mean that proton transfer was complete in the transition state of each. Since, however, general acid catalysis is definitely proved for some substrates, the change over from general to specific acid catalysis is a function of the reactivity of the substrate and the catalysing acid, *i.e.* position of the transition state along the reaction coordinate is important and for some substrates a value of the Brönsted parameter $\alpha$ of 1.0 should be observed. Whilst this argument is entirely theoretically reasonable, the observations of Kresge *et al.*[498] described above show it to be based upon experimental evidence which it is unsafe to interpret.

Gold[501] has suggested that, since the proton transfer will be much more rapid in strong than in weak acids, then if the proton switch between acid and base is fast compared with the rearrangements during the formation of the transition state, the chemical identity, *i.e.* origin, of the proton taken up by the substrate will be blurred and the structure of the transition state can be represented by a protonated substrate in association with several molecules of the base between which proton switch can occur. Under these conditions, the concentration of the transition state will not be related to the concentration of the catalysing acids, but may be controlled by the thermodynamic availability of protons in the system, *i.e.* by an acidity function. General acid catalysis will be observed when the proton remains attached to a molecule of the catalysing species HA for a period of time which is long compared with a time interval characteristic of the molecular motions during transition state formation. This argument, which is identical in effect to that of Melander[499] and is likewise theoretically reasonable, is unsatisfactory in that there is no reliable evidence for a dependence of rate upon acidity function $h_0$ in hydrogen exchange.

### 8.1.2 *Kinetic studies of the nature of the base undergoing hydrogen exchange*

The early work of Best and Wilson[451] and of Koizumi[454], in which exchange on aniline hydrochloride was shown to occur *via* the equilibrium concentration of free base has already been mentioned. Kendall *et al.*[502, 503] have made some more extensive studies on the tritiation of N,N-dialkylanilines by tritium-labelled ethanol in sulphuric acid. For *para*-substituted anilines two hydrogens exchanged and for the *meta*-substituted anilines three hydrogens exchanged. This latter implies that reaction is occurring on the free base, and the kinetic data was treated as though the reactivities of the *ortho* and *para* positions were equivalent. Second-order rate coefficients were determined over a range of temperatures (in some cases as little as 15 °C, however) and the rate coefficients extrapolated to 65 °C together with the derived kinetic parameters are given in Table 142, and these refer to reaction of dialkylaniline (0.015 *M*), with sulphuric acid (density, 1.84, 0.05 *M*) in a solution 95 % ethanol by volume. Rate coefficients were stated to be accurate to within $\pm 10 \%$, but the quoted data has a much wider variation and there are some unexplained and very large discrepancies in some of the parameters (see Table 142). Two sets of rate coefficients were quoted for the 3-nitro compound, due, it was believed, to non-equivalent reactions at the 2 (most rapid) and 4,6

## TABLE 142

RATE COEFFICIENTS AND KINETIC PARAMETERS FOR REACTION OF $R_2NAr$
WITH $H_2O$–$EtO[^3H]$–$H_2SO_4$ AT 65 °C[502, 503]

| R | Substituent in Ar | $10^7 k_1$ | $\Delta H^\ddagger$ | $\Delta S^\ddagger$ | $E_a$ | $\Delta F^\ddagger$ |
|---|---|---|---|---|---|---|
| Me | H | 6.21, 9.90 | 19.3, 16.6 | −18.3, −25.7 | 27.7 | 25.5, 25.3 |
| Et | H | 0.70 | 20.2 | −13.0 | 29.8 | |
| n-Pr | H | 2.08 | 18.1 | −21.9 | 26.9 | |
| n-Bu | H | 1.92 | 17.8 | −22.8 | 26.6 | |
| Me | 4-Me | 3.49 | 19.9 | −15.7 | 28.7 | |
| Me | 4-MeO | 0.570 | 20.4 | −16.0 | 29.5 | |
| Me | 4-F | 0.577 | 19.8 | −22.8 | 28.1 | |
| Me | 4-Cl | 4.32 | 16.1 | −32.5 | 23.9 | |
| Me | 4-Br | 4.11 | 15.8 | −35.8 | 23.3 | |
| Me | 3-Me | 79.4, 64.8 | 17.0, 14.0 | −31.4 | 25.4 | 23.8, 24.6 |
| Me | 3,5-Me$_2$ | 185 | 18.9 | −11.8 | 27.4 | |
| Me | 3-EtO | 185 | 18.1 | −10.4 | 27.1 | |
| Me | 3-NO$_2$ (1) | 2.05, 2.92 | 19.9, 8.8 | −29.9, −62.2 | 22.1 | 29.3, 29.8 |
| | (2) | 0.692 | | | 26.9 | 30.0 |
| Me | 2-F | 0.185 | 22.7 | −21.0 | 30.2 | |
| Me | 2-Cl | 3.12[a] | 32.9 | + 2.4 | 40.3 | |
| Me | 3-F | 126 | 17.3 | −24.1 | | 25.5 |
| Me | 3-Cl | 49.2 | 22.2 | − 8.9 | | 25.2 |
| Me | 3-Br | 45.8 | 16.4 | −28.5 | | 26.0 |
| Me | 3-MeO | 4,020 | 25.2 | +15.2 | | 20.1 |
| Me | 3-EtO | 3,740 | 19.6 | − 5.1 | | 21.3 |
| Me | 3-MeO$_2$C | 6.82 | 17.2 | −33.5 | | 28.6 |
| Me | 3-EtO$_2$C | 9.05 | 19.0 | −21.2 | | 26.2 |
| Me | 3-Pr$^i$O$_2$C | 7.55 | 22.1 | −15.3 | | 27.2 |
| Me | 3-Pr$^n$O$_2$C | 9.98 | 22.9 | − 9.6 | | 26.2 |
| Me | 3-CF$_3$ | 14.9 | 5.3 | −64.4 | | 27.1 |
| Me | 4-Me$_2$N[b] | 26.2 | 13.3 | −12.7 | | 17.6 |
| Me | 4-NO$_2$ | 12.3 | 28.0 | − 0.4 | | 28.15 |

The 2-Me-, 2-MeO-, 2,4-Me$_2$- and 2,6-Me$_2$-N,N-dimethylanilines underwent negligible exchange at *ca.* 90 °C.

[a] At 105.8 °C.

[b] In 90 vol. % ethanol as solvent.

(least rapid) positions. However, the very strong deactivation at a position *ortho* to nitro makes this highly unlikely. The most rapid reaction almost certainly occurs at the 6 position, reaction at the 2 and 4 positions being less rapid; the conclusions in the paper are not, however, altered since relative reactivities at the position *ortho* to the N,N-dialkyl group were under consideration.

As found by the earlier workers[451, 454], reaction takes place upon the small amount of free base present since *ortho*, *para*-substitution occurs. Observed second-order rate coefficients increased with increasing ethanol content of the medium, being approximately twice as large for 85 vol. % ethanol as for 20 vol. %, and this was found to derive from the change in concentration of free base with ethanol content of the medium for when this change is allowed for, the rate coefficient for reaction of the free base actually decreases with increasing ethanol content. This was attributed to lowering of the proton concentration of the medium through addition of ethanol which is more basic than the water it replaces (*cf.* the difference in rates of protodesilylation in methanol and ethanol, see p. 332). This experimental result was thought to argue in favour of a rate-determining proton transfer in the reaction, though, of course, if reaction took place upon a rapidly protonated intermediate (protonated at carbon), the higher concentration of this in the medium of higher free proton concentration would lead to the same result, so that this evidence for the A-$S_E2$ mechanism is inadmissible. Exchange rates became almost constant for acid : amine ratios > 1 (Table 143); this derives from the fact that

TABLE 143

SECOND-ORDER RATE COEFFICIENTS FOR REACTION OF PhNMe$_2$ (*ca.* 0.5 *M*) WITH H$_2$O–EtO[$^3$H]–H$_2$SO$_4$[502, 503]

| [H$_2$SO$_4$](M) | 0.5 | 1.0 | 1.5 | 2.0 | 2.5 | 3.0 | 0.06 | 0.125 | 0.25 | 0.50 | 1.0 |
|---|---|---|---|---|---|---|---|---|---|---|---|
| Ethanol (Vol. %) | 0 | 0 | 0 | 0 | 0 | 0 | 95 | 95 | 95 | 95 | 95 |
| 10$^6$k$_2$ (75 °C) | 1.78 | 1.98 | 1.95 | 1.76 | 1.66 | 1.55 | | | | | |
| 10$^6$k$_2$(83.9 °C) | | | | | | | 1.41 | 4.7 | 12.7 | 14.7 | 17.2 |

the free acid concentration then greatly exceeds that of the free base and first-order conditions prevail. First-order rate coefficients were determined by division of the observed second-order rate coefficients by the free base concentration (derived from the p$K_a$ value) and the logarithms of these coefficients correlated linearly with $-H_0$ values with a slope of 0.94.

From the rate data, kinetic parameters were derived for reaction of both the free base and its conjugate acid, and whereas the free energies of activation for the free base (Table 142) correlated satisfactorily with $\sigma^+$ values (correlation coefficient 0.935), those for the conjugate acid did not (correlation coefficient *ca.* 0.75, and 0.67 for the use of $\sigma$ values). (Note that the column headings $\Delta H^{\ddagger}$ and $\Delta S^{\ddagger}$ in the original paper[503] should be reversed.) In addition, no correlation existed between rate coefficients and activation energies, and this is because the activation energy includes the enthalpy of dissociation, and activation enthalpy of exchange. A further indication that reaction takes place on the free base is the $\rho$-factor for the reaction of $-3.54$, which is *less* than that obtained by Eaborn and Taylor[166, 464] for the detritiation of alkylbenzene in aqueous sulphuric acid; it follows that the form of the amine involved must be more reactive than the alkylbenzenes and this would not be true if it were the protonated species.

The fact that in this work a satisfactory linear free energy correlation was obtained for reaction at an *ortho* position again shows that hydrogen exchange is a reaction of very small steric requirement, as noted elsewhere[504].

Rates of exchange were either much slower, or negligibly slow for compounds with a substituent *ortho* to the N,N-dimethyl group, and this was attributed to inhibition of conjugation through this group being forced out of the plane of the aromatic ring. For the N,N-dialkyl compounds studied, differences in rate seemed to derive from the differences in entropy, probably associated with differential solvation at the nitrogen atom.

Ling and Kendall[505] have also studied isotope effects in the reaction (Table 144). Analysis of these data by the use of equation (226) gave values of

### TABLE 144

RATE COEFFICIENTS FOR REACTION OF ArNMe$_2$ (*ca.* 0.5 *M*) WITH H$_2$SO$_4$ (*ca.* 0.5 *M*) IN H$_2$O OR D$_2$O[505]

| Reaction | Solvent | Substituent in Ar | Temp. (°C) | $10^7 k_2$ | | | |
|---|---|---|---|---|---|---|---|
| Detritiation | H$_2$O or D$_2$O | H | 90 | 53.1[a] | 21.2[c] | | |
| Dedeuteration | H$_2$O | H | 90 | 89.9 | | | |
| Deuteration | D$_2$O | H | 90 | 74.7 | | | |
| Detritiation | H$_2$O | 3-Me | 60 | 11.4 | 23.8 | 9.43 | |
| or | or | 3-F | 40 | 7.77[a] | 16.5[b] | 4.98[c] | |
| Dedeuteration | D$_2$O | 3-MeO | 35 | 45.5 | 102 | 39.7 | |
| Tritiation | H$_2$O | H | 90 | 66.0 | | | |
| Tritiation | 20 % H$_2$O, 80 % D$_2$O | H | 90 | 63.0 | | | |
| Tritiation | 40           60 | H | 90 | 57.3 | | | |
| Tritiation | 60           40 | H | 90 | 56.1 | | | |
| Tritiation | 80           20 | H | 90 | 51.1 | | | |
| Tritiation | D$_2$O | H | 90 | 40.1 | | | |

[a] Detritiation in H$_2$O.
[b] Dedeuteration in H$_2$O.
[c] Detritiation in D$_2$O.

$k_1'(H)/k_1'(D) = 7.5$, $k_2'(H)/k_2'(D) = 1.69$ and $k_2'(H)/k_2'(T) = 4.1$; the former value may, however, be too high, uncertainty arising from lack of knowledge of the effect on the calculated rate coefficients of the difference in dissociation of the nitrogen protonated anilinium ion in H$_2$O and D$_2$O. On the other hand, the relatively small isotope effects for the loss of hydrogen from the intermediate (compare Table 139) leads one to expect a large isotope effect for attachment of hydrogen to the intermediate. For the 3-Me, 3-F, and 3-MeO compounds (which constitute a series of increasing reactivity, each compound being more reactive than the unsubstituted aniline), values of $k_2(H)/k_2(D)(T)$ were 2.09 and 6.3, 2.12 and 6.5, 2.24 and 7.2, respectively. Loss of hydrogen from the protonated intermediate appears to be more important the more reactive the compound (this is not completely certain in view of the different temperatures used for the kinetic studies), and formation of the intermediate therefore correspondingly less rate-determining, and this is as we would expect from the increasing electron-supplying effects of the substituents. The results overall also show that tritiation is faster than detritiation and this result (which is the opposite of that which may be predicted from the work of Challis and Long[497] relating to hydrogen exchange of azulene, p. 220) is entirely consistent with the first step of the reaction being principally rate-determining; it would be interesting to find the combination of base and acid which would give a perfectly symmetrical reaction path.

The possibility of an entropy–enthalpy relationship for the reaction was examined and found to give a correlation coefficient of only 0.727 which was however improved to 0.971 if only the external contributions to these parameters were used, *i.e.* these contributions arising from solvent interactions only. If compounds with substituents *ortho* to the amino group were excluded, this further improved to 0.996 and is likely therefore to be *real* (*cf.* the comments on p. 9). It was argued that the different amounts of desolvation of the aromatic on going to the transition state would depend upon the substituent, and that the resultant greater freedom for solvent molecules would mean decreased interaction energy or increased enthalpy so that the linear relationship follows.

Finally, these results clearly show yet again that attachment of hydrogen to the aromatic is not a fast step, thereby eliminating the A-1 mechanism; yet the data in mixed solvents followed the prediction of the Gross–Butler "cubic" equation, and once again this test turns out to be valueless.

Katritzky *et al.*[506–511, 513, 514] have made very extensive studies of the hydrogen exchange of anilines, phenols, and heterocyclic analogues in an attempt to determine the species undergoing exchange. They have made extensive use of the profiles of rate *versus* acid concentration, for if the former is relatively independent of the latter it suggests that the concentration of the reacting species must be decreasing with increasing acid concentration, and hence this species must be the free base; if, however, rates increase rapidly with increasing acid concentration, reaction almost certainly takes place upon the conjugate acid. Different substrates of the same general type will respond differently to changes in acid concentration. Thus an amine with a strongly electron-supplying substituent will protonate more readily than one without and the equilibrium concentration of the conjugate acid will be higher, hence it is more likely to undergo hydrogen exchange *via* this latter species than an amine with an electron-withdrawing substituent and examples of this behaviour are noted below. Furthermore, the activation energy for reaction of the conjugate acid in the presence of an electron-supplying substituent will be sufficiently low for reaction to be able to occur under moderate conditions, whereas electron-withdrawing substituents will cause the activation energy to be so high that reaction will preferentially occur with the small quantity of free base.

Bean and Katritzky[506] measured rates of deuteration of 4-substituted anilines in deuterated sulphuric acid at 107 °C. They confirmed the findings of Kendall *et al.*[502–505] that only two hydrogen atoms usually undergo exchange, but for 4-methoxyaniline four hydrogens exchanged and the two hydrogens *ortho* to the methoxy group gave rate coefficients which were strongly dependent upon acidity, $d \log k_2/d(-H_0) = 0.96$, so that the reaction occurs on the conjugate acid and this no doubt arises from the concentration of the lone pair on nitrogen though the effect of the methoxy substituent. For 4-aminopyridine and 4-aminopyridine-N-oxide, however, rate coefficients decreased with increasing acidity so that the free amine was the reacting species. (Since the free amines would be protonated at the ring nitrogen even at low acidities the reduction in the observed rate at higher acidities must arise from diprotonation.) The true acidity-independent rate coefficients for the reaction of the free amines were calculated from the equation

$$\log k_2 = \log k_{2\,obs} + pK - \log a_{H_2O} \tag{232}$$

and the values of $k_2$ were 0.43, 0.156, 0.00113, and 0.201 for reaction *ortho* to the amino group with 4-methyl, 4-chloro, 4-nitro and 4-methoxy substituents, respectively, present which gives $\rho = -3.51$ in good agreement with the data of Kendall *et al.* (above). From this the reactivity of the unsubstituted compound was interpolated (and not measured experimentally in view of the extra (*para*) position available for substitution) and comparison with the data for pyridine and pyridine-N-oxides yielded $\sigma^+$ values of 1.82 and 1.99 for $-NH^+-$ and $-N(OH)^+-$, respectively.

Another method of ascertaining the nature of the reacting species is to compare the reactivity of a heterocyclic compound with its derivative methylated at the heteroatom. Thus Katritzky and Ridgewell[507] measured first-order rate coefficients $(10^7 k_1)$ for reaction of 2,6-dimethoxypyridine, 2,4,6-trimethylpyridine and 1,2,4,6-tetramethylpyridinium sulphate with tritiated sulphuric acid over a range of temperatures (Table 145).

Rate coefficients were determined from 2–5 readings per run because of the difficulty of the analytical procedure so that the results are rather inaccurate. Acid concentrations were given only in the form of acidity function values so there is no means of accurately repeating this work. It is clear from the data that the increase in rate coefficient with increasing acidity is parallel for all three compounds, the slopes of the log rate coefficient *versus* $-H_0$ plots being about 0.5–0.8 depending on the temperature and compound, consequently reaction for each occurs *via* the conjugate acid. This is also confirmed by the large negative activation entropies which indicates reaction between two similarly charged ions. The increase in $\Delta S^{\ddagger}$ with increasing acidity has been attributed to more rapid desolvation of the singly charged reagents than the doubly charged transition state complex, with increasing acidity.

Katritzky *et al.*[508] have measured rates of deuteration of aminopyridine by deuterated sulphuric acid (Table 146), and for the 4-amino and 2-amino-5-methyl compounds, the general increase in rate with increasing acidity, $d \log k_1/d\,(-H_0) \approx 0.6$, shows reaction to be occurring on the conjugate acids. For the latter compound this is only true at acidities $> -H_0 = 4.0$, below which rates are relatively independent of acidity indicating reaction on the free base. For the 2,6-dichloro

TABLE 145

RATE COEFFICIENTS $(10^7 k_1)$ AND KINETIC PARAMETERS FOR REACTION OF ArH WITH $T_2O-H_2SO_4$[507].

ArH = 2,6-Dimethylpyridine

| Temp. (°C) | $-H_0$ | | | | | | | | | | | | | |
|---|---|---|---|---|---|---|---|---|---|---|---|---|---|---|
| | 8.58 | 8.59 | 8.64 | 8.68 | 8.86 | 8.87 | 8.96 | 8.97 | 9.01 | 9.11 | 9.16 | 9.32 | 9.52 | 9.66 |
| 182 | | 5.63 | | 5.55 | | | | | | | | | | |
| 203 | | | | | | | | | | | 41.1 | | | 75.0 |
| 204 | | 18.6 | | | | 26.7 | | 26.4 45.5 | 36.4 | 50.2 47.2 | | | 94.4 | |
| 222 | 71.3 | | | | 98.9 | | | | | | | 197 | | |

ArH = 2,4,6-Trimethylpyridine

| Temp. (°C) | $-H_0$ | | | | | | | | | | | | | | |
|---|---|---|---|---|---|---|---|---|---|---|---|---|---|---|---|
| | 6.22 | 6.30 | 6.51 | 6.75 | 6.86 | 6.96 | 7.00 | 7.05 | 7.11 | 7.35 | 7.48 | 7.61 | 7.68 | 7.95 | 8.13 |
| 182 | | | 13.9 | | | 25.0 | 22.2 | | 27.8 | | 66.7 | | | 108 | |
| 204 | 58.3 | | | | | | | 144 | | 250 | | | 258 | | 667 |
| 219 | | 211 | | 361 | 306 | 472 | | | | | | 777 | | | 1330 |

ArH = 1,2,4,6-Tetramethylpyridinium sulphate

| Temp. (°C) | $-H_0$ | | | |
|---|---|---|---|---|
| | 7.00 | 7.11 | 7.61 | 7.95 |
| 182 | 38.9 | 50.0 | 139 | 208 |

2,6-Dimethylpyridine

| | | | |
|---|---|---|---|
| $\Delta H^\ddagger$ | 36.5 $(H_0, -8.4)$ | | 32.5 $(H_0, -9.0)$ |
| $\log A$ | 11.3 | | 9.9 |
| $\Delta S^\ddagger$ | −48.5 | | −57.2 |

2,4,6-Trimethylpyridine

| | | | |
|---|---|---|---|
| $\Delta H^\ddagger$ | 37　$(H_0, -6.33)$ | 30.5 $(H_0, -7.40)$ | 23.5 $(H_0, -8.40)$ |
| $\log A$ | 12.1 | 9.8 | 7.2 |
| $\Delta S^\ddagger$ | −35 | −50.5 | −67 |

compound the rate–acidity profile shows reaction to be occurring on the free base, and this is confirmed by the much greater reactivity of this compound relative to the others (especially at low acidity); if the same species was undergoing reaction in each case the chloro compound would be very much less reactive. These results

TABLE 146

FIRST-ORDER RATE COEFFICIENTS FOR REACTION (AT THE 3 POSITIONS) OF X-SUB-
STITUTED PYRIDINES WITH $D_2O$–$D_2SO_4$ AT 107 °C[508]

| X = 4-NH₂ | | X = 2-NH₂-5-Me | | X = 4-NH₂-2,6-Cl₂ | |
|---|---|---|---|---|---|
| $-H_0$ | $10^7 k_1$ | $-H_0$ | $10^7 k_1$ | $-H_0$ | $10^7 k_1$ |
| 0.40 | 0.864 | 0.21 | 1.25 | 0.01 | 742 |
| 0.57 | 2.23 | 0.52 | 0.369 | 0.42 | 861 |
| 1.85 | 11.0 | 0.75 | 0.985 | 1.18 | 642 |
| 2.10 | 19.3 | 2.28 | 4.27 | 1.80 | 467 |
| 2.56 | 36.4 | 2.55 | 4.66 | 2.72 | 255 |
| 3.98 | 283 | 3.19 | 15.5 | 3.10 | 216 |
| 5.14 | 933 | 4.15 | 50.8 | 3.70 | 203 |
| 5.57 | 2,160 | 4.62 | 158 | 4.27 | 177 |
| 6.92 | 1,390 | 5.80 | 620 | 4.67 | 211 |
| 7.58 | 731 | | | 5.01 | 319 |
| | | | | 5.25 | 358 |
| | | | | 5.55 | 703 |

in particular demonstrate very nicely the fact that the electronic effects of substituents can dramatically modify the reaction mechanism (see p. 276).

Katritzky *et al.*[509] have also made a kinetic study of the deuteration of substituted pyridine-1-oxides (Table 147). For the 2,4,6-trimethyl compound, the
rate–acidity profile shows the conjugate acid to be reacting. The slope of the plot,
however, was less (0.33) than that (0.56) obtained for 2,4,6-trimethylpyridine

TABLE 147

FIRST-ORDER RATE COEFFICIENTS FOR REACTION OF X-SUBSTITUTED PYRIDINE-1-
OXIDES WITH $D_2O$–$D_2SO_4$[509].

| X = 2,4,6-Me₃ (202 °C) | | X = 3,5-Me₂ (230 °C) | | | X = 3,5-(MeO)₂ (118.6 °C) | |
|---|---|---|---|---|---|---|
| $-H_0$ | $10^7 k_1$ | pD | $10^7 k_1$ (4-H) | $10^7 k_1$ (2,6-H) | pD | $10^7 k_1$ (2,6-H) |
| 3.36 | 1.74 | 5.8 | | v. fast | 9.0 | ᵃ |
| 3.89 | 8.00 | 2.5 | | v. fast (200 °C) | 3.1 | 17.6 |
| 4.48 | 9.92 | 0.7 | | 2.100 (214 °C), | 1.9 | 82.9 |
| 5.60 | 32.0 | | | 811 (204 °), 301 (188 °) | 1.0 | 278 |
| 6.60 | 92.3 | $-H_0$ | | | $-H_0$ | |
| 7.64 | 264 | 0.20 | 91.7 | 160 | 0.32 | 512 |
| | | 1.73 | 52.7 | 63.9 | 1.10 | 356 |
| X = 3-OH (190 °C) | | 3.50 | 47.2 | 51.1 | 1.71 | 206 |
| | | 43.9 | 41.9 | 42.8 | 3.60 | 295 |
| $-H_0$ | $10^7 k_1$ | 8.11 | 20.5 | 15.1 | 4.32 | 697 |
| 2.13 | 201 | | | | 4.85 | 1,490 |
| 3.80 | 276 | | | | 6.84 | 6,230 |
| 4.63 | 505 | | | | | |
| 7.88 | 472 | | | | | |

ᵃ Rapid base-catalysed exchange at all positions.

TABLE 148

RATE COEFFICIENTS AND KINETIC PARAMETERS FOR REACTION OF ArH WITH $D_2O$–$D_2SO_4$[510]

ArH = 4-Pyridone

| $pD$ | 4.3 | 2.5 | 1.0[a] | | | | | | | |
|---|---|---|---|---|---|---|---|---|---|---|
| $-H_0$ | | | | 0.7 | 2.3 | 4.3[b] | 6.6 | 7.3 | 8.4 | 10.1[c] |
| $10^7 k_1$ (186.5 °C) | 428 | 668 | 776 | 374 | 262 | 199 | 190 | 211 | 226 | 218 |
| $10^7 k_1$ (170 °C) | 118 | 219 | 197 | 99.3 | 68.5 | 56.5 | 52.2 | 56.5 | 64.0 | 80.0 |

ArH = 1-Methyl-4-pyridone

| $pD$ | 5.2 | 3.0 | 2.3 | 0 | | | | | | |
|---|---|---|---|---|---|---|---|---|---|---|
| $-H_0$ | | | | | 1.1 | 2.6 | 4.2[d] | 6.8 | 9.0 | 10.2 |
| $10^7 k_1$ (170 °C) | <5 | 262 | 249 | 156 | 107 | 95.1 | 76.4 | 60.0 | 11.1 | 12.6 |

ArH = 2-Pyridone (P, 127.5 °C), 3-Methyl-2-pyridone (3-M, 102.8 °C) and 5-Methyl-2-pyridone (5-M, 120.4 °C)

| $pD$ | 1.5 | 1.0 | 0.7 | 0.5 | | | | | | | |
|---|---|---|---|---|---|---|---|---|---|---|---|
| $-H_0$ | | | | | 1.05 | 2.2 | 3.0 | 5.6 | 6.3 | 8.3 | 9.2 |
| $10^7 k_1$ (P) | | 160 | | | | 178 | | | 81.3 | | 29.8 |
| $10^7 k_1$ (3-M) | 37.4 | | 144 | | 112 | | 87.9 | 47.6 | 18.2 | | |
| $10^7 k_1$ (5-M) | | | 670 | | 671 | | 477 | 294 | 168 | | |

ArH = 4-Quinolone (Q) and 4-Methoxyquinoline (M), at 90.24 °C

| $pD$ | 0.5 | 0.0 | 2.5 | 4.5 | 5.8 | 6.4 | 7.5 | 9.2 | 10.4 |
|---|---|---|---|---|---|---|---|---|---|
| $-H_0$ | | | | | | | | | |
| $10^7 k_1$ (Q, 3 position) | 280 | 150 | 115 | 74.0 | | 15.7 | 32.2 | 1,240 | 5,830 |
| $10^7 k_1$ (Q, 6 position) | | | | | | | <0.3[e] | 54.8 | 34.7 |
| $10^7 k_1$ (M, 3 position) | | | | | 56.1 | | 91.2 | 600 | 2,310 |
| $10^7 k_1$ (M, 6 position) | | | | | | | | 40.7 | 36.3 |

[a] Rate coefficients at 159.3 and 150.2 °C were 115 and 52.1 giving $E_a$ = 28.1, log $A$ = 9.2 and $\Delta S^{\ddagger}$ = −6.

[b] Rate coefficients at 159.3 and 150.2 °C were 28.2 and 11.5 giving $E_a$ = 28.4, log $A$ = 9.4 and $\Delta S^{\ddagger}$ = −7.

[c] Rate coefficients at 159.3 and 150.2 °C were 43.3 and 22.4 giving $E_a$ = 25.8, log $A$ = 7.6 and $\Delta S^{\ddagger}$ = −13.

[d] Rate coefficients at 159.3 and 150.2 °C were 38.8 and 15.8 giving $E_a$ = 29.4, log $A$ = 9.4 and $\Delta S^{\ddagger}$ = −1.

[e] Wrongly quoted as >0.3.

at approximately the same temperature and this was attributed to departure from Hammett base behaviour of the carbon protonation. For 2,6-dimethylpyridine-1-oxide, side reactions permitted the measurement of the rate coefficient only at $H_0 = -8.7$, at 266 °C, the value of $201 \times 10^{-7}$ being 50 times less than that obtained for the trimethyl compound at $H_0 = -9.3$ and 258 °C. A value of 20 was obtained for $f_4^{Me}$ in the pyridine series and it was argued that this consistency would only be obtained if the 2,6 and 2,4,6 compounds reacted *via* the same species so that the former must also react *via* the conjugate acid. For the 3,5-dimethyl compound, reaction is restricted to the less reactive 2, 4, and 6 positions in the pyridine-1-oxide molecule and hence it is not surprising to find that the rate–acidity profile indicates reaction to occur on the free base.

For the 3,5-dimethoxy compound the rate–acidity profile shows the free base to be the reactive species at acidities $< -H_0 = 3.0$ and at higher acidities reaction occurs on the conjugate acid; surprisingly, no exchange was detected at the 4 position of this compound. Rates of exchange of the 3-hydroxy compound (at the 2 position) were relatively independent of a wide change in acidity of the medium so that exchange probably occurs on the free base.

A kinetic study of the deuteration of pyridones and quinolones by deuterated sulphuric acid yielded the data in Table 148[510]. For the 4-pyridones, the rapid rise in rate with increasing acidity in strongly basic solutions, and the levelling off in rate at about $H_0 = 0$ is consistent with reaction on the free base as is the small negative entropy of activation. The similarity in rate between 4-pyridone and its 1-methyl derivative shows reaction to take place on the form (XII) and not (XIII), *viz.*

(XII)          (XIII)          (XIV)

In addition, 4-methoxypyridine (XIV) did not exchange under the same conditions used for the pyridones, so this additionally argues against involvement of form (XIII). For the 2-pyridones the rate–acidity profile shows reaction to occur on the free base. A methyl substituent enhanced the reactivity of the position *meta* to it by a factor of about 2 for both the 3- and 5-methyl derivatives. Exchange at the 3 position of 4-quinolone shows a change in mechanism from reaction on the free base to reaction on the conjugate acid at about $-H_0 = 7$, and in this region the rates for the 4-methoxyquinoline are similar so that both species are of the form (XV) rather than (XVI). Exchange at the 6 position was very slow and similar for the methoxy and unsubstituted compounds so that both pieces of evidence indicate reaction on the conjugated acid (XV).

(XV)                    (XVI)

These workers also measured rates of exchange of phenol and phenoxide ion

TABLE 148a

FIRST-ORDER RATE COEFFICIENTS FOR REACTION OF PhOH WITH $D_2O-D_2SO_4$ AT 127.8 °C[510]

| pD | 4.54 | 4.0 | 2.00 | 1.65 | 1.45 | 1.08 | 0.76 |
|---|---|---|---|---|---|---|---|
| $10^7k_1$ | 16.6[a] | 19.5[b] | 25.7 | 57.8 | 93.8 | 176 | 343 |

[a, b] Obtained in acetate and formate buffers respectively by extrapolation to zero buffer acid concentrations.

(Table 148a) from which it is clear that except at the lowest acidities, exchange takes place on the conjugate acid.

Katritzky et al.[511] have measured rate coefficients for deuteration of 3,5-dimethylphenol and heterocyclic analogues. As in all of the deuteration work of this group, rates of exchange were measured by the NMR method, which is useful for following exchanges at more than one position in the molecule but is, of course, much less accurate than detritiation techniques. In this study, the chemical shift for the *ortho* and *para* protons for the parent compound was too small to allow separate integration, but it was apparent that rates of exchange at these two positions did not differ by a factor $> 4$. From the rate–acidity profile (Table 149) reaction clearly occurs on the neutral species at pD $< 3.5$ (the log $k_1$ *versus* pD slope was 0.96) and upon the anion at pD $> 3.5$ (slope zero), and the reactivity of the anion to the neutral molecule was estimated as $10^{7.8}$, close to the value of $10^7$ noted above.

The data in Table 149 show that 2,6-dimethyl-4-pyridone exchanges as the free base at $-H_0 < ca.$ 3.5 and as the conjugate acid at higher acidities. This result may be compared with the result for 4-pyridone which exchanges as the free base even at $-H_0 = 10$ (see above) and once again the effect of electron-supplying substituents in causing a change over to reaction on the conjugate acid is apparent. Likewise, the 1,2,6-trimethyl derivative shows the change over at even lower acidities $(-H_0 = 2.7)$. At high acidities, 2,6-dimethyl-4-pyridone, 1,2,6-trimethyl-4-pyridone, and 4-methoxy-2,6-dimethylpyridine (the latter two compounds being the N- and O-methyl derivatives, respectively) all react at a similar rate and have a similar dependence of rate upon acidity, so that the conjugate acids are involved as form (XVII) and not the alternative (XVIII). (Note that in

(XVII)          (XVIII)

the original paper, the figure demonstrating this is wrongly captioned, the O-methyl compound being incorrectly referred to as 4-methoxy-2,6-dimethyl-4-pyridone.) This situation is therefore, analogous to that found for the 4-quinolones

## TABLE 149

RATE COEFFICIENTS AND ARRHENIUS PARAMETERS FOR REACTION OF ArH WITH $D_2O$–$D_2SO_4$[511]

ArH = 3,5-Dimethylphenol at 100 °C

| $pD$ | 4.80 | 3.69 | 2.90 | 2.22 | 1.63 | 1.50 | 1.05 |
|---|---|---|---|---|---|---|---|
| $10^7 k_1$ | 38.0[a] | 38.0[b] | 71.2 | 180 | 611 | 862 | 193 |

ArH = 2,6-Dimethyl-4-pyridone at 107.8 °C

| $pD$ | 5.2 | 2.2 | 1.4 | | | | | | | | |
|---|---|---|---|---|---|---|---|---|---|---|---|
| $-H_0$ | | | | 0.8 | 2.3 | 4.3 | 5.6 | 6.3 | 7.5 | 8.4 | 10.7 |
| $10^7 k_1$ | 19.8 | 28.5 | 24.1 | 8.13 | 7.01 | 11.6 | 25.4 | 46.7 | 104 | 200 | 1,380 |

ArH = 1,2,6-Trimethyl-4-pyridone at 100 °C

| $pD$ | 4.1 | | | | | | | |
|---|---|---|---|---|---|---|---|---|
| $-H_0$ | | 0.9 | 2.2 | 3.9 | 4.5 | 6.3 | 8.2 | 9.7 |
| $10^7 k_1$ | 10.3 | 2.39 | 2.38 | 4.35 | 10.2 | 46.9 | 231 | 820 |

ArH = 4-Methoxy-2,6-dimethylpyridine at 100 °C

| $-H_0$ | 1.0 | 2.5 | 4.1 | 5.1 | 6.5 | 8.7 | 9.6 |
|---|---|---|---|---|---|---|---|
| $10^7 k_1$ | 0.148 | 0.148 | 0.83 | 2.77 | 15.4 | 270 | 767 |

ArH = 1-Hydroxy-2,6-dimethyl-4-pyridone at 100 °C

| $pD$ | 10.1 | 6.3 | 5.4 | 4.9 | 3.2 | 2.2 | 1.1 | | | | | | |
|---|---|---|---|---|---|---|---|---|---|---|---|---|---|
| $-H_0$ | | | | | | | | 1.2 | 2.5 | 3.3 | 6.1 | 6.9 | 9.3 |
| $10^7 k_1$ | <5 | 189 | 134 | 141 | 42.7 | 16.1 | 3.09 | 2.26 | 2.70 | 3.03 | 13.2 | 20.4 | 215 |

ArH = 4-Methoxy-2,6-dimethylpyridine-1-oxide at 100 °C

| $pD$ | 7.2 | 2.2 | 1.0 | | | | | | |
|---|---|---|---|---|---|---|---|---|---|
| $-H_0$ | | | | 1.1 | 2.3 | 4.3 | 6.7 | 7.9 | 10.3 |
| $10^7 k_1$ | <0.06 | <0.5 | <0.5 | <0.5 | <0.5 | <0.65 | 10.3 | 37.2 | 492 |

ArH = 2.6-Dimethyl-4-pyrone at 148.3 °C

| $pD$ | 2.0 | 0.9 | 0.5 | | | | | | | | | | |
|---|---|---|---|---|---|---|---|---|---|---|---|---|---|
| $-H_0$ | | | | 1.0 | 2.2 | 3.9 | 4.4 | 4.9 | 6.4 | 7.9 | 9.6 | 13.5 | 14.5 |
| $10^7 k_1$ | 330 | 503 | 792 | 623 | 378 | 205 | 174[c] | 147 | 92.9 | 63.7 | 29.8 | 186 | 164 |

ArH = 2.6-Dimethyl-4-thiapyrone at 148.3 °C

| $pD$ | 0.4 | | | | | | | | | |
|---|---|---|---|---|---|---|---|---|---|---|
| $-H_0$ | | 0.8 | 2.2 | 4.1 | 4.6 | 6.6 | 9.0 | 10.8 | 13.5 | 14.5 |
| $10^7 k_1$ | 142.5 | 35.0 | 26.5 | 18.8 | 17.8[d] | 14.5 | 13.6 | 12.9 | 123 | 56.5 |

[a,b] Obtained using acetate and formate buffers respectively by extrapolation to zero buffer acid concentration.

[c] Values at 157.5, 138.9, and 127.1 °C were 453, 70.0, and 24.6, respectively, giving $E_a = 32.0$ and $\log A = 11.9$.

[d] Values at 158.5, 138.9, and 127.1 °C were 59.8, 9.35, and 2.63, respectively, giving $E_a = 33.7$ and $\log A = 11.7$.

and is in contrast to that found for the unsubstituted pyridones and again derives from the effect of substituents. At low acidities, 2,6-dimethyl-4-pyridone and 1,2,4-trimethyl-4-pyridone have similar rates and therefore the former reacts as the pyridone analogous to (XII) and not the hydroxypyridine analogous to (XIII), which is the same result as found for the unsubstituted pyridines (which also reacted as the free base in this acid region). By contrast and in confirmation, at low acidities the O-methyl compound (4-methoxy-2,6-dimethyl-pyridine) was much less reactive than the other two compounds.

Similar reasoning was applied to determination of the reacting form of 1-hydroxy-2,6-dimethyl-4-pyridone, its rate being compared to that of 4-methoxy-2,6-dimethylpyridine-1-oxide. At high acidities, the rate coefficients for both compounds are similar and therefore, both must react as the conjugate acids (XIX, R = H or Me). In the intermediate acidity range $pD+1$ to $H_0 -2.5$, the rate–acidity profile slope was zero for the hydroxy compound corresponding to reaction on the neutral form (XX) which was preferred to (XXI) since 4-methoxy-2,6-dimethylpyridine-1-oxide was much less reactive. At lower acidities the rate

(XIX)          (XX)          (XXI)          (XXII)

increased and then levelled off with decreasing acidity and this was consistent with reaction upon the anion (XXII), which would be more reactive than the neutral species yet in equilibrium with it so that the rate–acidity profile would be zero.

For 2,6-dimethyl-4-pyrone and 4-thiapyrone, the kinetics were complicated by exchange in the methyl groups. Since the rates for both were relatively independent of acidity, reaction on the free base species was indicated.

From this work, relative reactivities were obtained by extrapolation to 100 °C (in the case of 2,6-dimethyl-4-pyridone by assuming an activation energy for the 7 °C temperature correction necessary) and to $H_0 = 0$. The temperature variation in $pK_a$ values was neglected and the results (Table 150) are highly dependent upon acidity, so they can only be regarded as very approximate. They thus add further confirmation to the growing body of data, which shows that electron-supplying substituents are greater in effect in less reactive systems[512], for substitution of $NH^+$ for CH *meta* to the reacting site in phenoxide ions lowers the rate by about $10^8$, whereas for trimethylpyridines a value of $10^{18}$ may be derived; however, for the 3,5-dimethylphenol neutral species a value of $10^4$ is obtained which is unexpectedly low. Values of $\sigma_m^+$ and $\sigma_p^-$ were derived for heterocyclic substituents as follows: $\geqslant \overset{+}{N}H(1.85, 3.4)$; $\geqslant \overset{+}{N}OH$ (2.1, 3.9); $\geqslant \overset{+}{N}\overset{-}{O}(0.8, 2.0)$; $>\overset{-}{O}-$ (3.0, 5.8); $>\overset{+}{S}-$ (3.2, 5.4).

TABLE 150

VALUES OF $\log k_1$ (EXTRAPOLATED TO 100° AND p$D$ = 0) FOR DEUTERATION OF ArH
BY $D_2O$–$D_2SO_4$[511]

| ArH | Position | Free base | Conjugate acid |
|---|---|---|---|
| Benzene[166] | 1 | −9.5 | |
| 1,3-Dimethylbenzene[477] | 4 | −5.0 | |
| Phenol anion[a] | 2, 4. 6 | 3.1–3.8 | − (3.9–4.4) |
| 3,5-Dimethylphenol anion | 2, 4, 6 | 5.3 | − 3.3 |
| 4-Nitrophenol anion[458] | 2 | | −(8.5–9.0) |
| 2-Pyridone | 3, 5 | −4.6 | ⩽−10 |
| 3-Methyl-2-pyridone | 5 | −4.2 | ⩽−10 |
| 5-Methyl-2-pyridone | 3 | −4.3 | ⩽−10 |
| 4-Pyridone | 3, 5 | − (3.5–4.0) | ⩽ − (12–14) |
| 1-Methyl-4-pyridone | 3, 5 | −4.0 | ⩽−14 |
| 2-Methyl-4-pyridone | 3, 5 | −5.2 | ⩽−12 |
| 3-Methyl-4-pyridone | 5 | −4.8 | ⩽−12 |
| 5-Methyl-4-pyridone | 3 | −4.9 | ⩽−12 |
| 2,6-Dimethyl-4-pyridone | 3, 5 | −1.5 | −7.5 |
| 4-Methoxy-2,6-dimethylpyridine | 3, 5 | | −9.3 |
| 4-Hydroxy-2,6-dimethylpyridine-1-oxide anion | 3, 5 | 2.3 | |
| 1-Hydroxy-2,6-dimethyl-4-pyridone | 3, 5 | −2.8 | −8.7 |
| 4-Methoxy-2,6-dimethylpyridine-1-oxide | 3, 5 | | −9.4 |
| 2,6-Dimethyl-4-pyrone | 3, 5 | −6.1 | ⩽−12 |
| 2,6-Dimethyl-4-thiapyrone | 3, 5 | −6.7 | ⩽−12 |
| 4-Quinolone | 3 | − (1.7–1.9) | −(7.6–8.7) |
| 4-Quinolone | 6 | | −( 11–13) |
| 4-Methoxyquinoline | 3 | | −(8.1–9.1) |
| 4-Methoxyquinoline | 6 | | −( 11–13) |

[a] Temperature dependence obtained from ref. 452.

Katritzky and Pojarlieff[513] studied the kinetics of deuteration of pyridazine derivatives by deuterated sulphuric acid. First-order rate coefficients for 4-aminopyridazine, pyradazin-4-one and -3-one are given in Table 151. Exchange at the 5 position of the former compound increased linearly with acidity, d log $k_1/\mathrm{d}(-H_0) = 0.60$, indicating reaction on the conjugate acid (XXIII below), though the positive entropy of activation tends to contradict this. However, the activation entropies for the other heterocycles were determined at higher acidities and it has been shown (Table 145) that the values became more positive for less acidic media. By comparison of the rates with the data for 4-amino-pyridine (Table 146) a value of 0.67 for $\sigma_p^+ (= N-)$, i.e. the *meta* position in pyridine, was determined, though this value is apparently severely affected by hydrogen bonding.

(XXIII)          (XXIV)

TABLE 151

RATE COEFFICIENTS AND KINETIC PARAMETERS FOR REACTION FOR REACTION OF PYRIDAZINE DERIVATIVES WITH $D_2O$–$D_2SO_4$[513]

ArH = 4-Aminopyridazine

| | | | | | | |
|---|---|---|---|---|---|---|
| $pD$ 8.36 | 7.97 | 7.59 | 6.80 | 6.35 | 6.00 | 5.75 |
| $10^7k_1$ (3H+6H, 127 °C) 308 | 257 | 188 | 118 | 50.3 | 27.0 | 12.0 |
| $pD$ | 0.06 | | | | | |
| $-H_0$ | | 0.84[a] | 1.32 | 2.15 | 3.33 | 7.20 |
| $10^7k_1$ (5H, 186 °C) 9.45 | | 26.3 | 64.1 | 234 | 927 | fast |

ArH = Pyridazin-4-one

| | | | | | | | | |
|---|---|---|---|---|---|---|---|---|
| $pD$ 2.40 | 2.10 | 1.70 | 1.32 | 0.70 | 0.20 | | | |
| $-H_0$ | | | | | | 0.23 | 3.35 | 6.61 |
| $10^7k_1$ (5H, 186 °C) 42.2 | 99.8 | 312 | 671 | 1,450 | 1,460 | 1,450 | 740 | 142 |
| $10^7k_1$ (3H, 186 °C) 412 | 392 | 339 | 255 | 97.0 | 28.2 | 15.7 | | |
| $10^7k_1$ (6H, 186 °C) 165 | 131 | 116 | 91.8 | 32.0 | 21.1 | 46.1 | | |

ArH = Pyridazine-3-one

| | | | | | | | |
|---|---|---|---|---|---|---|---|
| $pD$ 0.67 | 0.37 | | | | | | |
| $-H_0$ | | 0.43 | 0.87 | 1.24 | 2.30 | 3.65 | 7.50 |
| $10^7k_1$ (5H, 186 °C) 1.01 | 4.52 | 14.75 | 22.2 | 13.8 | 4.96 | 4.90 | 1.38 |

[a] Rates at 199.0, 208.3 and 215.4 °C were 98.9, 260 and 457, respectively, giving $\Delta H^{\ddagger} = 44$ and $\Delta S^{\ddagger} = 13$.

In neutral and alkaline media, the rate of exchange at the 3 and 6 position of 4-aminopyridazine is independent of acidity but decreases markedly when the media become more acidic. This was interpreted in terms of a rate-determining removal of the 6-proton by deuteroxide ion to give the ylid (XXIV), which reacts with deuterium oxide in a fast step. A similar result for the 3 and 6 positions of pyridazin-4-one suggests the same mechanism. For reaction at the 5 position, the rate–acidity profile indicated reaction on the free base as did that for the 5 position of pyridazin-3-one, though the appearance of a maximum in the rate at $-H_0 = 0.8$ was anomalous and suggested incursion of a further mechanism.

Finally, Katritzky et al.[514] have measured first-order rate coefficients for deuteration of pyrimidines by deuterated sulphuric acid (Table 152), and all pD and $-D_0$ values given in the Table refer, as in the earlier work, to a temperature of 20 °C. For 2-aminopyrimidine, reaction clearly occurs on the free base and comparison of the data with the earlier work on anilines and by making a number of assumptions, $\sigma_m^+$ for heterocyclic =N– was determined as 0.65. For the more reactive 6-amino-2,4-dimethyl-pyrimidine the incursion of reaction on the conjugate acid at higher acidities is apparent and this follows the previously established pattern. This work yielded a value of 0.55 for $\sigma_m^+$ and the range of

## TABLE 152

RATE COEFFICIENTS FOR REACTION OF PYRIMIDINES WITH $D_2O$–$D_2SO_4$ AT 107 °C[514]

ArH = 1,2-Dihydro-1,3-dimethyl-2-oxopyrimidinium salts [a]

| $pD$ | 3.00[b] | 2.90[b] | 2.55[b] | 1.25 | 1.10 | 0.50 | | | |
|---|---|---|---|---|---|---|---|---|---|
| $-D_0$ | | | | | | | 0.00 | 0.65 | 3.45 |
| $10^7 k_1 (5H)$ | 74.4 | 71.6 | 68.1 | 41.7 | 42.1 | 14.7 | 12.2 | 10.8 | 14.0 |

ArH = 2-Aminopyrimidine

| $pD$ | 4.75 | 4.50 | 4.02 | 2.95 | 1.85 | 1.18 |
|---|---|---|---|---|---|---|
| $10^7 k_1 (5H)$ | 2.28 | 4.08 | 10.3 | 17.7 | 19.4 | 23.3 |

ArH = 6-Amino-2,4-dimethylpyrimidine

| $pD$ | 7.40[b] | 6.90[b] | 6.30[b] | 5.60[b] | 3.70 | 1.88 | 1.80 | 0.95 | 0.27 |
|---|---|---|---|---|---|---|---|---|---|
| $10^7 k_1 (5H)$ | 3.07 | 3.06 | 3.58 | 3.47 | 3.86 | 21.7 | 25.9 | 128 | 361 |

ArH = Pyrimidin-2-one

| $pD$ | 5.00 | 4.95[b] | 4.65[b] | 4.30 | 2.80 | 2.20 | 1.40 | 1.30 | 0.27 | | | | | | |
|---|---|---|---|---|---|---|---|---|---|---|---|---|---|---|---|
| $-D_0$ | | | | | | | | | | 0.06 | 1.93 | 3.18 | 4.98 | 5.15 | 9.70 |
| $10^7 k_1 (5H)$ | 32.8 | 44.2 | 51.7 | 87.8 | 414 | 461 | 486 | 478 | 478 | 104 | 20.1 | 6.97 | 0.717 | 0.401 | 1.54 |

ArH = 4,6-Dimethylpyrimidin-2-one

| $pD$ | 4.70[b] | 4.60[b] | 4.40 | 4.20 | 2.90 | 2.20 | 1.65[c] |
|---|---|---|---|---|---|---|---|
| $10^7 k_1$ | 442 | 428 | 330 | 268 | 210 | 197 | 561[c] |

[a] Wrongly referred to as a pyridinium salt in ref. 514.
[b] In $CH_3COOD$.
[c] Refers to 1-methylpyrimidin-2-one.

these values (0.54–0.78) obtained from studies of electrophilic substitution of heterocycles (see also p. 18) seriously disagree with the value of 0.30 obtained in a study[376] in which direct measurement with no assumptions or approximations was made. The difference was attributed to hydrogen bonding in the former reactions though the effect seems too large to be accounted for by this alone.

The rate–acidity profile for pyrimidin-2-one indicated reaction on the free base but since the derived second-order rate coefficient is $10^4$ times greater than that for 2-pyridone, and the acidity dependence in the $H_0$ region was also greater, the slope of log $k_1$ versus $-H_0$ plot being 0.45, cf. 0.15 for 2-pyridone; reaction was, therefore, postulated as occurring via a covalent hydrate, hydration taking place at the 4 position. Methyl substitution increased the rate as expected and N-methyl substitution produced a larger effect than 4,6-dimethyl substitution and this may be due to alteration of the amount of covalent hydration at equilibrium. The data for the pyrimidinium salt indicated that reaction occurred on the base in equilibrium with the salt.

Paudler and Helmick[515] have measured half-lives for deuteration of some heterocycles by deuterated sulphuric acid at 100 °C. The equivalent first-order rate coefficients $(10^7 k_1)$ are as follows: imidazo[1, 2-a]pyridine, 427(3-H); imidazo[1,2-a]pyridine-N-methiodide, 62(3-H); imidazo[1,2-a]pyrimidine, 123-(3-H); imidazo[1,2-a]pyrimidine-N-methiodide, < 6.4 (3-H, 5-H); 1,2,4-triazolo[1,5-a]pyridimidine, 128(5-H); 1,2,4-triazolo[1,5-a]pyrimidine-N-methiodide, 11.7 (5-H). The lower reactivity in each case of the corresponding methiodides shows that the bases react as such and not as the conjugate acids.

### 8.1.3 Kinetic studies of the effects of catalysts on hydrogen exchange

It was appropriate to consider some of the experimental data relating to the effects of catalysts upon hydrogen exchange, previously and in another context (p. 207). Further data are now discussed.

Comyns et al.[516] first observed the rate enhancement by Lewis acid catalysts in hydrogen exchange in the tritiation of toluene by tritium chloride. Second-order rate coefficients were given approximately by

$$\text{Rate} = k_2[\text{TCl}][\text{SnCl}_4 \text{ or NO}_2] \tag{233}$$

for the effect of stannic chloride or nitrogen dioxide as catalysts. Reaction was complicated by the fact that tritium chloride underwent exchange with the glass walls of the reaction vessel (other workers have not detected or corrected for this) and the effect of nitrogen dioxide was unclear. With this catalyst, loss of tritium chloride occurred at the rate of 7 moles for each mole of catalyst, the fate of this reagent being unknown. Nitrogen dioxide catalysed the exchange with toluene to the extent of about 1.5–2.3 moles of tritium chloride per mole of catalyst. Second-order rate coefficients were $4 \times 10^{-4}$ for $\text{SnCl}_4$ as catalyst at 25 °C with $E_a = 10\pm4$, and $19 \times 10^{-3}$ for nitrogen dioxide as catalyst at 140 °C with $E_a = 7$; in the absence of these catalyst exchange between tritium chloride and toluene at 140 °C was negligible, and $10^4 k_1$ for reaction with mesitylene was 3–10. The mechanism of the catalytic action was not known, but the formation of a complex between toluene and stannic chloride was considered to be significant.

This study was extended by Satchell[517], who investigated the effect of stannic chloride with various Brönsted acids on the rate of tritiation of toluene (excess) at 25 °C. Using tritium–hydrogen chloride the data in Table 153 were obtained, and whereas the first-order rate coefficients were closely proportional to the stannic chloride concentration, they were relatively independent of the total concentration of hydrogen chloride at a constant tritium chloride concentration, the small rate

TABLE 153

RATE COEFFICIENTS FOR REACTION OF TCl–HCl–SnCl$_4$ WITH PhCH$_3$ AT 25 °C[517]

| $10[HCl]$ $(M)$ | 0.38 | 0.75 | 1.125 | 1.88 | 3.75 | 3.75 | 3.75 | 3.85 | 1.88 |
|---|---|---|---|---|---|---|---|---|---|
| $10[SnCl_4]$ $(M)$ | 1.66 | 1.66 | 1.66 | 1.66 | 1.66 | 3.32 | 0.83 | 0.42 | 3.22 |
| $10^6 k_1$ | 38 | 41 | 44 | 48 | 66 | 130 | 34 | 17 | 91 |

change observed being attributed to a salt effect. No exchange took place in the absence of stannic chloride and it was argued that the observed behaviour follows for a reaction first-order in both stannic chloride and tritium chloride, for at a given concentration of the former, the exchange rate would not depend upon the concentration of hydrogen chloride present since the latter would not increase the number of tritium–toluene collisions. The mechanism was proposed as equilibria (234)–(236)

$$SnCl_4 + ArH \rightleftharpoons ArHSnCl_4 \text{ (fast)} \tag{234}$$

$$ArHSnCl_4 + TCl \underset{k'_{-2}}{\overset{k'_2}{\rightleftharpoons}} Ar^+ HTSnCl_5{}^- \underset{k'_{-3}}{\overset{k'_3}{\rightleftharpoons}} ArTSnCl_4 + HCl \text{ (slow)} \tag{235}$$

$$ArTSnCl_4 \rightleftharpoons ArT + SnCl_4 \tag{236}$$

the evidence being, apart from the kinetics, the confirmation of the formation of a complex between stannic chloride and toluene as shown spectroscopically. The rate equation was, therefore,

$$\text{Rate} = \frac{K(k'_2 k'_3)}{(k'_{-2} + k'_3)} [ArH][SnCl_4][TCl] \tag{237}$$

which reduces to

$$\text{Rate} = k_2 [SnCl_4][TCl] \tag{238}$$

in the presence of excess toluene. Addition of 0.015 $M$ of water to the kinetic runs reported in Table 153 slowed the rates by only *ca.* 30 %, so that it seemed improbable that the results were affected by traces of water in the tritium chloride.

For reaction with tritiated acetic acid, the rate again depended upon the stannic chloride concentration and was relatively independent of the concentration of added acetic acid, except when this was large in which case a large rate decrease occurred (Table 154). Spectral evidence indicated the formation of a complex $[HSnCl_4(OAc)_2]^- [H_2OAc]^+$, the reaction of which with the aromatic was assumed to be rate-determining. At high acid concentrations, solvation of the complex by the excess acetic acid was thought to account for the lowering of its reactivity.

TABLE 154

RATE COEFFICIENTS FOR REACTION OF $CH_2COOT–CH_3COOH–SnCl_4$ WITH $PhCH_3$ AT 25 °C[517]

| $10[AcOH]$ $(M)$ | 2.16 | 2.16 | 2.16 | 1.08 | 4.30 | 8.60 | 18.0 |
|---|---|---|---|---|---|---|---|
| $10[SnCl_4]$ $(M)$ | 1.9 | 3.7 | 7.3 | 4.7 | 3.7 | 3.7 | 1.9 |
| $10^7k_1$ | 17 | 37 | 74 | 35 | 39 | 42 | 3.8 |

With trifluoroacetic acid, it was reported that very little exchange took place at 50 °C, even after a period of days, which is surprising in view of the fact that there was spectral evidence of complex formation between the acid and stannic chloride. The complex was assumed to be unreactive towards exchange and at the same time to compete in its formation with the formation of a complex between stannic chloride and the toluene solvent. In the absence of stannic chloride, about 20 % exchange would take place in seven days (see Table 159, p. 244).

From this work the relative efficiencies of Brönsted acids, in the presence of stannic chloride (0.166 $M$) in promoting hydrogen exchange at 25 °C can be ascertained from the rate coefficients ($10^6k_1$) as follows: HCl (44); $H_2O$ (27); AcOH (1.6); $CF_3COOH$ (very small). Thus the ability of the dual acid systems to transfer protons is not simply related to the conventional acid strength of the Brönsted component.

Shatenshtein et al.[518] also found a similar dependence of rate coefficient upon the concentrations of stannic chloride and acetic acid in the dedeuteration of $[1,4-^2H_2]$-durene in benzene. Rate coefficients increased linearly with increasing stannic chloride concentration and at a constant value of this the rate increased only slightly with increasing acetic acid concentration, except at high concentrations of the latter when the rate then decreased (Table 155). Cryoscopic mea-

TABLE 155

FIRST-ORDER RATE COEFFICIENTS FOR REACTION OF $[1,4-^2H_2]$-DURENE WITH $AcOH–SnCl_4$ IN $PhH$[518]

| $10[SnCl_4]$ $(M)$ | $10[AcOH]$ $(M)$ | $10^7k_1$ $(M)$ | $10[SnCl_4]$ $(M)$ | $10[AcOH]$ $(M)$ | $10^7k_1$ |
|---|---|---|---|---|---|
| 4 | 8 | 800(5°C) | 1 | 2 | 20(5 °C) |
| 4 | 2 | 400 | 2 | 4 | 70 |
| 2 | 2 | 200 | 2 | 4 | 80 |
| 2 | 4 | 200 | 4 | 8 | 200 |
| 1 | 0.5 | 40 | 0.29 | 0.57 | 20(50 °C) |
| 1 | 1.07 | 60 | 0.50 | 1 | 60 |
| 1 | 2 | 69 | 1.0 | 2 | 200 |
| 1 | 4 | 40 | 2.0 | 4 | 500 |
| 1 | 8 | 20 | 1 | 2 | 200(75 °C) |
| 0.5 | 1 | 30 | 2 | 4 | 800 |
| 0.29 | 0.57 | 9 | 2 | 4 | 1,000(100 °C) |

surements gave a maximum on the molecular weight–composition curve at a point which corresponds to $[SnCl_4\,2AcO]^{2-}\,2\,H^+$, and a plot of log $k_1$ *versus* the concentration of this was linear; the data of Satchell were shown to analyse in a similar manner. In the absence of benzene, log $k_1$ was linearly related to log $[SnCl_4]$ (Table 156), the rate of exchange being lower and the activation energy

TABLE 156

FIRST-ORDER RATE COEFFICIENTS FOR REACTION OF $[1,4\text{-}^2H_2]$-DURENE WITH AcOH–SnCl$_4$[518]

| $10[SnCl_4]$ $(M)$ | 1.0 | 2.0 | 5.0 | 10.0 | 2.0 | 4.93 | 9.87 | 2.0 | 2.0 |
|---|---|---|---|---|---|---|---|---|---|
| Temp. $(°C)$ | 25 | 25 | 25 | 25 | 49.7 | 50 | 50 | 75 | 95 |
| $10^7 k_1$ | 1 | 4 | 20 | 80 | 90 | 400 | 2,000 | 1,000 | 9,000 |

(24.3) higher because in excess of acetic acid the complex acid is converted to the less acidic onium salt $[SnCl_4AcO]^{2-}\,2AcOH_2^+$.

Shatenshtein *et al.*[519] have also measured the effect of boron trifluoride as a catalyst for hydrogen exchange in acetic acid and have compared it with stannic chloride (Table 157). The logarithm of the rate coefficient was linearly related to

TABLE 157

RATE COEFFICIENTS AND KINETIC PARAMETERS FOR REACTION OF $[^2H]$-ArH WITH BF$_3$ (OR SnCl$_4$)–HOAc[519]

| Catalyst | [Catalyst] $(M)$ | ArH | $10^7 k_1$ range[a] | $E_a$ | $\Delta S^{\ddagger}$ |
|---|---|---|---|---|---|
| BF$_3$ | 0.13 | $[^2H]$-C$_6$Me$_5$ | 1.3(25)–     550( 75) | 24.7 | −9.1 |
|  | 0.27 |  | 14  (25)–  1,700( 65) | 23.7 | −1.7 |
|  | 0.53 |  | 79  (25)–  7,300( 65) | 23.0 | −6.8 |
|  | 1.05 |  | 400 (25)–12,000( 55) | 22.2 | −6.1 |
|  | 1.93 |  | 3,500 (25)–13,000( 35) |  |  |
| SnCl$_4$ | 0.20 |  | 90 (15)– 3,000( 45) | 21.7 |  |
|  | 0.20 | $[2\text{-}^2H]$-1,4-Me$_2$C$_6$H$_4$ | 20 (85)–  200(110) | 23.4 |  |

[a] At temperature given in parentheses.

the concentration of boron trifluoride and to the acidity function $-H_0$ of the solution; with the latter the slope was 1.8, however, and no conclusions were arrived at. From this and the previously obtained data, the rate coefficients for reaction of hydrocarbons with a 0.2 $M$ solution of stannic chloride in acetic acid at 25 °C were $3 \times 10^{-9}$ (p-xylene), $4 \times 10^{-7}$ (durene) and $3 \times 10^{-5}$ (pentamethyl-benzene) by extrapolation of the observed data, and this rate spread is similar to that obtained with aqueous trifluoroacetic acid at 70 °C (see p. 250). The rate coefficients for reaction of pentamethylbenzene at 25 °C with a 0.2 $M$ solution of catalyst were $4 \times 10^{-7}$ (BF$_3$), $3 \times 10^{-5}$ (SnCl$_4$) and for a 1.0 $M$ solution, $4 \times 10^{-5}$ (BF$_3$) and $9 \times 10^{-4}$ (SnCl$_4$) though it was not clear how this latter value was obtained. The relative efficiency of stannic chloride to boron trifluoride was therefore in the range 100–20 and since previous work showed stannic chloride to be 200 times as effective as zinc chloride, the efficiency order was deduced as $SnCl_4 > BF_3 > ZnCl_2$.

Shatenshtein *et al.*, and Satchell[517], therefore attribute the increased reactivity in the presence of a catalyst to greater polarisation of the acid by the catalyst, but Satchell and Comyns *et al.*[516] seemed to favour reaction of this upon a complex formed between the aromatic and the catalyst. It is difficult, however, to envisage how this latter complex would have enhanced reactivity, since the catalysts are electron acceptors and it is possible that the catalyst enhances the reactivity of the Brönsted acid and lowers the reactivity of the aromatic, the former effect being

TABLE 158

FIRST-ORDER RATE COEFFICIENTS FOR REACTION OF [9-$^3$H]-PHENANTHRENE ($5.6 \times 10^{-3}$ $M$) WITH $CF_3COOH$ AND $CF_3COOAg$ AT 100 °C[520]

| $[CF_3COOAg]$ $(M)$ | 0 | 0.0025 | 0.005 | 0.01 | 0.02 | 0.05 | 0.1 | 0.2 |
|---|---|---|---|---|---|---|---|---|
| $10^7 k_1$ | 1,605 | 1,140 | 932.5 | 605[a] | 338 | 117 | 53.7 | 24.5 |

[a] Rate coefficients at 70 and 50 °C were 48.7 and 5.57 giving $E_a = 22.4$, *cf.* 20.3 in the absence of silver salt (see p. 245).

greater than the latter. Hence with trifluoroacetic acid in which the latter complex is less readily formed[517], significant rate enhancement is not observed. It is interesting to note that the rates of detritiation of [9-$^3$H]-phenanthrene by trifluoroacetic acid were reduced by added silver salts (Table 158)[520] this being shown spectroscopically to be due to the formation of silver ion–phenanthrene complexes. The presence of 0.05 $M$ of silver ion reduced the reactivity of [9-$^3$H]-phenanthrene by a factor of 13.7 and also that of [1-$^3$H]-naphthalene and [2-$^3$H]-$p$-xylene by factors of 8.5 and 6.6, respectively, and interestingly these reductions are paralleled by the equilibrium constants for argentation of these aromatics by silver nitrate.

The addition of Brönsted acids to trifluoroacetic acid causes a dramatic change in the ability of the latter to promote hydrogen exchange, and the sensitivity of the acid to added Brönsted acids is especially great at low added acid concentrations[477]. Consequently, it is of considerable importance to ensure absolute purity when carrying out kinetic studies with trifluoroacetic acid. The reason for the rate enhancement has not been investigated (and indeed it would be very difficult to do so), but exchange very probably occurs *via* the small quantity of the species $CF_3COOH_2^+$ that will be formed, though, of course, exchange may occur *via* the small concentration of added acid which may be exceptionally polarised by the trifluoroacetic acid medium (see also pp. 202, 255).

Lewis acids have been found to increase the rate of hydrogen exchange in halogen acids, and thus iodine increases the rate of hydrogen exchange in hy-

drogen iodide, and it has also been found that the rates increase approximately
linearly with aluminium bromide concentration in the reaction between deuterium
bromide and benzene[521]. The catalytic efficiencies were found to be: $AlBr_3$
$(5 \times 10^5) > GaBr_3(10^5) > FeBr_3(10^4) \gg BBr_3(30) > SbBr_3(6) > TiBr_4(1) \gg$
$SnBr_4$ which follows the usual order found for other electrophilic substitutions,
and the effects of the catalysts can be attributed to their ability to polarise the
hydrogen–halogen bond. The mean values of $10^6 k/10^2$ (moles catalyst/mole of
acid) were 100,000; 70,000; 2,000; 12; 2; 0.4; 0.

### 8.1.4 Kinetic studies of substituent effects in electrophilic aromatic hydrogen exchange

The advantages of the hydrogen exchange as a model electrophilic aromatic
substitution are now well recognised and have been emphasised[504, 522, 523], so
that a very considerable body of data is now to be found in the literature. In order
to simplify presentation of this, the data are considered under headings of the
acid medium employed for the studies.

#### (a) Kinetic studies with anhydrous trifluoroacetic acid solvent

This medium is undoubtedly the best for substituent effect studies for it combines good solvent properties with a convenient rate of exchange at accessible
temperatures with almost no side reactions. Most temperatures used are at or
above the b.p. (71.5 °C) of the acid and consequently sealed-tube techniques are
preferred. The disadvantages of the acid are sensitivity to impurities and relatively
high cost. The former can be overcome by fractionating either from concentrated
sulphuric acid and then from silver trifluoroacetate (to eliminate water and halogen
acids respectively these being the main impurities) or, more easily from silver
oxide and sulphuric acid in that order. The cost can be significantly reduced by
recovery and re-use, via neutralisation, evaporation of the salt to dryness and
heating the salt with concentrated sulphuric acid; this process removes the hydrogen isotopes acquired during kinetic studies.

Rate coefficients and thermodynamic data where available are gathered in
Table 159, which represents the largest body of kinetic data available under one
condition in an electrophilic aromatic substitution, and the absence of significant
steric effects has caused this data to be used for rigorously testing and modifying
existing theories of substituent effects in electrophilic substitution and for pro-
posing new ones. Discussion of this is, of course, inappropriate here and reference
to the original papers should be made. In one case, however, the data can be seen
to reinforce a recent theory of the effects of strain on substitution patterns which
was published subsequent to that containing the kinetic data. For deuteration of
1,2-dimethoxybenzene (XXV), benzo-1,3-dioxole (XXVI), and benzo-1,4-dioxan

## TABLE 159

RATE COEFFICIENTS AND ACTIVATION ENERGIES FOR REACTION OF [³H]-Ar (OR[²H]-Ar WHERE INDICATED) WITH ANHYDROUS $CF_3COOH$

| ArH | Position of ³H | $10^7 k_1$ at 70 °C or as indicated | $E_a$ |
|---|---|---|---|
| *Substituted benzenes* | | | |
| Benzene[520,524,525] | 1 | 0.095(70.1 °C), 3.64(110.15 °C), 15.36(128.25 °C) | 23.8 |
| Toluene[465] | 2 | 20.8(70.1 °C) | |
| Toluene[465] | 4 | 42.7(70.1 °C) | |
| Toluene[465] | 3 | 0.577 | |
| t-Butylbenzene[465] | 2 | 23.0(70.1 °C) | |
| t-Butylbenzene[465] | 4 | 50.9(70.1 °C) | |
| o-Xylene[527,528] | 3 | 127, 110[a] | |
| o-Xylene[527,528] | 4 | 180, 155[a] | |
| Indane[528] | 4 | 82 | |
| Indane[528] | 5 | 334 | |
| Tetralin[528] | 5 | 310 | |
| Tetralin[528] | 6 | 335 | |
| p-Xylene[529,540] | 2 | 139, 1,190(100.0 °C) | 18.1 |
| m-Xylene[529] | 2 | 3,780 °C | |
| m-Xylene[471,529] | 4 | 128(25.0 °C), 7,490 | 18.4 |
| 4-Chlorotoluene[526] | 2 | 0.037, 5.63(128.25 °C) | 23.8 |
| 4-Bromotoluene[526] | 2 | 6.06(128.25 °C) | |
| 2-Chloro-m-xylene[530] | 4 | 31.3 | |
| 4-Chloro-m-xylene[531] | 2 | 7.93 | |
| 5-Chloro-m-xylene[530] | 4 | 340 | |
| 4-Bromo-m-xylene[539] | 2 | 8.57 | |
| 2-Bromomesitylene[539] | 4 | 2,870 | |
| Anisole[526] | 2 | 6,945 | |
| Anisole[525] | 4 | 270(24.84 °C), 1,242(40.0 °C) | 18.6 |
| 4-Fluoroanisole[526] | 2 | 79.0 | |
| 2-Chloroanisole[539] | 4 | 192 | |
| 2-Chloroanisole[539] | 6 | 71.2 | |
| 3-Chloroanisole[539] | 2 | 1,190 | |
| 3-Chloroanisole[539] | 4 | 3,440 | |
| 3-Chloroanisole[539] | 6 | 2,975 | |
| 4-Chloroanisole[526] | 2 | 85.7 | |
| 4-Bromoanisole[526] | 2 | 89.7 | |
| 1,2-Dimethoxybenzene[532] | 4(²H) | 65.0(25.0 °C), 730(50.0 °C), 5,900(75.0 °C) | 18.6 |
| Benzo-1,3-dioxole[532] | 5(²H) | 101(25.0 °C), 1,090(50.0 °C), 8,400(75.0 °C) | 18.2 |
| Benzo-1,4-dioxane[532] | 6(²H) | 59.0(25.0 °C), 640(50.0 °C), 6,100(75.0 °C) | 19.2 |
| Diphenylmethane[533] | 2 | 4.4 | |
| Diphenylmethane[533] | 3 | 0.36 | |
| Diphenylmethane[525] | 4 | 11.05(70.1 °C) | |
| Diphenyl ether[534] | 2 | 654(70.1 °C) | |
| Diphenylether[535] | 2(²H) | 150(40.0 °C) | |
| Diphenylether[534] | 4 | 44.6(24.84 °C), 2,930(70.1 °C) | 18.9 |
| Diphenylether[535] | 4(²H) | 460(40.0 °C) | |
| Diphenylether[534] | 3 | 0.425(110.15 °C) | |
| Diphenyl sulphide[534] | 2 | 316(70.1 °C) | |
| Diphenylsulphide[534] | 4 | 934(70.1 °C) | |
| Thioanisole[525] | 4 | 124(24.84 °C), 538(40.0 °C) | 17.9 |

TABLE 159 (continued)

| ArH | Position of $^3H$ | $10^7 k_1$ at 70 °C or as indicated | $E_a$ |
|---|---|---|---|
| Biphenyl[536] | 2 | 9.31(70.1 °C) | |
| Biphenyl[536] | 4 | 13.45(70.1 °C) | |
| 4-Methylbiphenyl[537] | 4′ | 62(70.1 °C) | |
| 4-Methoxybiphenyl[537] | 4′ | 51(70.1 °C) | |
| 4-Chlorobiphenyl[537] | 4′ | 4.57(70.1 °C) | |
| 4-Bromobiphenyl[537] | 4′ | 3.90(70.1 °C) | |
| 4-Nitrobiphenyl[537] | 4′ | 0.573(110.15 °C) | |
| 3-Chlorobiphenyl[537] | 4′ | 1.91(70.1 °C) | |
| 2,4,6-Trimethoxybenzene[525] | 1 | 828(−10.0 °C), 27,250(24.84 °C) | 15.1 |
| N,N-Dimethylaniline[535, 572] | 4 | 0.01(30 °C), 8(75 °C), 60(90 °C), 800(115 °C) | 30.8 |
| N-Methyldiphenylamine[535, 572] | 4 | 10(30 °C) | |
| Triphenylamine[535, 572] | 4 | 20,000(30 °C) | |

*Polycyclics*

| ArH | Position of $^3H$ | $10^7 k_1$ at 70 °C or as indicated | $E_a$ |
|---|---|---|---|
| Biphenylene[538] | 1 | 9.8 | |
| Biphenylene[538] | 2 | 1,330 | |
| 9,10-Dihydrophenanthrene[526] | 2 | 260 | |
| Dihydroanthracene[527] | 1 | 42.2 | |
| Dihydroanthracene[527] | 2 | 61.8 | |
| Triptycene[527] | 1 | 4.84[b], 4.48 [539, c] | |
| Triptycene[527] | 2 | 34.4[b], 34.8 [539, c] | |
| Fluorene[524] | 2 | 1,600(70.1 °C) | |
| Fluorene[533] | 1 | 2.03 | |
| Fluorene[533] | 3 | 11.8 | |
| Fluorene[533] | 4 | 5.2 | |
| 9-Methylfluorene[524] | 2 | 1,660(70.1 °C) | |
| 9,9-Dimethylfluorene[524] | 2 | 1,900(70.1 °C) | |
| 2-Methylfluorene[537] | 7 | 6,170(70.1 °C) | |
| 2-Methoxyfluorene[535] | 7 | 4,840(70.1 °C) | |
| 2-Chlorofluorene[537] | 7 | 425 | |
| 2-Bromofluorene[537] | 7 | 340 | |
| 2-Carboxyfluorene[537] | 7 | 3.28 | |
| 2-Nitrofluorene[537] | 7 | 23.0 | |
| Phenanthrene[526] | 1 | 85.5, 924(100.0 °C)[540] | 20.2 |
| Phenanthrene[526] | 2 | 16.4 | |
| Phenanthrene[526] | 3 | 36.5 | |
| Phenanthrene[526] | 4 | 76.8 | |
| Phenanthrene[525, 540] | 9 | 23.3(50.0 °C), 154.5, 1,605(100.0 °C) | 20.3 |
| 9-Fluorophenanthrene[540] | 10 | 98.4(50.0 °C), 639.5, 5,420(100.0 °C) | 19.4 |
| 9-Fluorophenanthrene[540] | 1 | 16.2(50.0 °C), 116, 994(100.0 °C) | 19.8 |
| 9-Chlorophenanthrene[540] | 10 | 28.9, 337(100.0 °C), 1,400(120.0 °C) | 20.8 |
| 9-Bromophenanthrene[540] | 10 | 16.8, 205(100.0 °C), 798(120.0 °C) | 20.8 |
| 9-Iodophenanthrene[540] | 10 | 21.5, 234(100.0 °C), 907.5(120.0 °C) | 20.2 |
| 9-Cyanophenanthrene[540] | 10 | 6.5(100.0 °C) | |
| 9-Methylphenanthrene[540] | 10 | 30.4(0.0 °C), 539(25.0 °C), 5,310(50.0 °C), 28,250 | 18.0 |
| Naphthalene[541] | 1 | 110, 1,045(100.7 °C), 6,030(125 °C) | 18.8 |
| 2-Methylnaphthalene[541] | 1 | 811(24.94 °C), 1,980(35.04 °C), 4,750(45.05 °C) | 16.7 |
| 3-Methylnaphthalene[541] | 1 | 303 | |
| 4-Methylnaphthalene[541] | 1 | 189(24.94 °C), 447(35.04) °C), 1.180(45.05 °C) | 17.6 |

TABLE 159 (continued)

| ArH | Position of $^3H$ | $10^7 k_1$ at 70 °C or as indicated | $E_a$ |
|---|---|---|---|
| 5-Methylnaphthalene[541] | 1 | 240 | |
| 6-Methylnaphthalene[541] | 1 | 144 | |
| 7-Methylnaphthalene[541] | 1 | 319 | |
| 8-Methylnaphthalene[541] | 1 | 168 | |
| 2-Methoxynaphthalene[541] | 1 | 11,000(−11.95 °C), 16,850(−6.85 °C), 24,000(−24 °C) | 11.8 |
| 4-Methoxynaphthalene[541] | 1 | 5,410(−11.7 °C), 8,430(−6.2 °C), 14,800(1.0 °C) | 11.2 |
| 5-Methoxynaphthalene[541] | 1 | 426 | |
| 4-Phenylnaphthalene[541] | 1 | 1,750 | |
| 2-Fluoronaphthalene[541] | 1 | 188 | |
| 4-Fluoronaphthalene[541] | 1 | 505 | |
| 5-Fluoronaphthalene[541] | 1 | 4.73 | |
| 8-Fluoronaphthalene[541] | 1 | 5.84 | |
| 2-Chloronaphthalene[541] | 1 | 29.6 | |
| 3-Chloronaphthalene[541] | 1 | 0.256 | |
| 4-Chloronaphthalene[541] | 1 | 315(100.7 °C), 1,600(125 °C) | 19.8 |
| 5-Chloronaphthalene[541] | 1 | 3.11 | |
| 6-Chloronaphthalene[541] | 1 | 3.47 | |
| 7-Chloronaphthalene[541] | 1 | 15.2 | |
| 8-Chloronaphthalene[541] | 1 | 4.70 | |
| 2-Bromonaphthalene[541] | 1 | 18.2 | |
| 4-Bromonaphthalene[541] | 1 | 200(100.7 °C), 1,190(125 °C) | 21.8 |
| 5-Bromonaphthalene[541] | 1 | 2.77 | |
| 8-Bromonaphthalene[541] | 1 | 5.12 | |
| 2-Iodonaphthalene[541] | 1 | 25.3 | |
| 4-Iodonaphthalene[541] | 1 | 226(100.7 °C) | |
| 8-Iodonaphthalene[541] | 1 | 7.24 | |
| Naphthalene[541] | 2 | 14.3, 159(100.7 °C), 852(125 °C) | 20.3 |
| 1-Methylnaphthalene[541] | 2 | 3,180 | |
| 3-Methylnaphthalene[541] | 2 | 15.4 | |
| 4-Methylnaphthalene[541] | 2 | 42.8 | |
| 5-Methylnaphthalene[541] | 2 | 25.2 | |
| 6-Methylnaphthalene[541] | 2 | 277 | |
| 7-Methylnaphthalene[541] | 2 | 22.8 | |
| 8-Methylnaphthalene[541] | 2 | 44.5 | |
| 1-Methoxynaphthalene[541] | 2 | 7,120(−11.6 °C), 13,300(−5.7 °C), 23,400(0.0 °C) | 14.4 |
| 3-Methoxynaphthalene[541] | 2 | 515 | |
| 4-Methoxynaphthalene[541] | 2 | 22.6 | |
| 6-Methoxynaphthalene[541] | 2 | 1,770 | |
| 1-Fluoronaphthalene[541] | 2 | 7.74 | |
| 3-Fluoronaphthalene[541] | 2 | 8.53(100.7 °C), 50.1(125.0 °C) | 21.6 |
| 4-Fluoronaphthalene[541] | 2 | 3.38(100.7 °C), 18.4(125.0 °C) | 20.7 |
| 1-Chloronaphthalene[541] | 2 | 10.8(100.7 °C), 51.8(125.0 °C) | 19.1 |
| 3-Chloronaphthalene[541] | 2 | 6.99(100.7 °C), 40.0(125.0 °C) | 21.3 |
| 4-Chloronaphthalene[541] | 2 | 2.42(100.7 °C), 16.6(125.0 °C) | 23.5 |
| 5-Chloronaphthalene[541] | 2 | 2.64(100.7 °C), 18.1(125.0 °C) | 23.4 |
| 6-Chloronaphthalene[541] | 2 | 28.7(100.7 °C), 143(125.0 °C) | 19.6 |
| 7-Chloronaphthalene[541] | 2 | 5.12(100.7 °C), 28.9(125.0 °C) | 20.6 |
| 8-Chloronaphthalene[541] | 2 | 5.19(100.7 °C), 35.7(125.0 °C) | 23.5 |
| 1-Bromonaphthalene[541] | 2 | 6.38(100.7 °C), 29.8(125.0 °C) | 18.8 |

TABLE 159 (continued)

| ArH | Position of $^3H$ | $10^7k_1$ at 70 °C or as indicated | $E_a$ |
|---|---|---|---|
| 3-Bromonaphthalene[541] | 2 | 5.65(100.7 °C, 33.4(125.0 °C) | 21.1 |
| 4-Bromonaphthalene[541] | 2 | 2.26(100.7 °C), 15.0(125.0 °C) | 22.9 |
| 8-Bromonaphthalene[541] | 2 | 3.42(100.7 °C), 27.0(125.0 °C) | 25.2 |
| 1-Iodonaphthalene[541] | 2 | 12.6(100.7 °C) | |
| 3-Iodonaphthalene[541] | 2 | 9.49(100.7 °C), 56.0(125.0 °C) | 21.6 |
| 4-Iodonaphthalene[541] | 2 | 4.58(100.7 °C), 29.7(125.0 °C) | 22.8 |
| *Heterocyclic compounds* | | | |
| Thiophen[c,55] | 2($^2$H) | 15,000(0 °C), 8,000(−7 °C), 3,500(−14 °C) | 15±1 |
| Thiophen[525] | 2 | 2,360(−10.0 °C), 73,200(24.84 °C) | 15.4 |
| Thiophen[555] | 3($^2$H) | 44.0(25 °C) | |
| Thiophen[525] | 3 | 1,780(70.1 °C) | |
| 4-Methylthiophen[555] | 3($^2$H) | 18,000(25 °C) | |
| 4-Thiomethoxythiophen[555] | 2($^2$H) | 16,000(25 °C) | |
| 4-Thiomethoxythiophen[555] | 3($^2$H) | 2,700(25 °C) | |
| Benzothiophen[525] | 2 | 421(24.84 °C), 1,810(40.0 °C) | 17.5 |
| Benzothiophen[525] | 3 | 600(24.84 °C), 2,510(40.0 °C) | 16.8 |
| Benzofuran[525] | 2 | 1,615(70.1 °C) | |
| Dibenzofuran[534] | 1 | 12.8(70.1 °C) | |
| Dibenzofuran[534] | 2 | 34.9(70.1 °C) | |
| Dibenzofuran[534] | 3 | 29.7(70.1 °C) | |
| Dibenzofuran[534] | 4 | 15.2(70.1 °C) | |
| Dibenzothiophen[534] | 1 | 25.8(70.1 °C) | |
| Dibenzothiophen[534] | 2 | 174(70.1 °C) | |
| Dibenzothiophen[534] | 3 | 40.8(70.1 °C) | |
| Dibenzothiophen[534] | 4 | 34.4(70.1 °C) | |
| 5-Ethylcarbazole[525] | 1 | 4,000(−10.0 °C) | |
| 5-Ethylcarbazole[525] | 3 | 32,900(−10.0 °C) | |

ᵃ These lower values were obtained with a trifluoroacetic acid medium which gave rates for [4-$^3$H]-toluene 7 % lower than those given in this table.
ᵇ The rates for the 1 and 2 positions were incorrectly transposed in the original paper.
ᶜ These results obtained with a medium containing 2 % of carbon tetrachloride.

(XXVII) at the 4, 5, and 6 positions respectively (these being all *beta* to the side chain), initial rates ($10^7k_1$) at 50 °C were 400, 480, and 380, which are each about

(XXV)          (XXVI)          (XXVII)

1.8 times slower than for dedeuteration (Table 159) as expected[532]. The values of the rate coefficients decreased with time during a run due to slower deuteration at the α positions, the corresponding rates after an approximately equal interval of time being *ca.* 160, 70, and 170. It appears then that the α position of 1,3-dioxole is significantly less reactive than the α position of the other two molecules and also that the $\beta$ position is significantly more reactive than for the other $\beta$ positions (this is better and more reliably indicated by the data for dedeuteration in Table 159). This is exactly the same situation found for the methyl analogues, o-xylene, indane and tetralin, and may be attributed to the same cause, namely relief or enhancement of strain on going to the transition state of the reaction[542]. The same pattern is clearly discernible in the rate coefficients $(10^5 k_1)$ for deuteration at 25 °C at the positions indicated for the compounds (XXVIII)–(XXX)

One may also note that the dedeuteration and detritiation studies of thiophen agree very well both with regard to the activation energy for reaction at the 2 position, and the slightly greater reactivity in dedeuteration compared to detritiation. The relative reactivities deduced from these results for the 3 and 2 positions are also in fairly good agreement in view of the temperature difference, being 1,240 (70 °C, detritiation) and 3,300 (25 °C, dedeuteration).

(b) *Kinetic studies with trifluoroacetic acid in aprotic solvents*

A few kinetic studies have been carried out using mixtures of trifluoroacetic acid and an inorganic acid in aprotic solvents and description of these is deferred (see p. 257). There has been one study of the effect of such solvents on the rates of dedeuteration in pure trifluoroacetic acid (Table 160)[543]. The order with respect to the acid for reaction of pentamethylbenzene varied from 2.3 to 2.9 so that the acid dimers at least seem to be the exchanging entities involved. From log $k_1$ *versus* concentration plots the relative efficiencies of solvents containing 0.5 $M$ acid were interpolated as: hexane (1.0); carbon tetrachloride (1.2); benzene (5.1);

TABLE 160

FIRST-ORDER RATE COEFFICIENTS FOR THE REACTION OF $[^2H]$-$C_6Me_5$ WITH $CF_3COOH$ IN THE PRESENCE OF APROTIC SOLVENTS AT 25 °C [543]

| Solvent | $[CF_3COOH]$ (M) | $10^7 k_1$ | Solvent | $[CF_3COOH]$ (M) | $10^7 k_1$ |
|---|---|---|---|---|---|
| Benzene | 0.33 | 300 | Carbon tetrachloride | 0.45 | 16 |
| Benzene | 0.50 | 900 | Carbon tetrachloride | 0.60 | 30 |
| Benzene | 0.51 | 900 | Carbon tetrachloride | 0.75 | 60 |
| Benzene | 0.60 | 1,100 | 1,2-Dichloroethane | 0.32 | 70 |
| Benzene | 0.72 | 2,200 | 1,2-Dichloroethane | 0.49 | 200 |
| Hexane | 0.35 | 6 | 1,2-Dichloroethane | 0.72 | 520 |
| Hexane | 0.43 | 10 | Chlorobenzene | 0.23 | 90 |
| Hexane | 0.75 | 60 | Chlorobenzene | 0.29 | 180 |
| Hexane | 1.24 | 220 | Chlorobenzene | 0.41 | 410 |
|  |  |  | Chlorobenzene | 0.48 | 830 |

1,2-dichloroethane (13); chlorobenzene (42) and these differences were attributed to the specificity of solvation of trifluoroacetic acid by the solvents.

Hanstein and Traylor[544] used a 4.3 $M$ solution of trifluoroacetic acid in chloroform to measure the rates of deuteration of $(PhCH_2)_2Hg$ at 35 °C, for which $k_1 = 2,100 \times 10^{-7}$. Comparison with other deuteration rates (which, however, were neither quoted nor referred to) was said to give $\sigma_p^+(CH_2HgCH_2Ph)$ = $-1.14$ and the *ortho* : *para* ratio was 0.84. For the compound $PhCH_2B(OH)_3^-$, the rate of deuteration by 3 $M$ tartaric acid at 100 °C was found to be $5 \times 10^{-7}$ (*ortho* : *para* ratio = 1.1) and from this a value of $\sigma_p^+(CH_2B(OH)_3^-)$ of $-1.11$ was claimed, but without the relevant data used to obtain these $\sigma^+$ values cautious use of them seems appropriate.

Nesmeyanov et al.[545] used a mixture of ferrocene, deuterated trifluoroacetic acid and benzene in the molar ratios 1 : 2 : 20 in a preliminary investigation of the reactivity of ferrocene and its derivatives. At 25 °C, rate coefficients were $1,620 \times 10^{-7}$ (ferrocene) and $19.3 \times 10^{-7}$ (acetylferrocene). In a subsequent publication by Alikhanov and Shatenshtein[543] these values were altered to $1,600 \times 10^{-7}$ and $1.5 \times 10^{-7}$, respectively, and a value of $0.77 \times 10^{-7}$ added for 1,1-diacetylferrocene. Under the same conditions, toluene gave a value of $0.3 \times 10^{-7}$ so that the activating effects of these compounds relative to benzene can be approximately determined.

Nesmeyanov et al.[546] have also measured the effects of substituents in deuteration of ferrocene by deuterated trifluoroacetic acid in dichloromethane at 25 °C. Rate coefficients were measured for ferrocene and its derivative in a range of such acid mixtures, the composition of which was omitted, and in some cases the rate of exchange for ferrocene was calculated on the basis of a linear relationship between $\log k_1$ and $-H_0$. Results including the calculated $k_{rel}$ values are given in Table 161. It should be noted that, in discussing those results, the authors quoted the incorrect partial rate factors for dedeuteration of toluene arising from the use of the incorrect data for benzene (see p. 199). This should be taken into account

TABLE 161

RATE COEFFICIENTS AND RELATIVE RATES FOR THE REACTION OF X-FERROCENE WITH $CF_3COOD–CH_2Cl_2$ AT 25 °C[546]

| $X$ | $10^7k_1$ | $k_{rel}$[a] | $X$ | $10^7k_1$ | $k_{rel}$[a] |
|---|---|---|---|---|---|
| H | 9.1 | 1 | H | 1,750 | 1 |
| Me | 88 | 10.7 | Ph | 920 | 0.6 |
| Et | 88 | 10.7 | Cl | 40 | 0.25 |
| 1,1'-Et$_2$ | 370 | 50.8 | 1,1'-Ph$_2$ | 46 | 0.33 |
| 1,1'-(OMe)$_2$ | 170 | 23.3 | | | |
| | | | | | |
| H | 17,000 | 1 | H | 13,000 | 1 |
| 1,1'-Cl$_2$ | 10 | 0.0007 | 1-CO$_2$Me | 15 | 0.007 |

[a] Corrected for the different number of exchangeable hydrogen atoms.

in any quantitative evaluation of the substituent effect data presented in this paper.

Kinetic isotope effects have also been measured in trifluoroacetic acid–dichloromethane and are discussed under (d) below.

### (c) Kinetic studies using aqueous trifluoroacetic acid

A range of aromatic reactivities have been measured for this medium, mainly by Lauer et al.[547, 548, 550-552]. Some of this data was only published in the form of preliminary abstracts and unfortunately has been overlooked by later workers. Where comparison of substituent effect data with that obtained by later workers, who used the anhydrous medium, is possible, agreement is extremely good. In some of the work, relative rates rather than rate coefficients were given only and some data for deuteration are given in Table 162[547, 548]. It should be noted that the given relative reactivities for the polymethylbenzenes and which refer to a temperature of 70 °C were obtained by correcting the reactivities obtained at lower temperatures according to the activation energy (17.6) obtained between 29.65 and 70 °C for m-xylene. They may, therefore, be subject to considerable error and no quantitative conclusions may be derived from them; one may anticipate that all of the resultant "observed" reactivities will be too high and this has not been taken into account by the many workers who have quoted these results and the departures from additivity of substituent effects may be greater here than has hitherto been supposed (see, for example, ref. 549). The data in the Table show that the relative reactivities are insensitive to large changes in the medium composition, and this has been found to be true even for the addition of strong electrolytes (see p. 257).

Blackley[548] measured the rates of deuteration of biphenylene, fluorene, triphenylene, and phenanthrene relative to o-xylene as 6.15 : 5.85 : 1.08 : 1.32, which is in very good agreement with the values of 8.80 : 7.00 : – : 1.14 which may be deduced from the detritiation data in Table 159, obtained using anhydrous trifluoroacetic acid. Aqueous trifluoroacetic acid (with the addition in some cases of benzene to assist solubility) was used by Rice[550], who found that triptycene was 0.1 times as reactive per aromatic ring as o-xylene (cf. 0.13 derivable from Table 159) whereas the compound (XXXI) was 0.9 times as reactive as o-xylene. An exactly comparable measure is not available from Table 158, but dihydroanthracene (XXXII), which is similar, was 0.51 times as reactive as o-xylene and

(XXXI)          (XXXII)

the differing reactivities of (XXXI) and triptycene can be explained in terms of a new strain theory[527, 542]. Rice also found the reactivities of di-, tri-, and tetra-phenylmethanes (per aromatic ring relative to o-xylene) to be 37, 17 and 7 re-

TABLE 162

RELATIVE REACTIVITIES AND PARTIAL RATE FACTORS FOR REACTION OF ArH WITH $CF_3COOH-D_2O$ AT 70 °C[547, 548]

| ArH | Temp. (°C)[a] | Composition of medium (mole %) | | | Partial rate factors[b] | | |
|---|---|---|---|---|---|---|---|
| | | $CF_3COOH$ 50.5 $D_2O$ 10.1 $ArH$ 39.4 | 62.9 12.5 12.3+ 12.3PhMe | 81.5 16.3 1.8 | $f_o$ | $f_m$ | $f_p$ |
| Benzene | 70.1 | 6 | 6[c] | 6 | 1 | 1 | 1 |
| Toluene | 70.1 | 766 | 934[d] | 964[e] | 253 | 3.5–4 | 421 |
| Ethylbenzene | 70.1 | 625 | 970 | 1,088 | 259 | 5–6.9 (4.52)[f] | 449 |
| $n$-Propylbenzene | 70.1 | | 912 | | | 6.8 | |
| $i$-Propylbenzene | 70.1 | 550 | 1,017 | 1,145 | 259 (250) | (6.16) | 493 (483) |
| $n$-Butylbenzene | 70.1 | 453 | 1,044 | 1,146 | 202 | 8.4 | 470 |
| $s$-Butylbenzene | 70.1 | 422 | 889 | 1,043 | | | |
| $t$-Butylbenzene | 70.1 | 423 | 893 | e | | | |
| $i$-Pentylbenzene | 70.1 | | 1,074 | | | | |
| $p$-Xylene | 70.1 | | 4,930 (1,080) | | | | |
| $o$-Xylene | 70.1 | | 6,180 | | | | |
| $m$-Xylene | 70.1 | | $1.75 \times 10^5$ ($1.82 \times 10^5$) | | | | |
| 1,2,4-Trimethylbenzene | 40.1 | | $4.79 \times 10^5$ | | | | |
| 1,2,3-Trimethylbenzene | 40.1 | | $6.19 \times 10^5$ | | | | |
| 1,3,5-Trimethylbenzene | 0 | | $5.36 \times 10^7$ ($1.44 \times 10^7$) | | | | |
| 1,2,4,5-Tetramethylbenzene | 30.0 | | $1.51 \times 10^6$ | | | | |
| 1,2,3,5-Tetramethylbenzene | 0 | | $1.44 \times 10^8$ | | | | |
| Pentamethylbenzene | 0 | | $1.59 \times 10^8$ | | | | |

[a] Temperature at which relative reactivities were measured.
[b] Mostly approximate values.
[c] In approximately this medium, $10^7k_1 = 0.025$.
[d] In approximately this medium $10^7k_1(ortho) = 6.3$; $10^7k_1$ (para) = 10.0.
[e] In this medium $10^7k_1$(toluene) = 76.2 and $10^7k_1$(t-butylbenzene) = 82.0.
[f] Values in parentheses were obtained in ref. 548, the $f_m$ data being obtained from measurements at 100 °C.

spectively; from Table 159 the corresponding relative reactivity of diphenyl-methane is 37 so again the agreement is excellent. Partial rate factors for the 2, 3, and 4 positions of biphenyl were determined by Rice as 87, 4.9, and 130 respectively, and for the 4 position of 2'-, 3'-, and 4'-methylbiphenyls as 360, 5.2 and 490; again the effect of a 4'-methyl group (3.77) is in excellent agreement with the value of 4.00 obtained in detritiation (Table 158).

Evenson[551] used aqueous trifluoroacetic acid to determine some relative reactivities in dedeuteration as follows: benzcyclobutene < $o$-xylene < indane < tetralin > benzsuberone > benzcyclo-octene. The relative reactivities of the

4 : 5 positions in indane (0.23) and of the 5 : 6 positions in tetralin (1.15) is in excellent agreement with the values obtained by Vaughan and Wright[528] (0.246 and 1.08 respectively), who also found the same order of reactivity for o-xylene, indane, and tetralin.

Aqueous trifluoroacetic acid (87 mole %) containing deuterium oxide was used in an investigation of the kinetic effect of substitution of deuterium in the methyl group of toluene (methyl group composition, $CD_{2.7}H_{0.3}$). The rate at 70 °C was found to be only 5 % less than for light toluene[552]. This result may be compared with that obtained for tritiation of toluene (methyl group composition $CD_{2.97}H_{0.03}$) in 62 wt. % sulphuric acid at 25 °C. The rate obtained ($1,740 \times 10^{-7}$) was insignificantly different from that obtained from light toluene[553].

### (d) Kinetic studies in acetic acid–trifluoroacetic acid

For compounds which have very high reactivity, anhydrous trifluoroacetic acid may give inconveniently fast rates even at low temperatures and mixtures of acetic acid and trifluoroacetic acid have been used for these measurements, since the good solvent properties and freedom from side reactions are retained. Also, trifluoroacetic acid alone will hydrogen bond to aromatic ethers (and presumably phenols too) so that the observed rates in the pure acid give an erroneous measure of the substituent effect since the electron-releasing power of the substituent is reduced by this bonding. This was shown by Eaborn et al.[534, 537], who measured rate coefficients for detritiation in several acetic acid–trifluoroacetic media (Table 163). It is quite clear from these data that the reaction rates for the oxygen-containing compounds in the media containing acetic acid are reduced from those obtained with pure trifluoroacetic acid by a much smaller fraction than for the other compounds measured.

TABLE 163

RATE COEFFICIENTS ($10^7 k_1$) FOR REACTION OF [³H]-Ar WITH $CH_3COOH$–$CF_3COOH$ AT 70.1 °C[524, 537]

| ArH | Position of [³H] | $CH_3COOH$ in $CF_3COOH$ (Wt. %) | | | | |
|---|---|---|---|---|---|---|
| | | 0 | 1.68 | 13.2 | 19.3 | 22.8 |
| Diphenyl ether | 4 | 2,930 | | 1,015 | | 323 |
| Fluorene | 2 | 1,600 | | 382 | | 109 |
| 2-Methylfluorene | 7 | 6,120 | 1,800 | | | 49.4 |
| 2-Methoxyfluorene | 7 | 4,840 | | 2,370 | | 780 |
| 2-Bromofluorene | 7 | 340 | | | 35.8 | |
| 4-Methylbiphenyl | 4' | 62 | 43.5 | 23.1 | | 4.90 |
| 4-Methoxybiphenyl | 4' | 51 | 50.3 | 29.8 | | 8.15 |
| Dibenzofuran | 2 | 34.9 | | | 35.3 | |
| Dibenzothiophen | 2 | 174 | | | 14.3 | |
| Diphenylsulphide | 4 | 934 | | | 60.6 | |

## TABLE 164

RATE COEFFICIENTS ($10^7k_1$) FOR REACTION OF [2-$^3$H]-Ar WITH $CH_3COOH$–$CF_3COOH$ AT 24.84 °C[554]

| Ar | $CF_3COOH$ in mixture (mole %) | | |
|---|---|---|---|
| | 68.8 | 44.5 | 27.9 |
| Thiophen | 6,770 | 178 | 11,4 |
| 5-Chlorothiophen | 1,180 | | |
| 5-Bromothiophen | 880 | | |
| 5-Iodothiophen | 1,300 | | |
| 5-Methylthiophen | | 36,600 | 2,300 |
| 5-$t$-Butylthiophen | | 40,500 | 2,670 |
| 5-Phenylthiophen | | | 178 |
| 5-(2′-Thienyl)thiophen | | | 481 |
| 5-(5′-Methyl-2′-Thienyl)thiophen | | | 1,640 |

The reactivity of thiophen and its derivatives in detritiation in this medium have been measured (Table 164)[554], and a fairly good correlation of rates with $\sigma^+$ values for the benzene series was obtained with $\rho = -7.2$. Shatenshtein et al.[555] have also measured substituent effects in thiophen via dedeuteration in various acetic acid–trifluoroacetic acid media (Table 165) and for some substituents in the pure acids individually (Tables 159 and 165). The results are in good agreement with those of Butler and Eaborn[554] both with regard to the actual magnitude of the rates (bearing in mind the difference in medium and exchanging hydrogen isotope) and the activating effect of a 5-methyl group upon the 2 position (factors of ca. 200 being obtained in each case).

Hydrogen exchange in ferrocene and its derivatives has been investigated using this medium[556]. Some preliminary results for these compounds have been previously noted above under (b). Deuteration with a mixture of trifluoroacetic acid (25.1 mole %) in acetic acid at 25 °C gave the following average rate coefficients ($10^7k_1$): ferrocene, 9.1; methylferrocene, 88; 1,1′-diethylferrocene, 375; ethylferrocene, 88. These values were said to be obtained at an $-H_0$ value of 0.5 but so were the different values noted under (b) above, so, either the measurements are wrong, or, more likely, we have yet another example of the danger of relating rates at an acidity function value in one medium with those at a similar value in another medium. In a medium of $-H_0$ value 0.7 (28.3 mole % trifluoroacetic acid) ethylferrocene gave a rate coefficient of $128 \times 10^{-7}$.

A relative reactivity of ferrocene : benzene of $10^5$–$10^6$ has been quoted[557] following a kinetic study of the deuteration of ferrocene in acetic acid–trifluoroacetic acid mixtures at 25 °C, but the value is entirely in error, being based on two faulty assumptions. The data are given in Table 166 and a linear plot of log $k_1$ versus $-H_0$ was extrapolated to $-H_0 = 5.0$, a rate coefficient of $1.3 \times 10^{-1}$ being obtained. This was compared to the Gold and Satchell value for dedeuteration

TABLE 165

RATE COEFFICIENTS FOR REACTION OF $[^2H]$-Ar WITH $CH_3COOH$–$CF_3COOH$ AT 25 °C[555]

| ArH | Position of $[^2H]$ | Medium | $10^7k_1$ |
|---|---|---|---|
| Thiophen[a] | 2 | 10   mole % $CF_3COOH$ | 5.0 |
| Thiophen[a] | 2 | 50   mole % $CF_3COOH$ | 930 |
| Thiophen[a] | 2 | 33.3 mole % $CF_3COOH$ | 56 |
| 3-Methylthiophen | 2 | 33.3 mole % $CF_3COOH$ | 17,000 |
| 4-Methylthiophen | 2 | 33.3 mole % $CF_3COOH$ | 550 |
| 5-Methylthiophen | 2 | 33.3 mole % $CF_3COOH$ | 10,000 |
| 5-Methylthiophen | 2 | 2   mole % $CF_3COOH$ | 7 |
| 5-Methylthiophen | 2 | 20   mole % $CF_3COOH$ | 1,100 |
| 3-Methoxythiophen | 2 | HOAc | 530 |
| 5-Methoxythiophen | 2 | HOAc | 730 |
| 5-Methoxythiophen | 2 | 20   mole % $CF_3COOH$ | 60,000 |
| 5-Thiomethoxythiophen | 2 | 20   mole % $CF_3COOH$ | 6,600 |
| 4-Thiomethoxythiophen | 2 | 20   mole % $CF_3COOH$ | 2 |
| 3-Thiomethoxythiophen | 2 | 20   mole % $CF_3COOH$ | 24,000 |
| 5-Thiomethoxythiophen | 2 | 2   mole % $CF_3COOH$ | 30 |
| 5-Thiomethoxythiophen | 3 | 50   mole % $CF_3COOH$ | <0.2 |
| 5-Thiomethoxythiophen | 3 | 50   mole % $CF_3COOH$ | 3,800 |
| 4-Methylthiophen | 3 | 50   mole % $CF_3COOH$ | 9.6 |
| Thiophen | 3 | HOAc | 60 |

[a] The data for this compound is in good agreement with that which may be interpolated from Table 164.

of benzene (itself in error, see p. 199) and it was assumed that a linear extrapolation of the benzene data to this acidity could be made (also quite erroneous, see p. 198) and also that different acids of the same $-H_0$ value give the same exchange rate (there is not yet a single example where this has been observed whereas there are numerous examples which show the converse).

The kinetic isotope effect $k_D : k_T$ for hydrogen exchange of ferrocene in both trifluoroacetic acid–acetic acid and trifluoroacetic acid–dichloromethane has been measured[558]. In the former medium (1 : 1 molar ratio at 25 °C) $k_D : k_T$ was 1.2–1.3, which was less than that obtained for the 2 position of thiophene: $k_1(D) = 1,200 \times 10^{-7}$, $k_1(T) = 660 \times 10^{-7}$, $k_D : k_T = 1.9$. It is also lower than that obtained for pentamethylbenzene: $k_1(D) = 3,300 \times 10^{-7}$, $k_1(T) =$

TABLE 166

RATE COEFFICIENTS FOR REACTION OF FERROCENE WITH $CF_3COOH$–$CH_3COOD$ AT 25 °C[557]

| $CF_3COOH$ (g) | 44.5 | 48.17 | 65.38 | 91.5 | 79.39 | 93.21 | |
|---|---|---|---|---|---|---|---|
| $CH_3COOD$ (g) | 36.9 | 38.92 | 38.69 | 43.0 | 39.28 | 29.87 | |
| $-H_0$ | 1.12 | 1.16 | 1.9 | 2.18 | 2.33 | 2.51 | 5.00 |
| $10^7k_1$ | 180 | 210 | 910 | 2,200 | 2,400 | 4,600 | $1.3 \times 10^6$ |

$1,900 \times 10^{-7}$, $k_D : k_T = 1.8$, in a medium composed of trifluoroacetic acid and dichloromethane (composition not given); the lower value was attributed to metal participation in the reaction mechanism.

### (e) Kinetic studies using mineral acid–trifluoroacetic acid

The addition of mineral acid to trifluoroacetic acid brings about a dramatic increase in the rates of hydrogen exchange even when very small quantities of mineral acid are used. This brings the advantage that reaction rates can be measured at more convenient temperatures and also that the reactivities of strongly deactivated aromatics can be measured at accessible temperatures and reaction times: the need for very high purity in the trifluoroacetic acid is also eliminated. On the other hand, there are disadvantages in that halogen acids tend to come out of solution[471], so that reproducible rates are difficult to obtain with a given batch of acid, sulphuric acid causes sulphonation (see p. 60), and perchloric acid causes oxidation. Phosphoric acid may be more satisfactory but has not been used so far. Rate coefficients for reaction of a large number of compounds have been determined for sulphuric or perchloric acid solutions in trifluoroacetic acid and these are given in Table 167. Sufficient data are now available for these media such that it is possible to prepare a solution that will give fairly accurately a desired rate for a compound of known reactivity provided that it is appreciated that the media are much more sensitive to added acid than to added water. High concentrations of water reduce the solvent properties, but in media containing sulphuric acid, sulphonation decreases rapidly with increasing water content. In these media poor first-order plots are obtained for exchange if the water and sulphuric acid concentrations are low because sulphonation produces water as a by product and consumes sulphuric acid, both processes leading to a reduction in medium acidity and hence rate. The data in Table 167 do not show very good agreement between the dedeuteration and detritiation rate coefficients in the presence of aqueous perchloric acid, considering that dedeuteration should be faster than detritiation and that the temperature of measurement of the former reaction was higher. It seems doubtful if the difference in spread of rates is real; the rate coefficient quoted for the 9 position of anthracene is of very doubtful validity, since it corresponds to a reaction with a half-life of 25 sec and in the writer's experience the poor solubility of anthracene would require a much longer time than this even to dissolve sufficient quantity for measurement. Indeed the poor solubility could lead one to suppose that reaction was over (since there would be no spectral indication of the label) when in fact no anthracene had even gone into solution.

Rate coefficients have also been measured at a range of temperatures for some aromatics in aqueous perchloric acid–trifluoroacetic acid (Table 168)[468], and, surprisingly, the lower reactivity of benzene relative to toluene and $t$-butylbenzene appears to arise from a more negative activation entropy. This effect if real is

TABLE 167

RATE COEFFICIENTS ($10^7 k_1$) FOR REACTION OF [$^3$H]-Ar WITH $CF_3COOH$–$H_2O$–$H_2SO_4$ ($HClO_4$)

| | | | | | | | | | |
|---|---|---|---|---|---|---|---|---|---|
| $CF_3CO_2H$ (mole %) | | 95.31 | 83.40 | 93.00 | 92.04 | 99.96 | 89.50 | 96.77 | 91.80 |
| $H_2O$ (mole %)[a] | | 2.21 | 1.98 | 5.12 | 5.45 | 0.03 | 4.55 | 0.82 | 5.70 |
| $H_2SO_4$ (mole %)[a] | | 2.48 | 14.62 | 1.88 | | | | | |
| $HClO_4$ (mole %) | | | | | 2.51 | 0.01 | 5.95 | 2.41 | 2.50 |
| Temp. (°C) | | 25.0 | 25.0 | 25.0 | 25.0 | 25.0 | 55.0 | 55.0 | 31.0 |
| Ref. | | 147b 466 | 147c 559 | 465 | 147b 466 | 524 | 147b | 147b | 560 |
| ArH | Position of $^3$H | | | | | | | | |
| Benzene | 1 | 3.60 | 497 | | 14.3 | | 12,800 | 3,025 | 15.5[b] |
| Toluene | 3 | 33.0 | 2,610 | | 103 | | | | |
| Toluene | 2 | 1,950 | | 1,295[c] | 4,720[c, d] | | | | |
| Toluene | 4 | 2,530 | | 1,770[c] | 4,480[c, d] | 256 | | | |
| t-Butylbenzene | 3 | 144 | | | 339 | | | | |
| t-Butylbenzene | 2 | | | 1,565[c] | 5,620[c, d] | | | | |
| t-Butylbenzene | 4 | 3,110 | | 2,180[c] | 5,530[c, d] | | | | |
| Fluorobenzene | 3 | | | | | | 25.0 | | |
| Fluorobenzene | 2 | | | | | | 1,610 | 412 | |
| Fluorobenzene | 4 | | 860 | | | | | 5,420 | |
| Chlorobenzene | 3 | | 1.00 | | | | 16.0 | | |
| Chlorobenzene | 2 | | | | | | 465 | 107 | |
| Chlorobenzene | 4 | | 63.0 | | | | | 487 | |
| Bromobenzene | 3 | | | | | | 16.0 | | |
| Bromobenzene | 2 | | | | | | 350 | 83 | |
| Bromobenzene | 4 | | 36.0 | | | | 1,090 | 298 | |
| Iodobenzene | 3 | | | | | | 42.0 | | |
| Iodobenzene | 2 | | | | | | 525 | 129 | |
| Iodobenzene | 4 | | 43.0 | | | | | 339 | |
| Diphenyl | 3 | | | | 9.80 | | | | |
| Diphenyl | 2 | 477 | | | 748 | | | | 933 |
| Diphenyl | 4 | 515 | | | 748 | | | | 1.000 |
| Naphthalene | 1 | 3,890 | | | 5,300 | | | | 7,580 |
| Naphthalene | 2 | 456 | | | 885 | | | | 872 |
| Fluorene | 2 | | | | | 2,950 | | | |
| 9-Methylfluorene | 2 | | | | | 3,110 | | | |
| 9,9-Dimethyl-fluorene | 2 | | | | | 3,980 | | | |
| Phenanthrene | 9 | | | | | | | | 7,940 |
| Anthracene | 1 | | | | | | | | 29,500 |
| Anthracene | 9 | | | | | | | | 263,000 |

[a] Data in this medium is corrected for the sulphonation side reaction (for rates of these see pp. 60–62).

[b] Data in this column is for dedeuteration.

[c] Temperature of measurement incorrectly given as 70.1 °C in ref. 546.

[d] These rates were incorrectly reported in the form of partial rate factors in ref. 546.

TABLE 168

RATE COEFFICIENTS $(10^7 k_1)$ AND ARRHENIUS PARAMETERS[468] FOR REACTION OF $[^3H]$-Ar
WITH $HClO_4(0.05)$–$H_2O(1.80)$–$CH_3COOH(98.15)$[a]

| ArH | Temp. (°C) | | | | | $E_a$ | $log\ A$[b] |
|---|---|---|---|---|---|---|---|
|  | 24.22 | 39.25 | 50.30 | 65.75 | 90.8 |  |  |
| $[^3H]$-Benzene |  |  |  | 17.7 | 212 | 24.3 | 10.0 |
| $[4\text{-}^3H]$-Toluene | 34.1 | 254 | 970 | 4,435 |  | 23.4 | 11.8 |
| $[4\text{-}^3H]$-t-Butylbenzene | 45.1 | 322 | 1,246 | 5,860 |  | 23.5 | 12.0 |

[a] Mole % in parentheses.
[b] Pre-exponential factor from $k_1$ and $E_a$.

consistent with the transition state for the exchange of benzene being more
positively charged (as would be expected since it will be nearer to the Wheland
intermediate for substitution of a less reactive compound) and it follows that differential solvation between compounds in this medium is of considerable importance,
again consistent with the poor solvating ability of trifluoroacetic acid. Furthermore, the activation energy for this medium is insignificantly different from
that with anhydrous trifluoroacetic acid, yet the rates are 150-fold higher and this
indicates that added reagents (in this particular medium at least) increase the rate
principally through increased solvation of the transition states. This is entirely
in accord with the proposal (p. 202) advanced to account for the increased exchange rates in aqueous trifluoroacetic acid compared to the anhydrous medium,
and significantly the medium employed for the exchange measurements given in
Table 168 contained large quantities of water.

### (f) Kinetic studies with mineral acid–trifluoroacetic acid in aprotic solvents

Dallinga et al.[561] carried out a poorly executed and badly reported study of the
deuteration of some polycyclic hydrocarbons by sulphuric acid–deuterated trifluoroacetic acid in carbon tetrachloride (Table 169). Not only was the temperature
of measurement of rate coefficients not reported, but the relative rates were obtained by assuming a linear correlation of log rate coefficient with acidity function
$H_0$, it being claimed that this was so in the case of naphthalene, though in fact
the curve which was obtained (and in the *absence* of carbon tetrachloride, and at
20 °C), had a variation in slope of almost two. Consequently, little reliance can
be placed on these $k_{rel}$ values, which have been often quoted and a re-examination
of the reactivities of these compounds in a better solvent system (and preferably
with exchange from specifically labelled positions) would be desirable. The large
difference between the $k_{rel}$ values obtained here for naphthalene and and those
from other studies is believed to be an indication of the magnitude of the error in
this work. For a similar medium the rate coefficients for exchange of a number of

TABLE 169

RATE COEFFICIENTS AND RELATIVE RATES FOR REACTION OF ArH WITH
$D_2SO_4$–$CF_3COOD$–$CCl_4$, PROBABLY AT 40 °C[561]

| ArH | Position of exchange[a] | [$CF_3COOD$] (g) | [$H_2SO_4$] (g) | [$CCl_4$] (g) | $10^7k_1$ | $k_{rel}$[b] |
|---|---|---|---|---|---|---|
| Naphthalene | 1 | 10.098 | 0.103 | 40 | 308 | 178 |
| Naphthalene | 2 | 10.098 | 0.103 | 40 | 40.5 | 23.4 |
| Anthracene | 9 | 10.129 | 0.007 | 40 | 114,000[c] | $1.29 \times 10^6$ |
| Anthracene | 1 | 10.129 | 0.007 | 40 | 77.8 | 890 |
| Anthracene | 2 | 10.129 | 0.007 | 40 | 15.5 | 178 |
| Biphenyl | 2, 4 | 10.158 | 0.200 | 40 | 139 | 37.2 |
| Pyrene | 1 | 10.086 | 0.007 | 40 | 7,450 | 85,000 |
| Pyrene | 4 | 10.086 | 0.007 | 40 | >7 | >80 |
| Benz[a]anthracene | 7, 12 | 10.125 | 0.0065 | 40 | 35,000 | 40,000 |
| Perylene | 1, 3 | 10.268 | 0.0065 | 40 | 9,600 | 10,700 |
| Triphenylene | 1 | 10.721 | 0.200 | 40 | 116 | 30.9 |
| Triphenylene | 2 | 10.721 | 0.200 | 40 | 37,5 | 10.0 |
| Chrysene | 2 | 10.260 | 0.052 | 40 | 1,630 | 1,860 |
| Chrysene | 1, 3, 6 | 10.260 | 0.052 | 40 | 378 | 426 |

[a] These differ from those given in the paper which cannot be correct, *e.g.* different rates are quoted for the (identical) 1 and 3 position of pyrene.
[b] Approximate values (corrected for number of equivalent positions see text), and relative to benzene.
[c] Extrapolated from data at 20 °C.

methyl-substituted 1,2-benzanthracenes were measured but again the temperature of measurement was omitted (Table 170)[562].

Dallinga *et al.*[563] have also measured rates of deuteration of methylnaphthalenes by difluorohypophosphorous acid–deuterated trifluoroacetic acid in carbon tetrachloride, and the rate coefficients are recorded in Table 171. As with all deuteration studies one has to assign the observed rates to reaction at a particular position. Comparison of the data with the more rigorous detritiation results[541] shows that the assignment has been correct with the possible exception of the 2,6-compound for which there is some doubt. The detritiation data predict activation by the methyl groups to be greater at the 4 position than at the 3 position (after allowing for the greater basic reactivity of the $\alpha$ relative to the $\beta$ position) whereas the deuteration data predict the opposite. However, if the rates assigned to the 3 and 4 position are reversed one obtains the correct reactivity order but the difference in reactivities is far greater than predicted from detritiation. It seems probable that the assignments are correct but that there is an error in one of the rates which are obtained by the NMR method and, therefore, less accurate than those obtained by scintillation counting. In all other cases the reactivity orders are those which may be predicted from the detritiation data, though the difference between observed and calculated partial rate factors varies from 1.1 to 4.5 and this seems too great to arise from any breakdown of additivity of substituent effects. It would seem

TABLE 170

RATE COEFFICIENTS FOR REACTION OF BENZ[*a*]ANTHRACENES WITH $CF_3COOD$ (10 g)–$H_2SO_4$(1 g)–4-CHLOROTOLUENE (5 g) AT AN UNSPECIFIED TEMPERATURE[562]

| ArH[a] | $10^7k_1$ (12 position) | $'10^7k_1$ (7 position) |
|---|---|---|
| Benz[*a*]anthracene | 43 | 43 |
| 2-Me-benz[*a*]anthracene | 61 | 96 |
| 3-Me-benz[*a*]anthracene | 94 | 94 |
| 5-Me-benz[*a*]anthracene | 84 | 260 |
| 6-Me-benz[*a*]anthracene | 39 | 96 |
| 8-Me-benz[*a*]anthracene | 48 | 108 |
| 9-Me-benz[*a*]anthracene | 49 | 250 |
| 10-Me-benz[*a*]anthracene | 170 | 78 |
| 11-Me-benz[*a*]anthracene | 70 | 61 |
| 12-Me-benz[*a*]anthracene | | 550–1,100 |
| 7-Me-benz[*a*]anthracene | 610 | |

[a] Numbered according to I.U.P.A.C. nomenclature, contrary to the original publication.

wise to treat the data in Table 171 with some caution at present; it is quite clear, however, that acenaphthene has a very enhanced and so far unexplained reactivity relative to 1,8-dimethylnaphthalene.

Nesmeyanov *et al.*[564] have measured the rates of deuteration of benzene and cyclopentadienylmanganese tricarbonyl by deuterated sulphuric acid–trifluoro-

TABLE 171

RATE COEFFICIENTS FOR REACTION OF SUBSTITUTED NAPHTHALENES[a] WITH $CF_3COOD$–$HF_2PO_2$–$CCl_4$ AT 20 °C[563]

| Substituents | Position of exchange | $10^7k_1$ | Substituents | Position of exchange | $10^7k_1$ |
|---|---|---|---|---|---|
| H | 1 | 9.0 | 1,4-Me$_2$ | 2 | 889 |
| | 2 | 1.19 | | 5 | 83.5 |
| 2,3-Me$_2$[b] | 1 | 15,300 | | 6 | 15.4 |
| | 5 | 483 | 1,5-Me$_2$ | 4 | 3,450 |
| | 6 | 83.3 | | 2 | 462 |
| 2,6-Me$_2$ | 1 | 13,200 | | 3 | 18.6 |
| | 4 | 278 | 1,8-Me$_2$ | 4 | 2,180 |
| | 3 | 86.1 | | 2 | 2,140 |
| 2,7-Me$_2$ | 1 | 35,800 | | 3 | 7.5 |
| | 3 | 411 | 1,8-(CH$_2$)$_2$[c] | 4 | 43,000 |
| | 4 | 103 | | 2 | 18,800 |
| | | | | 3 | 160 |

[a] 2 mole of aromatic in 5 ml $CCl_4$ was reacted with 0.1 mole $HF_2PO_2$ in 1000 g $CF_3COOD$.
[b] Rates were extrapolated for this compound from data at 20 °C assuming the activation enthalpy (16.0) found for naphthalene.
[c] Acenaphthene.

*References pp. 388–406*

acetic acid in dichloromethane at 25 °C. With molar ratios of reagents of 1 : 0.2 : 3 : 3 the respective rate coefficients $(10^7 k_1)$ were 32 and 41, whereas with a molar ratio of 1 : 0.3 : 4.8 : 2 they were 60 and 133; competitive experiments under the former set of conditions, however, give a relative reactivity of 1 : 5. One may perhaps conclude that the tricarbonyl is more reactive than benzene.

## (g) Kinetic studies with mineral acid–acetic acid solutions

Cost apart, these media have no advantage over the corresponding trifluoro-acetic acid media, and the higher concentrations required to bring about convenient rates of exchange usually means that side reactions are more significant. Rate coefficients $(10^7 k_1)$ for detritiation at the 2 position of 9-X-fluorenes by acetic acid (20.83)–water(34.43)–sulphuric acid(44.74), mole % in parentheses, at 25 °C were: (X =)H, 3,900; Me, 3,790; Me$_2$, 3,580; and 581 for 4-[$^3$H]-toluene[525]. In this paper these values were wrongly reported as rates of protodesilylation. Rates of detritiation of alkylbenzenes in a similar medium (26.0 : 34.0 : 40.0 mole %) were: (X = ) H, 9.0; 3-Me, 44.8; 3-Bu$^t$, 84; 2-Me, 2,744; 4-Me, 2,816; 4-Bu$^t$, 2,067[464,466].

A number of workers have measured rates of dedeuteration of aromatic ethers in this medium (Table 172). The agreement between the different sets of work is poor in the case of anisole and good otherwise. The result of Lauer and Day[566] for dedeuteration at 80 °C is probably the most suspect; certainly one would not expect such large differences in activation energy for reaction at the *ortho* and

TABLE 172

RATE COEFFICIENTS AND KINETIC PARAMETERS FOR REACTION OF [$^2$H]-Ar WITH HOAc–H$_2$SO$_4$ (50 : 1 MOLAR RATIO)

| ArH | Position of [$^2$H] | Temp. (°C) | $10^7 k_1$ | $E_a$ | $\Delta H^{\ddagger}$ | $\Delta S^{\ddagger}$ | Ref. |
|---|---|---|---|---|---|---|---|
| Benzene | 1 | 100 | 0.163 | | | | 565 |
| Anisole | 4 | 60.8 | 36.1 | | | | 565 |
| | | 79.9 | 355 | | | | 565 |
| | | 80.0 | 880 | | | | 566 |
| | | 96.6 | 1,890 | | 26.3 | −4.4 | 565 |
| | | 100.0 | 3,000 | 16.0 | | | 566 |
| Anisole | 2 | 80.0 | 290 | | | | 566 |
| | | 100 | 1,500 | 21.4 | | | 566 |
| Thioanisole | 4 | 80.0 | 7.68 | | | | 565 |
| | | 99.3 | 63.5 | | | | 565 |
| | | 99.3 | 64.0 | | | | 567 |
| | | 119.9 | 436 | | 27.6 | −8.4 | 565 |
| Thioanisole | 2 | 99.3 | 18.0 | | | | 567 |
| Ethylphenyl ether | 4 | 100.0 | 4,000 | | | | 566 |
| n-Propyl ether | 4 | 100.0 | 4,050 | | | | 566 |
| i-Propyl ether | 4 | 100.0 | 7,450 | | | | 566 |

*para* positions in hydrogen exchange. Of interest also is the activation by the alkyl group in the side chain in the hyperconjugative order.

A mixture of sulphuric acid in tritiated acetic acid has been used in an investigation of hydrogen exchange of naphthalene and anthracene in excited states. Since the basicity of hydrocarbons is greater in the singlet and triplet states relative to the ground state one might expect hydrogen exchange to be faster under irradiation. These preliminary experiments showed the rates of exchange to be independent of the intensity of irradiation and the conclusion was that reaction occurs in the singlet state[568].

### (h) Kinetic studies in mineral acids

Much of the kinetic work in this category has already been described under the section relating to studies of mechanism. Additional data was obtained by Olsson[569], who measured rate coefficients $(10^7 k_1)$ for dedeuteration and detritiation of thiophen by 57.02 wt. % sulphuric acid at 24.6 °C as follows: [2-$^2$H], 3,890; [2-$^3$H], 2,000; [3-$^2$H], 3.72; [3-$^3$H], 2.20. The ratio of reactivities at the 2 and 3 positions (*ca.* 1,000) is in excellent agreement (bearing in mind the larger $\rho$-factor usually obtained with trifluoroacetic acid) with the value of *ca.* 1,250 which may be deduced from the data in Table 158. The ratio of dedeuteration to detritiation is 1.96 at the 2 position and 1.70 at the 3 position and thus decreases with decreasing reactivity of the reaction site.

Aqueous perchloric acid has been used in but one kinetic study, the dedeuteration of [4-$^2$H]-toluene at 22 °C in 70.8 wt. % acid, being found to be $38,600 \times 10^{-6}$ (after correcting for the solubility of toluene in the acid, the heterogeneous exchange rate being $94 \times 10^{-7}$)[570]. The relative reactivities of the *ortho : para* positions was found to be 1.09, and *meta : para* 0.065 so that perchloric acid, like sulphuric acid (see p. 199) gives high *ortho : para* ratios; however, the *meta : para* ratio is significantly different from the values with sulphuric acid. Comparison of these results with those obtained with aqueous sulphuric acid indicates that exchange rates in perchloric acid are considerably higher for an equivalent concentration.

Blackborow and Ridd[51] studied the deuteration of aryl quaternary ammonium ions by deuterated sulphuric acid under second-order conditions (Table 173), and the results showed that, contrary to earlier beliefs, the positive nitrogen pole has

### TABLE 173

SECOND-ORDER RATE COEFFICIENTS FOR REACTION OF $ArNR_3^+$ WITH 95 WT. % $D_2SO_4$ AT 35 °C[51]

| Substituent in Ar | | 3,5-Me$_2$ | | 2,6-Me$_2$ | 4-MeO | | 3-MeO |
|---|---|---|---|---|---|---|---|
| $R_3$ | Me$_3$ | Me$_2$H | | Me$_2$H | Me$_3$ | Me$_2$H | Me$_2$H |
| Position of exchange | 2   4 | 2   4 | | 3 | 3 | 3 | 2   4 |
| $10^5 k_2$ | 14   14 | 23.8  67.2 | | 6.02 | 168 | 196 | 112  112 |

no very marked directional effect, and also by virtue of the similarity of the rate coefficients in Table 173, that the protonated substrates are reacting as such and not as the corresponding free amines.

There has been one kinetic study using sulphuric acid in which methanol was employed as a co-solvent[571], the rate coefficients for dedeuteration of five-membered heterocyclics being measured (Table 174). Firstly, it should be noted

TABLE 174

RATE COEFFICIENTS ($10^7 k_1$) FOR THE REACTION OF [$^2$H]-Ar WITH $H_2SO_4$–$H_2O$–MeOH[a] AT 20 °C[571]

| Ar | Position of exchange | [$H_2SO_4$], (wt %) | | | | |
|---|---|---|---|---|---|---|
| | | 78.5 | 56.0 | 31.0 | 21.6 | 0.5 |
| Benzene | 1 | 9.16 | | | | |
| Thiophen | 3 | 86,100 | 5,0 | | | |
| Thiophen | 2 | | 9,720 | 28.6 | 4.69 | |
| Furan | 3 | | <500 | < 3 | | |
| Furan | 2 | | 4,720 | 28.3 | 5.0 | |
| Pyrrole | 3[b] | | | | | 8,000 |
| Pyrrole | 2[b] | | | | | 19,000 |

[a] Water–methanol wt % ratio of 1.00:1.63.
[b] Assumed positions of exchange.

that the reaction of the 3 position of thiophen in 78.5 wt. % acid has a half-life of just over a minute so this value must be very approximate indeed. Assuming a $\rho$-factor for the reaction of −8, one obtains $\sigma^+$ values for the 2 and 3 positions of thiophen of −0.91 and −0.495, respectively, which agree well with those values which have been firmly established from other reactions[572]; the choice of the value of −8 (which applies to aqueous sulphuric acid) may not be very accurate, however, since the rates of 2- to 3-substitution in the present medium (1,950) is rather larger that that obtained by Olssen[569] (1,000) in aqueous sulphuric acid (see above, p. 261). The results for furan, however, are obviously not reliable since they predict it to be less reactive than thiophen, which is certainly not true[572]. Protonation of the oxygen atom, or breakdown of the unstable furan ring in the strong acid media is likely to be responsible, and the changing relative reactivities of furan and thiophen with acidity shows that these results cannot be accepted at face value.

### (i) Kinetic studies with carboxylic acids other than trifluoroacetic acid

Anhydrous heptafluorobutyric acid has been used to catalyse hydrogen exchange, the interest in this acid being that it might show steric hindrance in the reaction; however, none has been observed. Tiers[570] used the acid to measure the rates of dedeuteration of [4-$^2$H]-toluene at 78 and 119 °C, the rate coeffi-

TABLE 175

RATE COEFFICIENTS ($10^7 k_1$) AND ACTIVATION ENERGIES FOR REACTION OF [4-²H]-PhNR₂ WITH XCOOH[535, 573]

| X | $R_2$ | | | | | | | | | | | | $E_a$ |
|---|---|---|---|---|---|---|---|---|---|---|---|---|---|
|   | $Me_2$ (30°C) | $MePh$ (30°C) | $Ph_2$ (30°C) | $Me_2$ (65°C) | $MePh$ (65°C) | $Ph_2$ (65°C) | $Me_2$ (75°C) | $Me_2$ (90°C) | $MePh$ (90°C) | $Me_2$ (115°C) | $MePh$ (115°C) | $Ph_2$ (115°C) | ($R_2 = Me_2$) |
| CH₃ |  |  |  | 30 | 10,000 | 600 | 400 | 1,600[a] | 180[b] | 10,000 | 1,000 | 30 | 22.4 |
| H |  |  |  | 40 | 40,000 | 1,000 | 50 | 300 |  | 4,000 |  |  | 29.7 |
| CH₂Cl |  |  |  | 20 | 20,000 | $10^5$ | 100 | 700 |  | 8,000 |  |  | 28.2 |
| CCl₃ |  |  |  |  |  |  | 50 | 300 |  | 4,000 |  |  | 29.2 |
| CF₃ | 0.01 | 10 | 20,000 |  |  |  | 8 | 60 |  | 8,000 |  |  | 30.8 |

[a] $k_1(ortho) = 2{,}300 \times 10^{-7}$.
[b] $k_1(ortho) = 240 \times 10^{-7}$.

cients being $9.9 \times 10^{-7}$ and $229 \times 10^{-7}$, respectively, so the acid is less effective than trifluoroacetic acid at catalysing exchange. The *ortho* : *para* ratio was found to be only 0.4 indicating a steric effect (and thereby also arguing against the A-1 mechanism), but later work gave the following rate coefficients $(10^7 k_1)$ for detritiation at 70.1 °C; [4-³H]-toluene, 7.08; [2-³H]-toluene, 2.79; [4-³H]-*t*-butylbenzene, 8.94; [2-³H]-*t*-butylbenzene, 3.34, and the similarity of *ortho* : *para* ratio for both aromatics showed the low ratio to arise from electronic and not steric effects[465]; likewise, a low ratio was obtained for detritiation of biphenyl, rate coefficients being $1.82 \times 10^{-7}$ and $4.35 \times 10^{-7}$ for the *ortho* and *para* positions[536].

Shatenshtein *et al.*[535, 573] have measured rates of dedeuteration of some aromatic amines by a range of carboxylic acids, the rate coefficients being given in Table 175, some of these being obtained by extrapolation from data at other temperatures. It can be seen from Table 175 that for triphenylamine, rates increase with increasing acid strength and may be interpreted to mean that reaction occurs on the free base under all conditions. For the compound N,N-dimethylaniline the converse is true, indicating that a change to reaction on the conjugate acid occurs at high acidities; this is also indicated by the activation energies which for the strong acids are too high for reaction on the free base. These results are exactly as expected for the greater availability of the nitrogen lone pair in N,N-dimethylaniline relative to triphenylamine would increase the opportunity for protonation and the situation is entirely analogous to that noted previously. For N-methyl-N-phenylaniline the rates pass through a maximum with increasing acid strength so that reaction occurs on the free base in weak acids and the conjugate acid in strong ones. These workers also found the rate coefficient for dedeuteration of the *para* position of diphenyl ether by acetic acid at 40 °C to be $460 \times 10^{-7}$.

### (j) Kinetic studies using halogen acids

Deuterium exchange studies using halogen acids have been carried out by Shatenshtein *et al.* Early work was carried out with deuterium fluoride which exchanged with benzene at rates $(10^7 k_1)$ of 9,000 (25 °C) and 30,000 (50 °C), the latter being reduced to 7,000 on addition of 0.2 g water per 20 g deuterium fluoride[574]. The observed rate coefficients for the pure acid were altered to 10,000 and 40,000 in a subsequent publication[575], and this was compared to a value of 0.5 for exchange with deuterium bromide at 20 °C. The rate difference was ascribed to the difference in the dielectric constant of the acid. Deuteration of anthracene gave a rate coefficient of 4,000 (average of all positions at 50 °C) and the low reactivity relative to benzene was attributed to protonation. Toluene gave a value of 200,000 (average of all positions at 25 °C).

Most work has, however, been carried out using liquid deuterium or hydrogen

bromide and the difficult experimental technique presumably accounts for the fact
that the quoted values tend to vary with succeeding publications; some rate coefficients are derived from two or three measurements only. The rate coefficients
and partial rate factors derived from them in this work have been often quoted
(*e.g.* ref. 504), but careful examination of the kinetic data shows that all of the
quoted partial rate factors may be in error by a factor of 2 or so (though this does
not alter any of the conclusions based upon comparison of this work with that
obtained by other workers). Now the rate coefficients for deuteration of benzene
was stated (above) and by Shatenshtein *et al.*[573] to be $0.5 \times 10^{-7}$ at 20 °C, but
it was revealed by Varshavskii and Shatenshtein[574] that the reaction of benzene
was carried out at *room temperature* for *42 weeks*, hardly rigorous or reliable kinetic
conditions. The rate of deuteration of toluene and biphenyl at 25 °C was found[576]
to be in the range $100–1,000 \times 10^{-7}$, and later the initial rate coefficient for
deuteration of biphenyl was found to be $700 \times 10^{-7}$ falling to $200 \times 10^{-7}$ (these
being attributed to the 2 and 4 positions) and for naphthalene the rate coefficient
fell from $20,000 \times 10^{-7}$ (1 position) to $600 \times 10^{-7}$ (2 position). Rates for naphthalene were apparently measured between $-50$ and $+34$ °C, $E_a$ being quoted as 3–4
which seems extraordinary. For *p*-terphenyl the initial rate coefficient $(100 \times 10^{-7})$
fell to a value quoted as $20 \times 10^7$ which is obviously in error; all of this data was
obtained at 25 °C[577a].

Subsequently[577b], the naphthalene values were changed to 30,000 and 600,
other values $(10^7 k_1)$ being: 1-nitronaphthalene, 0.7 falling to 0.6; 1-cyano-
naphthalene, 6 falling to 0.8; 2-naphthoic acid, 0.9 falling to 0.6; 1-chloronaph-
thalene, 850; and 1-bromonaphthalene, 700, though little can be deduced from
these results.

A kinetic study of dedeuteration at 25 °C yield the following rate coefficients for
$[^2\text{H}]\text{-}C_6H_4R$[578, 579]: (R = )4-Me, 1,900; 2-Me, 530; 3-Me, 2.7; 4-Ph, 1,400;
2-Ph, 260; 3-Ph, < 0.2; 2,3-benzo(1 position of naphthalene) 20,000–30,000 and
3,4-benzo (2 position of naphthalene) 500. Rather surprisingly, the rates of dedeuteration were little different from the rates of deuteration and it should be
noted that quoted partial rate factors in this work were obtained by dividing these
rates for *dedeuteration* at *25 °C* by the rates of *deuteration* of benzene at *20 °C*
and errors of a factor of 2 or more may be introduced by this.

Some other work of a semi-quantitative nature has been carried out using
hydrogen bromide but none of it yields usable rates. For example, the relative
rates of deuterium uptake in phenylated alkanes was measured and showed the
expected decrease in reactivity with increasing phenyl group substitution[580].

Hydrogen iodide has also been used as a catalyst for hydrogen exchange[581],
rate coefficients $(10^7 k_1)$ at 25 °C for reaction with $[^2\text{H}]C_6H_4R$ being: (R =)
2,3-benzo(1 position of naphthalene), 500; 4-Me, 40; 2-Me, 21; 4-Ph, 42; 2-Ph, 10.
The reaction was catalysed by added iodine, the rate being directly proportional
to the concentration of this, and for the latter compound rates were measured

in the presence of iodine, the quoted rate being obtained by extrapolation to zero iodine concentration.

The rate coefficients obtained with hydrogen iodide show that the effectiveness of halogen acids in catalysing hydrogen exchange is HF > HBr > HI and, as noted above, this was attributed to the difference in dielectric constant of the media.

## 8.2 BASE-CATALYSED HYDROGEN EXCHANGE

Kinetic studies of base-catalysed hydrogen exchange have been carried out by Roberts, by Shatenshtein, and by Streitweiser and their coworkers. In earlier work, potassium amide was used as base in liquid ammonia as solvent, whereas later workers used lithium and caesium cyclohexylamides in cyclohexylamine.

The reaction can be represented by equilibria (239) and (240)

$$ArH^* + MB \underset{k'_{-1}}{\overset{k'_1}{\rightleftharpoons}} Ar^-M^+ + BH^* \quad \text{(slow)} \tag{239}$$

$$Ar^-M^+ + BH \overset{k'_2}{\rightleftharpoons} ArH + MB \quad \text{(fast)} \tag{240}$$

and is subject to kinetic isotope effects the magnitude of which depend upon the extent to which the labelled solvent molecule diffuses away before the aryl metal reacts with solvent (see below). The experimental technique in these studies is difficult and consequently kinetic investigations have yielded rate coefficients which are subject to considerable experimental error.

The first studies were concerned with deuteration of aromatics by deuterated potassamide $(0.02\ M)$ in liquid ammonia (Table 176)[582, 583]. From this data it was not wholly apparent that electron-supplying substituents decrease the reaction rate and *vice versa* as has been subsequently confirmed. A further study of

### TABLE 176

FIRST-ORDER RATE COEFFICIENTS $(10^7 k_1)$ FOR REACTION OF ArH WITH $0.02\ M$ KN$(^2$H$)_2$ IN NH$_3$[582, 583]

| Temp. (°C) | PhH | PhMe | PhEt | Ph-i-Pr | Ph-t-Bu | Tetralin | Bibenzyl | Naphthalene | Biphenyl |
|---|---|---|---|---|---|---|---|---|---|
| 0 | 70 | 30 | 30 | | | | | | |
| 10 | 190 | | | 40 | | | | | |
| 25 | 820–850 | 210 | 200 | 170 | 140 | 40 | 450 | 5,900[a] | 2,800 |

[a] With $0.021\ M$ KN$(^2$H$)_2$; at a concentration of $0.01\ M$, rate coefficients at 10.5, 25, and 40 °C were 990, 3,400, and 13,000 giving $E_a = 14.2$.

## TABLE 177

FIRST-ORDER RATE COEFFICIENTS ($10^7 k_1$) FOR DEDEUTERATION OR DEUTERATION (ITALICISED) OF AROMATICS BY POTASSAMIDE IN LIQUID AMMONIA (TEMPERATURE °C, IN PARENTHESES)[535, 567, 573a, 579, 583-8]

| Aromatic | Position of exchange | [Potassamide] ±0.01 (M) | | | | | | | | $k_{rel}^{a}$ |
|---|---|---|---|---|---|---|---|---|---|---|
| | | 0.01 | 0.014 | 0.02 | 0.06 | 0.19 | 0.20 | 0.43 | 0.6 | |
| Benzene[b,c] | 1 | 40(0), 440(25), *1,600(40)* | 570(25), *3,100(40)* | 73(0), 860(25), 820(25) | 4(−30), *1,800(25)* | 4,200(25) | 4,900(25) | 890(0) | ~1(−33.4) | 1 |
| Toluene[d] | 2 | | | 10(0), 160(25) | | | 1,100(25) | | | 0.14–0.22 |
| | 3 | | | 23(0), 240(25) | | | 2,200(25) | | | 0.32–0.45 |
| | 4 | | | 16(0), 240(25) | | | 2,000(25) | | | 0.23–0.41 |
| Naphthalene | 1 | | | 8,000(25) | | | | | | 9.8 |
| | 2 | | | 3,600(25) | | | | | | 4.4 |
| Biphenyl[e] | 2 | | | 350(0), 820(10), 3,000(25) | | | | | | 3.7 |
| | 3 | | | 260(0), 750(10), 2,600(25) | | | | | | 3.2 |
| | 4 | | | 230(0), 540(10), 2,500(25) | | | | | | 3.0 |
| p-Xylene | 2 | | | | | | 490(25) | | | 0.1 |
| Mesitylene | 2 | | | | | | 74(25) | | | 0.015 |
| Durene | 3 | | | | | | 19(25) | | | 0.004 |
| Thioanisole | 2 | | | | 2,200(−30) | | | | | 550*[k] |
| | 3 | | | | 3,700(0) | | | | | 24* |
| | 4 | | | | 1,100(0) | | | | | 6.5* |

TABLE 177 (continued)

| Aromatic | Position of exchange | [Potassamide] ±0.01 (M) | | | | | | | | $k_{rel}^{a}$ |
|---|---|---|---|---|---|---|---|---|---|---|
| | | 0.01 | 0.014 | 0.02 | 0.06 | 0.19 | 0.20 | 0.43 | 0.6 | |
| Anisole | 2 | | | 700(−30) | | | | | 8,000(−33.4) | 365*(8,000) |
| | 3 | | | | | | | | ~10(−33.4) | (10) |
| | 4 | | | | | | | | ~0.1(−33.4) | 0.5*(0.1) |
| Diphenylether | 2 | | | 8,000(−40) | | | 3,000(25)$^{f}$ | | | 41* |
| | | | | 300(−20) | | | | | | |
| | 3 | | | 1,000(−10)*$^{g}$ | | | | | | |
| | | | | 3,000(0)*$^{g}$ | | | | | | |
| N-Methylaniline | 4 | | | | 500(0)$^{h}$ | | | | | 3.7* |
| | 2 | | | | 5,050(0)$^{i}$ | | | | | 33* |
| | 3 | | | | 460(0)$^{i}$ | | | | | 2.9* |
| | | | | | 190(0)$^{i}$ | | | | | 1.3* |
| N,N-Dimethylaniline | 2 | | | | 2,100(25) | | | | | 1.2* |
| | 3 | | | | 240(25) | | | | | 0.13* |
| | 4 | | | | 110(25) | | | | | 0.06* |
| Fluorobenzene | 2 | | | | | | | >4×10$^{6j}$ | | >4×10$^{6}$ |
| | 3 | | | | | | | 4,000 | | 4,000 |
| | 4 | | | | | | | 200 | | 200 |
| Benzotrifluoride | 2 | | | | | | | 6×10$^{5}$ | | 6×10$^{5}$ |
| | 3 | | | | | | | 10,000 | | 10$^{4}$ |
| | 4 | | | | | | | 10,000 | | 10$^{4}$ |
| 1,4-Dimethoxybenzene | 2 | | | 17,000(−20) | | | | | | 2,330* |
| 1,2-Dimethoxybenzene | 4 | | | 2,200(25)$^{l}$ | | | | | | 1.47* |
| 1-Methoxy-4-methyl-benzene | 2 | | | 14,000(0) | | | | | | 192* |
| 1,2-Dimethoxy-4,5-dimethylbenzene | 3 | | | 7,600(0)$^{m}$ | | | | | | 72.5* |
| 1-Methoxy-2,4,6-trimethylbenzene | 3 | | | | 72(25) | | | | | 0.04* |

TABLE 177 (continued)

| Aromatic | Position of exchange | [Potassamide] ±0.01 (M) | | | | | | | | $k_{rel}^{a}$ |
|---|---|---|---|---|---|---|---|---|---|---|
| | | 0.01 | 0.014 | 0.02 | 0.06 | 0.19 | 0.20 | 0.43 | 0.6 | |
| 1,4-Di-(N,N-dimethyl-amino)benzene | 2 | | | | 1,500(25) | | | | | 0.835* |
| N,N-Dimethylamino-2-methylbenzene | 4 | | | | 2,170(25) | | | | | 0.945* |

[a] Most of these values differ from those given in the original papers and which have been derived by the unjustifiable comparison of the benzene *deuteration* data with the *dedeuteration* data for the substituted compound; this same error has been made elsewhere by these Russian workers (see p. 265). In some cases there is no alternative to this approximation and the data so derived is marked with an asterisk. Some of the values differ very markedly from those given in the original papers and which seem to the reviewer to have been obtained by methods which defy the laws of simple arithmetic.

[b] $E_a = 15.8$.

[c] The value obtained for dedeuteration with 0.6 $M$ potassamide[588] represents a difference in rate from the deuteration data of an order of magnitude, and has been attributed[589] to experimental error in dedeuteration.

[d] $E_a = 17$.

[e] $E_a = ca. 15$.

[f] Obtained with 0.027 $M$ potassamide.

[g] With 0.022 $M$ amide.

[h] With 0.052 $M$ amide.

[i] With 0.058 $M$ amide.

[j] Reaction was complete in 10 sec.

[k] This was corrected to 330 in the original paper because "15±5 % of exchange occurred under conditions when only the exchange of the methylthio group is possible (0.01 $M$ KNH₂, −60 °C)". This may well have been true for the other *ortho* compounds (with a heteratom in the side chain) but no indication was given of examination for this possibility.

[l] Obtained with 0.05 $M$ potassamide.

[m] With 0.04 $M$ potassamide.

the effect of base concentration on rates of deuteration of benzene gave the data in Table 177, which show that although the rates increase with increasing concentration of the catalysing base they do not do so in direct proportion to this at higher base concentrations due, it is believed, to the formation of ion pairs.

Subsequent studies with the liquid ammonia–potassamide system have been carried out using dedeuteration, and the appropriate rate coefficients and relative rates (calculated from the Russian work) are given in Table 177. It is clear from the footnotes to the Table that a fairly wide interpretation of the relative rates is possible and since they are derived mostly from non-identical reactions and at different temperatures, finely detailed analysis of them should be avoided. This also applies to data derived from them by the Shatenshtein group, for in a number of publications (referring to the same work) the *ortho* : *para* ratios for substituents are miscalculated even from their own derivation of the partial rate factors (see, *e.g.* ref. 590). In addition, there is a tendency for these values to be changed with each succeeding publication in a manner which is most confusing to the reader (no reason is generally given for this) and it is recommended that Table 177 rather than the original paper be consulted for what are, in the reviewer's opinion, the more reliable values[754].

The data in the table show that the reaction is accelerated by $-I$ substituents and *vice versa*; consequently, substituent effects are most marked at the *ortho* position and Shatenshtein et al.[590] have shown that a correlation exists between the log rate of exchange and the $\sigma_I$ values for the *ortho* substituents. This suggests that steric hindrance is very slight in the reaction, and this is entirely consistent with the reaction mechanism in which rate-determining attack on hydrogen occurs.

The lack of steric hindrance is also shown by the kinetic data for *p*-xylene, mesitylene, and durene, the observed reactivities being close to those calculated by the additivity principle. The additivity principle has also been tested for the last seven compounds in Table 177, and for the first five of these it holds very well. If one assumes a value for $f_3^{MeO}$ of *ca.* 4.0 and takes the average of the values listed in the table for the methyl substituent partial rate factors, then the observed : calculated reactivity ratios are 1.6, 0.85, 0.75, 1.4 and 1.0. For the last two compounds in the table the ratios are 5.3 and 4.1, the reason for this being unknown.

Shatenshtein et al.[555, 591] have also measured rate coefficients for dedeuteration of thiophen derivatives by lithium or potassium *t*-butoxides in dimethyl sulphoxide or *t*-butyl alcohol (70 vol. %) in diglyme (Table 178). Interestingly, the 2 position is more reactive than the 3 position and this was reasonably attributed to the $-I$ effect of the hetero sulphur atom. The methyl substituent lowers the reactivity of the 2 position from each position in accord with its $+I$ effect and consequently the effect was greatest from the 3 position. However, the deactivation from the 5 position was greater than from the 4 position, and this was incorrectly attributed to the $+M$ effect of methyl group operating from the 5 position since

TABLE 178

RATE COEFFICIENTS FOR REACTION OF [²H]-THIOPHENES WITH $t$-BuO⁻ IN DIMETHYL-
SULPHOXIDE OR $t$-BuOH (70 VOL. %) IN DIGLYME 30 VOL. %)[555,591]

| Position of [$^2H$] | Substituent | [$t$-$BuO^-$] (M) | DMSO solvent | | $t$-BuOH-DG solvent $10^7k_1(25\ °C)^a$ |
|---|---|---|---|---|---|
| | | | $10^7k_1(25\ °C)$ | $10^7k_1(50\ °C)$ | |
| 2 | H | — | 1,000 | | 330 |
| 2 | 3-Me | — | 59 | | |
| 2 | 4-Me | — | 180 | | |
| 2 | 5-Me | — | 120 | | |
| 2 | 3-MeS | 0.86 | >110,000 | | 10,000 |
| 2 | 4-MeS | 0.86 | 30,000 | | 3,300 |
| 2 | 5-MeS | 0.54 | 50,000 | | 5,400 |
| 2 | 3-MeO | — | >100,000 | | |
| 2 | 5-MeO | — | 390 | | |
| 3 | H | — | 9 | 120[b] | |
| 3 | 4-Me | — | | 520[c] | |
| 3 | 2-MeS | 0.30 | 600 | 11,000[d] | |
| 3 | 4-MeS | 0.35 | 1,200 | 15,000[e] | |
| 3 | 5-MeS | 0.35 | | 1,100 | |

[a] [K-$t$-BuO] = 0.6 $M$.
[b] 1,400 at 75 °C, $E_a$ = 23.8.
[c] Value at 75 °C.
[d] $E_a$ = 20.6.
[e] $E_a$ = 18.0.
— = Not quoted

this can conjugate with the 2 position. However, the negative charge which develops in the transition state is in a non-conjugating $\sigma$ orbital and some other cause must be sought. The same result is found for the methoxy group in the 3 position, and a possible explanation is that the $p$ orbital density at the reaction site is increased by the mesomeric effect of this substituent and that this repels the attacking reagent to a significant extent; however, in the 3 position, the $-I$ effect of methoxy presumably predominates. In the same way, the greater rate enhancement from the 5 relative to the 4 position by a methylthio substituent can be attributed to a lowering of the $p$ orbital density at the 2 position, thereby favouring approach of the reagent and not, as stated by Shatenshtein *et al.*, to conjugation between the $d$ orbitals and the negative charge which is, of course, impossible. Alternatively, one can envisage that repulsion between the $p$ orbital electrons and the $\sigma$ orbital electrons will raise the energy of the transition state and this will be lowered by $p\pi$–$d\pi$ conjugation (and also increased by $+M$ substituents). However, so little is yet known about the transmission of electronic effects in the thiophen molecule that these proposals can only be very tentative at present.

Although no experimental details were given, it was stated that the relative reactivities of the 2 positions of furan, thiophene and selenophene were 1 : 500 :

$700^{555}$. This was again incorrectly attributed to conjugation between the carbanion and the unoccupied $d$ orbitals of sulphur and selenium, but can be explained in the manner described above.

For the exchanges carried out in liquid ammonia, kinetic isotope effects $k_D : k_T$ of 2.3–2.5 have been obtained for reaction of benzene, toluene, and naphthalene and for the reactions of the 2 positions of furan and thiophene with $t$-butoxide in dimethyl sulphoxide somewhat lower values, 1.5 and 1.3, respectively, were obtained[591], but whether this was a solvent or a substituent effect is not apparent from the data.

Sodium ethoxide in deuterated ethanol was used in a study of the rate of deuteration of anisolechromium tricarbonyl at 100 °C, the first-order coefficient $(2.65 \times 10^{-5})$ being reported as only 3 times greater than that for benzene, $cf.$ anisole in the reaction with potassamide in liquid ammonia (Table 177)[591a].

Streitweiser $et~al.$[592] have measured rates of base-catalysed dedeuteration and detritiation and have attempted to discover details of the reaction mechanism. Second-order rate coefficients for the reaction of some polycyclics with lithium cyclohexylamide in cyclohexylamine are given in Table 179, and it can be seen

TABLE 179

SECOND-ORDER RATE COEFFICIENTS FOR REACTION OF $[^2H(^3H)]$-Ar WITH $C_6H_{11}NHLi$ IN $C_6H_{11}NH_2$ AT 49.9 °C[592]

| ArH | Position of exchange | $10^6k_2(D)$ | $10^6k_2(T)$ | $k_{rel}$ |
|---|---|---|---|---|
| Benzene | 1 | 110 | 69 | 1 |
| Biphenyl | 2 | 128 | | 1.2 |
| | 3 | 396 | | 3.7 |
| | 4 | 249 | | 2.3 |
| Naphthalene | 1 | 720 | | 6.5 |
| | 2 | 440ᵃ | | 4.1 |
| Phenanthrene | 9 | 1,930 | | 17.9 |
| Anthracene | 1 | | 977 | 14.1 |
| | 9 | 4,600 | 4,900 | 42–71 |
| Pyrene | 1 | 2,870 | 2,230 | 26–32 |
| | 2 | | 1,860 | 27 |
| | 4 | | 2,800 | 40 |

ᵃ 89 at 25 °C, giving $E_a = 11.6$.

that the reaction is also best described in terms of a model based primarily on inductive effects and has a much lower activation energy than those carried out with $t$-butoxide ion in dimethyl sulphoxide. Furthermore, the results for the 1 position of naphthalene and the 2 position of biphenyl indicate fairly convincingly that the present reaction is subject to greater steric hindrance than the reactions in liquid ammonia and this is entirely consistent with the different sizes of the amide and cyclohexylamide ions. It should be noted that the $k_{rel}$ values in Table 179 refer to both dedeuteration and detritiation and are not therefore strictly comparable (in the sense of quantitative accuracy) since the latter reaction appears to have the greater selectivity as one would expect in view of the greater strength of the carbon–tritium bond relative to the carbon–deuterium bond, these being broken in the rate-determining step. This was also shown by the kinetic isotope effect which is for example 1.60 for the reaction of benzene.

The differences in rate for the two positions of naphthalene show clearly that an additional–elimination mechanism may be ruled out. On the other hand, the magnitude of the above isotope effect is smaller than would be expected for a reaction involving rate-determining abstraction of hydrogen, so a mechanism involving significant internal return had been proposed, equilibria (239) and (240), p. 266. In this base-catalysed (B-$S_E$2) reaction both $k'_1$ and $k'_2$ must be fast in view of the reaction path symmetry. If diffusion away of the labelled solvent molecule BH* is not rapid compared with the return reaction $k'_{-1}$ a considerable fraction of ArLi reacts with BH* rather than BH, the former possibility leading to no nett isotope effect. Since the diffusion process is unlikely to have an isotope effect then the overall observed effect will be less than that for the step $k'_1$.

Kinetic studies with caesium cyclohexylamide have also been performed[593]. For reaction of tritiated benzene, the kinetic order was one in both hydrocarbon and caesium cyclohexylamide ion-pair (Table 180) and rate coefficients were

TABLE 180

RATE COEFFICIENTS FOR REACTION OF [$^2$H($^3$H)]-Ph WITH $C_6H_{11}$NHCs IN $C_6H_{11}$NH$_2$ AT 25 °C[593]

| $10^2[C_6H_{11}NHCs]$ (M) | $10^4 k_1(D)$ | $10^4 k_1(T)$ | $k_1(D_1):k_1(T)$ | $10^2 k_2(T)$ |
|---|---|---|---|---|
| 2.3 | 32.1 | 12.0 | 2.67 | |
| 1.9 | 24.7 | 10.8 | 2.29 | |
| 4.8 | | 21.2 | | 4.4 |
| 0.47 | | 2.78 | | 6.0 |
| 1.3 | | 7.50 | | 5.8 |
| 2.0 | | 10.4 | | 5.2 |
| 0.79 | | 3.62 | | 4.6 |
| 3.6 | | 15.7[a] | | 4.3[a] |
| 3.1 | | 19.2[a] | | 6.2[a] |

[a] Data is for tritiation.

3,300 times greater than those obtained with the lithium compound. The isotope effect $k_D : k_T$ was significantly larger (ca. 2.4) in this reaction which suggests less participation of the internal return mechanism and consequently greater stability of the organocaesium intermediate compared to the corresponding organolithium intermediate.

A subsequent examination of the rates of detritiation and dedeuteration of toluene by lithium and caesium cyclohexylamides confirmed the larger isotope effect in the latter reaction (2.3 *cf.* 1.23), and a comparison of the relative reactivities (including those of Shatenshtein *et al.* for 0.2 *M* potassamide at 25 °C, Table 179) confirmed that the reaction with lithium cyclohexylamide is subject to greater steric hindrance and the overall spread of rates is similar in all three reactions[596] (Table 181); this table does not take into account the greater spread

TABLE 181

RELATIVE RATES FOR BASE-CATALYSED EXCHANGE OF ArH AT 25 °C[594]

| ArH | Position of exchange | Base | | |
|---|---|---|---|---|
| | | $C_6H_{11}NHLi$ | $C_6H_{11}NHCs$ | $NH_2K$ |
| Benzene | 1 | 1 | 1 | 1 |
| Toluene | 2 | 0.12 | 0.20 | 0.22 |
| | 3 | 0.54 | 0.59 | 0.45 |
| | 4 | 0.43 | 0.52 | 0.41 |
| Mesitylene | 1 | | 0.013 | 0.015 |

of rates in detritiation as opposed to dedeuteration. In this work, the activation enthalpy and entropy for reaction of the 2 position of toluene with caesium cyclohexylamide were found to be 15 and −17 respectively compared to the previously determined[593] values of 13 and −22 for benzene; the activation energies are therefore confirmed as being appreciably lower than for the potassamide system.

Although in the above work the rate spread for detritiation was greater than for dedeuteration, surprisingly, in the reaction of lithium cyclohexylamide at 25 °C with fluorobenzenes and benzotrifluorides, the relative rates for dedeuteration and detritiation were[595], respectively: 2-F, $6.3 \times 10^5$ and $3.4 \times 10^5$; 3-F, 107 and 86; 4-F, 11.2 and 9.1; 3-CF$_3$, 580 and 390, which therefore shows the reverse. The rate enhancement for these substituents were found to correlate reasonably with a field effect model for the inductive effect.

The importance of the inductive effect in controlling the reaction rates was further shown by Streitweiser and Humphrey[596], who measured the rates of dedeuteration of toluene ($\alpha$, $\alpha$-$d_2$), ($\alpha$, 2,4,6-$d_4$), and ($\alpha$, 2,3,4,5,6-$d_6$) by lithium cyclohexylamide at 50 °C and found the rate to be reduced by 0.4 %, 0.4 %, and 1.8 % for a deuterium atom in the *ortho*, *meta* and *para* positions respectively. The retardation is consistent with the +I effect of deuterium but the differential positional effect could not be rationalised in simple and general terms.

Rates of detritiation of the 1 and 2 positions of biphenylene and the 5 position of benzo[b]biphenylene by lithium cyclohexylamide at 50 °C relative to benzene have been determined as 490, 7.0, and 1,865, respectively and the enhanced reactivity of the position $\alpha$ to the strained 4-membered ring has been attributed to the enhanced electronegativity of the strained bridgehead carbon atom. The same explanation was advanced to account for the greater rate of detritiation of the 1 position of triptycene relative to the 2 position, the $k_{rel}$ values for detritiation by caesium cyclohexylamide at 25 °C being 20.8 and 2.77, respectively[539]. Relative rates were determined either by comparison of second-order rate coefficients with earlier data or by the competition technique.

Streitweiser *et al.*[597] have also measured second-order rate coefficients for hydrogen exchange of fluorobenzenes with sodium methoxide in methanol, Table 182. Nucleophilic displacement of fluoride ion by methoxide ion accompanies

TABLE 182

RATE COEFFICIENTS AND KINETIC PARAMETERS FOR REACTION OF ArH WITH NaOMe-[$^3$H]OMe[597]

| ArH | Temp. (°C) | $10^4 k_2$ | $\Delta H^{\ddagger}$ | $\Delta S^{\ddagger}$ |
|---|---|---|---|---|
| C$_6$F$_5$H | 0 | 10.3 | | |
| | 25 | 249 (242$^a$) (242$^b$) (562$^c$) (536$^d$) | | |
| | 40.1 | 1,360 | | |
| | 59.9 | 5,890 | 18.0 | −5.6 |
| 1,3-F$_2$C$_6$H$_4$ | 40 | 0.0061 | | |
| | 59.8 | 0.0927 | 27.9 | +1.9 |
| 1,2,3,4-F$_4$C$_6$H$_2$ | 25 | 0.00056 | | |
| | 40 | 0.0053 | 27.3 | −0.3 |
| 1,2,4,5-F$_4$C$_6$H$_2$ | 40 | 58.2 | | |
| 1,2,3,5-F$_4$C$_6$H$_2$ | 40 | 5.59 | | |

$^a$ For detritiation in methanol.
$^b$ For dedeuteration in methanol.
$^c$ For detritiation in deuterated methanol (containing a small amount of tritium).
$^d$ For deuteration in deuterated methanol.

the reaction but at a rate insignificant compared with the rate of exchange except for the 1,2,3,4-compound. Hence this side reaction was not a complication and it also follows that any mechanism which starts with nucleophilic addition to the ring can be ruled out. The logarithms of the partial rate factors were determined as $f_o^F$ 5.25; $f_m^F$, 2.03; $f_p^F$, 1.13; this confirms the general pattern noted previously.

The lack of a substrate isotope effect suggests very extensive internal return and is readily explained in terms of the fact that conversion of the hydrocarbon to the anion would require very little structural reorganisation. Since $k_{obs} = k_1' k_2'/(k_{-1}' + k_2')$ and $k_{-1}$ is deduced as $\gg k_2'$, then $k_{obs} = Kk_2'$, the product of the equilibrium constant and the rate of diffusion away of a solvent molecule, neither of the steps having an appreciable isotope effect. If the diffusion rates are the same for reactions of each compound then the derived logarithms of partial rate factors (above) become $pK$ differences between benzene and fluorobenzene hydrogens in methanol. However, since the logarithms of the partial rate factors were similar to those obtained with lithium cyclohexylamide, a Brönsted correlation with a proportionality constant $\beta$ of 1.0 is implied. This cannot be so, for the latter exchange reaction does not have a reverse step of carbanion reaction with solvent which is diffusion-controlled in view of the large observed isotope effect. However, these reactions involve formation of ion pairs and thus the solvated lithium cation, and the possible slowness of this desolvation in the reverse reaction may well account for the observed slowness of this reverse reaction, and also be independent of structure for a related series, giving rise to the apparent Brönsted correlation. It was pointed out that these arguments probably could not be applied to exchange reactions involving delocalized anions (derived from aliphatic hydrogens) since in this case the organic moiety also undergoes structural reorganisation.

The activation entropies were considerably different from the large negative values expected for a second-order reaction and this was attributed to the effect of the internal return mechanism.

Kinetic studies of base-catalysed hydrogen exchange of heterocyclic compounds have been carried out. Paudler and Helmick[515] measured second-order rate coefficients for deuteration of derivatives of imidazo[1,2-a]pyridine(XXXIII), imidazo[1,2-a]pyrimidine(XXXIV), and 1,2,4-triazolo[1,5-a]pyrimidine(XXXV)

(XXXIII)        (XXXIV)        (XXXV)

and their derivatives by deuterated sodium methoxide–methanol mixtures. Rate coefficients (where these were not too slow to measure) are given in Table 183. The principal resonance structures for these compounds have a positive charge on the 4-nitrogen atom; consequently, base-catalysed exchange at the adjacent 3 and 5 positions would be expected to be very rapid (relative to the other positions) and inspection of Table 183 confirms this. In addition, a negative charge resides on the 1-nitrogen but not in the principal resonance structure for the methiodide; consequently, significant exchange should occur at the 2 position of the methiodide and this is also observed. The methiodides are also more reactive than the unsubstituted compounds, and methyl substitution in the ring lowers the reactivity, both these observations being the expected ones.

The rate of deuteration of (XXXIII) at the 3 position by ca. 0.1 $M$ NaOD in $D_2O$ was 2.19 times faster than the dedeuteration of the same position by the same concentration of NaOH in $H_2O$ at 35 °C. This indicates that proton transfer takes place in a rate-determining step, but the result is not conclusive in view of the different reaction medium.

The intense activation of *ortho* positions by strongly electron-withdrawing groups has produced some interesting kinetic results in the base-catalysed exchange of pyridine and its derivatives. For the neutral molecule, exchange occurs

TABLE 183

SECOND-ORDER RATE COEFFICIENTS ($10^3 k_2$) FOR REACTION OF ArH WITH DEUTERATED NaOH–EtOH[515]

| Compound[a] | Substituent | Temp. (°C) | Position of deuteration | | | |
|---|---|---|---|---|---|---|
| | | | 2 | 3 | 5 | 6 |
| XXXIII | H | 65 | | 0.024[a] | 0.072[b] | |
| | 1-MeI | 35 | 6.1 | 9.4 | 0.033 | |
| | 1-MeI, 5Me | 35 | 5.2 | 5.2 | | |
| | 1-MeI, 6Me | 35 | 3.3 | 8.5 | | |
| | 1-MeI, 7Me | 35 | 3.9 | 7.8 | | |
| | 1-MeI, 8Me | 35 | 3.3 | 8.5 | | |
| XXXIV[c] | H | 65 | 0.025[b] | 0.15[b] | 1.5[b] | 0.048[b] |
| | H | 65 | | 0.145 | 1.64 | |
| XXXV | H | 35 | | | 10.9 | 0.0098 |
| | H | 65 | 0.037 | | >50 | 0.32 |
| | 2-Me | 35 | | | 6.1 | |
| | 7-Me | 35 | | | 3.8 | |
| | 1-MeI | 35 | >50 | Too unstable to measure | | |
| | 1-MeI, 7-Me | 35 | >50 | | | |

[a] For name see text.
[b] Using 0.5 $M$ NaOD, all other data obtained with 0.098 $M$.
[c] Corresponding methiodide unstable under reaction conditions.

rather more rapidly at the 3 and 4 positions than at the 2 position, but in weakly alkaline and neutral media, exchange occurs preferentially *via* the small equilibrium concentration of conjugate acid, which therefore gives extreme activation at the 2 position relative to the other positions. This has led to some apparent contradictions in the literature. For example, with sodamide in liquid ammonia at $-25$ °C, the relative rates of deuteration at the 4, 3, and 2 positions were[598] $10^3 : 10^2 : 1$, whereas in aqueous sodium hydroxide at 220 °C exchange was said to occur exclusively at the 2 position[599]. With 1 $M$ sodium hydroxide at 25 °C, the relative reactivity order was 3.0 : 2.3 : 1, whereas in very dilute sodium hydroxide exchange occurred principally at the 2 position[600]. The concentration of sodium hydroxide at which exchange occurred principally at the 3 rather than at the 2 position in 4-aminopyridine was higher than for pyridine, and higher still for 4-N,N-dimethyl-aminopyridine, and this follows from the increasing basicity of the heterocyclic nitrogen atom along the series[601].

## 9. Reactions involving replacement of a substituent by hydrogen

### 9.1 PROTODEMAGNESIATION

Only two kinetic studies[602, 603] have been carried out on this reaction

$$XC_6H_4{\cdot}MgBr + C_4H_9C{\equiv}CH \rightarrow XC_6H_5 + C_4H_9C{\equiv}CMgBr \qquad (241)$$

For the reaction of arylmagnesium bromide (0.15 $M$) and 1-hexyne (3.0 $M$) in ether Dessy and Salinger[602] found first-order rate coefficients (obtained by following the change in the dielectric constant of the medium) were linear to at least 90 % of reaction. With both reagents at 0.2 $M$ concentration, second-order kinetics were obtained to beyond 70 % of reaction indicating a first-order dependence on each reagent, though the rate coefficients under these conditions were 28 % greater than those determined under first-order conditions; this was attributed to a solvent effect of the excess of 1-hexyne in the latter determination. Second-order rate coefficients ($10^4 k_2$) at 31.5 °C were (X =) 3-CF$_3$, 0.22; 3-Cl, 0.33; 4-Cl, 0.6; H, 2.8; 4-CH$_3$, 6.22, and the logarithm of these values plotted against $\sigma$ values to give a $\rho$ value of $-2.5$ indicating a very reactive electrophile, which is not inconsistent with the acidic nature of the alkynyl hydrogen.

A subsequent investigation of the rates of reaction of phenylmagnesium bromide with 1-hexyne in which the progress of reaction was determined by a gas chromatographic analysis of the vapour in equilibrium with the reacting solution, showed that, with ether as solvent, the reaction was only first-order, and went over to second-order kinetics as tetrahydrofuran was added to the solution[603]. This change was also accompanied by a decrease in the reaction rate. The reason for this change in kinetic order and for the disagreement between the two sets of work has not been evaluated. The critical reagent in the first-order reaction was unfortunately not determined by varying the initial concentrations.

### 9.2 PROTODEMERCURATION

Protodemercuration was first studied kinetically by Kharasch et al.[604], who measured, qualitatively, the rates of cleavage of diarylmercury compounds by acids, viz.

$$ArHgAr' + HX \rightarrow ArH + Ar'HgX \qquad (242)$$

The relative case of cleavage of the aryl–mercury bond yielded a so-called scale of "electronegativities" (the group Ar in reaction (242) being considered the more "electronegative") though this scale is the exact reverse of the true electronegativity order and arose from a misunderstanding of the reaction mechanism.

The first quantitative study was made by Corwin and Naylor[605], who measured the rates of cleavage of diphenylmercury by formic acid (at approximately 30 °C) and acetic acid (at approximately 42 °C) in dioxan, dilatometrically. Formic acid cleaved the mercury compound more readily than acetic acid, and hydrochloric acid much more so than either. With acid in excess, good first-order kinetics were obtained in a given run, but runs carried out a different initial concentrations of mercury compound gave different rates, decreasing concentration yielding increasing rate coefficients; this abnormality was subsequently considered to arise from an unknown fault of dilatometry[606]. Neither chloride ions nor water had an appreciable effect on the rate, but increasing the concentration of acid produced a very large increase in the rate coefficient even though the acid is in large excess. This was thought to arise possibly from the existence of either polymolecular acid complexes or to hydrogen bonding of acid molecules with the dioxan solvent, so that the effective free acid concentration would be much less than the stoichiometric concentration.

A subsequent investigation using spectrophotometric and titration techniques indicated that the reaction order in carboxylic acid was approximately three and that first-order rate coefficients were, in fact, independent of the initial concentration of mercury compound[606]. The reaction mechanism was proposed as

$$\text{(243)}$$

In the reaction with perchloric acid, the reaction was strictly second-order, being first-order in both perchloric acid and aromatic, and in various mixtures of ethanol, dioxan, and water, the reaction rates correlated with the acidity function of the media, this being thought to indicate a reaction with a rate-determining proton transfer. Thus the increased reaction rate with increasing ethanol or dioxan content of the medium was attributed to protonated dioxan or ethanol being stronger attacking agents than hydronium ion, following a proposal by Braude[607]. Addition of perchlorate anion to the reaction with perchloric acid in 69.5 vol. % aqueous dioxan produced no rate change, consistent with the second step of the reaction being fast. However, addition of chloride ion produced a small rate acceleration, and since it produced no reaction in the absence of acid thereby ruling out direct nucleophilic attack on mercury to give the carbanion, the acceleration was attributed to the formation of a small equilibrium concentration of hydrochloric acid, which gives a high specific cleavage rate. The activation energy (determined only over a 10–15 ° range) for the reaction with acetic acid was unexpectedly lower (18.5) than for the reaction with perchloric acid (20.6, 95 % EtOH; 23.7, 97 % dioxan) but this was compensated by a lower log $A$ value (9.1 cf.

12.1 and 14.6), which indicated a specifically orientated transition state for the former reaction.

A further kinetic investigation of the rates of cleavage of diphenylmercury (and some dialkylmercurials) showed similar kinetic features[608]. The first-order rate for the reaction of diphenylmercury with acetic acid at 25 °C was $4.98 \times 10^{-4}$, which agreed quite well with the value from the above determination ($2 \times 10^{-4}$ at 42 °C with dioxan). In the presence of perchloric acid, second-order kinetics were found to be obeyed (for dineophyl mercury and presumably for diarylmercurials as well) for a twofold concentration change in both mercurial and perchloric acid. Two separate mechanisms were proposed for the reactions in the absence, and presence, of perchloric acid. Under the former conditions an $S_E i$ process (244) was

$$(244)$$

envisaged and this would be consistent with the lower log $A$ factor of the reaction, and the lack of effect of sodium acetate upon the reaction rate; this latter also ruled out the possibility of reaction with the conjugate acid of acetic acid. This reaction, involving assistance by nucleophilic attack on mercury, is unimportant in the presence of strong acid, when the electrophile is indicated as protonated acetic acid $AcOH_2^+$, the reaction mechanism being analogous to (243). Obviously this protonated acid would be less likely to carry out nucleophilic attack on mercury

A kinetic study of the cleavage of diphenylmercury with hydrochloric acid in dimethyl sulphoxide containing dioxan or water has been carried out, the progress of the reaction being followed conductimetrically[609]. The reaction was found to be strictly second-order for a 20-fold variation in the acid concentration. Most of the studies were carried out using a 10 : 1 by volume mixture of dimethyl sulphoxide and dioxan and rates ($k_2$) of cleavage of diphenylmercury were $9.3 \times 10^{-3}$ (25 °C) and $2.5 \times 10^{-2}$ (40 °C) giving $E_a = 12.2$ and $\Delta S^{\ddagger} = -2.9$. A study, at 32 °C, of the effect of increasing the dioxan content of the solvent gave a substantial rate increase (a tenfold increase in the mole fraction of dioxan increased the rate by 15 %), whereas replacing the dioxan by water decreased the rate. Addition of sodium sulphate had no effect, but addition of sodium chloride increased the rate. These observations were interpreted to mean that the attacking species was undissociated hydrogen chloride, for water would increase this dissociation, and chloride ion would decrease it and suppress the small quantity of reaction which must be occurring via the dissociated species (and which must, therefore, be a slower reaction). Addition of sulphuric acid gave only a 10 % increase in rate for a concentration equimolar to that of hydrogen chloride, and second-order kinetics were observed if mercurial and hydrochloric acid alone were used; the rate-determining step was, therefore, extremely unlikely to involve free protons. Confirmation of this was obtained through the observation that sulphuric acid on its own did not react with mercurials, *i.e.* bisulphate anion is not a suitable anion for attacking mercury. An attempt to provide further confirmation of these conclusions by using deuterium chloride gave inconclusive results, believed to be partly due to the breakdown (for reasons unknown) of second-order kinetics with this reagent. Confirmation was also obtained by the fact that the reaction of diphenylmercury with hydrogen bromide gave an activation energy of 17.4 and an entropy of $-11$, (though no rates were quoted). These large changes in thermodynamic parameters were considered to be too large to have arisen from a reaction involving attack of a solvated proton. Dessy *et al.*[609] therefore, presented evidence regarded by many as showing the reaction to involve a 4-centre reaction mechanism, but recently this view (in particular the experimental evidence leading to it) has been challenged[610].

The effects of substituents have been determined in the cleavage of diarylmercury compounds by hydrogen chloride in dimethyl sulphoxide–dioxan at 32 °C. Rates were measured within the temperature range 12.8–75.0 °C, though over a range of not more than 18 °C for each compound[611] (Table 184). The

TABLE 184

RATE COEFFICIENTS AND KINETIC PARAMETERS FOR REACTION OF $(XC_6H_4)_2Hg$ WITH HCl IN DMSO-DIOXAN[611]

| X | $10^3 k_2$ (32 °C) | $E_a$ | $\Delta S^{\ddagger}$ |
|---|---|---|---|
| 4-MeO | 460[a] | 9.2 | $-32$ |
| 4-Ph | 26.1 | 11.9 | $-30$ |
| H | 16.2 | 12.2 | $-29$ |
| 4-F | 13.6 | 11.4 | $-31$ |
| 4-Cl | 4.78 | 15.0 | $-22$ |
| 3-NO$_2$ | 0.14[a] | 20.8 | $-10$ |

[a] Extrapolation from data at other temperatures.

kinetic parameters, which must be treated cautiously in view of the very large possible error, indicate an unexpectedly large entropy factor for the nitro and possibly the chloro compound though the reason for this is not known. The log $k_{rel}$ values correlated with the Yukawa–Tsuno equation[612] with $r = ca.$ 0.5 and $\rho = -2.8$ and an apparent correlation of $E_a$ *versus* $\Delta S^{\ddagger}$ was considered as evidence for an isokinetic temperature and hence as confirmation of a single reaction mechanism for all the compounds. However, it is easily shown (see p. 9) that since $\Delta S^{\ddagger}$ is a linear function of $E_a$, a moderately linear plot of one against the other is inevitable for reactions with small $\rho$ factors and such plots are, therefore, meaningless.

Rate coefficients have been measured for the compound bis-o-phenylenedimercury under the same conditions and reaction proceeds as[613]

$$\text{(structure)} \xrightarrow[k']{HCl} \text{(structure)} \xrightarrow[k'']{HCl} \text{C}_6\text{H}_5\text{HgCl} \qquad (245)$$

The second-order rate coefficients ($k'$ and $k''$) were 0.315 and 0.048 respectively at 25 °C, showing that strain in the initial compound increases the reaction rate, and also that the *ortho*-chloromercuri substituent must be electron-supplying, otherwise cleavage of the first decomposition product would have yielded benzene and the bis-o-chloromercuribenzene, $C_6H_4(HgCl)_2$. Petrosyan and Reutov[613a] have, however, argued that the chloromercuri group is electron-withdrawing, the above results arising from steric hindrance to formation of bis-o-chloromercuribenzene.

The kinetics of the cleavage of a wide range of diarylmercurials by hydrochloric acid in 90 vol. % aqueous dioxan have been studied by Nesmeyanov *et al.*[614] The second-order rate coefficients are gathered in Table 185, and are correlated

TABLE 185

RATE COEFFICIENTS AND ACTIVATION ENERGIES[a] FOR REACTION OF $(XC_6H_4)_2Hg$ WITH HCl IN 90 VOL % AQUEOUS DIOXAN[614]

| X | $10^2k_2$ at | | | | X | $10^2k_2$ at | | | | |
|---|---|---|---|---|---|---|---|---|---|---|
| | 20 °C | 30 °C | 40 °C | 50 °C | | 30 °C | 40 °C | 50 °C | 60 °C | 70 °C |
| 4-EtO | 17.8 | 66.3 | | | 3-MeO | 0.552 | | 5.78 | 17.5 | |
| 4-MeO | 12.5 | 53.4 | | | 4-Cl | 0.123 | | 1.42 | 4.64 | |
| 2-Me | | 4.73 | | | 4-Br | 0.0846 | 0.316 | | | 3.65 |
| 4-Et | | 4.26 | | | 3-F | | | 0.417 | 1.36 | 4.78 |
| 4-Me | 1.00 | 4.14 | 14.4 | | 3-Cl | | | 0.315 | 1.14 | 3.56 |
| 3-Me | | 1.28 | | | 2-MeO | | | | | 0.275 |
| 2-MeO | 0.382 | 1.42 | 4.95 | | 4-CO₂Me | | | | | 2.48 |
| 2:3-Benzo | | 1.28 | 5.10 | 16.3 | 2-CO₂Me | | | | | 1.36 |
| H | 0.177 | 0.665 | 2.36 | | 2-Cl | | | | 0.185 | |
| 4-F | | 0.582 | 2.21 | 7.74 | | | | | | |
| (Aryl = α-thienyl) | 3.42 | 11.6 | | | | | | | | |

[a] Activation energies were within the range 23.0–26.8 and roughly decreased with increasing reactivity.

by the Yukawa–Tsuno equation[612] with $r = ca.$ 0.5 and $\rho = -3.78$ at 30 °C, and this is consistent with the data of Dessy and Kim[613] since the rate coefficients in this medium are lower than in dimethyl sulphoxide. Subsequently, Nesmeyanov *et al.*[615] measured the rates of cleavage, by hydrochloric acid, of diferrocenylmercury (and also of bis (cyclopentadienylmanganesetricarbonyl)mercury) in a range (80–92 vol. %) of aqueous dioxan mixtures and at four different temperatures (with a 15 °C range for each solvent mixture). The second-order rate coefficients ($10^2 k_2$) with 90 vol. % aqueous dioxan were 126 (25 °C) and 0.47 (40 °C) respectively from which the $k_{rel}$ values were calculated as 380 and 0.20. The activation energies and entropies determined in this study are gathered in Table 186. In

TABLE 186

VALUES OF $E_a$ AND $\Delta S^{\ddagger}$ FOR REACTION OF $R_2Hg$ WITH HCl IN AQUEOUS DIOXAN[615]

| Mercurial | Concentration of solvent (vol. %) | | | | | | | |
|---|---|---|---|---|---|---|---|---|
| | 80 | | 85 | | 90 | | 92.5 | |
| $(C_6H_5)_2Hg$ | 22.2, | −0.3 | 23.6, | 5.6 | 24.2, | 9.3 | 25.2, | 14.0 |
| $(C_5H_5FeC_5H_4)_2Hg$ | 16.9, | −6.2 | 18.0, | −1.3 | 18.7, | 2.7 | 19.5, | 6.6 |
| $[C_5H_4Mn(CO)_3]_2Hg$ | | | 23.0, | 0.3 | 22.9, | 1.9 | 22.4, | 1.4 |

general, the activation energy and entropy both decrease (within experimental error) with increasing water content. This was explained in terms of a more highly solvated transition state when the medium is more aqueous, leading to a decrease in the activation energy and restriction of the vibrations and rotations of the solvent molecules, hence the loss of entropy.

The rates of cleavage of diferrocenylmercury by aqueous perchloric acid in 90 vol. % aqueous dioxan were also measured. The rates were approximately half those obtained with hydrochloric acid at the same temperatures and gave values of $E_a$, 17.1 and $\Delta S^{\ddagger}$, −2.3. The decrease in both these parameters compared with the hydrochloric acid-catalysed cleavage was considered to arise from a more highly charged, and consequently more highly solvated transition state. The greater rates with hydrochloric acid were attributed to nucleophilic assistance as described by the earlier workers, and once again addition of chloride ion increased the rate for the reaction with hydrogen chloride and much more so for the reaction with perchloric acid (presumably through formation of a small equilibrium concentration of hydrogen chloride).

A comparison of the rate of hydrochloric acid-catalysed cleavage of diferrocenyl-mercury and ferrocenylmercuric chloride in 90 vol. % aqueous dioxan showed the latter to be 130 times less reactive than the former. The activation energies and entropies were 18.7 and +2.7, 19.1 and −5.6, respectively, so the difference in reactivity arises principally from the lower entropy for reactions of the latter compound and results for cleavage of (cyclopentadienylmanganesetricarbonyl)-mercuric chloride revealed the same pattern. No explanation was given, but nucleophilic attack on mercury and the need for a highly ordered transition state will obviously differ according to the electronegativity of the substituent on mercury.

The kinetics of the hydrochloric acid-catalysed cleavage of diarylmercurials in aqueous dioxan and tetrahydrofuran have also been measured by Nerdel and Makover[616], though satisfactory kinetic detail is not available. The reaction was second-order and the effect of aryl substituents was 4-Me > 2-Me > 3-Me > H.

The kinetics of the acid cleavage of arylmercuric halides, $viz.$

$$ArHgX + HY = ArH + HgXY \tag{246}$$

have also been investigated. Reutov $et$ $al.$[617] showed that with solvent dioxan containing 1–5 % water, the second-order rate coefficients decreased rapidly during a kinetic run and this was attributed to the mercuric halide formed, it being shown that addition of mercuric halide markedly lowered the reaction rate. Kinetics were, therefore, carried out in the presence of sodium iodide which complexes with the mercuric halide and with a mole ratio greater than 3.5 moles of iodide to one mole of phenylmercuric, second-order coefficients were maintained for a considerable portion of the reaction and were independent of a fourfold variation in concentration of hydrogen chloride; for subsequent kinetic studies a 7 : 1 molar ratio was employed. Addition of sodium iodide did, however, produce a marked rate acceleration ($ca.$ $10^4$) and this was the enhancement over the $initial$ rate of runs performed with no addition iodide. Consequently, this acceleration did not arise simply from removal of mercuric halide, nor was it considered to be simply a salt effect in view of the rate enhancement observed. The possibility that conversion of the phenylmercuric bromide to the iodide was causing the rate enhancement was eliminated by performing runs with the iodide which gave closely similar rates to those with the bromide, and the possibility that hydrogen iodide was formed and producing a large rate increase was eliminated by performing runs with hydrogen iodide which also gave similar rates. It was, therefore, proposed that a complex was formed between phenylmercuric bromide and sodium iodide, $PhHgBrI^-Na^+$, (spectrophotometric evidence suggesting this) and the rate coefficients indeed showed a dependence on the ratio of these two reagents; the increased reactivity of this complex then arises from increased polarisation of the carbon–mercury bond.

Further investigation of this reaction using dioxan solutions of various water content, indicated that the reaction was second-order in 90–95 vol. % aqueous dioxan, but the order decreased, along with the reaction rate, to one in 60–80 vol. % aqueous dioxan[618]. However, under the latter conditions the reaction rate was still dependent upon the concentration of hydrochloric acid, and it was proposed that here the mechanism is $S_E1$, equilibria (247) and (248), $viz.$

$$[ArHgCl]^-Na^+ \rightleftharpoons Ar^- + [HgCl]Na^+ \quad \text{(slow)} \tag{247}$$

$$Ar^- + HCl \rightleftharpoons ArH + Cl^- \quad \text{(fast)} \tag{248}$$

Tortuous explanations were involved in order to account for the effects of changing water concentration, the addition of chloride ions, and of trichloroacetic and phosphoric acids, in terms of this mechanism; explanation of the specific effects observed on carrying out the reaction in chloroacetic, trichloroacetic, and phosphoric acids proved to be equally difficult and are obviously quite irreconcilable with the mechanism. A detailed account of these effects is considered inappropriate at this time by virtue of the fact that Kreevoy and Hansen[619] found that acid cleavage of alkylmercuric chlorides is profoundly accelerated by the presence of oxygen, that Jensen and Heyman[620] found that kinetic studies on dialkylmercurials are very dependent upon the presence of oxygen, and that Hegarty et al.[621] showed that the $S_E1$ mechanism proposed by Reutov et al.[622] for the cleavage of dibenzylmercury by hydrochloric acid in dioxan arises from an unsatisfactory kinetic investigation which included neglect of the effect of oxygen. In the absence of oxygen satisfactory second-order kinetics were obtained for the reaction in contrast to the first-order kinetics obtained by Reutov et al., confirming the results of an earlier kinetic investigation[616]. It seems very probable, therefore, that oxygen is interfering in the hydrochloric acid cleavage of phenylmercuric bromide and, until this is shown not to be the case, the data of Reutov et al. must be treated with caution.

Reutov et al.[622a] have also examined the effects of anions upon kinetics of the acid-catalysed cleavage of phenylmercuric bromide in dimethylformamide at 17 °C. In the presence of 0.1 $M$ sodium bromide the reaction is again second-order, and is faster with hydrogen chloride ($k_2 = 7.9 \times 10^{-2}$ for a 3-fold variation in acid concentration) than with hydrogen bromide ($k_2 = 6.5 \times 10^{-2}$). Bromide ion is, however, eight times as effective in catalysing the reaction, so that these effects of anions cannot arise from formation of the corresponding acids; this was confirmed by the lack of catalytic effect of picrate ion. The relative reactivities of the halogen acids in this reaction was again regarded as evidence for nucleophilic participation of the acid anion.

Brown et al.[623] have determined the effects of a range of substituents in protodemercuration of arylmercuric chlorides in 10 vol. % aqueous ethanol by hydrochloric acid at temperatures between 50 and 100 °C. Kinetics were second-order overall, being first-order in aromatic and in hydrogen ion. Rates were measured under first-order conditions (acid in excess) and the derived second-order rate coefficients are given in Table 187. The data at 70 °C correlated with $\sigma^+$ values, $\rho$ being $-2.44$; the similarity between this and the factor for acid cleavage of diarylmercurials (p. 282) is a strong indication of a closely similar transition state, hence it is not surprising that both sulphuric and perchloric acids also failed to cleave the arylmercuric chlorides, though addition of chloride or bromide ions produced reaction. As in the case of the cleavage of diarylmercurials it is possible to speculate that reaction occurs via a small equilibrium concentration of halogen acid and the fact that addition of sodium fluoride produced no reaction is not inconsistent, since the hydrogen–fluorine bond length may be too small to facilitate

TABLE 187

RATE COEFFICIENTS ($10^5 k_2$) AND KINETIC PARAMETERS FOR REACTION OF
$XC_6H_4HgCl$ WITH HCl IN 10 VOL. % EtOH[623]

| Temp. (°C) | X | | | | | | |
|---|---|---|---|---|---|---|---|
| | H | 4-Me | 3-Me | 4-Cl | 3-Cl | 4-OMe | 3-OMe |
| 50 | | | | | | 313 | |
| 60 | 5.76 | 31.6 | 8.81 | 2.96 | 1.45 | 822 | 3.24 |
| 70 | 12.0 | 83.6 | 28.3 | 8.04 | 3.13 | 1,770 | 8.53 |
| 80 | 37.2 | 248 | 79.4 | 20.3 | 9.61 | | 19.6 |
| 90 | 95.0 | | | | | | |
| 100 | 256 | | | | | | |
| $E_a$ | 23 | 24 | 25 | 23 | 23 | 19 | 21 |
| $\Delta S^{\ddagger}$ at 70 °C | −12 | −5 | −3 | −14 | −16 | −14 | −18 |

concurrent electrophilic and nucleophilic attack and fluorine is also a weaker nucleophile. A different view has been taken by Brown et al.[623], who envisaged reaction occurring as

$$ArHgCl + H_3O^+ \underset{slow}{\rightleftharpoons} ArHHgCl^+ \underset{fast}{\overset{Cl^-}{\rightleftharpoons}} ArH + HgCl_2 \qquad (249)$$

a reaction which would be zeroth-order (as found) in chloride ion. It was considered that the chloride or bromide-ion catalysis would produce largely undissociated mercuric halide as the reaction product, whereas other ions would produce more dissociated salts and consequently mercury cations which would promote the reverse reaction, mercuration (see p. 186); hence under the latter conditions no reaction would occur. The inability to detect a first-order dependence on chloride ion concentration when the latter is low was attributed to the inability to achieve complete protonation of the aromatic even at high acidities, because of the low basicity of the arylmercuric chloride.

In a subsequent investigation[624], the rates of protodemercuration of the 2 positions of furan, thiophene, and selenophene, and the 3 position of furan by hydrochloric acid in 10 vol. % aqueous ethanol were measured at temperatures between 50 and 80 °C, the kinetic data being given in Table 188. At low chloride ion concentration ($0.01\ M$) and a hydrogen ion concentration of $0.10\ M$ the rates of cleavage of the 3-furyl compound at 80 °C were independent of variations in the chloride ion concentration, but at a lower hydrogen ion concentration of $2 \times 10^{-3}$ $M$, rates of cleavage of the 2-furyl and 2-thienyl compounds at 50 °C varied as the chloride ion concentration varied from 0.002 to 4.00 $M$. This was partly due to salt effect, for addition of sodium perchlorate also caused a significant but much smaller rate increase (approximately one-third of that of the equivalent amount of chloride), and the remaining acceleration was attributed to either reaction oc-

TABLE 188

RATE COEFFICIENTS ($10^3 k_2$) AND KINETIC PARAMETERS FOR REACTION OF RHgCl WITH HCl IN 10 VOL. % EtOH[624]

| Temp. (°C) | R | | | | |
|---|---|---|---|---|---|
| | Phenyl[a] | 2-Furyl[b] | 3-Furyl[c] | 2-Thienyl[b] | 2-Selenophenyl[b] |
| 50 | | 80 | 3.7 | 52 | 125 |
| 55 | | 118 | | 75 | |
| 60 | 0.0576 | 205 | 9.0 | 106 | 252 |
| 65 | | 309 | | 143 | |
| 70 | 0.12 | 494 | 18.0 | 206 | 456 |
| 75 | | 511 | | 280 | |
| 80 | 0.372 | 956 | 36.1 | | 834 |
| $E_a$ | 23 | 18 | 16 | 15 | 14 |
| $\Delta S^{\ddagger}$ | −12 | −9 | −21 | −20 | −18 |

[a] At ionic strength 0.40.
[b] At ionic strength 0.002.
[c] At ionic strength 0.10.

curring *via* hydrogen chloride ion pairs as proposed by Dessy *et al.* (p. 281), or *via* complex anions of the type $RHgCl_2^-$ which would be much more susceptible to electrophilic attack than the compound RHgCl, though the failure to observe a kinetic dependence on chloride ion concentration when this is low seems to argue against this.

The wide variation in the entropy factors for both the substituted phenyl and heterocyclic compounds and in particular for the methoxyphenyl and furan derivatives was considered to be strong evidence for solvent effects being predominant in determining the activation entropy. Consequently, discussion of the substituent effects in terms of electronic factors alone requires caution in this reaction. Caution is also needed since rates for the substituted phenyl compounds were only determined over a 20 °C range. The significance of entropy factors has also been indicated by the poor correlation of the data of the electrophilic reactivities of the heterocyclic compounds, as derived from protodemercuration, with the data for other electrophilic substitutions and related reactions[572].

## 9.3 PROTODEBORONATION

The kinetics of protodeboronation are rather complex, probably due to the occurrence of more than one reaction mechanism depending to some extent on the medium involved. Much of the work has been devoted to showing that a linear log rate *versus* $H_0$ plot does *not* mean that the A-1 mechanism applies.

Kuivila and Nahabedian[625] showed that with aqueous sulphuric acid the logarithms of the first-order rate coefficients for the protodeboronation of 4-

TABLE 189

VARIATION OF RATE COEFFICIENTS WITH ACIDITY FOR REACTION OF 4-MeOC$_6$H$_4$B(OH)$_2$ WITH H$_2$SO$_4$, H$_3$PO$_4$ AND HClO$_4$[625]

| $H_2SO_4$ (wt %) | 16.1 | 29.4 | 40.6 | 50.4 | 54.6 | 20.1 | 24.7 | 30.1 |
|---|---|---|---|---|---|---|---|---|
| $-H_0$ | 0.67 | 1.56 | 2.37 | 3.21 | 3.63 | 0.95 | 1.28 | 1.62 |
| $10^6 k_1$ (25 °C) | 1.18 | 12.4 | 119 | 931 | 2,530 | | | |
| $10^6 k_1$ (40 °C) | | | | | | 13.5 | 29.9 | 77.8 |

| $H_2SO_4$ (wt %) | 3.11 | 5.14 | 10.1 | 20.3 | 29.6 |
|---|---|---|---|---|---|
| $-H_0$ | −0.37 | −0.11 | 0.37 | 1.00 | 1.60 |
| $10^6 k_1$ (60 °C) | 4.11 | 7.57 | 23.0 | 115 | 560 |

| $HClO_4$ (wt %) | 30.3 | 40.7 | 44.0 | 50.5 | 56.2 | 18.6 | 25.5 | 30.95 | 33.85 |
|---|---|---|---|---|---|---|---|---|---|
| $-H_0$ | 1.37 | 2.23 | 2.58 | 3.48 | 4.40 | 0.72[a] | 1.10[a] | 1.43[a] | 1.63[a] |
| $10^6 k_1$ (25 °C) | 6.53 | 46.5 | 92.7 | 470 | 2,790 | | | | |
| $10^6 k_1$ (60 °C) | | | | | | 8.22 | 22.2 | 51.1 | 76.6 |

| $H_3PO_4$ (wt %) | 36.0 | 45.5 | 53.6 | 60.9 | 61.2 | 67.9 |
|---|---|---|---|---|---|---|
| $-H_0$ | 0.40 | 0.82 | 1.18 | 1.56 | 1.58 | 2.06 |
| $10^6 k_1$ (25 °C) | 4.10 | 20.7 | 101 | 753 | 910 | 5,720 |

| $H_3PO_4$ (wt %) | 10.6 | 27.1 | 39.9 | 49.2 | 57.5 |
|---|---|---|---|---|---|
| $-H_0$ | 0.0 | 0.0 | 0.50 | 0.82 | 1.22 |
| $10^6 k_1$ (60 °C) | 3.40 | 30.8 | 195 | 910 | 4,640 |

[a] Values at 25 °C, those at 60° being unavailable.

methoxybenzene-boronic acid (Table 189) correlated with $H_0$ with a slope of −1.15[625]. The rate coefficients were higher than those obtained for hydrogen exchange in the same media (see p. 203) and this is consistent with the lower strength of the carbon–boron relative to the carbon–hydrogen bond.

Measurement of the rates of protodeboronation of 4-methoxybenzeneboronic acid by aqueous perchloric and phosphoric acids (Table 189) revealed that the data at 25 °C gave moderately linear plots of log $k$ versus $-H_0$ with slopes of 0.85 and 1.85, respectively, but at 60 °C the data for phosphoric acid gave a distinct curve, as did that for perchloric acid. (This plot was not shown but is curved in the same direction as the corresponding plot at 25 °C). The different slope obtained with different acids, and the markedly different reaction rates obtained for different acids of the same acidity function serve to show that apparently linear correlations of logarithms of rate coefficients with acidity function (with approximately unit slope) is a poor indication of specific acid catalysis (see also hydrogen exchange pp. 196–222). In addition, the curve plot obtained with phosphoric acid at 60 °C had a decreasing slope with decreasing acidity, which was more consistent with general acid catalysis by $H_3O^+$ and $H_3PO_4$, and this is considered further below.

The kinetics of protodeboronation of a range of substituted benzeneboronic acids in aqueous sulphuric acid mixtures were also examined. Good first-order kinetics were obtained in all cases except for the 3-fluoro compound (due to the sulphonation side reaction) and the 3-trifluoromethyl compound, which hydrolysed, hydrogen fluoride being produced; the rates with this latter compound were only followed to 30 % reaction. The kinetic details are summarised in Table 190, from

TABLE 190

VARIATION IN SLOPE OF LOG $k_1$ *versus-*$H_0$ PLOTS WITH ACIDITY, TEMPERATURE AND SUBSTITUENT IN THE REACTION OF $XC_6H_4B(OH)_2$ WITH $H_2SO_4$[625]

| X | Temp. (°C) | $H_2SO_4$ (wt %) | Slope $log\ k_1/-H_0$ | X | Temp. (°C) | $H_2SO_4$ (wt %) | Slope $log\ k_1/-H_0$ |
|---|---|---|---|---|---|---|---|
| 4-MeO | 60 | 3–30 | 1.10 | 4-Br | 60 | 48–70 | 0.87 |
|  | 40 | 20–30 | 1.15 |  | 60 | 70–84 | 1.00 |
|  | 25 | 16–55 | 1.15 | 3-F | 79.4 | 57–65 | 0.71 |
| 4-Me | 60 | 29–56 | 1.03 |  | 69.4 | 57–65 | 0.69 |
|  | 40 | 43–54 | 1.06 |  | 60 | 55–70 | 0.72 |
|  | 25 | 43–54 | 1.05 |  | 60 | 70–84 | 1.00 |
| 4-F | 60 | 50–65 | 0.86 |  | 40 | 80–84 | 1.14 |
|  | 40 | 53–62 | 0.85 |  | 25 | 79–83 | 1.35 |
|  | 25 | 53–63 | 0.86 | 3-Cl | 60 | 59–70 | 0.65 |
| H | 60 | 41–75 | 0.90 |  | 60 | 70–80 | 1.00 |
|  | 40 | 71–74 | 1.10 | 3-$NO_2$ | 60 | 75–97 | 0.84 |
|  | 25 | 70–74 | 1.16 |  |  |  |  |

which it is immediately obvious that no satisfactory linear free energy correlation of this data is possible in view of the random correlations of logarithms of rates with acidity functions; a correlation would only be possible if the variation in slope with reactivity was regular, the least reactive compound having the highest slope. For the 3-chloro and 3-fluorobenzeneboronic acids, the correlation of log rate against $H_0$ suggested two straight lines with a distinct change of slope indicating a change of reaction mechanism and this is discussed further below.

It is, however, very difficult to decide whether two apparently straight lines are, in fact, merely a curve and this can only be satisfactorily resolved by measurements over an extended range of acidity which is usually precluded by the extremely high or low rates that would result. Support for a change in mechanism was considered to be provided by the kinetic parameters given in Table 191, some of which is derived from rates obtained by extrapolation of the *assumed linear* log $k_1$ *versus* $-H_0$ plots. For example, the activation energy for the 3-fluoro compound is considerably lower for the > 70 % acid media than for the < 70 % acid media. Now whilst these changes in kinetic parameters may indicate changes in mechanism, they also indicate that unsatisfactory linear free-energy correlations

TABLE 191

VARIATION IN KINETIC PARAMETERS WITH SUBSTITUENT AND ACID FOR REACTION
OF $XC_6H_4B(OH)_2$ WITH $H_2SO_4$, $H_3PO_4$, OR $HClO_4$[625]

| $X$ | Medium (wt %) | $E_a$ | $\Delta S^{\ddagger}$ | $X$ | Medium (wt %) | $E_a$ | $\Delta S^{\ddagger}$ |
|---|---|---|---|---|---|---|---|
| 4-MeO | 30 % $HClO_4$ | 23.6 | − 5.2 | 4-F | 30 % $H_2SO_4$ | 22.3 | −22.4 |
| | 30 % $H_2SO_4$ | 21.1 | −12.0 | | 55 % $H_2SO_4$ | 21.7 | −16.2 |
| | 38 % $H_3PO_4$ | 18.1 | −23.9 | 3-F | 55 % $H_2SO_4$ | 23.1 | −19.5 |
| | 48 % $H_3PO_4$ | 17.5 | −22.6 | | 64 % $H_2SO_4$ | 23.0 | −18.7 |
| | 58 % $H_3PO_4$ | 15.0 | −26.1 | | 81 % $H_2SO_4$ | 18.4 | −20.3 |
| 4-Me | 30 % $H_2SO_4$ | 21.1 | −22.6 | | 83 % $H_2SO_4$ | 17.1 | −22.9 |
| | 55 % $H_2SO_4$ | 20.2 | −15.5 | H | 72 % $H_2SO_4$ | 18.3 | −19.8 |
| | 58 % $H_3PO_4$ | 18.2 | −23.8 | | 74 % $H_2SO_4$ | 18.1 | −18.7 |
| | 68 % $H_3PO_4$ | 17.1 | −22.3 | | | | |
| | 72 % $H_3PO_4$ | 15.7 | −25.1 | | | | |

will be obtained. This is, in fact, observed, for whilst the data for 74.5 wt. %
sulphuric acid at 60 °C gives a good correlation with $\sigma^+$ values ($\rho = -5.2$), the
data for 55.4 wt. % sulphuric acid gives a curve, the more deactivated compounds
reacting at a faster rate than would be expected for a linear correlation, and as
expected as a consequence of the relatively low slopes of the log $k_1$ *versus* $-H_0$
plots for the less reactive compounds in the weaker acid media.

To examine more fully the possibility of the general acid catalysis noted above.
the kinetics of protodeboronation in aqueous phosphate buffers were measured[625],
A more reactive substrate, 2,6-dimethoxybenzeneboronic acid was necessary and a
preliminary examination using perchloric acid at 25 and 60 °C (Table 192) showed

TABLE 192

RATE COEFFICIENTS FOR REACTION OF $2,6\text{-}(MeO)_2C_6H_3B(OH)_2$ WITH $HClO_4$[625]

| Temp. (°C) | $10^6k_1$ | $10^3k_2$ | $-log\ [H^+]$ | $-H_0$ | Ionic strength |
|---|---|---|---|---|---|
| 25 | 7.75 | | −0.009 | 0.20 | |
| | 14.7 | | −0.167 | 0.51 | |
| | 24.0 | | −0.292 | 0.76 | |
| | 39.5 | | −0.389 | 0.99 | |
| 60 | 1.53 | 2.23 | 2.166(2.13)[a] | | 0.50 |
| | 2.27 | 2.32 | 2.011(2.03)[a] | | 0.50 |
| | 4.85 | 2.47 | 1.709(1.70)[a] | | 0.50 |
| | 12.1 | 2.47 | 1.311(1.33)[a] | | 0.50 |
| | 23.5 | 2.41 | 1.011(1.00)[a] | | 0.50 |
| | 22.7 | 2.30 | 1.011 | | 0.30 |
| | 22.1 | 2.26 | 1.011 | | 0.20 |
| | 21.4 | 2.19 | 1.011 | | 0.10 |

[a] Measured pH.

that in dilute solution at 60 °C, the second-order rate coefficients (obtained by dividing the first-order rate coefficients by the hydrogen ion concentration) gave a constant value of *ca.* $2.38 \times 10^{-3}$ (at constant ionic strength of 0.50 obtained by the addition of sodium perchlorate), *i.e.* the rate depended on the hydronium ion concentration. At a constant hydronium ion concentration, the rate increased only 10 % for a 5-fold increase in ionic strength, as might be expected for a reaction of similar polarity in ground and transition states. At acid concentrations above 1 *M*, rates (25 °C) correlated with acidity function rather than with hydronium ion concentration, the slope of the log $k_1$ *versus* $-H_0$ plot being 0.90. With aqueous buffer solution (constant phosphoric acid concentration of 0.1021 *M*, and variable dihydrogen phosphate concentration) rate coefficients (60 °C; $\mu = 0.50$), were not linearly correlated with hydronium ion concentration, and the difference in this rate and that expected from the measurements with perchloric acid, when divided by the actual concentration of phosphoric acid, gave reasonably constant values (Table 193) indicating general acid catalysis by molecular phosphoric acid.

TABLE 193

RATE COEFFICIENTS FOR REACTION OF $2,6\text{-}(MeO)_2C_6H_3B(OH)_2$ WITH PHOSPHATE BUFFERS AT 60°C[625]

| $[NaH_2PO_4]$ (M) | pH | $10^6 k_1$ | $[H_3PO_4]^b$ | $k_2(H_3PO_4)^a$ |
|---|---|---|---|---|
| 0.100 | 2.09 | 3.16 | 0.094 | 1.30 |
| 0.200 | 2.32 | 2.24 | 0.097 | 1.13 |
| 0.200 | 2.32 | 2.14 | 0.097 | 1.03 |
| 0.400 | 2.69 | 1.82 | 0.100 | 1.33 |
| 0.500 | 2.73 | 1.90 | 0.100 | 1.46 |
| 0.500 | 2.78 | 1.81 | 0.100 | 1.41 |

$^a$ $k_2(H_3PO_4) = [k_1 - 2.38 \times 10^{-3} \text{ antilog } (-pH)]/[H_3PO_4)]$.
$^b$ Stoichiometric concentration of phosphoric acid $[H_3PO_4] = [0.1021 - \text{antilog}(-pH)]$.

Extension of these studies to formic acid media (containing 4 vol. % ethylene glycol and 1.3 vol. % water) showed that for protodeboronation of 4-methoxybenzeneboronic acid at 25 °C) rates were invariant of a tenfold variation in acidity produced by adding sodium formate (0.05–0.20 *M*) to the medium (Table 194), and in this range the concentration of molecular formic acid is essentially constant. This was, therefore, assumed to be the reactive species. At higher acidities the rate increased, which was attributed to the increase in concentration of hydronium ions and protonated formic acid ions which bring about reaction more readily[625].

Thus the above observations indicate that the reaction mechanism is not A-1 and confirmation was provided by measurement of the solvent isotope effect $k_H/k_D$ for protodeboronation of 4-methoxybenzeneboronic acid in 6.3 *M* sulphuric

TABLE 194

RATE COEFFICIENTS FOR REACTION OF $4\text{-MeOC}_6\text{H}_4\text{B(OH)}_2$ WITH
HCOOH (94.7)–$H_2O$ (1.3)–$C_2H_4(OH)_2$(4.0)[a], AT 60 °C[625]

| $[HCO_2Na]$ (M) | $[H_2SO_4]$ (M) | $-H_0$ | $10^5 k_1$ |
|---|---|---|---|
| 0.1963 | | −0.84 | 1.86 |
| 0.0982 | | −0.14 | 1.78 |
| 0.0491 | | 0.17 | 1.89 |
| 0.0295 | | 0.33 | 2.12 |
| 0.00982 | | 0.72 | 3.06 |
| 0.00491 | | 0.92 | 3.61 |
| | | 1.31 | 4.91 |
| | | 1.31 | 4.67 |
| | | 1.31 | 4.58 |
| | 0.01032 | 2.00 | 7.87 |
| | 0.0258 | 2.34 | 11.8 |
| | 0.0413 | 2.51 | 14.5 |
| | 0.0482 | 2.58 | 17.9 |

[a] Vol. %.

acid at 25 °C (3.7) and 2,6-dimethoxybenzeneboronic acid in 0.1 $M$ perchloric
acid at 60 °C (1.65 ≡ 2.0 at 25 °C), the rate coefficients being given in Table
195[625]. Since a substrate will be converted to its conjugate acid to a greater extent
in $D_2O$ than in $H_2O$, then according to the A-1 mechanism the reaction should
go faster in the former medium and since it does not this mechanism may be
eliminated. The data are, however, consistent with the A-$S_E$2 mechanism, since
this predicts a faster rate in $H_2O$ through the proton being more loosely bound

TABLE 195

SOLVENT ISOTOPE EFFECTS FOR REACTION OF ARYLBORONIC ACIDS WITH ACID IN
$D_2O$ OR $H_2O$[625]

4-Methoxybenzene-
boronic acid in 6.31 $M$
$H_2SO_4$ at 25 °C

2,6-Dimethoxybenzeneboronic acid in $HClO_4$ at 60 °C

| $D/(H+D)$ | $10^4 k_1$ | $[HClO_4](M)$ | $D/(H+D)$ | $10^5 k_1$ | $10^3 k_2$[a] | $\mu$ |
|---|---|---|---|---|---|---|
| 0 | 3.45 | 0.0975 | 0 | 23.5 | 2.41 | 0.50 |
| 0.221 | 3.02 | 0.104 | 0.198 | 24.4 | 2.35 | 0.50 |
| 0.443 | 2.42 | 0.104 | 0.396 | 23.8 | 2.29 | 0.50 |
| 0.598 | 1.99 | 0.104 | 0.595 | 22.3 | 2.15 | 0.50 |
| 0.799 | 1.50 | 0.104 | 0.793 | 19.8 | 1.91 | 0.50 |
| 0.995 | 0.94 | 0.1013 | 0.99 | 14.2 | 1.42 | 0.50 |
| | | 0.0507 | 0.99 | 7.12 | 1.42(1.43)[b] | 0.46 |
| | | 0.0203 | 0.99 | 2.95 | 1.47(1.49)[b] | 0.43 |

[a] Misreported as $10^{-3}k_2$.
[b] Values obtained by extrapolation to $\mu = 0.50$.

in the transition state than in the ground state. The greater isotope effect observed in sulphuric acid was attributed to the species $HSO_4^-$ being a significant (and less reactive) protonating species in this medium. This means that the transition state for such a reaction will be nearer to the Wheland intermediate. Consequently, the C–H bond formation will be greater, resulting in the larger isotope effect as observed. However, an additional contributing factor must be the lower reactivity of the 4-methoxy compound. Consequently the transition state for reaction of this is likely to be nearer to the Wheland intermediate than for reaction of the 2,6-dimethoxy compound.

A plot of the rate coefficients (abscissa) for the 4-methoxy compound against the isotopic composition of the medium (ordinate) gives a straight line, whereas the Gross–Butler theory (see p. 6) predicts a curve, convex upwards for the A-1 and A-2 mechanisms. The validity of this prediction is now very suspect, but since in the present case a straight line was obtained these mechanisms may be eliminated for this reaction. The greater reactivity of the 2,6-dimethoxy compound permitted the use of more dilute acid solutions and the calculation of second-order rate coefficients. A plot of these coefficients against the isotopic composition of the medium gave a curve, convex upwards, but the values of the coefficients were all less than predicted by the Gross–Butler theory, but could, however, be predicted by a modification[626] of this theory applicable to reactions with a rate-determining proton transfer, i.e. the A-$S_E$2 mechanism was again indicated.

The data in Table 191 show a large difference in entropy of activation for reaction of the 4-methoxy compound in the three acids. The difference in the data for perchloric and sulphuric acids was again thought to arise from a significant amount of reaction in the latter occurring via bisulphate anion. This anion was proposed as being insufficiently electrophilic to react except by coordination with the side chain, as in scheme (250)

$$\underset{\text{(fast)}}{\overset{B(OH)_2}{\bigcirc} + HSO_4^- \rightleftharpoons} \quad \underset{\text{(slow)}}{\overset{\overset{OH}{|}}{HO-B-OSO_3H}{\bigcirc} \rightleftharpoons} \quad \overset{HO\quad OH}{\underset{H}{B-OSO_3^-}} \qquad (250)$$

and this more highly ordered transition state results in a more negative entropy of activation. The less negative entropy of activation (for the 4-methyl compound) observed with 55 % than with 30 % sulphuric acid points to the lesser importance of the bisulphate ion mechanism at high acidities. The increasingly negative activation entropies observed for a given concentration of sulphuric acid on going to the less reactive aromatics (compare the data for the 4-methoxy and 4-methyl compounds in 30 % sulphuric acid) was likewise regarded as a consequence of the increasing importance of this A-$S_E i$ mechanism (250) for the latter. It also follows that for the less reactive compounds the log $k$ versus $-H_0$ plots will have linear slopes at lower acidities (relative to the more reactive compounds) and this was observed (Table 190).

Measurements of the solvent isotope effect for different aromatic substrates in a range of sulphuric acid media (Table 196) showed the values to increase as

TABLE 196

SOLVENT ISOTOPE EFFECTS FOR REACTION OF $XC_6H_4B(OH)_2$ WITH $H_2SO_4$ IN $D_2O$ OR $H_2O$[625]

| $X$ | $H_2SO_4$ (wt %) | $k_H/k_D$ | $X$ | $H_2SO_4$ (wt %) | $k_H/k_D$ |
|---|---|---|---|---|---|
| 4-MeO | 22.4 | 2.02 | 4-F | 60.6 | 2.16 |
|  | 32.5 | 1.93 |  | 65.4 | 2.27 |
| 4-Me | 40.0 | 1.60 | 3-F | 64.0 | 2.40 |
|  | 45.0 | 1.64 |  | 68.0 | 2.48 |
|  | 50.0 | 1.69 |  | 72.0 | 2.68 |
|  | 55.0 | 1.74 |  | 76.0 | 2.99 |
|  |  |  |  | 80.0 | 2.99 |

the reactivity of the substrate decreases and as expected since the transition state will be nearer to the Wheland intermediate and the degree of proton transfer to the aromatic increased[625]. The levelling off of the isotope effect for the 3-fluoro compound in sulphuric acid of concentration > 76 % was regarded as additional evidence for the described change in mechanism, it being argued that the A-$S_E$i mechanism would produce a smaller isotope effect. Once the bisulphate anion is bonded to boron, the ring carbon to which boron is attached becomes much more nucleophilic, consequently transfer of the side-chain proton in the rate-determining step takes place more readily and the decreased isotope effect results. Likewise, the smaller isotope effect for the 4-methyl relative to the 4-methoxy compound would follow from the postulated greater incursion of the A-$S_E$i mechanism in the reaction with the former[625].

A third mechanism of protodeboronation has been detected in the reaction of benzeneboronic acids with water at pH 2–6.7[625]. In addition to the acid-catalysed reaction described above, a reaction whose rate depended specifically on the concentration of hydroxide ion was found. In a preliminary investigation with aqueous malonate buffers (pH 6.7) at 90 °C, 2-, 4-, and 2,6-di-methoxybenzeneboronic acids underwent deboronation and followed first-order kinetics. A secondary reaction produced an impurity which catalysed the deboronation, but this was unimportant during the initial portions of the kinetic runs.

That the reaction was not catalysed by the buffer anion is shown by the data in Table 197, which gives the rate coefficients observed $k_{1\ obs}$ and the rate coefficients $k_{1\ corr}$ corrected for difference in pH and ionic strength to values of 6.70 and 0.14 respectively. The existence of the general acid-catalysed mechanism for the reaction was demonstrated by the data in Table 198, which gives the rate coeffi-

TABLE 197

RATE COEFFICIENTS FOR REACTION OF $2,6\text{-}(MeO)_2C_6H_3B(OH)_2$ WITH MALONATE BUFFERS AT 90 °C[626]

| $10^3[CH_2(COO^-)COONa]$ $M$ | $\mu$ | $pH$ | $10^6k_{1obs}$ | $10^6k_{1corr}$[a] |
|---|---|---|---|---|
| 0.155 | 0.04 | 6.70 | 22.0 | 22.0 |
| 0.543 | 0.14 | 6.70 | 18.1 | 18.1 |
| 0.764 | 0.20 | 6.70 | 30.2 | 18.8 |
| 1.55 | 0.40 | 7.15 | 53.4 | 17.9 |

[a] See text.

cients observed $k_{1\,obs}$ for the hydrolysis of 2,6-dimethoxybenzeneboronic acid at a buffer ratio $(H_2A/HA^-) = 0.155$, and those $k_{1\,corr}$ corrected to pH 3.60 and $\mu = 0.14$; a plot of $k_{1\,corr}$ (not $k_{1\,obs}$ as stated in the paper) *versus* [malonic acid] gave a reasonable correlation line which was slightly curved for an unidentified

TABLE 198

VARIATION IN RATE COEFFICIENTS FOR REACTION OF $2,6\text{-}(MeO)_2C_6H_3B(OH)_2$ WITH CONCENTRATION OF MALONIC ACID IN MALONATE BUFFERS AT 90 °C[626]

| $10^3[CH_2(COOH)_2]$ $(M)$ | $pH$ | $10^6k_{1obs}$ | $10^6k_{1corr}$[a] | $10^3[CH_2(COOH)_2]$ $(M)$ | $pH$ | $10^6k_{1obs}$ | $10^6k_{1corr}$ |
|---|---|---|---|---|---|---|---|
| | 3.60 | | 5.13 | 6.00 | 3.39 | 18.1 | 15.3 |
| 0.27 | 3.70 | 5.11 | 5.87 | 6.00 | 3.55 | 12.9 | 12.5 |
| 0.58 | 3.67 | 5.12 | 6.67 | 6.00 | 3.58 | 11.8 | 11.6 |
| 1.17 | 3.67 | 8.60 | 9.15 | 11.7 | 3.61 | 21.9 | 22.0 |
| 1.38 | 3.59 | 8.92 | 8.84 | 18.0 | 3.65 | 29.9 | 30.3 |
| 1.90 | 3.54 | 9.64 | 9.11 | 47.4 | 3.62 | 55.7 | 54.2 |
| 2.74 | 3.59 | 12.5 | 12.4 | 50.4 | 3.50 | 57.5 | 54.6 |
| 3.16 | 3.72 | 7.25 | 8.14 | 56.0 | 3.55 | 46.9 | 45.1 |
| 3.66 | 3.56 | 14.0 | 13.7 | 74.6 | 3.58 | 76.6 | 75.0 |
| 3.90 | 3.76 | 9.27 | 10.5 | | | | |

[a] See text.

reason. It is, however, apparent from comparison of this plot with the data in Table 198 that many of the less satisfactory points were omitted from the plot and in view of the very large experimental error that the data indicate, it could well be that the correlation line is, in fact, straight.

Examination of the effect of pH on the rates of protodeboronation of the 2,6-dimethoxy compound at 90 °C in malonic acid–sodium malonate buffer solutions of ionic strength 0.14 gave the data in Table 199. A plot of these data revealed the curve shown in Fig. 3 (one of the points was misplotted on the original) and the linear portions of the plot were attributed to acid and base catalysis as shown on Fig. 3, and since the rates in the region of pH 4–5 are higher than would be

TABLE 199

VARIATION IN RATE COEFFICIENTS WITH pH FOR REACTION OF $2,6\text{-}(MeO)_2C_6H_3B(OH)_2$ WITH MALONATE BUFFERS AT 90 °C $(\mu = 0.14)$[626]

| pH | 6.72 | 6.42 | 6.42[a] | 6.05 | 5.85 | 5.45 | 4.42 | 3.81 | 3.55 | 3.00[b] | 2.70[b] | 2.40[b] | 2.00[b] |
|---|---|---|---|---|---|---|---|---|---|---|---|---|---|
| $10^6k_1$ | 18.1 | 9.36 | 9.20 | 4.25 | 3.05 | 2.04 | 1.78 | 3.66 | 5.96 | 20.7 | 34.5 | 90.2 | 230 |

[a] In presence of $3.45 \times 10^{-3}$ $M$ boric acid.
[b] $HClO_4$ used for these runs.

expected from a combination of the acid and base-catalysed reactions, it was assumed that a pH-independent reaction also operates and the first-order rate of this reaction was quoted as $0.63 \times 10^{-6}$. This value is, however, clearly wrong since

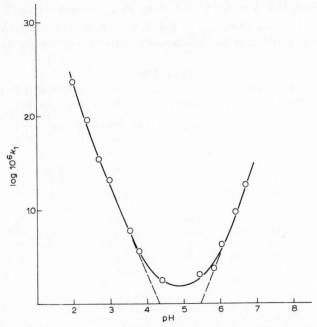

Fig. 3. Variation of rate with pH for protodeboronation of 2,6-dimethoxybenzeneboronic acid.

subtraction of this rate from the observed rates does not give values which give linear log rate *versus* pH plots as required; linear plots are, however, obtained if the pH-independent reaction has a first-order rate of $1.1 \times 10^{-6}$. Even so it is not clear how good first-order kinetics could have been obtained throughout the pH range if two simultaneous reactions of dissimilar rate were taking place.

The mechanism of the base-catalysed reaction was proposed as equilibria (251) and (252)

$$ArB(OH)_2 + H_2O \overset{K_a}{\rightleftharpoons} ArB(OH)_3^- + H^+ \qquad \text{(fast)} \qquad (251)$$

$$ArB(OH)_3^- + H_2O \rightleftharpoons ArH + B(OH)_3 + OH^- \quad \text{(slow)} \tag{252}$$

since the equilibrium concentration of the boronate anion would be proportional to pH. However, one would have expected addition of boric acid to have reduced the reaction rate if this were so, but the data in Table 199 have shown this not to be the case. An alternative mechanism would be the A-$S_E i$ mechanism previously proposed, with $HO^-$ replacing $HSO_4^-$ as the proton source.

In this reaction the effect of changing the ionic strength of the medium was also studied by addition of sodium perchlorate which caused the rate coefficient to *decrease*. However, this was shown to arise from the resultant change in pH of the solution, for when this was allowed for, the normal positive salt effect was observed.

Rate coefficients for protodeboronation of a number of substituted benzene-boronic acids were measured by Kuivila *et al.*[627] at pH 6.70 and 6.42 at 90 °C, $\mu = 0.14$. The relative rates at the two pH values were reasonably constant, indicating that the same reaction was being studied for each compound (Table 200). The results indicate that all substituents increase the rate of reaction but

TABLE 200

RATE COEFFICIENTS FOR THE REACTION OF $XC_6H_4B(OH)_2$ WITH MALONATE BUFFERS AT 90 °C[627]

| X | $10^6 k_1$ | | $10^6 k'_2$ | X | $10^6 k_1$ | | $10^6 k'_2$ |
|---|---|---|---|---|---|---|---|
| | *pH 6.70* | *pH 6.42* | | | *pH 6.70* | *pH 6.42* | |
| H | 0.145 | 0.08 | 0.688 | 2-F | 11.1 | | 7.14 |
| 4-MeO | 0.610 | 0.322 | 8.54 | 3-F | 0.332 | | 0.208 |
| 2-MeO | 1.53 | 0.772 | 25.8 | 4-Cl | 0.187 | 0.10 | 0.322 |
| 4-Me | 0.260 | 0.145 | 3.10 | 2-Cl | 8.60 | 4.76 | 12.0 |
| 2-Me | 0.360 | 0.208 | 12.6 | 3-Cl | 0.217 | | 0.152 |
| 3-Me | 0.294 | | 1.50 | 2,6-(MeO)$_2$ | 18.1 | 9.36 | 2,200 |
| 4-F | 0.250 | 0.141 | 0.966 | | | | |

this is because the observed rates are the product of the equilibrium constant for equilibrium (251) multiplied by the forward rate of equilibrium (252). Accurate values of $K_a$ are not known for this solvent system but using the values for 25 % methanol at 25 °C, the values of $k'_2$ were calculated as given in Table 200. These values now plot against the Yukawa–Tsuno equation[612] with $r = ca.$ 0.5 and $\rho = -2.32$. The point for the unsubstituted compound lies well off the straight line, but in view of the estimations made in deriving the values of $k'_2$ this may not be real. The smaller rho factor than was found for the acid-catalysed reaction was attributed to the aromatic being negatively charged in the rate-determining step and thus less stabilisation or destabilisation of the transition state by substituents in the aromatic ring would be required. Finally, this reaction gives rise to very high *ortho* : *para* ratios in marked contrast to the acid-catalysed reaction. Release of steric strain in the rate-determining transition state would obviously be a major factor for the proposed mechanisms, and electrostatic and hydrogen bonding interactions have also been involved to help explain the results, since the expected order of relief of steric strain does not parallel the decrease in the *ortho* : *para* ratio observed.

The effect of metal-ion catalysis (especially that of cadmium ion) in the above reaction has been studied[628], and in Table 201 are listed the first-order rate coefficients for protodeboronation of 2,6-dimethoxybenzeneboronic acid in malonic acid–sodium malonate buffer or perchloric acid, observed in the absence ($k_1$) or presence ($k_1^{Cd}$) of cadmium ion, together with the second-order rate coefficients ($k_2$) obtained by dividing the difference of these values by the cadmium ion concentration. The data of the first ten rows of Table 201 are plotted in Fig. 4 and the

TABLE 201

RATE COEFFICIENTS[628] FOR REACTION OF $2,6\text{-}(MeO)_2C_6H_3B(OH)_2$ WITH MALONATE BUFFER OR $HClO_4$ IN THE PRESENCE OF $Cd^{2+}$ AT 90 °C

| $10^3[Cd^{2+}]$ (M) | pH | $10^6 k_1$ | $10^6 k_1^{Cd}$ | $10^3 k_2$ |
|---|---|---|---|---|
| 0.10 | 2.2 | 151 | 176 | 250 |
| 0.10 | 2.47 | 67.4 | 82.4 | 150 |
| 0.10 | 2.91 | 35.1 | 42.1 | 70 |
| 0.10 | 3.10 | 18.2 | 19.5 | 13 |
| 0.10 | 3.18 | 10.4 | 15.4 | 50 |
| 0.10 | 3.62 | 5.10 | 13.4 | 83 |
| 0.10 | 4.42 | 1.78 | 11.8 | 100 |
| 0.10 | 5.45[a] | 2.04 | 47.5 | 455 |
| 0.10 | 6.05[a] | 4.25 | 140 | 1,360 |
| 0.10 | 6.70[a] | 18.1 | 698 | 6,800 |
| 1.00 | 2.24[a] | 168 | 190 | 22 |
| 1.00 | 3.18 | 10.4 | 32.2 | 22 |
| 5.00 | 3.19 | 10.3 | 66.2 | 11 |
| 10.0 | 3.13 | 13.6 | 98.4 | 8.5 |
| 50.0 | 3.17 | 17.2 | 312 | 6.0 |
| 100 | 3.16 | 16.1 | 442 | 4.3 |
| 10.0 | 2.05 | 200 | 214 | 1.4 |
| 1.0 | 2.01 | 223 | 240 | 17 |
| 0.10 | 1.91 | 245 | 246 | |

[a] Malonic acid–sodium malonate buffer used in these experiments only.

curve of the reaction in the absence of cadmium ion is virtually superimposable upon that in Fig. 3. Analysis of the curve for the cadmium ion-catalysed reaction into two straight lines as before indicates the presence of a catalysed pH-independent reaction, the first-order rate coefficient of which was estimated[628] as $6 \times 10^{-6}$, but in fact a better fit is obtained with a value of $7 \times 10^{-6}$ compared

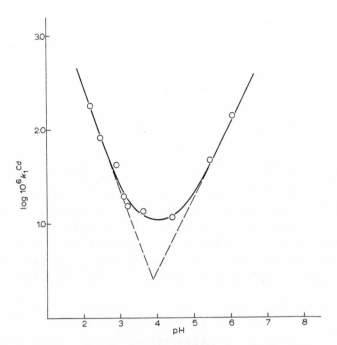

Fig. 4. Variation of rate with pH for cadmium ion-catalyzed protodeboronation of 2,6-dimethoxy-
benzeneboronic acid.

with the value of $1 \times 10^{-6}$ obtained without the catalyst. Thus the pH-independent
reaction is catalysed by cadmium ion, and it is obvious from Fig. 4 that the base-
catalysed reaction is strongly catalysed by cadmium ion whilst the acid-catalysed
reaction appears to be barely so. Conversion of the first-order rate coefficient for
the pH-independent reaction to the second-order coefficient gives a value of $70 \times 10^3$
and subtraction of this from each value in the last column of Table 201 gives the
true indication of the catalysis by cadmium ion which is relatively insignificant
for the acid-catalysed reaction compared to the base-catalysed reaction. (The fact
that this subtraction renders the values negative at pH 3.10 and 3.18 is almost
certainly due to erroneously low rate coefficients obtained for the cadmium ion-
catalysed reaction, and this is apparent from the graph.) Confirmation of the
marked rate acceleration in the region of higher pH was obtained with 2-methoxy-
benzeneboronic acid, which at pH 6.70 gave first-order rate coefficients of
$42.4 \times 10^{-6}$ and $203 \times 10^{-6}$ for addition of $1 \times 10^{-4}$ and $5 \times 10^{-4}$ $M$ respectively of
cadmium ion, whereas in the absence of this a value of $1.53 \times 10^{-6}$ was obtained.

Rows 12–16 in Table 201 show that the acceleration by cadmium ion is not
linear with concentration, becoming proportionally less as the concentration is
increased. The last entry shows the magnitude of the experimental error in the
acid-catalysed portion of the reaction for this result now indicates that cadmium
ion has no effect on the acid-catalysed reaction.

No firm conclusions were drawn with regard to the mechanism of the catalysed reaction, but the proposal of Ainley and Challenger[629], equilibria (253) and (254),

$$ArB(OH)_2 + CdX_2 \overset{H_2O}{\rightleftharpoons} ArCdX + B(OH)_3 + HX \qquad \text{(slow)} \qquad (253)$$

$$ArCdX + H_2O \rightleftharpoons ArH + CdXOH \qquad \text{(fast)} \qquad (254)$$

was considered to be not unreasonable in view of the fact that mercury(II), silver, and thallium(III) salts react with areneboronic acids to form arylmetals. The alternative possibility, namely, that the protodeboronation product is formed directly through proton transfer to the aromatic from one of the ligand water molecules of the metal ion, was considered to be less likely. The fact that the acid-catalysed reaction rate was unaffected by the presence of cadmium ion strongly suggests that whatever the catalysed mechanism is, it does not operate in media of low pH because this marked change in mechanism would undoubtedly produce an observable change in rate.

The effects of a range of substituents have been measured in this reaction, with

TABLE 202

RATE COEFFICIENTS FOR THE $Cd^{2+}$-CATALYSED REACTION OF $XC_6H_4B(OH)_2$ WITH MALONATE BUFFER AT 90 °C[628].

| $X$ | $10^3[Cd^{2+}]$ (M) | $10^6k_1$ | $10^6k_1{}^{Cd}$ | $10^6k_2$ | $10^6k'_2$ |
|---|---|---|---|---|---|
| H | 0.1 | 2.39 | 0.145 | 22.4 | 0.121 |
| 4-MeO | 0.1 | 5.46 | 0.610 | 48.5 | 0.680 |
| 4-MeO | 1.0 | 50.3 | 0.610 | 49.7 | 0.699 |
| 4-Me | 0.1 | 2.66 | 0.260 | 24.0 | 0.286 |
| 4-Me | 1.0 | 25.3 | 0.260 | 25.0 | 0.297 |
| 4-Cl | 0.1 | 5.19 | 0.187 | 50.0 | 0.086 |
| 4-Cl | 1.0 | 48.3 | 0.187 | 48.2 | 0.083 |
| 4-F | 0.1 | 4.86 | 0.250 | 46.1 | 0.178 |
| 4-F | 1.0 | 50.0 | 0.250 | 49.7 | 0.196 |
| 3-MeO | 0.5 | 14.5 | 0.343 | 28.4 | 0.062 |
| 3-MeO | 1.0 | 32.2 | 0.343 | 31.9 | 0.069 |
| 3-Me | 0.5 | 10.1 | 0.294 | 19.6 | 0.100 |
| 3-Me | 1.0 | 20.8 | 0.294 | 20.5 | 0.104 |
| 3-Cl | 0.5 | 19.8 | 0.217 | 39.4 | 0.027 |
| 3-Cl | 1.0 | 35.2 | 0.217 | 35.0 | 0.025 |
| 3-F | 0.5 | 6.49 | 0.322 | 12.3 | 0.0076 |
| 3-F | 1.0 | 11.3 | 0.322 | 11.0 | 0.0070 |
| 2-MeO | 0.1 | 42.5 | 1.53 | 410 | 6.89 |
| 2-Me | 0.1 | 5.81 | 0.360 | 54.5 | 1.93 |
| 2-Cl | 0.1 | 83.8 | 8.60 | 752 | 1.00 |
| 2-F | 0.1 | 204 | 11.1 | 1,930 | 1.23 |
| 2,6-(MeO)$_2$ | 0.1 | 638 | 18.1 | 6,800 | 800 |

the conditions, pH = 6.70 and $\mu$ = 0.14, and temperature 90 °C[628]. The kinetic data are gathered in Table 202, where $k_2$ is the rate coefficient of the second step of the reaction, i.e. $k_2$ is corrected for the equilibrium of the first step as before. However, if the first step of the reaction is now supposed to be slow and the second step fast, treatment of the data in this way seems inappropriate. The fact that the data so treated give a reasonable correlation with $\sigma$ values with $\rho$ = $-1.2$ suggests that the base-catalysed reactions in the presence and absence of cadmium ion must have greater similarities than is indicated by equilibria (251) and (252) on one hand and equilibria (253) and (254) on the other. The rather poor correlation with $\sigma$ values here is not surprising, since we are dealing with the difference of two numbers which magnifies the experimental error. As in the case of the previous base-catalysed reaction, high *ortho : para* ratios are observed which does not seem consistent with the mechanism given by equilibria (253) and (254) since, if anything, the rate-determining step involves an increase in size on going through the transition state.

Finally, the relative effectiveness of metal ions in catalysing this reaction has been measured[628] and the first-order coefficients together with the derived second-order coefficients are given in Table 203 for the protodeboration of 2,6-dimethoxybenzeneboronic acid at 90 °C, $\mu$ = 0.14 and pH = 6.70.

TABLE 203

CATALYTIC EFFECT OF METAL IONS ON THE REACTION OF $2,6\text{-}(MeO)_2C_6H_3B(OH)_2$
WITH MALONATE BUFFERS AT 90 °C AND pH 6.70[628]

| Metal ion | $10^3[M^{2+}]$ (M) | $10^6k_1$ | $k_2$ |
|-----------|--------------------|-----------|-------|
| $Ni^{2+}$ | 1.0 | 23.0 | 0.0049 |
| $Mg^{2+}$ | 1.0 | 30.6 | 0.013 |
| $Co^{2+}$ | 1.0 | 51.3 | 0.0332 |
| $Zn^{2+}$ | 100 | 35.2 | 1.67 |
| $Cd^{2+}$ | 100 | 70.3 | 5.22 |
| $Pb^{2+}$ | 100 | 236 | 21.8 |
| $Cu^{2+}$ | 100 | 416 | 39.8 |
| $Cu^{2+a}$ | 100 | 421 | 40.3 |

[a] As chloride. All other salts were nitrates or perchlorates.

Rate coefficients have been determined for protodeboronation of benzene- and thiophene-boronic acids in a range of aqueous perchloric acid mixtures at temperatures between 25 and 90 °C[630]. First-order rate coefficients are given in Table 204, but not all of the rates were measured at the acidities indicated (see Table footnote), but were corrected to these making use of the linear relationship which was found, in preliminary experiments, to exist between log rate and $H_0$. Similarly, the relative reactivities to thiophene-2-, thiophene-3-, and benzene-boronic acid ($8.5 \times 10^5$, $7.1 \times 10^3$, and 1.0, respectively) were obtained at an $H_0$

TABLE 204

RATE COEFFICIENTS ($10^6 k_1$) AND KINETIC PARAMETERS FOR REACTION OF ArB(OH)$_2$ WITH HClO$_4$[630]

| Temp. (°C) | Benzeneboronic acid 8.5 M HClO$_4$ $H_0 = -4.69$ | Thiophene-2-boronic acid | | Thiophene-3-boronic acid 6.0 M HClO$_4$ $H_0 = -2.83$ |
|---|---|---|---|---|
| | | 2.07 M HClO$_4$ $H_0 = -0.81$ | 1.0 M HClO$_4$ $H_0 = -0.22$ | |
| 25 | | 100 | | |
| 35 | | 245 | | |
| 40 | | | 157 | 87.3 |
| 45 | | 595 | | |
| 50 | 5.08[a] | | 230 | 218 |
| 55 | | 1,380 | | |
| 60 | 15.3[b] | | 931 (59.5 °C) | 662 |
| 70 | 58.0 | | 1,230 | 1,770 |
| 90 | 162 | | | |
| 90 | 424 | | | |
| $E_a$ | 26 | 16 | | 22 |
| $\Delta S^{\ddagger}$ | −5, −26[c] | | −26 | −11, −21[c] |

[a] Calculated from the value obtained at $H_0 = -5.95$.
[b] Calculated from the value obtained at $H_0 = -5.73$.
[c] Value depended on the method of calculation.

value of $-0.22$ (1.0 $M$ perchloric acid) at 70 °C by application of the linear log $k$ versus $-H_0$ relation.

The relative reactivities were qualitatively in accord with the values obtained in other electrophilic substitutions, but quantitatively there are quite large deviations in that the $\sigma^+$ values (which can be calculated assuming that the $\rho$ factor for the perchloric acid-catalysed reaction is not very different from that obtained[625] with the sulphuric acid-catalysed reaction) are $-1.20$ and $-0.80$ for the 2 and 3 positions of thiophene respectively. These are much greater than obtained in other reactions (especially those obtained in a solvent-free system)[572], so the implication here is that solvent effects are very significant in determining these reactions as indicated by the very large differences in entropies of activation. Brown et al.[630] suggest that an additional reaction mechanism may be

$$(255)$$

in which the electron deficiency at boron created by the initial rapid protonation is made good by coordination with the adjacent water molecule, producing a favourable situation for the subsequent bond-breaking and formation. Such a process would require a large negative entropy of activation, as observed, and will also depend on the electron availability at boron and consequently the reactivity of the aromatic; hence the observed relative reactivities may be significantly dependent upon the entropy of activation and may not truly represent the reactivities of the aromatic ring positions.

### 9.4 PROTODECARBONYLATION

This description covers the reactions

$$ArCOX + H^+ = ArH + XCO^+ \tag{256}$$

in which a substituent containing a carbonyl group is replaced in the aromatic ring by hydrogen; X may either be OH (decarboxylation) or hydrogen or alkyl (deacylation).

#### 9.4.1 Protodecarboxylation

The kinetics of this reaction can be very conveniently followed by measuring the volume of carbon dioxide liberated in a given time, and the earlier studies employed this technique, though more recently spectrophotometric (UV) methods have been favoured. Most of the kinetic studies have been directed towards determining whether the $S_E1$ or A-$S_E2$ mechanisms, equilibria (257) and (258) or (259) and (260), respectively,

$$ArCOOH \rightleftharpoons Ar^- + COOH^+ \quad \text{(slow)} \tag{257}$$

$$Ar^- + H^+ \text{ (from } COOH^+) \rightleftharpoons ArH \quad \text{(fast)} \tag{258}$$

$$ArCOOH + H^+ \rightleftharpoons HAr^+COOH \quad \text{(slow)} \tag{259}$$

$$HAr^+COOH \rightleftharpoons ArH + CO_2 + H^+ \quad \text{(fast)} \tag{260}$$

were applicable to the acid-catalysed reaction, the latter mechanism being unanimously favoured. However, the evidence to determine the exact details of the A-$S_E2$ reaction has been somewhat conflicting and it has proved difficult to reach firm conclusions as to the reaction mechanism until recently.

An early qualitative study of this reaction[631] revealed that it tends to take place at a measurable rate only if bulky substituents are adjacent to the carbonyl group, so that it is likely that the carbon–carbon bond is considerably weakened in the rate-determining step (see, however, p. 306).

The first quantitative study of the reaction was carried out with anthracene-9-carboxylic acid (which possesses the necessary steric requirement by virtue of the *peri*-hydrogen atoms, and is very reactive at the 9 position towards electrophilic substitution). Schenkel[632] found that the decarboxylation rate was increased in the presence of acid, and the first-order rate coefficients (believed to be in sec$^{-1}$) are given in Table 205. It was subsequently concluded that in the absence of acid,

TABLE 205

RATE COEFFICIENTS FOR DECARBOXYLATION OF ANTHRACENE-9-CARBOXYLIC ACID[632]

| No added acid | | Added picric acid | | Added dichloroacetic acid | |
|---|---|---|---|---|---|
| Temp. (°C) | $10^5 k_1$ | Temp. (°C) | $10^5 k_1$ | Temp. (°C) | $10^5 k_1$ |
| 226.6 | 15.4 | 132.7 | 0.192 | 117.4 | 14.8 |
| 226.7 | 14.3 | 146.8 | 0.65 | 132.7 | 35.0 |
| 258.5 | 19.7 | 157.1 | 1.4 | 146.7 | 70.0 |
| 281.6 | 29.0 | | | | |

decarboxylation occurred *via* an $S_E1$ mechanism and in the presence of acid *via* an A-$S_E2$ mechanism[633].

The kinetics of the sulphuric acid-catalysed decarboxylation of a range of alkyl substituted benzoic acids have been measured by Schubert *et al.*[634, 635]. The variation of rate coefficient with temperature for mesitoic acid is given in Table 206 and the value for the methyl ester shows that, at this acid concentration, the

TABLE 206

RATE COEFFICIENTS FOR REACTION OF 2,4,6-Me$_3$C$_6$H$_2$COOH WITH H$_2$SO$_4$[634, 635]

| Temp. (°C) | 50 | 60 | 70 | 80 | 90 |
|---|---|---|---|---|---|
| $10^3 k_1$ (83.3 wt % $H_2SO_4$) | 0.121 | 0.484 | 1.54 | 4.93(4.79)$^a$ | 13.85 |

$^a$ Value for the methyl ester $E_a = 27.4$, log $A = 14.6$.

ester decarboxylates *via* initial rapid formation of the carboxylic acid. The variation of first-order rate coefficients (average values) with compound and acid concentration (at 80 °C) is shown in Table 207 and it is apparent that there is a fall-off in rate coefficient as the sulphuric acid concentration approaches 100 wt. %. The data for mesitoic acid showed no linear correlation between log rate coefficient and $H_0$, hence any possibility of the A-1 mechanism was eliminated, but in view of the first-order dependence of rate upon mesitoic acid concentrations, and a fairly linear dependence of rate upon the hydronium ion concentration, the A-2 mechanism was thought (incorrectly) to be appropriate, *i.e.* the rate-determining step is between a water molecule and the aromatic which has been proton-

TABLE 207

RATE COEFFICIENTS ($10^3 k_1$) FOR REACTION OF $2,6\text{-}X_2\text{-}4\text{-}YC_6H_2COOH$ WITH $H_2SO_4$[634, 635]

| $H_2SO_4$ (wt %) | $X = Y = Me$ 80 °C | $H_2SO_4$ (wt %) | $X = Me$ $Y = Et$ 80 C° | $Y = i\text{-}Pr$ $X = Me$ 80 °C | $H_2SO_4$ (wt %) | $X = Y = Et$ 60 °C | $X = Y = i\text{-}Pr$ 60 °C |
|---|---|---|---|---|---|---|---|
| 81.8 | 4.63 | 82.5 | 5.06 | 4.02 | 88.7 | 1.89 | |
| 83.3 | 4.93 | 84.9 | 5.38 | 5.27 | 89.9 | 1.73 | |
| 84.1 | 4.95 | 87.5 | 5.13 | 5.13 | 91.4 | 1.34 | |
| 86.1 | 4.77(4.95)[a] | 90.1 | 4.06 | 4.23 | 93.0 | 0.96 | 3.03 |
| 88.2 | 4.38 | 93.1 | 2.63 | 2.53 | 94.5 | 0.55 | 1.25 |
| 91.3 | 3.06 | 96.0 | 0.85 | 0.89 | 96.0 | 0.21 | 0.61 |
| 94.4 | 1.38 | 100.0 | 0 | 0 | 97.4 | | 0.26 |
| 96.9 | 0.184 | | | | 100.1 | | |
| 100.0 | 0 | | | | | | |

[a] In the presence of anhydrous sodium sulphate.

ated in a rapid-equilibrium according to

$$(261)$$

This may be ruled out not only by the unlikely resonance structure (XXXVI), but by the fact that the bisulphate anion (much more basic than water) produced no rate acceleration (Table 207) so that the hydrogen which is eliminated along with carbon dioxide is not transferred to base in a rate-determining step. The decrease in rate towards 100 wt. % sulphuric acid was even more rapid than expected from the decrease in hydronium ion concentration in this direction, and was initially attributed to a decrease in the concentration of $[ArH \cdot COOH]^+$ through formation of the acylium ion $ArCO^+$. However, the concentration of this latter ion and also of its precursor $ArCOOH_2^+$ were subsequently determined *via* a spectrophotometric method, which showed that above a concentration of 85 wt. % sulphuric acid, the $ArCOOH_2^+$ ion is predominant relative to $ArCO^+$ and $ArCOOH$, and that the concentration of acylium ion only becomes significant in solutions of *ca.* 95 wt. % sulphuric acid[635]. Consequently, although the rates reach a maximum in about 85 wt. % sulphuric acid, in which the concentration of hydronium ion is near its maximum, the levelling-off in rate arises substantially through the conversion of the carboxylic acid to the unreactive protonated species $ArCOOH_2^+$.

The reaction was confirmed as being straightforward $A\text{-}S_E2$ by the observation of general acid catalysis (although this conclusion concerning the mechanism

## TABLE 208

SECOND-ORDER RATE COEFFICIENTS FOR REACTION OF $2,4,6\text{-}Me_3C_6H_2COOH$ WITH $H_2SO_4$ AT 60 °C[635]

| $H_2SO_4$ (wt %) | $10^4 k_{obs}$ | $[ArCOOH]:[ArCO_2H_2^+]$ | $[H_2SO_4]$ (M) | $k_2^{H_2SO_4}$ |
|---|---|---|---|---|
| 72.0 | 2.075 | 100:3 | 0.206 | 104 |
| 76.2 | 3.27 | 100:11 | 0.336 | 108 |
| 80.5 | 6.675 | 100:39 | 0.70 | 133 |
| 84.8 | 7.30 | 51:100 | 2.1 | 103 |
| 90.3 | 4.90 | 12:100 | 7.1 | 64 |
| 95.5 | 2.5 | 2:100 | 13.3 | 138 |

was not reached by Schubert *et al.*[634, 635]). Catalysis by molecular sulphuric acid was indicated by the data in Table 208 which give the first-order rate coefficients, the relative actual concentrations of ArCOOH and $ArCOOH_2^+$, and the concentrations of *molecular* sulphuric acid in a range of sulphuric acid solutions at 60 °C; the rates attributable to molecular sulphuric acid, calculated from

$$k^{2H_2SO_4} = k_{obs}([ArCO_2H] + [ArCO_2H_2^+])/([ArCO_2H][H_2SO_4]_{molec})$$

are seen to be reasonably constant over a wide acid range, but since the rate is still very significant when the molecular sulphuric acid concentration becomes very small, catalysis by hydronium ion was thought to be significant at the lower acid concentration.

The data in Table 207 show that whilst compounds with the same *ortho* substituents but different *para*-alkyl substituents react at much the same rate, bulkier *ortho* substituents produce a large rate enhancement which was attributed to destabilisation of the ground states of these compounds, *i.e.* steric acceleration. This does not entirely prove that carbon–carbon bond breaking occurs in the rate-determining step of the reaction, for if in the initial slow step the proton becomes attached to the ring carbon bearing the carboxylic group (as is customary for electrophilic substitution) then the change in hybridisation from $sp^2$ to $sp^3$ will relieve the steric strain. Alternatively, the electron density at this ring carbon may be enhanced (relative to that in benzoic acid) through the carbonyl group being forced out of coplanarity with the aromatic ring and hence being unable to exert its full electron-withdrawing effect.

The rates of decarboxylation of a range of 3-substituted mesitoic acids in 82.1 wt. % phosphoric and 83.0 wt. % sulphuric acids have been measured[636] over a range of temperatures as indicated in Table 209, which gives the first-order rate coefficients together with the log $A$ values and the energies, enthalpies, and entropies of activation calculated at 80 °C for sulphuric acid, and 119 °C for phosphoric acid; these kinetic parameters are somewhat different from those

TABLE 209

RATE COEFFICIENTS AND KINETIC PARAMETERS FOR REACTION OF
3-X-2,4,6-Me$_3$C$_6$HCOOH WITH H$_2$SO$_4$ AND H$_3$PO$_4$[636]

| Temp. (°C) | X | $10^3 k_1$ | $E_a$ | log A | $\Delta S^{\ddagger}$ | $\Delta H^{\ddagger}$ |
|---|---|---|---|---|---|---|
| | *83.0 wt % H$_2$SO$_4$* | | | | | |
| 64.2 | H | 0.974 | 27.7 | 14.9 | 9.4 | 27.0 |
| 70.5 | | 1.86 | | | | |
| 79.2 | | 5.13 | | | | |
| 85.8 | | 11.4 | | | | |
| 64.2 | OH | 0.270 | 21.3 | 10.2 | −12.1 | 20.6 |
| 70.5 | | 0.456 | | | | |
| 79.2 | | 1.05 | | | | |
| 85.8 | | 1.74 | | | | |
| 48.6 | Me | 0.559 | 23.7 | 13.85 | 4.6 | 23.0 |
| 54.8 | | 1.13 | | | | |
| 61.4 | | 2.14 | | | | |
| 68.5 | | 5.12 | | | | |
| 73.2 | | 7.95 | | | | |
| 60.6 | Et | 2.39 | 25.9 | 14.35 | 7.8 | 25.2 |
| 64.0 | | 3.71 | | | | |
| 60.9 | | 6.74 | | | | |
| 77.0 | | 8.04 | | | | |
| | *82.1 wt % H$_3$PO$_4$* | | | | | |
| 103.6 | H | 0.448 | 28.5 | 13.2 | 1.3 | 27.7 |
| 109.3 | | 0.717 | | | | |
| 116.2 | | 1.44 | | | | |
| 127.0 | | 3.97 | | | | |
| 103.6 | OH | 0.833 | 15.9 | 6.7 | −28.4 | 15.9 |
| 116.2 | | 1.67 | | | | |
| 120.4 | | 2.08 | | | | |
| 127.0 | | 3.04 | | | | |
| 134.9 | | 5.58 | | | | |

given in the paper which are almost entirely in error. The errors in the activation energy and entropy were stated to be ±0.8 and ±2.8 respectively, but since the error will largely depend upon the range of temperatures employed it is clear that there will be a considerable variation in these values which are almost certainly too small, for example, for the ethyl compound. The decarboxylation rates roughly followed the order expected for an electrophilic substitution and significantly the 3-bromo- and 3-nitro-compounds did not decompose in sulphuric acid below 140 and 170 °C, respectively. Little can be deduced about the reactivity of the 3-hydroxy compound which clearly has a very significant entropy effect; in phosphoric acid its reactivity is greater or less than the unsubstituted compound according to the acid strength. It is probable that a major factor is increasing protonation of the hydroxyl group as the acid strength increases, resulting in increasing deactivation in that direction.

Evidence that the second step in the reaction, the breaking of the carbon–carbon bond is partially rate-determining has been provided by Bothner-By and Bigeleisen[637], who found that the decarboxylation of mesitoic acid in 88.75 wt. % sulphuric acid at 61.2 and 92 °C resulted in enrichment of carbon-12 in the liberated carbon dioxide such that the respective $k_{C_{12}} : k_{C_{13}}$ values were $1.037 \pm 0.003$ and $1.032 \pm 0.001$, the smaller value being obtained at the higher temperature as expected. This result was subsequently confirmed by Stevens et al.[638], who decarboxylated mesitoic acid in 86 wt. % sulphuric acid. At 99 °C, over 99 % reaction readily occurred and the $C_{12} : C_{13}$ and $C_{12} : C_{14}$ ratios in the liberated carbon dioxide and hence the original acid were determined as 131.96 and 86.00, respectively[638]. By comparison with the isotopic content of the carbon dioxide liberated at 60 °C after 13.7 % and 17.5 % reaction, the $k_{C_{12}} : k_{C_{13}}$ and $k_{C_{12}} : k_{C_{14}}$ values are determined as 1.037 and 1.108 (13.7 %) and 1.039 and 1.094 (17.5 %), respectively.

The kinetics of the decarboxylation of hydroxybenzoic acids have been studied. Brown et al.[639] used resorcinol as an acidic solvent, since it was liquid over a wide range of temperatures and was similar to the reaction products; the first-order rate coefficients are given in Table 210. The significant decrease in the log $A$ factor

TABLE 210

RATE COEFFICIENTS $(10^4 k_1)$ AND ARRHENIUS PARAMETERS FOR REACTION OF $XC_6H_4COOH$ WITH RESORCINOL[639]

| Temp. (° C) | X = 2-OH | Temp. (° C) | X = 2,4-(OH)₂ | Temp. (° C) | X = 2,4,6-(OH)₃ |
|---|---|---|---|---|---|
| 206.5 | 1.32 | 180.4 | 5.65 | 109.5 | 15.4 |
| 216.3 | 2.53 | 181.0 | 6.61 | 110.0 | 17.9 |
| 218.3 | 3.15 | 182.2 | 7.41 | 110.9 | 18.3 |
| 219.5 | 3.25 | 192.0 | 13.1 | 117.8 | 24.2 |
| 219.8 | 3.53 | 195.2 | 16.2 | 123.5 | 29.5 |
| 223.0 | 3.96 | 197.9 | 20.4 | 124.5 | 36.7 |
| 224.0 | 5.02 | 205.0 | 31.1 | | |
| 225.0 | 8.95 | 207.1 | 36.7 | | |
| 226.0 | 10.5 | | | | |
| $E_a$ | 33.6 | | 29.2 | | 13.6 |
| log A | 11.5 | | 10.9 | | 5.0 |

for the di-ortho substituted compound compared to the other acids is very probably a result of steric congestion at the reaction site. The activation energy of decarboxylation of benzoic acid in this medium was estimated to be at least 39, based upon the fact that no decomposition could be detected at 250 °C, and thus the reaction rate is increased by substituents which usually enhance electrophilic substitutions. No firm conclusion was reached regarding the reaction mechanism except to note that the acceleration by electron-supplying groups is consistent with the $S_E2$ mechanism but not the $S_E1$ mechanism.

A subsequent kinetic investigation of the effect of added hydrochloric acid upon the rate of decarboxylation of 2,4,6-trihydroxybenzoic acid was analysed in terms of three possible mechanisms[640]:

(i) The unimolecular decomposition of the acid anion, the $S_E1$ mechanism, leading to a dependence of rate upon [ArCOO$^-$]. The rate of this reaction should obviously be retarded by added acid since the concentration of this ion should decrease.

(ii) The bimolecular attack of hydronium ion upon the carboxylic acid, the A-$S_E2$ mechanism leading to a dependence upon [H$_3$O$^+$][ArCOOH]. The rate here should rapidly increase with increased added acid concentration, since the concentration of both these species should be increased.

(iii) The bimolecular attack of hydronium ion upon the carboxylate anion, leading to a rate dependent upon [H$_3$O$^+$][ArCOO$^-$], which is indistinguishable from the (unlikely) decomposition of the carboxylic acid by interaction with water molecules (excess) and whose rate will depend upon [ArCOOH].

In Table 211 some rate coefficients calculated according to these dependencies

## TABLE 211

RATE COEFFICIENTS[a] FOR REACTION OF 2,4,6-(HO)$_3$C$_6$H$_2$COOH WITH HCl[640]

| Temp. (°C) | [HCl] (M) | $10^3k'$ | $10^3k''$ | $10^3k'''$ |
|---|---|---|---|---|
| 49.8 | | [b] | 135 | 1.09 |
| | 0.0458 | 2.94 | 23.9 | 1.19 |
| | 0.1783 | 16.4 | 10.0 | 1.76 |
| 45.0 | | [b] | 68.3 | 0.613 |
| | 0.346 | 1.45 | 19.4 | 0.750 |
| | 0.1278 | 5.65 | 7.7 | 0.997 |
| 39.8 | | [b] | 42.2 | 0.356 |
| | 0.0356 | 0.80 | 11.1 | 0.421 |
| | 0.126 | 3.90 | 4.8 | 0.608 |

[a] See text for definition of the coefficients.
[b] In the absence of acid, poor first-order kinetics were obtained assuming this mechanism.

are listed, the values $k'$, $k''$, and $k'''$ being the first-order, second-order, and apparent first-order coefficients (including the effect of H$_2$O), respectively. Clearly, the $S_E1$ mechanism cannot be correct and neither of the bimolecular mechanisms seem satisfactory either. The kinetics obtained assuming reaction of the hydronium ion upon carboxylate ion are the most consistent with prediction, even so the rate coefficients increase significantly with increasing acid concentration, suggesting an incursion of mechanism (ii), and further more complex analysis of the data indicated that this was so, the reaction upon the acid being about five times more important, at 40 °C. The activation energies for these reactions were 15.2 (acid) and 21.5 (anion) the corresponding log $A$ factors being 7.82 and 13.3, but no reasons were advanced for the apparent lower reactivity of the anion.

Since the above investigation was somewhat inconclusive, the kinetics of the decarboxylation of 2,4,6-trihydroxybenzoic acid were further investigated[641] using aqueous sulphuric and perchloric acids as catalysts and by following the reaction rate spectrophotometrically rather than gasometrically as above. It was hoped also, that with this more reactive acid (compared to the tri-alkyl acids) decarboxylation could be studied in an acidic medium in which formation of the acylium ion was unimportant, the ratio of carboxylate anion to carboxylic acid was also studied spectrophotometrically. First-order rate coefficients at 30 °C are given in Table 212. The results obtained in the absence of added acid confirmed the conclusions of the earlier workers in eliminating the $S_E 1$ mechanism, but difficulty was experienced in reconciling the levelling off in rate at 10.7 wt. % perchloric acid, with any reasonable acid-catalysed mechanism. Spectrophotometric analysis showed that only one ionisation equilibrium occurs in the 0–38 % acid region and since $ArCOO^-$ is present in water, the equilibrium must be that to form the carboxylic acid, and not protonation of the latter to give $ArCOOH_2^+$; hence any mechanism involving the latter could be eliminated and the formation of this ion cannot be invoked to account for the levelling-off of rate coefficient observed. There was, however, some spectral evidence that this protonation does occur in 56.6 wt. % perchloric acid and 75.2 wt. % sulphuric acid and in the latter acid in particular, the rate coefficient is particularly small. No firm conclusion as to the reaction mechanism was reached, but from a consideration of the equilibria present in the reaction medium and the constancy of the $k_1'$ values (see Table 212)

TABLE 212

RATE COEFFICIENTS FOR REACTION OF 2,4,6-$(HO)_3C_6H_2COOH$ WITH $HClO_4$ AT 30 °C[641]

| $HClO_4$ (wt %) | 0 | 0.139 | 0.278 | 0.556 | 1.09 | 2.01 | 5.70 | 10.7[a] | 18.1 | 21.5 | 26.1 | 30.6 | 38.2 | 45.7 | 56.6 |
|---|---|---|---|---|---|---|---|---|---|---|---|---|---|---|---|
| $10^4 k_1$ | 0 | 0.74 | 1.07 | 1.51 | 1.91 | 2.24 | 2.66 | 2.78 | 2.86 | 2.86 | 2.86 | 2.88 | 2.74 | 2.36 | 1.12 |
| *$10^4 k_1'$ | | 2.69 | 2.59 | 2.66 | 2.73 | 2.77 | 2.82 | 2.78 | 2.86 | 2.86 | 2.86 | 2.88 | 2.74 | | |

[a] $E_a$ was quoted in this medium as 19.6, "determined over a 40° range". However, it appears from the figure given that this value refers to 21.5 wt % acid, determined over a 30° range, the rates at 20, 40 and 50 °C being 0.92, 8.13, and 22.2, respectively.

In 75.4 wt % sulphuric acid, $10^4 k_1 = 0.15$.

* See text.

where [$k_1' = k_1([ArCOOH] + [ArCOO^-]/[ArCOOH])$], the rate-determining step of the reaction was thought to be either a first-order decomposition of ArCOOH or the reaction of a proton with $ArCOO^-$, which seems the more likely. It was emphasized that since hydroxy acids have an extra site for protonation compared to the alkyl acids, the reaction mechanisms may not be the same for both types of compound.

This study was extended to decarboxylation of 2,4,6-trimethoxybenzoic acid in aqueous sulphuric, perchloric and hydrochloric acids, the average values of the first-order rate coefficients obtained at 10 °C being listed in Table 213[642]. In the

TABLE 213

RATE COEFFICIENTS FOR REACTION OF 2,4,6-$(MeO)_3C_6H_2COOH$ WITH ACID AT 10 °C[642]

| Acid (wt %) | $10^3k_1$ | Acid (wt %) | $10^3k_1$ | Acid (wt %) | $10^3k_1$ |
|---|---|---|---|---|---|
| 10.3 $HClO_4$ | 0.216 | 49.8 $HClO_4$ | ~17 | 58.5 $H_2SO_4$ | "fast" |
| 19.9 $HClO_4$ | 0.795 | 59.7 $HClO_4$ | 0.675 | 10.1 HCl | 1.2 |
| 29.6 $HClO_4$ | 2.75 | 18.2 $H_2SO_4$ | 0.94 | 19.3 HCl | 12 |
| 39.0 $HClO_4$ | 9.15 | 39.4 $H_2SO_4$ | "fast" | | |

acid range studied, spectral studies indicated that formation of the acylium ion is significant and is probably complete in 64 wt. % perchloric acid, so that this may account for the fall-off in rate coefficients above 50 wt. % acid. Between 5 and 60 wt. % acid there was some evidence of the existence of another derivative of the carboxylic acid, perhaps $ArCOOH_2^+$, but it may also have arisen from a medium effect upon the spectrum of the free acid. If the spectrum does arise through formation of $ArCOOH_2^+$ which was assumed complete below 47 wt. % acid, it was possible to show that $k_{obs}$ increased at a faster rate with increasing acid concentration than is required by both the unimolecular mechanism, and the bimolecular mechanism with specific hydronium ion catalysis. Hence the general acid-catalysed mechanism would seem to be confirmed. Activation energies were determined (between 0 and 30 °C) for reaction with both 10.3 and 59.7 wt. % perchloric acids, giving values of 18.4 and 20.6 respectively, the corresponding entropies being $-12.4$ and $-2.1$. Since these values included all the parameters for the pre-equilibrium no conclusions were drawn from them.

The decarboxylation of 2,4-dimethoxybenzoic acid by 2.5–8.0 $M$ sulphuric acid at 60.5 °C has also been studied and division of the first-order rate coefficient by the molar concentration of sulphuric acid gave constant values $(2.94 \pm 0.25 \times 10^{-6})$ throughout the acid range so that the A-$S_E2$ reaction was again indicated[643].

Further evidence regarding the mechanism was provided by Lynn and Bourns[643a], who found a pH-dependent carbon-13 isotope effect in the decarboxylation of 2,4-dihydroxybenzoic acid in acetate buffers. The dependence was interpreted in favour of the A-$S_E2$ mechanism, for an increase in acetate concentration would increase $k'_{-1}$ and hence partitioning of the intermediate so that $k'_2$ becomes more rate-determining.

Support for the A-$S_E2$ mechanism comes also from the measurement[643b] of the solvent isotope effect in the perchloric acid-catalysed decarboxylation of 2,4-dihydroxybenzoic acid at 50.03 °C, $\mu = 0.1$. Table 214 shows the observed first-order rate coefficients $k_1$, and coefficients $k'_1$ for the decomposition of the acid calculated from the equilibrium constants $K_1(\text{H}) = 2.18 \times 10^{-4}$ and $K_1(\text{D})$

TABLE 214

RATE COEFFICIENTS FOR REACTION OF 2,4-$(\text{HO})_2\text{C}_6\text{H}_3\text{COOH}$ IN $\text{HClO}_4$ ($\mu = 0.1$) AT 50 °C[643b]

| $10^2$[perchloric acid] (M) | [KCl] (M) | $10^7 k_1$ | $10^7 k'_1$ |
|---|---|---|---|
| 0.97 ($\text{HClO}_4$) | 0.090 | 2.34 | 2.40 |
| 1.75 ($\text{HClO}_4$) | 0.082 | 2.41 | 2.44 |
| 2.24 ($\text{HClO}_4$) | 0.077 | 2.49 | 2.52 |
| 1.00 ($\text{DClO}_4$) | 0.090 | 14.1 | 15.2 |
| 1.80 ($\text{DClO}_4$) | 0.082 | 14.5 | 15.1 |

$7.75 \times 10^{-4}$; the second-order rate coefficients $k'_2$ were determined as $1.97 \times 10^{-3}$ (H) and $1.12 \times 10^{-3}$ (D) leading to the solvent isotope effect of 1.76. This is entirely consistent with the more rapid transfer of a proton from the weaker base $\text{H}_2\text{O}$ than from $\text{D}_2\text{O}$ in the first and principally rate-determining step of the reaction.

The mechanism of decarboxylation of acids containing an amino substituent is further complicated by the possibility of protonation of the substituent and the fact that the species $\text{NH}_2\text{ArCOOH}$ is kinetically equivalent to the zwitterion $\text{NH}_3^+\text{ArCOO}^-$. Both of these species, as well as the anion $\text{NH}_2\text{ArCOO}^-$ and even $\text{NH}_3^+\text{ArCOOH}$ must be considered. Willi and Stocker[644] investigated by the spectroscopic method the kinetics of the acid-catalysed decarboxylation of 4-aminosalicylic acid in dilute hydrochloric acid, (ionic strength 0.1, addition of potassium chloride) and also in acetate buffers at 20 °C. The ionisation constants $K_0 = [\text{HA}][\text{H}^+][\text{H}_2\text{A}^+]^{-1}$ (for protonation of nitrogen) and $K_1 = [\text{A}^-][\text{H}^+][\text{HA}]^{-1}$, were determined at $\mu = 0.1$ and 20 °C. The kinetics followed equation (262)

$$\text{Rate} = k'_2[\text{HA}] + k''_2[\text{H}_2\text{A}^+] \tag{262}$$

where $k'_2$ and $k''_2$ are second-order coefficients and the terms in the equation were, therefore, interpreted as representing simultaneous second-order reaction of hydrogen ion with 4-aminosalicylate ion, and 4-ammonium salicylate zwitterion, respectively. The respective rate coefficients were $4.18 \times 10^{-6}$ and $4.1 \times 10^{-7}$, and this is qualitatively but certainly not quantitatively consistent with retardation of an A-$S_E2$ reaction rate by the $\text{NH}_3^+$ substituent.

Subsequently, rate coefficients were determined for the decarboxylation of 4-amino-, (and also 4-hydroxy-, 4-methoxy-, and 4-methyl)-salicyclic acids in aqueous solution at various temperatures[645]. From the rate equations

$$\text{Rate} = k'_1[\text{HA}] = k'_2[\text{A}^-][\text{H}^+] \tag{263}$$

$$\text{Rate} = k_1[\text{HA}]_{\text{stoich}} = k_1[\text{HA}]+[\text{A}^-]) \tag{264}$$

where $k_1$ is the observed rate coefficient, the first-order rate coefficient for de-carboxylation of the unionized acid $k'_1$ may be calculated from $k'_1 = k_1(1+K_1/[\text{H}^+])$. Values of $k'_1$ so obtained were found to be essentially independent of the hydrogen ion concentration (Table 215). The second-order coefficients $k'_2$ for

TABLE 215

RATE COEFFICIENTS FOR REACTION OF 4-X-2-OH-C$_6$H$_3$COOH WITH AQUEOUS HCl
($\mu = 0.1$)[645]

| 4-X | Temp. (°C) | $10^3[H^+]$ (M) | $10^6k_1$ | $10^6k'_1$ |
|---|---|---|---|---|
| OH | 50.03 | 19.8 | 1.50 | 1.56 |
|  |  | 9.81 | 1.41 | 1.53 |
|  |  | 2.40 | 1.13 | 1.49 |
|  |  | 1.10 | 0.909 | 1.55 |
| OMe | 50.03 | 19.8 | 0.870 | 0.912 |
|  |  | 9.82 | 0.838 | 0.920 |
|  |  | 4.82 | 0.769 | 0.925 |
|  |  | 2.32 | 0.613 | 0.871 |
|  |  | 1.03 | 0.448 | 0.872 |
| Me | 85.07 | 24.8 | 0.787 | 0.816 |
|  |  | 19.84 | 0.781 | 0.818 |
|  |  | 9.82 | 0.769 | 0.843 |

reaction of the carboxylate anion were obtained by dividing $k'_1$ by $K_1$ and both first- and second-order coefficients for a range of temperatures are listed in Table 216 together with the calculated Arrhenius parameters. From the empirical rate law it was concluded that the reaction is either a unimolecular decomposition of HA or a bimolecular reaction of $H^+$ with $A^-$. Significantly, the second-order rate coefficients at 50 °C (data for methyl extrapolated) correlate against $\sigma^+$ values ($\rho = -4.38$), which is further evidence for a rate-determining attack of a proton upon the aromatic ring, and the activation energies are decreased, as expected, by increased electron supply. A unimolecular decomposition of HA to $A^-$ and $H^+$ would however, be retarded by electron supply.

The kinetics of decarboxylation of 4-aminosalicylic acid in some buffer solutions at 50 °C were studied. The first-order rate coefficients increased with increasing buffer concentration, though the pH and ionic strength were held constant (Table 217). This was not a salt effect since the rate change produced by substituting potassium chloride for the buffer salt was shown to be much smaller. It follows from the change in the first-order rate coefficients ($k_1$) with

TABLE 216

RATE COEFFICIENTS AND ARRHENIUS PARAMETERS FOR REACTION OF 4-X-2-OH-$C_6H_3COOH$ WITH AQUEOUS HCl ($\mu = 0.1$)[645]

| 4-X | Temp. (°C) | $10^6k'_1$ | log A | $E_a$ | $10^3k'_2$ | log A | $E_a$ |
|------|-----------|-----------|-------|-------|-----------|-------|-------|
| NH$_2$ | 20.07 | 41.8 | 10.81 | 21.7 | 1.80 | 13.22 | 20.05 |
| | 36.87 | 34.7 | | | 13.1 | | |
| | 50.03 | 130 | | | 43.0 | | |
| OH | 50.03 | 1.53 | 10.48 | 24.1 | 0.197 | 13.06 | 23.3 |
| | 70.04 | 13.2 | | | 1.63 | | |
| | 85.07 | 60.4 | | | 6.86 | | |
| OMe | 50.03 | 0.907 | 10.56 | 24.5 | 0.093 | 14.00 | 25.2 |
| | 70.04 | 8.28 | | | 0.77 | | |
| | 85.07 | 38.3 | | | 4.43 | | |
| Me | 70.04 | 0.155 | 10.50 | 27.2 | 0.0136 | 15.28 | 30.0 |
| | 85.07 | 0.825 | | | 0.0865 | | |

buffer ratio that the absolute increase of $k_1$ is proportional to [BH$^+$] and not [B], so that the reaction is subject to general acid catalysis. The A-1 mechanism is then ruled out since the rate could not be affected by the concentration changes of the weak acid in the buffer if the buffer ratios (pH) and ionic strength were held constant. Furthermore, the A-2 and A-S$_E$2 mechanism can be distinguished by virtue of the fact that steric hindrance to the attack by base should hinder the A-2 mechanism but not the A-S$_E$2 mechanism; no evidence of a retardation in rate parallelling the nature of the base component of the buffers was observed, except for a very small effect with 2,6-lutidine.

TABLE 217

RATE COEFFICIENTS FOR DECARBOXYLATION OF 4-NH$_2$-2-OH-C$_6$H$_3$COOH IN BUFFER SOLUTIONS AT 50 °C[645]

| Buffer | $10^2$[B] (M) | [KCl] (M) | [B] [BH$^+$] | $10^6k_1$ |
|--------|-----------|-----------|--------------|-----------|
| Aniline | 0.520 | 0.095 | 1.000 | 23.8 |
| | 1.029 | 0.090 | 1.000 | 25.8 |
| | 1.537 | 0.085 | 1.000 | 27.2 |
| Pyridine | 0.495 | 0.095 | 0.988 | 5.24 |
| | 0.990 | 0.090 | 0.988 | 5.50 |
| | 1.483 | 0.085 | 0.988 | 5.91 |
| | 1.980 | 0.080 | 0.988 | 6.16 |
| 2,6-Lutidine | 0.789 | 0.092 | 0.978 | 0.254 |
| | 1.577 | 0.084 | 0.978 | 0.263 |
| | 2.367 | 0.076 | 0.978 | 0.281 |
| Acetic acid | 0.983 | 0.090 | 0.069 | 12.0 |
| | 1.967 | 0.080 | 0.969 | 12.8 |
| | 2.945 | 0.070 | 0.969 | 13.4 |

The kinetics of the decarboxylation of anthranilic acid have recently been examined. Earlier, an investigation of the decarboxylation of anthranilic acid in aqueous or acidic solution at 100 °C gave a $C_{12} : C_{13}$ value of 108.02 after 72 % reaction in 1.0 M sulphuric acid compared with 108.5 from complete decarboxylation, so that there is virtually no kinetic isotope effect for this compound[646]. First-order rate coefficients are given in Table 218 and from the variation of rate

TABLE 218

RATE COEFFICIENTS FOR REACTION OF ANTHRANILIC ACID WITH $H_2SO_4$ AT 100 °C[646]

| $H_2SO_4$ (wt %) | 2.4 | 4.8 | 7.0 | 9.3 | 17.4 | 25.4 |
|---|---|---|---|---|---|---|
| $10^5 k_1$ | 3.75 | 3.80 | 3.92 | 3.86 | 3.22 | 2.36 |

with acid concentration, reaction was thought not to occur on the acid species $NH_3^+ RCOOH$; this may be concluded also from the magnitude of the rate coefficients which, by comparison with the other data, would be expected to be very much lower if the substituent was a positive pole. Likewise the ion $NH_2RCOO^-$ was not considered to be reacting since its concentration should remain roughly constant throughout the acid range studied, and consequently so should the rate. According to Stevens et al.[646], the reaction rate increased with increasing acid concentration up to 1 M (9.3 wt. %). Their data show no evidence of this and their arguments against involvement of the carboxylate anion are, therefore, irrelevant. In attempting to decide whether reaction occurs on the zwitterion or the neutral molecule, Stevens et al. argued that since the ratio of neutral molecules to zwitterions is greater in para than in ortho-aminobenzoic acid, then if reaction occurs via the neutral molecule then it should be faster for the para compound, whereas it is slower; it was concluded, therefore, that reaction occurs on the zwitterion, but this argument is quite fallacious as we should expect the ortho compound to be more reactive by virtue of the well-established effect of steric acceleration. Consequently, no satisfactory conclusion was reached as to the form of the reacting anthranilic acid.

Dunn et al.[646a] examined the rates of decarboxylation of 4-methoxyanthranilic acid in media of pH 1–5 at 25 and 60 °C and of 4-methylanthranilic acid at pH 1–3 at 25 °C. They found the rate to pass through a maximum with increasing acidity and these maxima did not correspond with those acidities at which the concentration of the various anions and zwitterions would be maximal. Likewise, Los et al.[646b] observed similar trends in the decarboxylation of 2-substituted-4-aminobenzoic acids and concluded that if an intermediate such as $HAr^+COOH$ is formed by ring protonation then it cannot decarboxylate without first losing the carboxylic proton. Willi et al.[647] have studied the decarboxylation of anthranilic acid and have concluded that the maximum in the rate arises through a change in the rate-determining step of the reaction. In weakly acidic (pH 4.7–1.4) solutions at $\mu = 0.1$, first-order rate coefficients increased along with the increasing concentration of the cation $H_2A^+$ and the reaction kinetics could be expressed in terms of equation (262) with similar conclusions to those apertaining to decarboxylation of 4-aminosalicyclic acid. The activation energy for reaction with the acid anion (*i.e.* nitrogen not protonated) was 22.9 (determined, however, only between 70.1 and 85.0 °C) *cf.* 22.8 for 4-aminobenzoic acid and 23.3 for 2,4-dihydroxybenzoic acid; the corresponding log $A$ values were 12.8, 12.5, and 13.1. For reactions upon the zwitterions of the former two compounds the activation energies were reported as 22.4 and 21.7 respectively, the corresponding log $A$ values being 10.3 and 10.3 so that reaction on the zwitterion turns out to be slower, as expected, but by not nearly a large enough factor in the writer's view if the reacting species is $NH_3{}^+C_6H_4COO^-$ rather than $NH_2C_6H_4COOH$. For anthranilic acid, kinetic studies with acetate buffers showed no evidence of general acid catalysis (*cf.* 4-aminosalicyclic acid, see p. 312), but it was argued that general acid catalysis may remain undetectable if the Bronsted coefficient $\alpha$ is close to unity. With 0.1–3.9 $M$ hydrochloric acid solutions, the first-order coefficients $10^6k_1$ at 85.0 °C and at the hydrogen ion concentration, *i.e.* [HCl] $(M)$, indicated in parentheses, were: 7.28 (0.100); 7.54 (0.478); 6.58 (0.991); 4.00 (1.92); 2.16 (2.94); 1.11 (3.94), and again the maximum in rate occurred at an acid concentration considerably different from that at which there are maxima in the concentrations of the reacting species. From a complex analysis of the experimental data this behaviour was shown to be reconcilable with a change in mechanism such that at higher acid concentrations $k'_{-1} > k'_2$ in the usual $S_E2$ mechanism. This was argued to be not inconsistent with the carbon isotope data of Stevens *et al.*[646] since in 1 $M$ sulphuric acid, carbon–carbon bond cleavage would not be sufficiently rate-determining to be measurable outside the limits of experimental error.

### 9.4.2 Protodeacylation

If measurable rates are to be obtained at accessible temperatures, protodeacylation, reaction (256, X = H or R), has the same steric requirement as protodecarboxylation namely the presence of *ortho* substituents. This was first indicated by the study by Louise[648], who found that reaction of benzoylmesitylene with hot phosphoric acid gave mesitylene and benzoic acid, and Klages and Likroth[631] showed that alkylaromatics could be readily deacylated by hot phosphoric acid if alkyl substituents were in the *ortho* position. Subsequently, Arnold and Rondestvedt[649] found that in the reaction with 85 % phosphoric acid, 4-acetyloctahydroanthracene (XXXVII) gave a higher yield of deacylated product in a given time than did 4-acetylhydrindacene (XXXVIII), and this probably arose (in part at least) from the greater steric hindrance in the ground state of the former, though it is likely that electronic effects are significant here in view of the relative reactivities of indane and tetralin at the α positions[542].

The first kinetic study was carried out with 2,6-dimethyl- and 2,4,6-trimethyl-acetophenone in sulphuric acid, the rate of the deacylation was followed by measuring the rate of formation of acetic acid, and was first-order in aromatic[650]. Since a linear correlation of log rate *versus* $H_0$ was observed over a wide acid range, the rate-determining step of the reaction was considered to be a first-order decomposition of the conjugate acid of the ketone. Consequently, the observed relative reactivity of the tri- to di-methyl compound (102 in 80.4 wt. % acid at 40 °C) was modified for the differences in p$K$ for the fast equilibrium to give the conjugate acid, resulting in a difference in reactivity of 66 for the step believed to be rate-determining; since the mechanism has subsequently been shown to be incorrect this value should be ignored. Sulphonation was a significant side reaction for the dimethyl compound, and accounted for the decrease in yield of acetic acid with increasing sulphuric acid concentration through conversion of the aromatic to the less reactive sulphonic acid, prior to deacylation. Consequently, the production of acetic acid ceased to be related to the amount of aromatic taken for a given run, and the observed rates had to be corrected for this side reaction. The first-order rate coefficients (observed and corrected) are given in Table 219 together with the observed rate coefficients for the 2,4,6-trimethyl compound.

Subsequent kinetic studies have indicated that the linear log $k$ *versus* $H_0$ plot

TABLE 219

RATE COEFFICIENTS FOR REACTION OF ArCOCH₃ WITH H₂SO₄[650]

| | $Ar = 2,6\text{-}Me_2C_6H_3$ at 40 °C | | | $Ar = 2,4,6\text{-}Me_3C_6H_2$ at 30 °C | |
|---|---|---|---|---|---|
| $H_2SO_4$ (wt %) | $10^5k_{1obs}$ | $10^5k_{1sulph}$ | $10^5k_{1deacyl}$ | $H_2SO_4$ (wt %) | $10^5k_{1obs}$ |
| 72.5 | 0.348 | | 0.348 | | |
| 75.0 | 0.926 | | 0.926 | | |
| 77.5 | 2.35 | | 2.35 | | |
| 80.0 | 5.25 | | 5.25 | 80.4 | 278(633, 40 °C) |
| 82.5 | 11.3 | | 11.32 | 83.5 | 516 |
| 85.0 | 21.8 | | 21.8 | 54.8 | 678 |
| 87.5 | 38.7 | 1.9 | 36.8 | 87.1 | 802 |
| 90.0 | 56.0 | 5.0 | 51.0 | | |
| 93.0 | 73.0 | 11.0 | 52.0 | 90.4[a] | 0.546 (20 °C) |
| 96.0 | 95.0 | 26.1 | 68.9 | 99.0[a] | 4.60 (20 °C) |

[a] In methanesulphonic acid.

obtained in the above study is misleading as a diagnosis of mechanism and have all indicated *general* rather than *specific* acid catalysis, so that the A-$S_E$2 (rather than A-1 mechanism is considered to be appropriate, *viz.* equilibria (265, 266),

$$ArCHO + HA \underset{k'_{-1}}{\overset{k'_1}{\rightleftharpoons}} Ar^+H.CHO + A^- \tag{265}$$

$$Ar^+H.CHO + A^- \overset{k'_2}{\rightleftharpoons} ArH + CO + HA \tag{266}$$

and studies have been principally directed towards determining which of these steps is rate-determining.

The kinetics of deacylation of 2,4-6-trimethyl, 2,4,6-triethyl and 2,4,6-tri-*iso*-propylbenzaldehyde have been determined over a wider acid range (50–100 wt. % sulphuric) than used in the above study, the reaction rates being determined either by a spectrophotometric or gasometric method, the latter being suitable here

TABLE 220

RATE COEFFICIENTS AND KINETIC PARAMETERS FOR DEACYLATION OF 2,4,6-$R_3C_6H_2CHO$ BY $H_2SO_4$[651]

| Me (100 °C) | | | | Et (100 °C) | | | i-Pr (80 °C) | | | |
|---|---|---|---|---|---|---|---|---|---|---|
| $H_2SO_4$ (wt %) | $10^3k_1$ (gas) | $H_2SO_4$ (wt %) | $10^3k_1$ (spec) | $H_2SO_4$ (wt %) | $10^3k_1$ (gas) | $10^3k_1$ (spec) | $H_2SO_4$ (wt %) | $10^3k$ (gas) | $H_2SO_4$ (wt %) | $10^3k$ (spec) |
| 70.0 | 0.47 | 51.5 | 0.035 | 75.0 | 1.32 | | 82.5 | 2.1 | 70.6 | 2.33 |
| 72.7 | 0.705 | 60.3 | 0.21 | 77.5 | 2.88 | | 84.9 | 2.7[e] | 74.9 | 3.16 |
| 75.0 | 0.93 | 70.6 | 0.75 | 80.1 | 3.85 | | 90.1 | 2.5 | 79.9 | 3.71 |
| 77.7 | 1.10 | 74.6 | 1.21 | 82.5 | 4.45 | | 93.0 | 2.0 | 84.9 | 3.33[f] |
| 80.1 | 1.25 | 74.7 | 1.32 | 84.9 | 4.73[c] | | 96.0 | 1.3 | 90.3 | 2.70 |
| 82.9 | 1.32 | 85.2 | 1.46[b] | 85.2 | | 0.918[d] | 100.1 | 0.495 | 96.0 | 1.11 |
| 84.9 | 1.38[a] | | | 87.5 | 4.73 | | | | 98.1 | 1.01 |
| 87.6 | 1.44 | 96.0 | 1.46 | 90.1 | 4.6 | | | | 100.1 | 0.43 |
| 90.1 | 1.48 | 98.1 | 1.14 | 93.0 | 3.4 | | | | | |
| 93.6 | 1.57 | 100.1 | 0.63 | 96.0 | 2.7 | | | | | |
| 96.0 | 1.42 | | | 99.7 | | 0.33[h] | | | | |
| 98.3 | 1.05 | 92.0 | 0.03[g] | 100.1 | 1.1 | | | | | |
| 100.1 | 0.63 | 99.7 | 0.045[g] | | | | | | | |
| 100.4 | 0.70 | | | | | | | | | |
| 100.8 | 1.3 | | | | | | | | | |
| $k_{rel}$ | 1 | | 1 | | 4.1 | 4.5 | | 19.5 | | 20.6 |

[a] 0.147 (80 °C), 0.462 (90 °C), $E_a = 29.1$, $\Delta S^\ddagger = 3.9$.

[b] 0.518 (90 °C), 0.166 (80 °C), 0.05 (70 °), $E_a = 28.6$, $\Delta S^\ddagger = 2.8$.

[c] 1.79 (90 °C), 0.608 (80 °), $E_a = 26.9$, $\Delta S^\ddagger = 0.7$.

[d] At 90 °C. Also 0.743 (80 °), 0.274 (70 °), 0.185 (60 °), $E_a = 26.1$, $\Delta S^\ddagger = -1.3$.

[e] 7.4 (90 °) C, 1.07 (70 °), $E_a = 23.8$, $\Delta S^\ddagger = -5.1$.

[f] 1.33 (70 °C) 0.48 (60 °), 0.15 (50 °), $E_a = 24.3$, $\Delta S^\ddagger = -3.4$.

[g] With methanesulphonic acid.

[h] At 80 °C.

since aldehydes give carbon monoxide rather than formic acid[651]. Where both methods were used, rate coefficients were in good agreement at the higher acid concentrations, but diverged at the lower concentrations, this being attributed to poorer solubility of the aldehyde in the weaker sulphuric acid media, so that some extraction into the separated reaction product hydrocarbon layer occurred in the gasometric method; for the spectrophotometric method the small quantities of aromatic used eliminated this possibility. The kinetic data are given in Table 220, some of the coefficients being the *average* of the values observed. The effect of solubility upon the rate coefficients (gasometric) is indicated by the fact that increasing the initial concentration of the *iso*propylaldehyde from 0.06 to 0.28 $M$ decreased the first-order rate coefficient from 2.9 to $2.1 \times 10^{-3}$. The rate of deacylation of mesitaldehyde-3-sulphonic acid in 100 wt. % sulphuric acid was $0.08 \times 10^{-3}$, *i.e.* 8 times less than for the unsulphonated compound, and the effects of salts and ionic solvents on the deacylation of the methyl and *iso*propyl compound are given by the representative data in Table 221.

TABLE 221

RATE COEFFICIENTS FOR DEACYLATION OF 2,4,6-$R_3C_6H_2CHO$ BY 100.1 WT. % $H_2SO_4$[651]

| Added substance | $R = Me(100 °C)$ | | $R = i\text{-}Pr(80 °C)$ | |
|---|---|---|---|---|
| | $(M)$ | $10^3k_1$ | $(M)$ | $10^3k_1$ |
| None | | 0.63 | | 0.495 |
| $(NH_4)_2SO_4$ | 0.28 | 0.80 | | |
| $(NH_4)_2SO_4$ | 0.84 | 1.10 | | |
| $(NH_4)_2SO_4$ | 1.41 | 1.37 | 1.40 | 1.76 |
| $(NH_4)_2SO_4$ | 2.82 | 1.38 | 3.10 | 3.75 |
| $Na_2SO_4$ | 1.41 | 1.36 | 1.40 | 1.60 |
| $NaH_2PO_4$ | 1.51 | 1.37 | | |
| $PhNO_2$ | 2.8 | 0.50 | | |

It is apparent from the data in Tables 220 and 221 that a plot of rate coefficients *versus* wt. % sulphuric acid passes through a maximum and this completely rules out specific acid catalysis. In addition, the maximum occurs at a different acid strength for each compound, which implies a variety of acid-catalysing species whose importance varies according to the aromatic substrate. Spectroscopic studies showed that over the whole range in which a decline in rate coefficient is observed, the aldehydes are present entirely as their conjugate acids. Consequently, formation of these alternative less reactive species cannot account for the gradual decline in rates observed. Neither can the formation of the less reactive sulphonic acid account for the results since the rate of deacylation of this compound is 8 times less than that of the observed rate, and below 90 wt. % acid, it was shown to desulphonate before deacylating. Added salts produced substantial rate acceleration and this seemed to derive from the added anion since addition of a polar solvent, nitrobenzene, produced a rate decrease. The reaction is clearly subject to steric acceleration as observed in protodecarboxylation and as indicated by the differences in entropy since the difference in rigidity of the ground and transition states will vary according to the steric hindrance in the ground state; the differences in the other kinetic parameters are most unlikely to be real and are well within the experimental errors expected in kinetic studies of this kind.

The mechanism of the reaction was interpreted in terms of equilibria (265) and (266) with the second step at least partially rate-determining in view of the acceleration observed with added anions, the effect of base being envisaged as in (XXXIX).

$$Ar \overset{+}{\underset{\underset{H}{\overset{\phantom{x}}{|}}}{\overset{H}{\diagup}} \overset{\phantom{x}}{\underset{\phantom{x}}{C}} \overset{O}{\diagdown_H} \quad \frown_{A^-}$$

(XXXIX)

The species ArCHO is present only in very low concentration since it is in equilibrium with the conjugate acid.

Determination of the solvent and substrate isotope effects in the deacylation of mesitaldehyde yielded the coefficients in Table 222[652]. The substrate isotope

TABLE 222

RATE COEFFICIENTS ($10^3 k_1$) FOR REACTION OF 2,4,6-Me$_3$C$_6$H$_2$CHO WITH SULPHURIC ACID AT 80 °C[652]

| $H_2SO_4$ (wt %) | ArCHO in $H_2SO_4$ | ArCDO in $H_2SO_4$ | $k_H/k_D$ | ArCHO in $D_2SO_4$ | $k_{H_2SO_4}/k_{D_2SO_4}$ (= X) | $\dfrac{X[BD^+]}{[BH^+]}$ |
|---|---|---|---|---|---|---|
| 59 | 0.018 | | | 0.032 | 0.56 | 0.98 |
| 59.88 | 0.022 | 0.012 | 1.8 | | | |
| 65 | 0.054 | | | 0.075 | 0.72 | 1.03 |
| 70 | 0.085 | | | 0.100 | 0.85 | 0.98 |
| 85 | 0.182 | 0.065 | | 0.119(0.051)[a] | 1.5(1.3)[a] | 1.5(1.3)[a] |
| 85.21 | 0.182 | 0.065 | 2.8 | | | |
| 96 | 0.192 | | | 0.091 | 2.1 | 2.1 |
| 96.31 | 0.192 | 0.070 | 2.8 | | | |
| 99.5 | 0.091 | | | 0.038 | 2.4 | 2.4 |
| 100.04 | 0.083 | 0.049 | 1.8 | | | |

[a] For ArCDO.

effect shows clearly that the second step of the reaction, (266), must be at least partially rate-determining, whilst the solvent isotope effect indicates that the first step (265) must be likewise. The existence of a negative solvent isotope effect in media of low acidity was attributed to the difference in p$K$ of the conjugate acids (oxygen protonation) producing a difference in concentration of the unprotonated and reactive species ArCHO in the two media; the first three results in the last column of Table 222 shows that the ratio of carbon to oxygen basicity is the same in each case, but in stronger acid media, oxygen protonation is complete in both solvents and the equilibrium concentrations of ArCHO are then the same in both media. Thus the true solvent isotope effect is *ca.* 1.0 up to 70 % acid and $> 1$ above so that the first step, protonation of the aromatic ring, becomes increasingly rate-determining relative to the second step as the acid concentration increases, *i.e.* in the stronger media $k'_{-1} \sim k'_2$ and in the weaker media $k'_{-1} \gg k'_2$. Hence $k'_2$ becomes greater relative to $k'_{-1}$ as the acid strength increases and the most effective base changes from $H_2O$ to $HSO_4^-$; this was, therefore, attributed to steric hindrance to $HSO_4^-$ in the reverse step of equilibrium (265) since the position of base attack in this step is different (and more hindered) to that in equilibrium (266). Schubert *et al.* argued that if this theory was correct then one would expect this effect to be even more important for 2,4,6-*tri*-isopropylbenzaldehyde. Consequently, the rate coefficients for deacylation of this compound and its deuterated isomer were measured along with its solvent isotope effect for the protium compound[653], the results (some rate coefficients averaged) being summarised in Table 223. The results are more consistent, in that as the isotope effect $k_H : k_D$ decreases, indicating that $k'_2$ is becoming larger relative to $k'_{-1}$ the solvent isotope effect increases showing that the first step is becoming more rate-determining. Whereas with mesitaldehyde the substrate isotope effect was positive at all acid

TABLE 223

RATE COEFFICIENTS ($10^3 k_1$) FOR REACTION OF 2,4,6-$Pr^i_3C_6H_2CHO$ WITH SULPHURIC ACID AT 80 °C[653]

| $H_2SO_4$ (wt %) | ArCHO in $H_2SO_4$ | ArCDO in $H_2SO_4$ | $k_H/k_D$ | ArCHO in $D_2SO_4$ | $k_{H_2SO_4}/k_{D_2SO_4}$ (= X) | $\dfrac{X[BD^+]}{[BH^+]}$ |
|---|---|---|---|---|---|---|
| 70.7 | 2.33 | | | 0.752 | 3.2 | 2.0 |
| 72.5 | 2.70 | 2.12 | 1.27 | | | |
| 77.4 | 3.45 | 2.93 | 1.18 | | | |
| 79.0 | 3.60 | 2.82 | 1.27 | | | |
| 80.0 | 3.70 | | | 0.952 | 3.9 | 1.95 |
| 84.9 | 3.24 | 2.45 | 1.32 | | | |
| 86.6 | 3.05 | | | 0.772 | 4.0 | 2.0 |
| 89.8 | 2.33 | 2.325 | 1.01 | | | |
| 90.9 | 2.30 | 2.20 | 1.04(1.03)[a] | 0.558(0.542)[b] | 4.1(4.1)[b] | 2.05 |
| 93.4 | 2.01 | 1.87 | 1.07 | | | |
| 93.5 | 2.01 | | | 0.478 | 4.2 | 2.15 |
| 95.5 | 1.51 | | | 0.306 | 4.9 | 2.45 |
| 96.8 | 1.305 | 1.30 | 1.01 | | | |
| 99.0 | 0.723 | | | 0.146 | 5.0 | 2.5 |

[a] Data for $D_2SO_4$.
[b] Data for ArCDO.

concentrations, indicating that the second step was partially rate-determining, for tri-*iso*propylbenzaldehyde the isotope effect is much smaller, especially at the higher acid concentrations, indicating that the first step is now almost entirely rate-determining, which would follcw from the greater steric hindrance to ring protonation.

From the shapes of the rate *versus* acid strength graphs obtained for mesit-aldehyde and triisopropylbenzaldehyde it was concluded that although neither compound was specific acid-catalysed, the latter compound showed the nearer tendency to this catalysis at the higher acid concentrations; again this may be a manifestation of the greater steric hindrance to protonation by $H_3^+SO_4$ than by $H_3O^+$.

Deacylation has been studied with 2,4,6-trimethoxybenzaldehyde so that the increased aromatic reactivity would permit behaviour in weaker acid media to be examined[654]. First-order rate coefficients obtained at 80 °C are given in Table 224. Whilst the rate coefficients are identical in each acid up to 2 $M$, there-

TABLE 224

RATE COEFFICIENTS FOR REACTION OF 2,4,6-$(MeO)_3C_6H_2CHO$ WITH ACID AT 80 °C[654]

| $[HClO_4]$ (M) | $10^3k_1$ | $[HCl]$ (M) | $10^3k_1$ | $[HBr]$ (M) | $10^3k_1$ |
|---|---|---|---|---|---|
| 0.103 | 0.052 | 0.36 | 0.21 | 0.24 | 0.13 |
| 1.03 | 0.60 | 1.41 | 1.08 | 0.54 | 0.34 |
| 2.32 | 1.62 | 2.36 | 2.45 | 1.25 | 0.92 |
| 3.83 | 2.99 | 2.98 | 3.87 | 2.01 | 1.79 |
| 4.60 | 3.22 | 4.15 | 7.11 | 2.53 | 2.61 |
| 5.38 | 2.71 | 5.01 | 10.9 | 3.36 | 4.29 |
| 5.43 | 2.66 | 6.10 | 16.2 | 4.16 | 6.46 |
| 6.71 | 1.10 | 7.06 | 18.5 | 4.54 | 7.87 |
| 6.79 | 1.09 | 7.13 | 19.0 | 5.64 | 9.69 |
| 7.94 | 0.45 | 8.75 | 18.5 | 6.60 | 10.2 |
| | | 9.14 | 17.9 | 7.94 | 9.42 |
| | | 9.80 | 15.2 | 8.81 | 7.59 |

after there are significant departures so that even in these dilute acids the A-1 mechanism does not apply. As in the case of the other aldehydes, it was concluded that reaction does not occur with the oxygen protonated species since this is rapidly formed and thus the rate-determining step of the reaction would have to be de-composition of this species as it would be unlikely to protonate further. Since the earlier work ruled out a unimolecular decomposition of this protonated species and it is difficult to envisage how base attack on this species would give the required product, reaction was assumed to occur on the neutral molecule. Reac-tion of the methoxy compound is different from that of the alkylated compounds in that formic acid is produced instead of carbon monoxide, hence it was proposed that here base attacks the carbonyl carbon and not the aldehydic hydrogen, which was concluded to be hydrogen-bonded to the methoxyl oxygen. From an examination of the correlation of $k_{obs}$ with the ratio $[ArCHO]/[ArCHO]_{stoich}$ it appeared that for each medium a measure of specific molecular acid catalysis was obtained, and which decreased along the series HCl > HBr > HClO$_4$.

## 9.5 PROTODEALKYLATION

Although there are a number of reports in the literature of this process, (see ref. 1) only one of these relates to a kinetic study of the reaction[168]. The ease with which the reaction takes place depends upon the stability of the leaving carbonium ion. Consequently, de-$t$-butylation is most frequently observed and in a kinetic study of the sulphonation of $t$-butylbenzene in aqueous sulphuric acid (see p. 72) this side reaction was sufficiently prominent for rates to be easily measurable (Table 225)[168]. Comparison of the rates with those in Table 42 shows that de-

TABLE 225

RATE COEFFICIENTS ($10^6 k_1$) AND ACTIVATION ENERGIES FOR REACTION OF PhCMe$_3$
WITH AQUEOUS H$_2$SO$_4$[168]

| $H_2SO_4$ (wt %) | Temp. °C) | | | $E_a$ |
|---|---|---|---|---|
| | 5 | 25 | 35 | |
| 77.5 | 0.0083 | 0.57 | 3 | 34.0 |
| 79.8 | 0.096 | 3.98 | 17.0 | |
| 84.0 | 2.82 | 64 | 164 | |
| 85.8 | 14.2 | 140 | 430 | 19.5 |
| 89.1 | 77 | | | |

$t$-butylation occurs 4–5 times slower than sulphonation and also that the activation energy is higher. As in the case of sulphonation this decreases with increasing acidity and the decrease is parallel for the two reactions, it being reasonably suggested that this probably follows from changes in solvent polarity and activity coefficients. Interestingly, the dependence of log rate upon the acidity function $H_0$ gives a slope of ca. 2.0 and thus close to the values observed for hydrogen exchange in this medium; both reactions almost certainly involve rate-determining transfer of a proton to the aromatic.

## 9.6 PROTODESILYLATION

Early qualitative studies on this reaction (267)

$$ArSiR_3 + HX = ArH + SiR_3X \tag{267}$$

indicated that electron-supplying substituents in the aromatic ring increased the rate of reaction and *vice versa*, but no quantitative kinetic studies were made[655].

Most of the kinetic studies have been carried out by Eaborn and by Benkeser and their coworkers, and the first study was that of Eaborn[656], who measured, spectrophotometrically, the rates of cleavage of 4-methoxyphenyltrimethylsilane by hydrochloric and perchloric acids in methanol, and by hydrochloric acid in aqueous dioxan, both at 47.9 °C (Table 226). The reaction was first-order in silane

TABLE 226

RATE COEFFICIENTS[656] FOR REACTION OF 4-MeOC$_6$H$_4$SiMe$_3$ WITH ACID[a] AT 49.7 °C

9 Vol. % methanol

| $[HCl]_4$ (M) | 11.68 | 7.65 | 6.43 | 5.74 | 4.27 | 3.38 | 2.31 | 1.08 | 0.498 |
|---|---|---|---|---|---|---|---|---|---|
| $10^5 k_1$ | 113 | 36.5 | 29.2 | 23.5 | 13.9 | 9.18 | 5.49 | 2.30 | 0.70 |
| $[HClO_4]$ (M) | 5.74 | 4.12 | 2.17 | 0.918 | | | | | |
| $10^5 k_1$ | 335 | 188 | 67.3 | 20.7 | | | | | |

27 Vol. % methanol

| $[HCl]$ (M) | 7.63 | 6.43 | 5.74 | 3.38 |
|---|---|---|---|---|
| $10^5 k_1$ | 28.3 | 20.8 | 18.2 | 8.45 |

Aqueous dioxan

| $[H_2O]$ (M) | 8.12 | 10.11 | 12.64 | 15.17 | 20.22 | 25.28 | 25.28 | 25.28 | 25.28 | 26.90 | 28.77 |
|---|---|---|---|---|---|---|---|---|---|---|---|
| $[HCl]$ (M) | 0.278 | 0.526 | 0.664 | 0.789 | 1.045 | 0.526 | 0.794 | 1.055 | 1.324 | 0.870 | 0.973 |
| $10^5 k_1$ | 2.78 | 6.51 | 8.45 | 10.5 | 20.0 | 0.534 | 1.27 | 3.85 | 23.8 | 16.2 | 24.0 |

[a] 1 ml of which was added to a solution of the silane in 10 ml of the stated solvent.

and a linear relationship existed between the logarithms of the first-order rate coefficients and $H_0$ for aqueous methanol and 45.5 vol. % aqueous dioxan (25.28 $M$ in water) solutions, but with less aqueous dioxan media departure from this linearity was observed. The slopes of the log $k$ *versus* $-H_0$ relationships were approximately unity, but this was fortuitous since it was subsequently shown that the slope varies according to the base chosen to measure the acidity function[657], and arises from the fact that $f_{BH^+}/f_B$ values are not independent of the base in solutions of low dielectric constant. On the basis of the correlation of rate with acidity function, the A-2 mechanism was thought to apply, nucleophilic attack on silicon probably participating in the second slow step *viz.* attack of water on the protonated intermediate.

Further mechanistic evidence was provided by Benkeser and Krysiak[658], who determined the effects of added salts and water on the rates of cleavage of xylyltrimethylsilanes by *p*-toluenesulphonic acid in acetic acid at 25 °C, the progress of the reaction being followed by dilatometry; the first-order rate coefficients are given in Table 227. Clearly the addition of water retards the reaction, as

TABLE 227

RATE COEFFICIENTS FOR REACTION OF 4-MeC$_6$H$_4$SO$_3$H (0.80 $M$) WITH 3,4-Me$_2$C$_6$H$_3$SiMe$_3$
IN HOAc AT 25 °C[658]

| $[H_2O]$ ($M$) | 3.0 | 4.0 | 6.0 | 6.0 | 6.0 | 6.0 | 6.0 |
|---|---|---|---|---|---|---|---|
| [*Added salt*] ($M$) | | | | 0.3, LiClO$_4$ | 0.3, KCl | 0.3, HCl | 1.0, LiCl |
| $10^5 k_1$ | 60.7 | 36.0 | 19.8 | 27.2 | 31.4 | 39.2 | 417 |

was apparent from the data in Table 226, and the addition of salts accelerates it. The effect of salts was attributed to their being solvated, which reduces the effective water concentration and which would, therefore, be most effective with the highly solvated lithium ion. Retardation by added water obviously arises from reduction in the effective acidity of the medium by providing an alternative base for protonation.

A similar effect of added water and salts was also found in the cleavage of alkylsilicon hydrides RSiH$_3$ by aqueous acid and for which the mechanism is probably somewhat similar[657]; this reaction also showed a solvent isotope effect of 2.25 ($k_H/k_D$), and a similar value (1.35) was observed for the cleavage of 4-methoxyphenyltrimethylsilane by hydrogen chloride (4.71 $M$) in 25 mole % aqueous dioxan at 50 °C, the rate coefficients being $64.2 \times 10^{-5}$ (H) and $41.3 \times 10^{-5}$ (D)[659]; the true isotope effect was probably somewhat greater since the presence of hydrogen chloride would have reduced the deuterium content of the added deuterium oxide. Rather greater values were obtained for the cleavage of other trialkylmetal groups from aromatics, but since less hydrogen chloride was required for these reactions the above deuterium concentration effect may be partly responsible. The magnitude of the isotope effect shows that proton transfer occurs in a step which is at least partially rate-determining, and the original mechanistic proposal was replaced by the normal A-S$_E$2 mechanism appropriate to other acid-catalysed aromatic substitutions, and this work, therefore, provided further evidence that linear log $k$ *versus* $-H_0$ plots are a poor diagnosis of reaction mechanisms.

The above results do not rule out the possibility that nucleophilic attack on silicon assists the cleavage, a transition state such as (XL), below being involved,

(XL)

this being originally proposed because the ease of cleavage of the $MR_3$ group, *viz.* $CR_3 \ll SiR_3 < GeR_3 \ll SnR_3 \ll PbR_3$[660], did not appear to correlate with the contemporary values for the electronegativities of the metals. However, later measurements[661-3] showed that the above order is the order of electron release by $MR_3$ and hence it is this release which stabilises the electron-deficient transition state for the reaction. The 4-centre mechanism is thus no longer necessary but cannot be discarded in the absence of evidence to the contrary.

A further examination of the solvent isotope effect in 65–100 mole % $H_2O$–$CF_3COOH$ or $D_2O$–$CF_3COOD$ at 25 °C showed a maximum in the rate with both media at *ca.* 85 mole % acid (Table 228), as was also observed for

TABLE 228

RATE COEFFICIENTS AND SOLVENT ISOTOPE EFFECT FOR REACTION OF 4-$ClC_6H_4SiMe_3$
WITH AQUEOUS TRIFLUOROACETIC ACID AT 25 °C[471]

| $CF_3COOH$ in $H_2O$ (mole %) | 100 | 99.2 | 95.4 | 94.4 | 89.6 | 85.2 | 81.6 | 77.6 | 76.2 |
|---|---|---|---|---|---|---|---|---|---|
| $10^5k_1$ | 102.5 | 107 | 125 | 123 | 131 | 133 | 133 | 124 | 122 |

| $CF_3COOH$ in $H_2O$ (mole %) | 72.15 | 70.7 | 66.4 | 63.6 | 50.9 | 100[a] | 92.2[b] | 99.2[c] | 99.2[d] |
|---|---|---|---|---|---|---|---|---|---|
| $10^5k_1$ | 108 | 100 | 80 | 69 | 22 | 114 | 116 | 700 | 495 |

| $CF_3COOD$ in $D_2O$ (mole %) | | 99.6 | 88.0 | 83.2 | 78.8 | 64.7 |
|---|---|---|---|---|---|---|
| $10^5k_1$ | | 16.8 | 18.0 | 17.8 | 17.0 | 11.7 |

[a] 0.1$M$ Lithium trifluoroacetate added.
[b] 0.1$M$ Sodium trifluoroacetate added.
[c] 0.1 $M$ Lithium chloride added.
[d] 0.1 $M$ Lithium perchlorate added.

detritiation (p. 202), and an isotope effect $(k_H/k_D)$ which ranged from 6.2 in 100 mole % acid through 7.4 at 80 mole % acid to 7.0 at 65 mole % acid[471]. The maximum was attributed to competition between lowering of the rate as the base water is added, thereby lowering the acidity probably by replacing the dimeric $CF_3COOH$ by the more weakly acidic monohydrate $CF_3COOH.H_2O$, and increasing the rate through better solvation of the transition state by water; in this respect the smaller increase in rate on adding deuterium oxide compared with protium oxide follows from the poorer solvating ability of the former. The magnitude of the isotope effect is such that the proton transfer must be about half complete by the time the transition state is reached.

The reactivity of 4-chlorophenyltrimethylsilane towards trifluoroacetic acid in this study was comparable to that of 2,4-dimethyltritiobenzene (for hydrogen exchange) and the similarity of the reactivity *versus* medium composition profiles argues strongly for a similar transition state in each reaction. However, substituent effects are much larger in hydrogen exchange than in protodesilylation and it was proposed that this arises through ground state stabilisation or destabilisation by the substituent of the $p_\pi - d_\pi$ bonding in the arylsilicon compounds so that the net effect of the substituent upon a transition state of similar charge to that in hydrogen exchange would thus be less.

The final aspect of the mechanism, namely the effect of different electron supply in the alkyl groups of $ArSiR_3$ has now been settled. From the ease of cleavage of $MR_3$ groups (M = metal) noted above, one would expect that increased electron supply from R would increase the reaction rate. The first kinetic studies[664] in fact indicated the opposite, as shown by the data in Table 229, and although the

TABLE 229

RATE COEFFICIENTS ($10^5 k_1$) FOR REACTION OF $3\text{-}RC_6H_4SiX_3$ WITH ACID AT 25 °C[664]

| R | $SiMe_3$[a] | $k_{rel}$ | $SiEt_3$[a] | $k_{rel}$ | $Si(i\text{-}Pr)_3$[b] | $k_{rel}$ |
|---|---|---|---|---|---|---|
| H | 6.4 | 1 | 2.74 | 1 | 0.257 | 1 |
| Me | 13.8 | 2.2 | 6.45 | 2.4 | 1.39 | 5.4 |
| Et | 14.2 | 2.2 | 7.82 | 2.9 | 1.65 | 6.4 |
| *i*-Pr | 15.1 | 2.4 | 8.57 | 3.1 | 2.03 | 7.9 |
| *t*-Bu | 18.0 | 2.8 | 9.84 | 3.6 | 2.38 | 9.3 |

[a] In acetic acid, 2.35 $M$ in HCl and 7.23 $M$ in $H_2O$.
[b] In acetic acid, 1.78 $M$ in $HSO_4^-$ and 0.44 $M$ in $H_2O$ (average rates).

rate coefficients for the tri-*iso*propyl compounds are not directly comparable with the others, the much enhanced $k_{rel}$ values indicate that the reactivity is markedly diminished in this series. The same decrease was noted in the reaction of $PhSiMe_3$, $PhSiMe_2Et$, $PhSiEt_3$, and $PhSi(i\text{-}Pr)_3$ with acetic acid 1.2 $M$ in *p*-toluenesulphonic acid and 4.0 $M$ in water, at 25 °C, the corresponding rate coefficients ($10^5 k_1$) being 0.210, 0.198, 0.1035, and 0.0133; the very marked decrease on going to the *iso*propyl compound strongly suggests a steric effect[665]. A similar effect was noted by Eaborn and Pande[666] for the cleavage of 4-methoxyphenyltrialkylsilanes in ethanol (5 vol.) containing aqueous perchloric acid (1 vol. )at 50 °C, the rate coefficients ($10^5 k_1$) being: 8.01 $M$ $HClO_4$, ethyl 43.8, cyclohexyl 4.22; 4.10 $M$ $HClO_4$, ethyl 5.05, cyclohexyl 0.467; 2.05 $M$ $HClO_4$, ethyl 1.23. The steric effect has been eliminated by kinetic studies of compounds of the type $4\text{-}MeOC_6H_4Si$ $(C_6H_4X)_3$, where variation in X would have little effect on approach of the reagent to the reaction site but would have a measurable electronic effect[667]. The data are given in Table 230 (along with data for other $SiR_3$ substituents). Comparison

## TABLE 230

RATE COEFFICIENTS ($10^5 k_1$) FOR REACTION OF 4-MeOC$_6$H$_4$SiR$_3$ WITH HClO$_4$ IN MeOH AT 50 °C[667]

| $R_3$ | 0.293 | 2.55 | 3.05 | 3.12 | 3.68 | 4.26 | 4.58 | 6.0 | 6.37 | 7.50 | 9.14 | 9.20 | 9.48 | 11.96 | 12.19 | 12.9 |
|---|---|---|---|---|---|---|---|---|---|---|---|---|---|---|---|---|
| | | | | | | | [HClO$_4$] (M)[a] | | | | | | | | | |
| Me$_3$ | 3.70 | | | 98.3 | 180 (66.7)[b] | | 273 (112) | | | | | | | | | |
| Et$_3$ | | | | | 86.7 (32.0) | | 138 (55.0) | | | | | | | | | |
| n-Pr$_3$ | | | | | 75.0 (28.5) | | 118 (46.3) | | | | | | | | | |
| i-Pr$_3$ | | | | | | | | | | | 203 (80.0) | | | | | |
| Me$_2$Ph | | 21.5 | 31.1 | 30.0 | | 67.5 | 92.3 | | 255 | 492 | | | | | | |
| MePh$_2$ | | | | | | | | 38.0 | 57.5 | 98.0 | 275 | | | | | |
| Ph$_3$ | | | | | | | | | 26.2 | | 59.2 (21.5) | 62.8 | | 285 | | |
| (4-MeOC$_6$H$_4$)$_3$ | | | | | | | | | | | | | | | | |
| (4-Me$_3$SiCH$_2$C$_6$H$_4$)$_3$ | | | | | | | | | | | | | | | | ~4.25[c] |
| (4-MeC$_6$H$_4$)$_3$ | | | | | | 16.5 | | | | | 132 | 83.1 | 98.0 | 154 | | |
| (3-MeC$_6$H$_4$)$_3$ | | | | | | | | | | | | | 22.5 | | 31.0 | |
| (4-ClC$_6$H$_4$)$_3$ | | | | | | | | | | | | | 7.97 | 96 | | |
| (3-ClC$_6$H$_4$)$_3$ | | | | | | | | | | | | | | ~1.3 | ~1.0 | |
| (2-MeC$_6$H$_4$)$_3$ | | | | | | | | | | | | | | | | |
| (PhCH$_2$)$_3$ | | | | | | 19.7 | | 75.3 | | | | | | | | |
| Me$_2$(ClCH$_2$) | | | | | | | | | 11.8 | | | | | 16.7 | | |
| Me(ClCH$_2$)$_2$ | | | | | | | | | | | | | ~167 | | | |
| Me$_2$(CH$_2$NH$_2$Et$^+$) | | | | | | | | | | | ~54.3[d] | | | ~60.2 | | |
| (EtO)$_3$ | | | | | | | | | | | | | | | | |

[a] Concentration of perchloric acid, 2 ml of which was added to a solution of the organosilane in methanol (5 ml).

[b] Rate coefficients in parentheses were determined at 40 °C.

[c] This rate is approximate since several Si-C bonds are being broken simultaneously and cannot be compared directly with the other data without making a correction for this[668].

[d] This rate almost certainly refers to the hydrolysed product[668].

of the effects of the 4-methyl and 4-chloro substituents on the phenyl ring show
that electron supply does indeed increase the rate of cleavage (of the 4-methoxy-
phenyl group) but by a much smaller factor than when these groups are in the
aromatic ring being substituted, which is to be expected since in the latter case the
effect does not have to be transmitted through the silicon atom. The variation
in rate coefficient on changing R from methyl to *iso*propyl confirms the result of
Benkeser *et al.*[665] both in direction and in being a steric effect, and this is also
indicated by the marked reduction in rate which arises from the presence of an
*ortho*-methyl substituent in the phenyl ring which not only shields the reaction
site but hinders solvation in the 4-methoxy-substituted phenyl ring; the steric
effect here and also for some of the other cases is twofold. Activation energies
determined from the above data did not vary by more than the experimental error
from 19.5 so that no conclusions are appropriate, but they did appear to be larger
for the less acidic media as expected.

   The ease with which both the arylsilicon compounds can be made and the
kinetics of subsequent protodesilylation measured, has led to the very intensive
use of this reaction as a means of determining substituent effects in aromatic
substitution, the major bulk of the work having been carried out by Eaborn *et al.*
In order to cover a wide range of substituent effects at measurable rates, the media
were varied to give convenient rates, it being assumed (since the evidence indicates
this) that the relative reactivities are not much altered by these changes. The
majority of these studies have employed methanol (5 vol.)+aqueous perchloric
acid (2 vol.) as the reaction medium, the strength of the added acid being given
along with the rate coefficients (determined spectrophotometrically) in Table
231[525, 663, 669–678, 682]. The range of media in which rates have been measured
is such that it is possible to obtain the reactivity of any one compound relative
to another from the data in the Table. Where it is impossible to obtain a relative
rate directly, and comparison with some intermediate rate is necessary, the
resultant $k_{rel}$ value may differ slightly from the literature value since it will depend
to some small extent on the intermediate rate or rates selected; variations are insufficient to alter any of the conclusions regarding substituent effects and which
have been drawn from this work.

   A discussion of this enormously extensive range of substituent effects is inappropriate here but certain of the kinetic features require comment. Firstly,
the rate coefficients increase regularly with increasing acid concentration except
in one or two instances. The rate coefficients in 12.07–12.5 $M$ perchloric acid for
phenyltrimethylsilane do not increase regularly as expected, but this is almost
certainly due to inaccuracies resulting from pipetting the 2 : 5 volume mixtures
which will be magnified by both the viscosity and strength of these the most concentrated acids (see also later). The rate coefficients for the dimethylamino compound increase then decrease as the acid strength is increased and this clearly
arises from a decrease in the concentration of the neutral amine (which must,

## TABLE 231

RATE COEFFICIENTS ($10^5 k_1$) FOR REACTION OF ArSiMe$_3$ COMPOUNDS WITH METHANOL (5 VOL.)+ PERCHLORIC ACID (2 VOL.) AT 50, 50.15 (ITALICISED) OR AT 51.2 °C, (IN PARENTHESES)[525, 663, 669–678, 682]

| Ar | [HClO$_4$](M) | | | | | | | | | | |
|---|---|---|---|---|---|---|---|---|---|---|---|
| | 0.001056 | 0.01056 | 0.103 | 0.1056 | 0.200 | 0.44 | 0.56 | 0.62 | 0.823 | 2.35 | 2.86 |
| 4-NH$_2$C$_6$H$_4$ | (76.7) | (175) | (193.5) | | | | | | (175) | | |
| 2,4,6-Me$_3$C$_6$H$_2$ | | 35.6 | (40.3) | 67.3 | (193.5) | | | 293 | (430) | | |
| 3-(9-ethylcarbazolyl) | | | | 63.5 | | | | | | | |
| 4-HOC$_6$H$_4$ | | (8.91) | | | | (39.2) | | | (85.3) | | |
| 2-furyl | | | | | | | 70.4 | | | | |
| 2-thienyl | | | | | | | 19.7 | | | | 287 |
| 2,6-Me$_2$C$_6$H$_3$ | | | | | | | | 19.5 | | | |
| 4-MeOC$_6$H$_4$ | | | | | | | | | (12.0) | (66.3) | 90.0 |
| 4-SiMe$_3$CH$_2$C$_6$H$_4$ | | | | | | | | | (13.85) | | |

| Ar | [HClO$_4$] (M) | | | | | | | | | | |
|---|---|---|---|---|---|---|---|---|---|---|---|
| | 3.06[a] | 3.08 | 4.53 | 4.68 | 6.05 | 6.16 | 6.21 | 6.76 | 7.34 | 7.54 | 7.60 |
| 4-MeOC$_6$H$_4$ | | | 303 | | | | | | | | |
| 4-SiMe$_3$CH$_2$C$_6$H$_4$ | 1.04 | | 59.2 (67.5) | | | | | 195 | 312 | | 355 |
| 4-GeMe$_3$CH$_2$C$_6$H$_4$ | 15.2 | | | | | | | 336 | 525 | | |
| 4-SiMe$_3$CH.Pr$^n$C$_6$H$_4$ | | | 56.7 | | | | | | | | |
| 2-MeOC$_6$H$_4$ | | | (71.7) | | | | | 423 | | | |
| 2-MeO-5-F-C$_6$H$_3$ | | | | | | | | 9.00 | | | |
| 2-MeO-5-Cl-C$_6$H$_3$ | | | | | | | | 5.09 | | | |
| 2-MeO-5-Br-C$_6$H$_3$ | | | | | | | | 4.75 | | | |
| 2-MeO-5-I-C$_6$H$_3$ | | | | | | | | 7.17 | | | |
| 4-PhOC$_6$H$_4$ | | | | | | | | 45.9 | | | |
| 3-pyrenyl | | | 52.5 | 116 | | | | | | | |
| 2-benzthienyl | | | | 20.8 | | | | | | | |
| 4-MeSC$_6$H$_4$ | | | | 15.2 | | | | | | | |
| 2,6-Me$_2$C$_6$H$_3$ | | 230 | | | | | | | | | |
| 2,4-Me$_2$C$_6$H$_3$ | | 28.0 | | | 194 | | | | | | |
| 2,3-Me$_2$C$_6$H$_3$ | | 4.77 | | | 33.3 | | | | | | |
| 3,4-Me$_2$C$_6$H$_3$ | | 3.65 | | | 26.0 | | | | | | |
| 2,5-Me$_2$C$_6$H$_3$ | | 2.80 | | | 19.9 | | | | | | |
| 3,5-Me$_2$C$_6$H$_3$ | | | | | 2.78 | | | | | | |
| 4-MeC$_6$H$_4$ | | | 4.50 | | 10.6 | | | | 21.2 | | 23.2 |
| 4-Bu$^t$CH$_2$C$_6$H$_4$ | | | (4.53) | | | | | | | | 22.3 |
| 2-MeC$_6$H$_4$ | | | | | | | 8.47 | | 19.2 | | |
| 3-MeC$_6$H$_4$ | | | | | | | | | 2.5 | | |
| C$_6$H$_5$ | | | | | | | | | 1.05 | | |

TABLE 231 (continued)

| Ar | [HClO₄] (M) | | | | | | | | | | |
|---|---|---|---|---|---|---|---|---|---|---|---|
| | 7.66 | 9.09 | 9.16 | 9.22 | 9.4 | 9.45 | 10.6 | 12.05 | 12.07 | 12.4 | 12.5 |
| 4-NMe$_2$C$_6$H$_4$ | | (89.2) | | | | | | | | | |
| 4-PhOC$_6$H$_5$ | 118 | | | | | | | | | | |
| 2-benzthienyl | | | | | | 127 | | | | | |
| 2-MeC$_6$H$_4$ | | 44.1 | | (53.4) | | | | | | | |
| 2,3-Me$_2$C$_6$H$_3$ | | | | | | | 322 | | | | |
| 3,4-Me$_2$C$_6$H$_3$ | | | | | | | 255 | | | | |
| 4-MeC$_6$H$_4$ | 24.5 | 54.1 | | (63.3) | 60.3 | 64.0 | | | | 209 | |
| 4-Bu$^t$CH$_2$C$_6$H$_4$ | | | | | 58.0 | | | | | | |
| 3-MeC$_6$H$_4$ | | | | (6.83) | | | | 30.8 | | 28.8 | |
| C$_6$H$_5$ | | 2.60 | 3.0$^b$ | (3.0) | | 3.2 | | 12.15 | 13.1 | 12.6 | 11.1 |
| 4-EtC$_6$H$_4$ | | | | (58.3) | | | | | | | |
| 4-Pr$^i$C$_6$H$_4$ | | | | (51.6) | | | | | | | |
| 4-Bu$^t$C$_6$H$_4$ | | | | (46.7) | | | | | | | |
| 4-SiMe$_3$C$_6$H$_4$ | | | | (7.56) | | | | | | | |
| 3-SiMe$_3$CH$_2$C$_6$H$_4$ | | | | (18.5) | | | | 80.0 | | | |
| 3-SiMe$_3$CHPr$^n$C$_6$H$_4$ | | | | | | | | 9.92 | | | |
| 3-Bu$^t$CH$_2$C$_6$H$_4$ | | | | | | | | | | 51.0 | |
| 4-PhC$_6$H$_4$ | | | | (10.65) | | | | | | | |
| α-naphthyl | | | 24.2 | | | | | | | | |
| β-naphthyl | | | 6.5 | | | | | | | | |
| 2-fluorenyl | 57.2 | | | | | 146 | | | | | |
| 2-(9-methylfluorenyl) | 52.7 | | | | | 135 | | | | | |
| 2-(9,9-dimethyl-fluorenyl) | 43.7 | | | | | 114 | | | | | |
| 5-tetralinyl | | | | | | | 392 | | | | |
| 6-tetralinyl | | | | | | | 302 | | | | |
| 4-indanyl | | | | | | | 128 | | | | |
| 5-indanyl | | | | | | | 350 | | | | |
| 3-benzcyclobutyl | | | | | | | 26.8 | | | | |
| 4-benzcyclobutyl | | | | | | | 257 | | | | |
| 2-PhOC$_6$H$_4$ | | | | | | | | | 114 | | |
| 2-PhSC$_6$H$_4$ | | | | | | 34.2 | | | | | |
| 3-benzthienyl | | | | | | 130 | | | | | |
| 1-benzfuranyl | | | | | | | | | 8.54 | | |
| 2-benzfuranyl | | | | | | 61.5 | | | | | |
| 3-benzfuranyl | | | | | | 7.63 | | | 31.5 | | |
| 4-benzfuranyl | | | | | | | | | 12.1 | | |
| 1-dibenzthienyl | | | | | | | | | 72.7 | | |
| 2-dibenzthienyl | | | | | | 20.0 | | | | | |
| 3-dibenzthienyl | | | | | | | | | 26.2 | | |
| 4-dibenzthienyl | | | | | | | | | 15.0 | | |
| 2-(9,10-dihydrophenan-thryl) | | | | | | | | | 168 | | |
| 9-phenanthryl | | | | | | 14.3 | | | 58.3 | | |
| 3-phenanthryl | | | | | | | | | 27.3 | | |
| 2-phenanthryl | | | | | | | | | 23.0 | | |
| 9-anthracyl | | | | | | | | | 53.8 | | |
| 4-FC$_6$H$_4$ | | | | (2.25) | | | | | | | |
| 4-ClC$_6$H$_4$ | | | | (0.40) | | | | | | | |

TABLE 231 (continued)

| Ar | [HClO₄] (M) | | | | | | | | | | |
|---|---|---|---|---|---|---|---|---|---|---|---|
| | 7.66 | 9.09 | 9.16 | 9.22 | 9.4 | 9.45 | 10.6 | 12.05 | 12.07 | 12.4 | 12.5 |
| 4-BrC₆H₄ | | | | (0.30) | | | | | | | |
| 3-Buᵗ.C₆H₄ | | | | | | | | | | | 43.0 |
| 3-SiMeC₆H₄ | | | | | | | | | | | 36.5 |
| 3-HOC₆H₄ | | | | | | | | | | | 9.69 |
| 3-PhOC₆H₄ | | | | | | | | | | | 3.97 |
| 3-MeSC₆H₄ | | | | | | | | | | | 2.08 |
| 2-SiMe₃CH₂C₆H₄ | | 80.8 | | | | | | | | | |
| 2-SiMe₃(CH₂)₂C₆H₄ | | 45.3 | | | | | | | | | |
| 2-SiMe₃(CH₂)₃C₆H₄ | | 31.2 | | | | | | | | | |
| 2-SiMe₃(CH₂)₄C₆H₄ | | 35.0 | | | | | | | | | |
| 3-SiMe₃(CH₂)₂C₆H₄ | | | | | | | | 44.2 | | | |
| 3-SiMe₃(CH₂)₃C₆H₄ | | | | | | | | 46.9 | | | |
| 3-SiMe₃(CH₂)₄C₆H₄ | | | | | | | | 42.9 | | | |
| 4-SiMe₃(CH₂)₂C₆H₄ | | 73.4 | | | | | | | | | |
| 4-SiMe₃(CH₂)₃C₆H₄ | | 56.7 | | | | | | | | | |
| 4-SiMe₃(CH₂)₄C₆H₄ | | 61.7 | | | | | | | | | |

ᵃ This medium is very probably 2.06 M (see text).
ᵇ Incorrectly quoted as 0.30 in ref. 670.

therefore, be the reactive species) through protonation in the stronger acid media. The apparent decrease in the rate for the 4-trimethylsilylmethylphenyl compound in 3.06 $M$ acid (relative to that for 2.35 and 4.53 $M$ acid) probably arises from an error either from measuring the acid strength, or typographically and it should be noted that an acid strength of 2.06 $M$ would give approximately the observed rate; the value is stated as 3.06 $M$ in the original thesis[676].

One further kinetic feature emerges from these data, and these is that proto-desilylation rates are considerably less in ethanol than in methanol. The rate coefficient obtained[666] for the 2,4,6-trimethylphenyl compound in ethanol (5 vol.) and 0.97 $M$ perchloric acid (1 vol.) is $34.8 \times 10^{-5}$, whereas in methanol (5 vol.) and 0.44 $M$ perchloric acid (2 vol.) the rate coefficient from Table 231 is $193.5 \times 10^{-5}$, and these acid strengths are roughly equivalent (in view of the volume difference). Thus the rate in methanol is approximately six times greater and this can probably be attributed to the acidity of $MeOH_2^+$ being greater than $EtOH_2^+$ but even so the magnitude of this difference is quite surprising and it would be useful to have a more accurate comparison of the reactivities.

A very wide range of substituent effects have also been measured using acetic acid (4 vol.) and aqueous sulphuric acid (3 vol.), the rate coefficients for the corresponding strengths of the added sulphuric acid being given in Table 232 (refs. 537, 673, 679–685). It will be noted that the rate coefficients do not always increase with an increase in the strength of the sulphuric acid (as stated) and this arises from the same reason noted above except that the problem becomes more magnified by the use of the more viscous sulphuric acid and attention has been drawn to this[682]. As in the previous case, this is not important insofar as the relative reactivities are determined at a particular acid concentration, and the true concentration does not matter since relative reactivities are virtually independent of the acid concentration. It is, however, recommended that the range of rate coefficients be consulted in any attempted repetition of these kinetic studies.

Activation energies and log $A$ values have been determined for some compounds over the temperature range 40.06–50.18 °C but the range of the former is barely outside the possible experimental error of $\pm 1.5$ kcal.mole$^{-1}$ for rates reproducible to $\pm 1.5\%$ (as quoted) for a 10 °C measurement range, and similar conclusions apply to the log $A$ values, so that discussion of the variations is inappropriate, especially since the values depend upon the medium composition[679, 680]. The activation energies averaged 21.0 and the log $A$ values *ca.* 11.5 (after correction of rates to sec$^{-1}$) so that a concerted reaction (proposed earlier) would seem to be quite possible since the entropy of activation will be of the order of 7 e.u.

The kinetics of cleavage of some of these compounds have been measured by dilatometry and in other acidic media, and the rate coefficients and relative rates are given in Table 233[658, 673, 686, 687]. It can be seen that there is relatively little variation in the spread of rate coefficients with change in the acid and this argues against nucleophilic participation of the acid in the rate-determining step

TABLE 232

RATE COEFFICIENTS ($10^5 k_1$) FOR REACTION OF ArSiMe$_3$ COMPOUNDS WITH ACETIC ACID (4 VOL.)+SULPHURIC ACID (3 VOL.) AT 50 °C OR AT 50.2 °C (ITALICISED)[537, 673, 679–685]

| Ar | [$H_2SO_4$] (M) | | | | | | | |
|---|---|---|---|---|---|---|---|---|
| | 1.15 | 5.65 | 6.41 | 7.6 | 8.1 | 8.75 | 9.5 | 9.6 |
| 2-thienyl | 570 | | | | | | | |
| 3-thienyl | *131* | 699 | | | | | | |
| 4-MeOC$_6$H$_4$ | 115.5 | | | | | | | |
| 4-Me$_3$SiCH$_2$C$_6$H$_4$ | 23.0 | | | | | | | |
| 4-MeC$_6$H$_4$ | | *109* | 193.5 | | | | | |
| 2-SiMe$_3$C$_6$H$_4$ | | | | | 833 | | | |
| 2-Bu$^t$C$_6$H$_4$ | | | | | 392 | | | |
| 2-PhC$_6$H$_4$ | | | | | | 485 | | |
| 4-PhC$_6$H$_4$ | | | *30.3* | 88.4 | | 235 | | |
| 4-(4'-MeOC$_6$H$_4$)C$_6$H$_4$ | | | | 278 | | | | |
| 4-(4'-MeC$_6$H$_4$)C$_6$H$_4$ | | | | 158.5 | | | | |
| 4-(4'-SiMe$_3$C$_6$H$_4$)C$_6$H$_4$ | | | | 118 | | | | |
| 4-(4'-ClC$_6$H$_4$)C$_6$H$_4$ | | | | 48.3 | | | | 305 |
| 4-(4'-BrC$_6$H$_4$)C$_6$H$_4$ | | | | | | | | 272 |
| 4-(4'-MeO$_2$CC$_6$H$_4$)C$_6$H$_4$ | | | | | | | | 122 |
| 4-(4'-HO$_2$CC$_6$H$_4$)C$_6$H$_4$ | | | | | | | | 107 |
| 4-(4'-NO$_2$C$_6$H$_4$)C$_6$H$_4$ | | | | | | | | 51.7 |
| C$_6$H$_5$ | | | | 30.8 | 48.9 | | 183 | |
| 3-ClC$_6$H$_4$ | | | | | | | 2.18 | |

TABLE 232 (continued)

| Ar | [$H_2SO_4$] (M) | | | | | | | |
|---|---|---|---|---|---|---|---|---|
| | 16.0 | 16.3 | 16.4 | 18.3ᵃ | 18.4 | 18.7 | 19.0 | 19.1 |
| 3-ClC$_6$H$_4$ | | | *533* | | | | | |
| | | | *639* | | | | | |
| 2-BrC$_6$H$_4$ | | 1,190 | | | | | | |
| 3-CF$_3$C$_6$H$_4$ | | 140 | | | | | | |
| 4-HO$_2$CC$_6$H$_4$ | | | *78.7* | | 315 | | | |
| 4-HO$_3$SC$_6$H$_4$ | | 51.5 | | | | 183 | | 156 |
| 2-HO$_2$CC$_6$H$_4$ | | 223 | | | | | | |
| 2-HO$_3$SC$_6$H$_4$ | | 114 | | | | | | |
| 2-NO$_2$C$_6$H$_4$ | | | | | | | | 9.07 |
| 3-NO$_2$C$_6$H$_4$ | | | | | | 74.9 | | |
| 4-NO$_2$C$_6$H$_4$ | | 4.80 | | | 23.0 | 28.5 | | 23.0 |
| | | | | | *26.0* | | | |
| 4-Me$_2$HN$^+$C$_6$H$_4$ | | | | | | | | 158 |
| 4-Me$_3$N$^+$C$_6$H$_4$ | | | | *81.0* | 55.0 | | | |
| | | | | | *82.0* | | | |
| 2-Me$_3$HN$^+$C$_6$H$_4$ | | | | | | | | 9.03 |
| 4-Me$_3$P$^+$C$_6$H$_4$ | | | | | 18.3 | | | |
| 4-(HO)$_2$(PO)C$_6$H$_4$ | | | | | 55.0 | | | |
| 4-(EtO)$_2$(PO)C$_6$H$_4$ | | | | | 30.7 | | | |
| 4-Me$_2$(PO)C$_6$H$_4$ | | | | | 15.7 | | | |
| 4-Ph$_2$(PO)C$_6$H$_4$ | | | | | 10.7 | | | |
| 3-(HO)$_2$(PO)C$_6$H$_4$ | | 73.3 | | | 250 | | | |
| 3-(EtO)$_2$(PO)C$_6$H$_4$ | | 55.0 | | | 150 | | | |
| 3-Ph$_2$(PO)C$_6$H$_4$ | | | | *73.3* | | | | |
| 2-CH$_3$CO.C$_6$H$_4$ | | | | | | ~0.042 | | |
| 2-PhCO.C$_6$H$_4$ | | | | | | ~0.25 | | |
| 3-CH$_3$CO.C$_6$H$_4$ | ~17 | | | | | | | |
| 3-PhCO.C$_6$H$_4$ | ~5.8 | | | | | | | |
| 4-CH$_3$CO.C$_6$H$_4$ | ~1.7 | | | | | | | |
| 4-PhCO.C$_6$H$_4$ | ~3.3 | | | | | | | |
| C$_6$F$_5$ | | | | *3.10* | | | | |
| 3-Me$_3$N$^+$CH$_2$C$_6$H$_4$ | | *550* | | | | | | |
| 4-Me$_3$N$^+$CH$_2$C$_6$H$_4$ | | *242* | | | | | | |
| 3-Me$_3$As$^+$C$_6$H$_4$ | | | *28.0* | | *134* | | | |
| 3-Me$_3$P$^+$C$_6$H$_4$ | | | | | 67.5 | | | |
| 4-Me$_3$As$^+$C$_6$H$_4$ | | | | | 55.0 | | | |
| 4-Me$_3$N$^+$C$_6$H$_4$ | | | | | 50.0 | | | |
| 3-Me$_3$N$^+$C$_6$H$_4$ | | | | | 49.2 | | | |
| 4-Me$_3$P$^+$C$_6$H$_4$ | | | | | 18.5 | | | |

ᵃ In ref. 683 it was incorrectly stated that equal volumes of aqueous sulphuric acid and acetic acid were used; the volume ratios were, in fact, 3:4 as used for the rest of the studies reported in this table.

ᵇ Probably in error in view of the rate coefficients for the 3-ClC$_6$H$_4$ substituent.

TABLE 232 (continued)

| Ar | [H₂SO₄] (M) | | | | | | | | | |
|---|---|---|---|---|---|---|---|---|---|---|
| | 9.9 | 10.0ᵃ | 10.1 | 10.2 | 11.7 | 12.2 | 12.3 | 12.7 | 13.0ᵇ | 15.9 |
| 4-PhC₆H₄ | | | 769 | | | | | | | |
| C₆H₅ | 288 | 217 | 272 | 272 | | | | 700 | | |
| 3-ClC₆H₄ | 3.75 | | | | 14.8 | | 25.7 | 29.0 | 7.75 | 642 |
| 3-(EtO)₂(PO)OC₆H₄ | | | | | 13.3 | | | | | |
| 3-(EtO)₂(PO)CH₂C₆H₄ | 61.7 | | | | | | | | | |
| 3-(HO)₂(PO)CH₂C₆H₄ | 9.7 | | | | | | | | | |
| 4-(EtO)₂(PO)OC₆H₄ | 102 | | | | | | | | | |
| 4-(EtO)₂(PO)CH₂C₆H₄ | 202 | | | | | | | | | |
| 4-(HO)₂(PO)CH₂C₆H₄ | 223 | | | | | | | | | |
| 4-FC₆H₄ | | 208 | 260 | | | | | | | |
| 4-ClC₆H₄ | | | 57.5 | | | | | | | |
| 4-BrC₆H₄ | | | 28.3 | | | | 252 | | | |
| 4-IC₆H₄ | | | 27.5 | | | | | | | |
| 2-FC₆H₄ | | 15.75 | | 19.7 | 129 | | | | | |
| 2-ClC₆H₄ | | | | | | 66.6 | | | | |
| 2-BrC₆H₄ | | | | | | 44.0 | | | | |
| 2-IC₆H₄ | | | | | | 67.3 | | | | |
| 3-MeOC₆H₄ | | | 116.5 | | | | | | | |
| 3-PhC₆H₄ | | | 92.7 | | | | | | | |
| 3-BrC₆H₄ | | | | | | | 25.2 | | | |
| 3-CF₃C₆H₄ | | | | | 2.92 | | 4.45 | | | 122.5 |
| 3-HO₃SC₆H₄ | | | | | | | | | | 86.0 |
| 4-HO₃SC₆H₄ | | | | | | | | | | 53.3 |
| 4-MeO₂CC₆H₄ | | | | | | | | | | 112 |
| 4-HO₂CC₆H₄ | | | | | | | 4.66 | | | 76.0 |
| 3-MeO₂CC₆H₄ | | | | | | | 24.0 | | | |
| 3-HO₂CC₆H₄ | | | | | | | 19.0 | | | 312 |

of the reaction, especially since increase in nucleophilicity of the acid anion does not seem to produce a parallel decrease in the spread of rates, though it must be noted that the order of effectiveness of acids in cleaving phenylsilane at −78 °C is HBr > HCl > HF[688]. The effect of reduction in the steric crowding at the reaction site on going to the protonated transition state, i.e. on changing the hydridisation from $sp^2$ to the less-crowded $sp^3$, is very apparent from the reactivities of the 2,3 and 2,6-dimethyl compounds, hindrance in the former compound arising from a buttressing effect, (and is also marked in the 2,4,6-trimethyl compound[675], Table 231); this effect, steric acceleration, was also proposed as the cause of the unexpected rate enhancement for protodesilylation of the 2 position of biphenyl[684] (Table 232). Steric acceleration of rate does not, however, seem to be important for most singly ortho-substituted compounds, since substituent effects at the ortho-position can be satisfactorily correlated with those at the para[681].

A few substituent effects have been determined with aqueous hydrochloric

TABLE 233

RATE COEFFICIENTS ($10^5 k_1$) AND $k_{rel}$ FOR PROTODESILYLATION OF $XC_6H_4SiMe_3$ WITH ACID AT 25 °C[658,673,686,687]

| X | Medium A | $k_{rel}$ | Medium B | $k_{rel}$ | Medium C | $k_{rel}$ | Medium D | $k_{rel}$ | $k_{rel}$ (Table 231) | $k_{rel}$ (Table 232) |
|---|---|---|---|---|---|---|---|---|---|---|
| H | 0.148 | 1 | 2.43 | 1 | 6.75 | 1 | 0.56 | 1 | 1 | 1 |
| 3-Me | 0.325 | 2.2 | 5.22 | 2.15 | 21.7 | 3.22 | | | 2.37 | |
| 3-Bu$^t$ | | | 7.29 | 3.00 | 43.3 | 6.42 | | | 3.88 | |
| 2-Me | 2.33 | 15.7 | | | | | | | 17.7 | |
| 4-Me | 2.12 | 14.3 | | | 218(372) | 32.3 | 142 | 25.3 | 20.3 | 18.0 |
| 4-Bu$^t$ | | | | | 183(298) | 27.1 | | | 15.6 | |
| 3,5-Me$_2$ | 0.51 | 3.44 | | | | | | | 5.8 | |
| 2,5-Me$_2$ | 5.15 | 34.8 | | | | | | | 41.5 | |
| 3,4-Me$_2$ | 4.93 | 33.3 | | | | | | | 54.3 | |
| 2,3-Me$_2$ | 6.95 | 47 | | | | | | | 69.5 | |
| 2,4-Me$_2$ | 41.0(10.5) | 277 | | | | | | | 405 | |
| 2,6-Me$_2$ | 500 (122) | 3,380 | | | | | | | 3,330 | |
| 4-MeO | | | | | | | 1,240 | 2,210 | 1,280 | 1,010 |

Medium A: *p*-Toluenesulphonic acid (0.80 *M*) and water [3.2 *M* (6.0 *M*)] in glacial acetic acid.

Medium B: Hydrochloric acid (1.17 *M*) and water (3.62 *M*) in glacial acetic acid.

Medium C: Trifluoroacetic acid (3 vol.) and acetic acid (2 vol.) (or 1:1 vol. ratio).

Medium D: Perchloric acid in acetic acid, medium composition not quoted[687]; rate coefficients obtained by extrapolation from values at other temperatures.

acid in glacial acetic acid and the appropriate rate coefficients are given in Table 234[431, 665, 686]. The main kinetic features which emerge are the following:

(a) The spread of rates is relatively independent of the concentration of aqueous hydrochloric acid and is similar to that obtained in other media. Bearing in mind the small differences that do exist, there is reasonable agreement between the $k_{rel}$ values for the 3-Br and 3-MeO compounds (0.088 and 0.63) and those derived from Table 232 (0.117–0.128 and 0.43). The $k_{rel}$ values for the phenyl (3.24), 4-methylphenyl (5.56), 4-chlorophenyl (1.56), $\alpha$-naphthyl (14.3) and $\beta$-naphthyl (3.23) substituted compounds are in fair agreement with those in Tables 232 and 233 (3.55, 5.15, 1.57, 8.07, and 2.17) the values for the polycyclic compounds

TABLE 234

RATE COEFFICIENTS ($10^5 k_1$) FOR REACTION OF ArSiMe$_3$ WITH AQUEOUS HCl–HOAc AT 25 °C[431, 665, 686]

| [HCl] (M) | 2.48 | 3.1 | 0.587 | 2.35 | 1.17 | 0.62 | 0.00256ᵃ |
|---|---|---|---|---|---|---|---|
| [H$_2$O] (M) | 8.20 | 9.62 | 1.81 | 7.23 | 3.62 | 2.05 | 0.323 |
| Ar | | | | | | | |
| C$_6$H$_5$ | 6.60 | 10.4 | 1.30ᵇ | 6.40 | 2.44 | 1.43 | |
| 3-MeC$_6$H$_4$ | | | | 13.85 | 5.22 | | |
| 3-EtC$_6$H$_4$ | | | | 14.2 | | | |
| 3-$i$-PrC$_6$H$_4$ | | | | 15.1 | | | |
| 3-$t$-BuC$_6$H$_4$ | | | | 18.0 | 7.29 | | |
| 3-BrC$_6$H$_4$ | 0.058 | | | | | | |
| 3-MeOC$_6$H$_4$ | | 6.58 | | | | | |
| 3,5-Me$_2$C$_6$H$_3$ | | | | 25.9 | 10.5 | | |
| 3,5-Et$_2$C$_6$H$_3$ | | | | 30.5 | | | |
| 3,5-($i$-Pr)$_2$C$_6$H$_3$ | | | | 32.8 | | | |
| 3,5-($t$-Bu)$_2$C$_6$H$_3$ | | | | 48.5 | 19.3 | | |
| 3,5-(MeO)$_2$C$_6$H$_3$ | | | 3.62ᶜ | | | | |
| 2,3-Me$_2$C$_6$H$_3$ | | | | 122ᵈ | | | |
| 4-(4'-MeC$_6$H$_4$)C$_6$H$_4$ | | | | | | 7.95 | |
| 4-(4'-ClC$_6$H$_4$)C$_6$H$_4$ | | | | | | 2.33 | |
| 4-(2'-MeC$_6$H$_4$)C$_6$H$_4$ | | | | | | 10.8 | |
| 4-(2'-ClC$_6$H$_4$)C$_6$H$_4$ | | | | | | 1.10 | |
| 4-Ph.C$_6$H$_4$ | | | | | | 4.62 | |
| 1-naphthyl | 94.3ᵉ | | | | | | |
| 2-naphthyl | 21.3ᵉ | | | | | | |
| ferrocenyl | | | | | | | 4.68 |
| 1'-Me-ferrocenyl | | | | | | | 12.4 |
| 1'-$i$-Pr-ferrocenyl | | | | | | | 7.52 |
| 1'-$t$-Bu-ferrocenyl | | | | | | | 3.44 |

ᵃ Data in this column is for detriethylsilylation and was determined using GLC analyses.
ᵇ Not quoted in the original paper[686], but derived from the quoted $k_{rel}$ value and rate coefficient for the 2,3-Me$_2$ compound.
ᶜ This rate was quoted for a different medium in the original paper[686], which must be in error if the reported $k_{rel}$ value is correct as appears to be the case.
ᵈ Value may be in error since $k_{rel}$ is rather higher than those given in Table 233.
ᵉ The original paper[665] quotes the water content for measurement of these rates as 3.2 M which must be incorrect.

differing most. The latter two values may be compared with those obtained by Nasielski and Planchon[687], *viz.* 10.5, rate coefficient $5.89 \times 10^{-5}$, and 0.69, rate coefficient $3.90 \times 10^{-5}$ in medium D (Table 233); the latter value for the $\beta$ position of naphthalene is difficult to interpret. This work also gives values for the 3 position of pyrene (647, rate coefficient $3.64 \times 10^{-5}$) and the 9 position of phenanthrene (12.6, rate coefficient $7.08 \times 10^{-5}$) which do not agree well with the values from Table 232 (221 and 4.45 respectively). Possibly the mechanism for desilylation of polycyclics is not as straightforward as for other compounds less able to form $\pi$ complexes. In this respect the abnormally low $k_{rel}$ values of the 9 position of anthracene (4.1, Table 232) is in disagreement with expectation based on other electrophilic substitutions.

(*b*) Additivity of substituted effects holds well for the *m*-dialkyl and *m*-dimethoxy compounds so that buttressing across a 1,3 position is unimportant even for the *t*-butyl substituent.

(*c*) The effect of alkyl substituents on protodesilylation of ferrocene follows the hyperconjugative order, but is here clearly due to a steric effect since the *t*-butyl group deactivates. This was interpreted as possibly meaning that the *t*-butyl and triethylsilyl groups block the protonation of the iron atom which some believe to be the first step in cleavage reactions of ferrocene derivatives, though the possibility of reduced reactivity due to direct steric interaction of these groups could not be ruled out.

TABLE 235

RATE COEFFICIENTS $(10^5 k_1)$ AND KINETIC PARAMETERS FOR REACTION OF ArSiMe$_3$ WITH HCl IN 20 VOL. % AQUEOUS MeOH AT 50.1 °C[689]

| Ar | [HCl] (M) | | | | | |
|---|---|---|---|---|---|---|
| | 0.097 | 0.121 | 0.170 | 0.243 | 0.292 | 0.485 |
| Ferrocenyl | 4.05 | 5.09 | 7.83 | 14.3 | 19.3 | 43.9 |
| 4-Methoxyphenyl | 2.72 | 3.53 | 4.85 | 7.02 | 9.28 | 17.3 |
| 2,4-Dimethylphenyl | | | 1.38 | 2.07 | 2.72 | 5.55 |

Variation in rate coefficient $(10^5 k_1)$ with water content at constant [HCl] $= 0.17\ M$

| Vol. % | 20 | 9 | 7 | 2.3 |
|---|---|---|---|---|
| Ferrocenyl | 7.83 | 11.0 | 13.2 | 30.7 |
| 4-Methoxyphenyl | 4.83 | 6.18 | 7.17 | 14.3 |

Variation in rate coefficient $(10^5 k_1)$ with temperature in 20 vol. % aq. MeOH.

| Temp. (°C) | 60.0 | 55.0 | 50.1 | 40.0 | 30.0 | $\Delta H\ddagger$ | $\Delta S\ddagger$ |
|---|---|---|---|---|---|---|---|
| Ferrocenyl (0.292 M HCl) | 61.3 | 31.7 | 19.3 | 61.1 | | 22.1 | $-7$ |
| 4-Methoxyphenyl (0.485 M HCl) | | | 17.3 | 61.1 | 21.7 | 19.7 | $-15$ |

Initial protonation of iron in protodesilylation of trimethylsilylferrocene was not, however, favoured as a mechanism by Marr and Webster[689], who measured rates by the spectroscopic method using hydrochloric acid in 20 vol. % aqueous methanol (Table 235) and found that the rate of desilylation of the ferrocene compound was little more than that for the 4-methoxyphenyl and 2,4-dimethyl compounds. The similarity of the spread of rates in the different media and the similar activation energies and entropies were considered as evidence that the transition states for reaction of all three compounds were similar. The lower activation energy obtained for the 4-methoxyphenyl relative to the ferrocene compound may arise from the different media involved; the difference in entropy seems, however, to be rather larger than one might have expected even allowing for the solvent differences.

Hydrochloric acid in anhydrous methanol (1 : 4 volume ratio) has been used to desilylate the trimethylsilyl derivatives of ferrocene, ruthenocene, and osmocene, the rate coefficients ($10^5 k_1$) being; ferrocene (5.60, 4.08), ruthenocene (261, 182), and osmocene (104, 80.2) for 0.596 $M$ and 0.477 $M$ hydrochloric acid, respectively, (temperature not quoted)[690].

## 9.7 PROTODEGERMYLATION

The first kinetic study of the acid-catalysed cleavage of the aryl–germanium bond, reaction (268)

$$ArGeR_3 + HX = ArH + XGeR_3 \qquad (268)$$

was carried out by Eaborn and Pande[660], who measured the rate of cleavage of the triethyl and tricyclohexylgermanium groups from the *para* position of anisole by aqueous perchloric acid (1 vol.) in ethanol (5 vol.) at 50 °C. The spectrophotometric method was identical to that used for protodesilylation and the reaction gave first-order kinetics; rate coefficients ($10^5 k_1$) were R = cyclohexyl 172 (8.01 $M$ $HClO_4$), 18.9 (4.1 $M$ $HClO_4$); R = ethyl, 77.8 (4.1 $M$ $HClO_4$), 19.8 (2.05 $M$ $HClO_4$). Thus, as for protodesilylation, the rate decreases with decreasing concentration of added acid, and comparison of the kinetic data with that obtained in protodesilylation (p. 327) showed that the trialkylgermanium group is cleaved 15.5 (ethyl) to 40.5 (cyclohexyl) times more readily than the corresponding trialkylsilicon group from the *para* position of anisole and 35.6 (ethyl) times more readily from benzene. The difference in the ethyl and cyclohexyl data and the observation that the tricyclohexylmetal group is cleaved less readily than the triethylmetal group for both germanium and silicon confirms that these effects arise from steric hindrance to protonation of the aromatic ring. Obviously with the larger germanium atom, the alkyl groups are situated further from the aromatic ring and their size difference becomes less important. The greater rate difference in the case of cleavage from benzene indicates that cleavage of germanium is a less selective reaction whose transition state must be nearer to the ground state than for cleavage of silicon; the greater reactivity of the germanium compounds is entirely consistent with this.

The greater reactivity of the germanium compounds compared to the corresponding silicon compounds is considered to arise from the greater electron-releasing ability of germanium compared to silicon[660, 661], which facilitates attachment of the proton to the aromatic ring. The earlier proposals that attack of the proton at the ring is aided by nucleophilic attack of the acid anion upon the metal is, as is the case of protodesilylation, no longer necessary to explain the results, though it cannot be excluded as a possibility. A kinetic study of the solvent isotope effect in the protodegermylation of $4\text{-}XC_6H_4GeEt_3$ compounds by hydrochloric acid (1 vol.) in 25 mole % aqueous dioxan (1 vol.) at 50 °C (Table 236)[659] indicated that the first step in the reaction is the rate-determining transfer

TABLE 236

RATE COEFFICIENTS FOR REACTION OF $4\text{-}XC_6H_4GeEt_3$ WITH HCl–AQUEOUS DIOXAN AT 50° C[659]

| $X$ | $[HCl]$ (M) | $10^5k_1$ ($H_2O$) | $10^5k_1$ ($D_2O$) | $k_{H_2O}/k_{D_2O}$ |
|---|---|---|---|---|
| MeO | 4.35 | 233 | 137 | 1.71 |
|  | 3.23 | 98.3 | 57.5 | 1.71 |
|  | 3.18 | 95.8 | 55.0 | 1.74 |
|  | 2.36 | 51.5 | 29.3 | 1.76 |
|  | 1.80 | 29.5 | 16.8 | 1.75 |
| $CH_2SiMe_3$ | 4.71 | 49.4 | 30.9 | 1.60 |

of a proton from solvent to aromatic. Significantly, the isotope effects are greater than obtained in protodesilylation (1.55) and it has been argued that, since the observed effect is a balance of solvent–hydrogen bond breaking and carbon–hydrogen bond making, then, since the latter (which decreases the overall observed effect) will be more nearly formed in desilylation through the transition state being closer to the Wheland intermediate, the experimental result follows.

The effects of a wide range of substituents have been measured for protodegermylation $ArGeEt_3$ compounds by perchloric acid (2 vol.) in methanol (5 vol.) at 50 °C (Table 237)[666]. A plot of the log $k_{rel}$ values *versus* $\sigma^+$ values gave a very good straight line correlation with $\rho = -3.9$, but a correlation against the more precise Yukawa–Tsuno equation[612] was not attempted. The reaction is clearly, therefore, an electrophilic substitution with the usual A-$S_E2$ mechanism as indicated by the kinetic data. As in the case of desilylation, rate coefficients increase with increasing acid concentration except for the 4-$NMe_2$

TABLE 237

RATE COEFFICIENTS ($10^5 k_1$) FOR REACTION OF ArGeEt$_3$ WITH HClO$_4$ IN MeOH[a] AT 50 °C[666]

| Ar | [HClO$_4$] (M)[a] | | | | | | | | | | |
|---|---|---|---|---|---|---|---|---|---|---|---|
| | 0.00527 | 0.0527 | 0.229 | 0.495 | 0.970 | 2.80 | 4.01 | 6.21 | 7.34 | 9.15 | 11.91 |
| 4-Me$_2$NC$_6$H$_4$ | 135.5 | 166 | 167.5 | | | | | | | | |
| 2,4,6-Me$_3$C$_6$H$_2$ | | 62.2 | 299 | | | | | | | | |
| 4-HOC$_6$H$_4$[b] | | | 60 | 147 | | | | | | | |
| 4-MeOC$_6$H$_4$ | | | | 29 | 75.4 | | | | | | |
| 2-MeOC$_6$H$_4$ | | | | | 29.2 | 215 | | | | | |
| 4-SiMe$_3$CH$_2$C$_6$H$_4$ | | | | | 22.2 | 169 | | | | | |
| 4-PhC$_6$H$_4$ | | | | | | 40.8 | 88.3 | | | | |
| 4-MeC$_6$H$_4$ | | | | | | 14.7 | | 152 | 247 | | |
| 2-MeC$_6$H$_4$ | | | | | | 12.8 | 32.0 | 128 | | | |
| 4-EtC$_6$H$_4$ | | | | | | | | 142 | | | |
| 4-i-PrC$_6$H$_4$ | | | | | | | | 131 | | | |
| 4-t-BuC$_6$H$_4$ | | | | | | | | 125 | | | |
| 2-PhC$_6$H$_4$ | | | | | | | | | 55.7 | | |
| 4-PhC$_6$H$_4$ | | | | | | | | | 46.5 | | |
| α-naphthyl | | | | | | | | | 107 | | |
| β-naphthyl | | | | | | | | | 82.2 | | |
| 3-t-BuC$_6$H$_4$ | | | | | | | | | 58.7 | 145.5 | |
| 3-MeC$_6$H$_4$ | | | | | | | | | 37.0 | 96.0 | |
| C$_6$H$_5$ | | | | | | | 2.57 | 10.9 | 17.3 | 46.0 | |
| 4-FC$_6$H$_4$ | | | | | | | | | 15.3 | 44.0 | |
| 3-MeOC$_6$H$_4$ | | | | | | | | | 9.95 | 26.8 | |
| 4-ClC$_6$H$_4$ | | | | | | | | | | 7.67 | 35.4 |
| 4-BrC$_6$H$_4$ | | | | | | | | | | 5.70 | 27.5 |
| 4-IC$_6$H$_4$ | | | | | | | | | | | 28.7 |
| 3-ClC$_6$H$_4$ | | | | | | | | | | | 3.48 |

[a] Concentration of perchloric acid, 2ml of which was added to a solution of organogermane in methanol (5 ml).
[b] Measurements actually carried out on the 3-Me$_3$SiOC$_6$H$_4$ compound which rapidly solvolyses to the 4-HOC$_6$H$_4$ compound.

compound which gives rates relatively insensitive to acid concentration. Likewise, this was attributed to protonation of the substituent to give the unreactive conjugate acid, so that the increase with increasing acidity in the rate at which the neutral amine reacts is compensated by the decrease in the concentration of the neutral amine.

Finally, rate coefficients for the protodegermylation of compounds ArGeEt$_3$ (where Ar contains mostly strongly electron-withdrawing substituents in acetic acid (4 vol.) containing sulphuric acid (3 vol.) have been measured at 50 °C; these coefficients and the strength of the added acid are given in Table 238. The log $k_{rel}$ values correlated precisely with the Yukawa–Tsuno equation[612] with $\rho = -4.4$ and $r = 0.6$ which implies that the electrophile is more selective in this medium. However, the correlation log $k_{rel}$ (acetic acid–water–sulphuric

TABLE 238

RATE COEFFICIENTS ($10^5 k_1$) FOR REACTION OF $XC_6H_4GeEt_3$ WITH HOAc–AQUEOUS $H_2SO_4$ AT 50 °C[666a]

| X | $[H_2SO_4]$ $(M)$[a] | | | | | | | | |
|---|---|---|---|---|---|---|---|---|---|
|  | 0.295 | 1.42 | 3.62 | 5.55 | 6.45 | 9.65 | 11.69 | 13.85 | 14.20 |
| 4-MeO | 92.3 | | | | | | | | |
| 2-MeO | 43.3 | 318 | | | | | | | |
| 4-Me | | 21.2 | 132 | | | | | | |
| 4-Ph | | | 25.9 | 115.5 | | | | | |
| 3-Me | | | | 84.7 | | | | | |
| H | | | | 47.5 | 115 | | | | |
| 3-MeO | | | | | 53.5 | | | | |
| 4-Cl | | | | | 17.7 | | | | |
| 4-Br | | | | | 14.0 | 157 | | | |
| 3-F | | | | | | 37.3 | | | |
| 3-Br | | | | | | 22.8 | | | |
| 3-Cl | | | | | | 22.7 | 127.5 | | |
| 3-HOOC | | | | | | | 118 | | |
| 3-CF$_3$ | | | | | | | 36.1 | 123 | |
| 4-CF$_3$ | | | | | | | 16.4 | 60.1 | |
| 4-HOOC | | | | | | | 34.4 | | 206 |
| 3-NMe$_3^+$ | | | | | | | | | 50.3 |
| 4-NMe$_3^+$ | | | | | | | | | 42.1 |
| 3-NO$_2$ | | | | | | | | | 31.8 |
| 4-NO$_2$ | | | | | | | | | 15.0 |

[a] Concentration of sulphuric acid, 3 vol. of which was added to 4 vol. of acetic acid.

acid) = 0.95 log $k_{rel}$ (methanol–water–perchloric acid) applied (as it did in protodesilylation thereby confirming the mechanistic similarity of the two reactions), and the fact that the $\rho$ factor appears to be greater for cleavage by sulphuric acid arises only through use of the Yukawa–Tsuno equation for the data with this medium, as it can easily be shown that use of a diminished $r$ value results in an increased $\rho$ value.

The log $k_{rel}$ values for protodegermylation were shown to give an excellent correlation against the log $k_{rel}$ values for protodesilylation, which may be regarded as confirmation that the overlap kinetic technique is satisfactory.

## 9.8 PROTODESTANNYLATION

Kinetic studies of the acid-catalysed cleavage of aryltrialkylstannanes, reaction (269)

$$HX + ArSnR_3 = ArH + XSnR_3 \tag{269}$$

were first made by Eaborn and Pande[660], who measured the rates by the spectroscopic method using a medium composed of aqueous perchloric acid (1 vol.)
and ethanol (5 vol.) at 50 °C. As in the case of aryl–silicon and aryl–germanium
bond cleavage, acid catalysis was indicated by a parallel between increase of rate
and increase of concentration of added acid (concentration in parentheses) as
follows: $4\text{-MeOC}_6\text{H}_4\text{SnEt}_3$, $32.7 \times 10^{-5}$ (0.01 $M$), $17.7 \times 10^{-5}$ (0.057 $M$) and the
electrophilic nature of the reaction was indicated by a rate increase by electron-
supplying substituents, $e.g.$ $\text{PhSnEt}_3$, $107 \times 10^{-5}$ (0.97 $M$) $cf.$ the above value for
the 4-methoxy compound in a much weaker acid. The rate was also decreased by
an increase in size of the alkyl group, $e.g.$ $4\text{-MeOC}_6\text{H}_4\text{Sn(cyclohexyl)}_3$, $16.1 \times 10^{-5}$
(0.01 $M$), and the trend noted with the silicon and germanium compounds (see
pp. 327, 339) is continued here as the ratio of ethyl : cyclohexyl rates is now only
2.0, which further supports the theory that this effect is steric in origin (see p. 339).

Comparison of the cleavage rates with those obtained for the corresponding
silicon and germanium compounds showed that the aryl–tin bond is cleaved very
much more rapidly than the aryl–silicon or aryl–germanium bonds, $e.g.$, a rate
factor of $10^4$ applies between the corresponding tin and germanium compounds.
This was considered so out of line with the reported differences in electronegativities that solvent attack on the tin atom was considered as a participating step
in the reaction mechanism. However, subsequent and more reliable measures of the
difference in the electron-releasing ability of the tin relative to silicon and germanium[661] made it clear that the order is not unexpected so that nucleophilic
assistance by attack at tin is less likely, but by no means impossible.

Measurement of the solvent isotope effect for the cleavage of phenyltriethyl-
stannane by hydrochloric acid (1 vol.) in 25 mole % aqueous dioxan (1 vol.
containing either $H_2O$ or $D_2O$) gave the following rate coefficients (concentration
of acid in parentheses): $k_{H_2O}$, $260 \times 10^{-5}$, $k_{D_2O}$, $106 \times 10^{-5}$ [0.903 $M$]; $k_{H_2O}$,
$66 \times 10^{-5}$, $k_{D_2O}$ $25 \times 10^{-5}$ [0.507 $M$], the respective isotope effects being 2.45 and
2.64. These values are rather larger than obtained with germanium and silicon
compounds of similar reactivity, due partly to the smaller amount of added hydrochloric acid and thus to a greater isotopic purity of the deuterium-containing
solvent, but attributed also to the transition state for the reaction being nearer
to the ground state so that compensation for the isotope effect in solvent O–H
bond breaking by C–H bond making is reduced[659] (see also p. 340).

The rates of cleavage of a range of aryltricyclohexylstannanes by ethanol
(50 vol.) and perchloric acid (2 vol.) at 50 °C have been measured by Eaborn
and Waters[691], the rate coefficients and strengths of added acid being given in
Table 239. The logarithms of the $k_{rel}$ values give a very good correlation against
the Yukawa–Tsuno equation[612] with $\rho = -3.8$ and $r = 0.4$, thereby confirming
the evidence of the solvent isotope studies, $viz.$ that the transition state is nearer
the ground state than are the transition states for protodesilylation and protodegermylation. A significant difference between these reactions and protodestan-

TABLE 239

RATE COEFFICIENTS ($10^5k_1$) FOR REACTION OF $XC_6H_4Sn(C_6H_{11})_3$ COMPOUNDS WITH $HClO_4$–EtOH AT 50 °C[691]

| X | [$HClO_4$] $(M)$[a] | | | | | | | | | |
|---|---|---|---|---|---|---|---|---|---|---|
| | 0.001 | 0.002 | 0.0283 | 0.099 | 0.313 | 0.571 | 2.35 | 3.41 | 4.48 | 9.15 |
| 4-NMe$_2$ | 128 | 258 | | | | | | | | |
| 4-MeO | | | 14.6 | 56.7 | | | | | | |
| 4-t-Bu | | | | | 21.0 | 40.6 | 213 | | | |
| 4-i-Pr | | | | | 20.3 | 40.0 | 207 | | | |
| 4-Et | | | | | 17.0 | | 157 | | | |
| 4-Me | | | | 4.95 | 17.3 | 34.0 | 159 | | | |
| 3-Me | | | | | | 10.8 | 55 | | | |
| 2-Ph | | | | | | 11.7 | | | | |
| 4-Ph | | | | | | 10.4 | | | | |
| H | | | | 0.90 | | 5.88 | 29.6 | 48.7 | 76.6 | |
| 3-MeO | | | | | | | 26.5 | | 58.0 | |
| 4-F | | | | | | | | | 47.5 | |
| 4-Cl | | | | | | | | | 4.4 | |
| 4-Br | | | | | | | | | 11.1 | |
| 3-Cl | | | | | | | | | 2.97 | 9.86 |
| 4-CO$_2$H | | | | | | | | | 7.67 | |
| 4-NMe$_3$$^+$ | | | | | | | | | | 1.73 |

[a] Concentration of perchloric acid, 2 vol. of which were added to 50 vol. of ethanol containing the organostannane.

nylation is the fact that 4-NMe$_2$-substituted compound shows a substantial rate increase with increase in concentration of added acid and this was attributed to freedom of the free base from protonation at the low acid concentrations employed for the protodestannylation.

Kinetic studies[662, 683, 691, 692] have also been made of protodestannylation of aryltrimethylstannanes by ethanol containing aqueous perchloric acid at 50 °C. Rate coefficients are given in Table 240. Some significant features of these results are, firstly, that the effects of substituents in the reaction (which occurs more readily than cleavage of the cyclohexyl compounds because of the steric factor) are smaller as might be expected since the transition state is nearer to the ground state. Secondly, the pentafluorophenyl compound is less deactivating than expected from the product of the individual effects of fluorine at the *ortho*, *meta*, and *para* positions of benzene and as found also in protodesilylation (see Table 232, p. 334). Thirdly, the rate coefficients for the Me$_3$MCH$_2$ substituents show quite unambiguously, that electron release increases along the series Me$_3$Si < Me$_3$Ge < Me$_3$Sn, and it is this evidence which is regarded as the principal reason for the increased rates of cleavage of the aryltrialkyl metal compounds along this series. Fourthly, the 4-methoxyphenyltriphenylstannane can be shown, by comparison with the data in Table 239, to be 35.5 times less reactive than the somewhat

TABLE 240

RATE COEFFICIENTS ($10^5 k_1$) FOR REACTION OF $ArSnMe_3$ WITH $HClO_4$-EtOH(MeOH) AT 50 °C[662, 683, 691, 692]

| Ar | $[HClO_4]$ $(M)^a$ | | | | | | |
|---|---|---|---|---|---|---|---|
| | 0.099* | 2.35* | 9.40** | 0.48*** | 0.60*** | 1.07*** | 2.10*** |
| 4-MeC$_6$H$_4$ | 17.8 | | | | | | |
| C$_6$H$_5$ | 5.17 | 150 | 557 | 380 | 500 | | |
| 4-ClC$_6$H$_4$ | | 57.5 | 205 | | | | |
| 3-ClC$_6$H$_4$ | | 26.3 | 78.9 | | | | |
| 3-BrC$_6$H$_4$ | | | 79.1 | | | | |
| 4-HC≡CC$_6$H$_4$ | | | 203 | | | | |
| 3-HC≡CC$_6$H$_4$ | | | 132 | | | | |
| 2-FC$_6$H$_4$ | | | | 155 | 198 | | |
| 3-FC$_6$H$_4$ | | | | 85 | 118 | | |
| 4-FC$_6$H$_4$ | | | | 358 | 460 | | |
| C$_6$F$_5$ | | | | 168 | 210 | | |
| 4-SiMe$_3$CH$_2$C$_6$H$_4$ | | | | | | 56.7 | 93.4 |
| 4-GeMe$_3$CH$_2$C$_6$H$_4$ | | | | | | 76.7 | 127 |
| 4-SnMe$_3$CH$_2$C$_6$H$_4$ | | | | | | 182 | 300 |
| 4-MeOC$_6$H$_4^b$ | | 52.5 | | | | | |

$^a$ Concentration of perchloric acid, 2 vol. of which was added to the following volumes of solvent: * 50 vol. of EtOH; ** 100 vol. of EtOH (at 50.2 °C); *** 5 vol. of MeOH.
$^b$ Protodestannylation of 4-MeOC$_6$H$_4$SnPh$_3$.

similarly sized tricyclohexyl compound and this further confirms that electron supply or withdrawal in the aryltrimethyl metal compounds increases or retards respectively, aryl–metal bond cleavage.

The rates of protodestannylation of a range of aryltrialkyl compounds in solutions of hydrochloric acid in methanol have been measured at various temperatures and the second-order rate coefficients extrapolated or interpolated to 25 °C, and at a standard concentration of acid, are given in Table 241[693]. Again an increase in the size of the alkyl group causes a decrease in the rate, and the data for R = Me correlates with $\sigma^+$ values with $\rho = -2.17$ and the reaction appears, therefore, to be very much less selective than for cleavage of the R = cyclohexyl compounds, and this accords with the findings of Eaborn and Waters[691]. Difficult to explain, however, are the relative reactivities of the 1- and 2-naphthyl compounds which are quite out of line with the observation on all other known electrophilic substitutions of these two aromatic positions; possibly steric hindrance to solvation by the peri-hydrogen is a significant factor for 1 substitution, but one would have expected a significant if not greater compensation from steric acceleration. Further investigation of these compounds is clearly desirable. Arrhenius parameters were determined for the reaction, the activation energies ranging from 9–16, but the quoted accuracy of these values, which were determined over a 20–30 °C range, does not correspond with that quoted in another paper by the same group. Thus

TABLE 241

RATE COEFFICIENTS ($10^5k_2$) FOR REACTION OF $ArSnR_3$ WITH HCl–MeOH AT 25 °C[693]

| Ar | R | | |
|---|---|---|---|
| | Me | i-Pr | n-Bu |
| 4-$MeOC_6H_4$ | 158 | | |
| 4-$MeC_6H_4$ | 18.6 | | |
| 3-$MeC_6H_4$ | 5.62 | | |
| 9-Phenanthyl | 4.47 | | |
| 1-Naphthyl | 6.31 | 2.24 | 3.80 |
| 2-Naphthyl | 6.31 | 1.07 | 2.76 |
| $C_6H_5$ | 3.24 | 0.576 | 1.70 |
| 4-$BrC_6H_4$ | 1.80 | | |

[a] Corrected to a standard concentration of HCl in MeOH, the value of which was not quoted.

for protodestannylation of $PhSnMe_3$[694] second-order rate coefficients ($10^2k_2$) were quoted as 1.24 (15.8 °C), 2.79 (25.0 °C), 3.53 (28.2 °C), 6.18 (34.6 °C), and 6.47 (35.0 °C) leading to $E_a = 15.37\pm0.04$; however, the data in Table 241 give a value of 16.8 and the difference between these values is probably a more realistic measure of the experimental error.

An interesting kinetic study has been made of the cleavage of aryltrimethyl-stannanes by aqueous methanolic alkali at 50 °C[695]. This reaction is base-cat-alysed as shown by the following rates ($10^5k_1$) for cleavage of the 3-trifluoro-methylphenyl compound in methanol (3 vol.) at the concentrations (in paren-theses) of sodium hydroxide (2 vol.): 148 (3.805 $M$), 173.5 (5.33 $M$), 207 (7.71 $M$). The mechanism approximates to that given in equilibria (270) and (271)

$$ArSnMe_3 + MeO^- \rightleftharpoons Ar^- + SnMe_3OMe \quad \text{(slow)} \qquad (270)$$

$$Ar^- + MeOH \rightarrow ArH + MeO^- \quad \text{(fast)} \qquad (271)$$

in which hydrogen attacks the ring in a second and fast step and is, therefore, analogous to base-catalysed hydrogen exchange (see p. 266).

Further aspects of the mechanism are revealed by an examination of the sub-stituent effects in this reaction, and the first-order rate coefficients[695] are given in Table 242. Firstly, the spread of rates is much less than in base-catalysed hy-drogen exchange, and this arises from the much greater reactivity of the tin compounds so that the transition state lies closer to the reactants. It was argued that, since the hydrogen exchange reaction must involve synchronous breaking of the C–H bond and attack of the base (through symmetry), the synchronous mechanism was possible for destannylation.

Secondly, destannylation appears to be more sterically hindered than the hy-drogen exchange reaction, since in the latter the $\alpha$-position of naphthalene is rate-

TABLE 242

RATE COEFFICIENTS ($10^5 k_1$) FOR REACTION OF $XC_6H_4SnMe_3$ COMPOUNDS WITH NaOMe–MeOH AT 50 °C[695]

| X | $10^5 k_1$ | X | $10^5 k_1$ | X | $10^5 k_1$ | X | $10^5 k_1$ | X | $10^5 k_1$ |
|---|---|---|---|---|---|---|---|---|---|
| 2-F | 582 | 2-Cl | 152 | 3-HC≡C | 76.7 | 4-MeO | 33.0 | 2,3-C₄H₄[a] | 18.7 |
| 4-Me₃N⁺ | 208 | 2-F₃C | 142 | 4-F | 58.3 | 4-Me₂N | 32.8 | 4-Me | 17.3 |
| 3-F₃C | 206 | 3-F | 110 | 4-MeS | 48.3 | 3-Me₃Si | 28.0 | 3-Me | 16.7 |
| 3-Br | 183 | 4-Cl | 93.3 | 4-Ph | 36.3 | 4-Me₃Si | 22.8 | 3-H₂N | 13.2 |
| 4-F₃C | 165 | 4-Br | 91.7 | 3-MeO | 35.8 | H | 20.1 | 2-MeO | 12.3 |
| 3-Cl | 163 | 4-HC≡C | 78.3 | 3-Ph | 34.0 | 4-t-Bu | 14.2 | 2-Me | 10.0 |

[a] $XC_6H_4$ = 1-naphthyl.

enhancing but not in the former, and this is true also for the effects of other substituents, *e.g.* 2-MeO. Such behaviour can be rationalised in the writer's view by postulating attack at tin by the base, such that inversion occurs (XLI), in which case this process will be hindered by *ortho* substituents whereas hydrogen exchange will not.

(XLI)

Thirdly, the electronic effects of some substituents in the two reactions differ. The logarithms of the $k_{rel}$ values for *meta* substituents and for normally electron-withdrawing substituents correlate satisfactorily with $\sigma$ values with $\rho = +2.18$. For substituents which are usually electron-supplying, *e.g.* 4-NMe₂ and 4-OMe, no correlation was possible because these substituents now acted as fairly strongly electron-*withdrawing* substituents though in the *meta* positions their effect was normal. This anomaly could be partly explained by assuming that inductive effects only operate (since the negative charge on the ring cannot be conjugated with the $\pi$ system) but even so the correlation between log $k_{rel}$ and $\sigma_I$ was poor (and significantly worse for the *meta* substituents than for the correlation with $\sigma$). It was pointed out that the discrepancy may arise from our lack of knowledge of the importance of field effects when they operate, as here, through the substrate cavity.

The mechanism of the reaction was postulated as being either synchronous attack of the $RO^-$ ion (R = Me or H) at tin and breaking of the aryl–tin bond, or rapid formation of a pentacoordinate tin intermediate [$ROMe_3SnPh^-$] followed by rate-determining breaking of the aryl–tin bond. In either case the phenyl-carbanion could separate as such and then react rapidly with solvent, or it could react with solvent as it separated.

One possible explanation for the abnormal results noted above comes from considering the $\pi$-electron density distribution when $+M$ substituents are present. This is indicated on the structure (XII) such that a solid constituent $p$ orbital is a region of higher

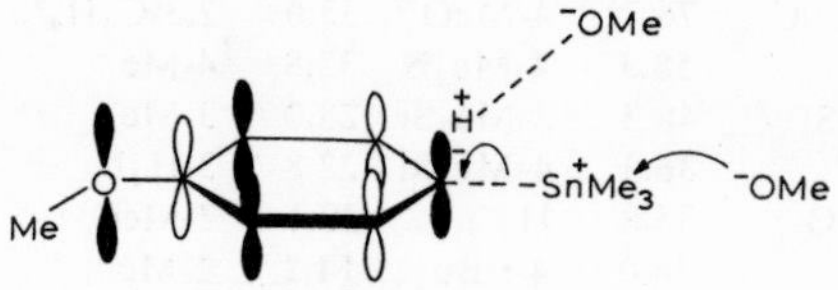

(XLII)

electron density; the full charges indicated may in fact be only partial. It seems possible that the positive end of the attacking solvent molecule (the proton) will be attracted to the $\pi$ cloud region of high electron density, *i.e.* the hydrogen will be very favourably aligned for subsequent substitution, and this can thus result in a more rapid reaction than in those cases where the $\pi$ cloud at the position of substitution is of low density. It should be noted that in this latter situation the hydrogen atom will not necessarily be repelled, consequently strongly electron-withdrawing substituents should not produce a converse effect, in agreement with the experimental observation. Similarly, an $+M$ substituent in the *meta* position will only raise the electron density at the $p$ orbitals adjacent to the position of the substitution; consequently, no abnormal effect will be detected. It follows from this argument that the carbanion does not become completely free, otherwise the subsequent substitution of the proton would be very fast and therefore non-rate-determining*.

### 9.9 PROTODEPLUMBYLATION

The acid-catalysed cleavage of aryltrialkylplumbanes, reaction (272)

$$ArPbR_3 + HX = ArH + XPbR_3 \qquad (272)$$

has been subjected to fewer kinetic studies than the other cleavages of the Group IV metals from the aromatic ring. The method of kinetic study differs somewhat in that the lead atom has the strongest absorption in the UV spectrum and it was the change in this spectrum with time, rather than that of the aromatic ring, which has been utilised in rate measurements.

---

* This prediction has now been confirmed by the observation (R. Alexander, C. Eaborn and T. G. Traylor, *J. Organometal. Chem.*, 21 (1970) P65) that a solvent-isotope effect is obtained in base cleavage of benzyl- and aryl–tin bonds by NaOH in MeOH–MeOD.

The first kinetic study was that of Eaborn and Pande[660], who found that the first-order rate coefficient for cleavage of phenyltrimethylplumbane by 0.0057 $M$ perchloric acid (1 vol.) in ethanol (5 vol.) was $168 \times 10^{-5}$ at 50 °C. Thus, the plumbane is about 10 times more reactive than the 4-methoxyphenyl tin compound and, hence, about 650 times as reactive as the corresponding phenyl tin compound (see pp. 343, 344). The increasing reactivity of the aryl–metal bond with increasing atomic number of the Group IV metal is thus uniformly maintained and may be assumed to arise from the increasing availability of electrons from the metal along the series. Likewise, the increase in the solvent isotope along the series is maintained here, since phenyltriethylplumbane gave a rate coefficient of $58.5 \times 10^{-5}$ ($H_2O$) and $19.2 \times 10^{-5}$ ($D_2O$) at 50 °C in a solution of 0.00304 $M$ hydrochloric acid in 25 mole % aqueous dioxan. The same consideration with regard to isotopic purity of the solvent referred to in the other demetallations (pp. 340, 343) apply here but a substantial part of the increase in the isotope effect probably arises from the transition state being nearer to the ground state; evidence for this is the fact that a kinetic study of substituent effects gave a correlation with the Yukawa–Tsuno equation[612] with $\rho = -2.5$ and $r = \sim 0.4$ so that the $\rho$ factor is the smallest observed for the demetallations, consistent with the transition state being nearer the ground state. Rate coefficients for this study using perchloric acid and ethanol at 25 °C are given in Table 243.

TABLE 243

RATE COEFFICIENTS ($10^5 k_1$) FOR PROTODEPLUMBYLATION OF $XC_6H_4PbMe_3$ WITH $HClO_4$–EtOH AT 25 °C[660]

| $X$ | $[HClO_4]$ $(M)^a$ | | | | |
|---|---|---|---|---|---|
|  | 0.0022 | 0.01 | 0.022 | 0.053 | 0.097 |
| 4-MeO | 88.4 |  |  |  |  |
| 4-Me |  | 65.8 | 147 |  |  |
| H |  | 19.7 | 42.7 | 106 |  |
| 4-Cl |  |  |  | 33.8 | 61.3 |
| 3-Cl |  |  |  |  | 24.0 |

$^a$ Concentration of perchloric acid, 1 vol. of which was added to a solution of organoplumbane in ethanol (10 vol.).

## 9.10 PROTODESULPHONATION

Kinetic studies of protodesulphonation, equilibrium (273)

$$ArSO_3H + HX \rightleftharpoons ArH + XSO_3H \tag{273}$$

have mostly been initiated by the desire to elucidate the mechanism of the Jacobsen reaction, *i.e.* the isomerisation of arylsulphonic acids in sulphuric acid. The main problem has been to determine whether or not isomerisation proceeds *via* protodesulphonation followed by sulphonation. Whilst some workers favour this mechanism[696], there is evidence that it cannot be wholly correct, since Syrkin *et al.*[697] have observed incomplete equilibration of isotopic label in the reaction of *ortho* and *para* toluenesulphonic acids in $^{35}$S-sulphuric acid.

Whatever the mechanism of protodesulphonation may be, it is likely to be the exact reverse of that appropriate for sulphonation (pp. 56–77). The first kinetic study seems to have been that of Crafts[698], who measured the rates of desulphonation of the sulphonic acids derived from toluene, *meta* and *para*-xylene, mesitylene, 1,2,4-trimethylbenzene, 2,4,6-trichloro- and 2,4,6-tribromo-benzenes, amino acids and naphthalene, with sulphuric and halogen acids as catalysts, and between temperatures of 80–218 °C; no conclusions as to the reaction mechanism were obtained. A more useful study was that of Pinnow[150], who determined the equilibrium concentrations of quinol and quinolsulphonic acid after reacting mixtures of these with 40–70 wt. % sulphuric acid at temperatures between 50 and 100 °C, and hence the first-order rate coefficients for sulphonation and desulphonation (Tables 34 and 35, p. 62,63) since clearly the equilibrium concentration is a result of the balance of the forward and backward reactions. As described also on p. 62, this work demonstrated what has been subsequently confirmed many times, namely that rates of desulphonation increase less readily with increasing acid concentration than do rates of sulphonation and one would indeed expect that desulphonation would be more favoured by more aqueous media. The variation in sulphonation and desulphonation rates with temperature show clearly that the activation energy for desulphonation is higher than that for sulphonation and that this difference is greater the more concentrated the acid.

Rates of desulphonation of $\alpha$- and $\beta$-naphthalene sulphonic acids by 45.1–70.3 wt. % sulphuric acid at 140 °C and 180 °C were determined by Lantz[699] from the proportion of compound desulphonated, though rate coefficients were only quoted in a graphical form. For the $\alpha$ compound, the logarithm of the rate was linearly proportional to the acid concentration, but for the $\beta$ acid, the rates were maximal at about 54 wt. % acid (at 180 °C). This decrease in rate with increasing acid concentration subsequently led Lantz[700] to propose that desulphonation involved reaction of the sulphonic acid anion with sulphuric acid since the equilibrium concentration of the anion would decrease with increasing acid concentration. Cowdrey and Davis[152] later showed that first-order rate coefficients derived from the data of Pinnow and of Lantz on desulphonation of naphthalene-$\alpha$-sulphonic acid, and of Crafts on the desulphonation of xylene sulphonic acids in 5–15 $M$ hydrochloric acid could be satisfactorily correlated by means of equation (274)

$$\text{Rate} = k_1[\text{H}_3\text{O}^+][\text{HSO}_4^-][\text{H}_2\text{O}]^{-1} = kK[\text{H}_2\text{SO}_4] \tag{274}$$

where

$$[H_3O^+] \text{ and } [HSO_4^-] = [H_2SO_4]_{stoich} \text{ and } [H_2O] = [H_2O]_{stoich} - [H_2SO_4]_{stoich};$$

discrepancies between theory and observation were explained in terms of ionic interactions and conversion of $RSO_3^-$ to $RSO_3H$ at high acid concentration.

The kinetics of desulphonation of sulphonic acid derivatives of *m*-cresol, mesitylene, phenol, *p*-cresol, and *p*-nitrodiphenylamine by hydrochloric or sulphuric acids in 90 % acetic acid were investigated by Baddeley *et al.*[701], who reported (without giving any details) that rates were independent of the concentration of sulphuric acid and nature of the catalysing anion, and only proportional to the hydrogen ion concentration. The former observation can only be accounted for if the increased concentration of sulphonic acid anion is compensated by removal of protons from the medium to form the undissociated acid; this result implies, therefore, that reaction takes place on the anion and the mechanism was envisaged as rapid protonation of the anion (at ring carbon) followed by a rate-determining reaction with a base.

A kinetic study of the desulphonation of 2-naphthylamine-1-sulphonic acid by hydrochloric, sulphuric and phosphoric acids showed the rate to be proportional to the concentration of hydrogen ions and the aromatic and a mechanism involving the formation of 1-naphthylsulphamic acid was proposed[702].

Gold and Satchell[160] pointed out that a linear relationship existed between the logarithms of the rate coefficients obtained by Crafts[698], and the $-H_0$ acidity function over the range $-0.4-2.4$, assuming that the values at 140 °C, the desulphonation rate-measurement temperature, were proportional to those at 25 °C. The slope of the plot was 1.05, and they interpreted this to mean that there is a rapid reversible attachment of a proton to the aromatic ring giving a $\pi$ complex which rearranges in the slow step to a second $\pi$ complex in which H and $SO_3$ are interchanged, $SO_3$ being subsequently lost in a fast step; this is thus the exact reverse of their mechanism for sulphonation, which is no longer accepted (see p. 67). It is, therefore, relevant that a later analysis[171] of Pinnow's data showed a log rate *versus* $-H_0$ correlation with a slope of 0.6 which is very different from the value of 1.0 ideally required by this theory.

Kilpatrick *et al.*[171] measured the rate of desulphonation of mesitylenesulphonic acid in excess of 12–13.5 $M$ sulphuric acid. The rate coefficients are given in Table 43 (p. 73), and these were determined by spectrophotometrically measuring the concentration of sulphonic acid present at equilibrium, after reacting the sulphonic acid with the sulphuric acid of appropriate concentration; again the increase in rate of desulphonation with increase of acid strength is much less than the increase in the rate of sulphonation. The increase in rate of desulphonation of mesitylene-sulphonic acid with increasing acid concentration was less than that for 1,2,4,5-tetra-, 1,2,3,5-tetra-, and penta-methylbenzenesulphonic acids[703] (Table 244) the slopes of the plots of log $k_1$ against acidity function being respectively 0.59, 0.99, 0.85, and 0.95; no definite reason was advanced for this, but one can reasonably speculate that steric acceleration is a primary factor.

TABLE 244

RATE COEFFICIENTS ($10^5k_1$) FOR REACTION OF ArSO$_3$H WITH H$_2$SO$_4$ AT 12.3 °C[703]

| [H$_2$SO$_4$] (M) | Ar | | |
|---|---|---|---|
| | 1,2,4,5-Me$_4$C$_6$H | 1,2,3,5-Me$_4$C$_6$H | Me$_5$C$_6$ |
| 8.73 | | | 2.15 |
| 9.86 | | | 7.73 |
| 10.13 | | | 10.3 |
| 10.22 | | | 11.3 |
| 11.55 | | 3.77 | |
| 11.99 | | 6.23 | |
| 12.65 | | 11.4 | |
| 12.90 | | 14.4 | |
| 13.08 | 1.66 | | |
| 13.50 | | 24.8 | |
| 13.95 | 4.77 | | |
| 14.40 | 8.00 | | |
| 14.86 | 13.7 | | |

A smaller dependence of rate upon acidity for desulphonation compared to sulphonation was also observed by Muramoto[704], the log rate coefficient *versus* $-H_0$ slopes being respectively 0.7 and 1.67 (*m*-xylene, 100 °C), 0.6 and 1.5 (*p*-xylene, 130 °C).

Spryskov and Ovsyankina[705] measured rates of desulphonation of α- and β-naphthalenesulphonic acids, and benzenesulphonic acids containing the following substituents: 2-, 3-, and 4-Cl, NH$_2$, and COOH, 3-OH, 4-OH, 4-Br, and 3-CO$_2$H-4-OH, in terms of the amount of hydrolysis for a given strength of catalysing acid in a given time at a given temperature. Catalysing acids were water, sulphuric, hydrochloric, phosphoric, and acetic acids, the order of effectiveness being HCl > H$_2$SO$_4$ > H$_3$PO$_4$ > H$_2$O. Interestingly, *ortho* compounds were always the most reactive, which strongly suggests that steric acceleration is a very important factor here; this is not surprising in view of the steric hindrance to sulphonation and is doubtless the factor giving rise to the lower thermodynamic stability of α compared to β-naphthalenesulphonic acid. They regarded the mechanism as involving complex formation between ArSO$_3$H and H$_3$O$^+$ in a pre-equilibrium followed by rate-determining reaction of the complex with any available base, such as the acid anion, and thus differ from Baddeley *et al.*[701] in that the sulphonic acid and not the anion is involved. The rate retardation observed on adding acetic acid to one of the other strong acids was interpreted as arising from conversion of the sulphonic acid to $RSO_3^-$ and $AcOH_2^+$, thereby reducing the concentration of the species they considered to be the reactive one. This mechanism was supported by Leitman and Pevsner[706], who studied the desulphonation of xylene- and ethylbenzene-sulphonic acids in excess 50–75 wt. % sulphuric acid at 90–145 °C, and found a maximum in the rate at 60–70 wt. % sulphuric acid, this was considered to be related to the observation by Feilchenfeld[707] of a maximum separation of the phase $ArSO_3H.H_2O$ in *ca.* 65 wt. % sulphuric acid (spent acid composition) for toluene- and xylene-sulphonic acids. On addition of water at 30 °C, the solubility of the sulphonic acid decreases, and this was attributed to conversion of the sulphonic acid anion to the molecular acid. At higher acid concentrations (65–80 wt. %) the solubility increases, and this was attributed to formation of $ArSO_3H_2^+$. Spryskov and Ovsyankina[705] thus considered reaction of this intermediate (or $ArSO_3H.H_3O^+$) with $HSO_4^-$ as the rate-determining step of the reaction, and this would yield the aromatic and two molecules of sulphuric acid; hence the reverse reaction would be expected to be second-order in sulphuric acid, but this is not found.

An extensive study of desulphonation has been made by Cerfontain *et al.*[696, 708]. The earlier study[696] appeared to show that the rate coefficients for isomerisation of sulphonic acids can be correlated with desulphonation rate coefficients and sulphonation isomer distribution, suggesting that isomerisation is an intermolecular process. However, the results of Syrkin *et al.*[697] described above show that this is not entirely true (if at all). On the assumption that the process is intermolecular, rate coefficients ($10^4 k_1$) were obtained for the desulphonation of *para* and *meta-t*-butylbenzenesulphonic acids as follows: 4.9 and 0.23 (79.9 °C), 140 and 4.4 (109.8 °C) 2,700 and 63 (141.2 °C) from which activation energies of $30\pm3$ and $27\pm3$ were obtained, respectively[696]. The rate coefficient for desulphonation of *p*-toluenesulphonic acid was found to be 3,310 (141.2 °C) and thus the hyperconjugative order of alkyl group activation was followed. These rate coefficients may be subject to large errors, however, not only because of the mechanistic assumption, but because it was also assumed that the ratio of dealkylation rate coefficients (required for the calculation) is independent of acid concentration.

The most recent and perhaps the most extensive study of desulphonation is that of Wanders and Cerfontain[708], who avoided the trouble caused by resulphonation of the desulphonation product by passing a stream of nitrogen through the reaction mixture so that the aromatics produced were removed from solution. The nitrogen stream was moistened in order to prevent any nett uptake of water from the reaction solution and in this way that acid strength was maintained constant to within $\pm0.5$ wt. %; the magnitude of this variation, regarded in this context as satisfactory, serves to illustrate the difficulties attendant upon sulphonation and desulphonation studies. Benzene- and toluene-sulphonic acids in 65–85 wt. % acid were studied at temperatures between 120 and 155 °C. First-order rate coeffi-

TABLE 245

RATE COEFFICIENTS ($10^6 k_1$) FOR REACTION OF ArSO$_3$H WITH AQUEOUS H$_2$SO$_4$[708]

| Temp. (°C) | H$_2$SO$_4$ (wt %) | Ar | | | |
|---|---|---|---|---|---|
| | | Ph | 2-MeC$_6$H$_4$ | 3-MeC$_6$H$_4$ | 4-MeC$_6$H$_4$ |
| 120.1(±0.2) | 69.5(±0.2 %) | | 28.9 | | 15.3 |
| 120.1(±0.2) | 79.3 | 1.36 | 207 | 3.05 | 94.0 |
| 120.1(±0.2) | 84.8 | 0.85 | | 4.54 | |
| 139.6 | 64.4 | | 62.1 | | 42.3 |
| 140.3 | 69.8 | | 190 | | 116 |
| 140.3 | 74.4 | | | 9.3 | 333 |
| 140.3 | 74.8 | 4.00 | 461 | 11.7 | 306 |
| 140.3 | 75.2 | | 456 | | |
| 140.3 | 80.0 | 6.27 | 1,200 | 20.8 | 417 |
| 140.3 | 84.8 | 4.11 | | 30.0 | |
| 154.4 | 69.5 | | 556 | | 364 |
| 154.4 | 79.7 | 16.4 | ~3,800 | | ~3,400 |
| 154.4 | 85.0 | 8.61 | | | |

cients are given in Table 245, and plots of logarithms of these rates *versus* the acidity function $-H_0$ was considered to give straight lines except for the 4-methyl compound; however, two of the lines involved two points only. Plots of log rate coefficient *versus* log $[H_3O^+]^2/a_{H_2O}$ were moderately linear and this was interpreted as indicating reaction upon the sulphonate anion, rather than the acid. For the 4-methyl compound, there was considerable curvature at high acid concentrations; hence it was suggested that reaction might be occurring on the free acid at these concentrations for this compound.

Partial rate factors were calculated for the reaction and the ratio log $f_p$/log $(f_p/f_m)$ varied from 1.24 to 1.32 compared with the range $1.31 \pm 0.10$ found for most electrophilic substitutions, thereby giving some reliability to the data. However, the partial rate factors vary markedly with the medium and in an ill-defined manner; thus for example $f_m^{Me}$ is 3.3 in 80.0 wt. %, acid and 7.3 in 84.8 wt. % acid (both at 140.3 °C) due to the rates for the benzene compound passing through a maximum at *ca.* 80.0 wt. % acid, and the rates for the 4-methyl compound also show a tendency to pass through a maximum. This apparent increase in the selectivity of the reaction may be indicating a change in mechanism from reaction on the anion to reaction on the acid, as indicated also by the rate maxima.

In summary, and in view of the reaction mechanism being necessarily the reverse of that appropriate to sulphonation, it can probably best be summarised as consisting of equilibria (275)–(277) or (275), (276) and (278), *viz.*

$$ArSO_3H \ (+H_2O) \ \rightleftharpoons \ ArSO_3^- \ + \ H^+(H_2O) \quad (fast) \tag{275}$$

$$ArSO_3^- \ + \ H_2SO_4 \ \rightleftharpoons \ Ar\underset{SO_3^-}{\overset{+}{<}}^{H} \ + \ HSO_4^- \quad (slow) \tag{276}$$

$$Ar\underset{SO_3^-}{\overset{+}{<}}^{H} \ + \ H_3O^+ \ \rightleftharpoons \ ArH \ + \ H_3SO_4^+ \quad (fast) \tag{277}$$

$$Ar\underset{SO_3^-}{\overset{+}{<}}^{H} \ + \ H_2SO_4 \ \rightleftharpoons \ ArH \ + \ SO_3(H_2SO_4) \quad (fast) \tag{278}$$

for it seems most probable that reaction occurs on the sulphonic acid anion rather than on the free acid, except perhaps at higher acid concentrations.

## 9.11 PROTODEHALOGENATION

The first kinetic studies of this reaction, *viz.*

$$ArHal + HX = ArH + XHal \tag{279}$$

were of a fairly qualitative nature, and consisted of measuring the percentage decomposition at given times in the reaction of halogenated phenols with hydriodic acid in acetic acid at 25 and 78 °C[709]. Substituents activated the removal of iodine in the order: 2,4-(HO)$_2$ > 4-(HO) > 2-HO > 3-HO, and 2-Me > 4-Me > 3-Me > H, whilst from phenol a 4-iodo substituent was removed more readily than a 4-bromo substituent, which was in turn removed more readily than 2-bromo substituent. Clearly, the reaction is an electrophilic substitution.

The removal of iodine from 2-iodo-4-nitroaniline in refluxing hydrochloric acid (*ca.* 100 °C), had a first-order rate coefficient of $4.7 \times 10^{-5}$, and for 2-iodo-1,3-diaminobenzene, 2-iodoaniline, and 2,5-diiodo-4-aminotoluene in hydrochloric acid containing stannous chloride the respective rate coefficients were $11.7 \times 10^{-5}$, $3.3 \times 10^{-4}$, and $2.5 \times 10^{-4}$, respectively, these results being interpreted as indicating an electrophilic substitution[710].

The first kinetic study of the reaction aimed at elucidating details of the reaction mechanism was that of Gold and Whittaker[711], who measured the rates of de-iodination of substituted 4-iodophenols by hydriodic acid in aqueous acetic acid at temperatures ranging between 20 and 50 °C. The relative rates of reaction of the phenols for the following substituents were: H (1.0); 2-Me (2.4); 2-Cl (0.08); 3-Me (48); 3-Me-4-*i*-Pr (160), which demonstrated the electrophilic nature of the reaction, and the corresponding activation energies were: 22.4, 22.2, 15.7, 20.7, 19.0; the low value for the least reactive chloro compound is unexpected. The kinetic form of the reaction was very complex, being first-order in aromatic, third-order in acid, and inverse seven-halves-order in water; consequently, no firm conclusions as to the mechanism could be deduced.

A more successful kinetic study has been made by Choguill and Ridd[712], who studied the rates of deiodination of 4-iodoaniline by aqueous mineral acids. The reaction is first-order in aromatic, the rate being independent of the concentration of hydrogen and iodide ions. A first-order rate coefficient of *ca.* $70 \times 10^{-6}$ at 98 °C was obtained with sulphuric, perchloric and hydrochloric acids, and was independent of the concentration of the latter over the range 0.025–6.0 *M*. All of the rates were obtained in the presence of 0.25 *M* iodide ion and were unchanged for a fourfold increase in iodide concentration. Under the conditions of the experiments, 0.25 *M* of iodide ion was sufficient to displace equilibrium (280)

$$\text{NH}_3^+\text{PhI} + \text{H}^+ + 2\,\text{I}^- \rightleftharpoons \text{NH}_3^+\text{PhH} + \text{I}_3^- \tag{280}$$

to complete deiodination, and it is this mass law effect which accounts for the superiority of hydiodic acid in the reaction, since iodide ion itself has no effect. The kinetic equation for the reaction was thus

$$\text{Rate} = k[\text{NH}_3^+\text{C}_6\text{H}_4\text{I}]_{\text{stoich}} \tag{281}$$

and since the concentration of the free amine (which must almost certainly be the reacting species) in the acid range covered is equal to that of the conjugate acid, reaction was envisaged as occurring between the free amine and a proton.

By the principle of microscopic reversibility, it follows that protodeiodination must in all steps be the reverse of iodination, and since this latter reaction is partly rate-determining in loss of a proton (see pp. 94–97, 136) it follows that attachment of a proton should be rate-determining in the reverse reaction; this was found to be the case, the first-order rate coefficients for reaction in $\text{H}_2\text{O}$ and 97.5 % $\text{D}_2\text{O}$ being 76.6 and $13.1 \times 10^{-6}$ respectively, so that $k_{\text{H}_2\text{O}}/k_{\text{D}_2\text{O}} = 5.8$.

Since the rate was independent of acidity even over the range where $H_0$ and pH differ, and the concentration of free amine is inversely proportional to the acidity function $h_0$ it follows that the rate of substitution is proportional to $h_0$. If the substitution rate was proportional to $[\text{H}_3\text{O}^+]$ then a decrease in rate by a factor of 17 should be observed on changing $[\text{H}^+]$ from 0.05 to 6.0. This was not observed and the discrepancy is not a salt effect since chloride ion had no effect. Thus the rate of proton transfer from the medium depends on the acidity function, yet the mechanism of the reaction (confirmed by the isotope effect studies) is A-$S_E2$, so that again correlation of rate with acidity function is not a satisfactory criterion of the A-1 mechanism.

## 10.  Reactions involving replacement of a substituent X by a substituent Y

### 10.1 MERCURIDEMERCURATION

The exchange of mercury between alkyl and aryl mercury compounds equilibria (282)–(284)

$$RHg^*X + HgX_2 \rightleftharpoons RHgX + Hg^*X_2 \qquad (282)$$

$$RHgX + RHgX \rightleftharpoons RHgR + HgX_2 \qquad (283)$$

$$RHgX + RHg^*R \rightleftharpoons RHgR + RHg^*X \qquad (284)$$

has been studied fairly intensively from the kinetic viewpoint and the reaction is an electrophilic substitution. It is difficult to divorce the observations and conclusions from the studies of the alkyl from those of the aryl compounds and where necessary the two are considered together in the following. Equilibria (282), (283), and (284) have been termed one, two, and three alkyl (or aryl) exchanges, respectively, by Ingold et al.[713] who demonstrated the unambiguous existence of the former and latter and determined the kinetic form of the other.

The first observations relevant to the kinetics and mechanism of the reaction were those of Wright[714], who found that in the reaction of bis-cis-2-methoxy-cyclohexylmercury with mercuric chloride, only the cis product was obtained, and of Winstein et al.[715], who found that in the reaction of cis-2-methoxycyclo-hexylneophylmercury with mercuric ($^{203}$Hg) chloride, the label became almost equally distributed in both reaction products.

The first extensive kinetic study of these reactions was carried out by Ingold et al.[713b], who demonstrated that for the two-alkyl exchange reaction, i.e. the reverse direction of equilibrium (283), the kinetics in acetone and ethanol were cleanly second-order, viz.

$$\text{Rate} = k_2[\text{R}_2\text{Hg}][\text{HgBr}_2] \qquad (285)$$

the reaction occurred with retention of configuration, and the rates increased with increasing ionicity of X in $HgX_2$ ($NO_3 > OAc > Br$); the reaction could be reversed (the so-called symmetrisation reaction) by continuously removing $HgBr_2$. It was concluded that the reaction proceeded by a bimolecular $S_E2$ mechanism, the alternative $S_Ei$ mechanism being eliminated by virtue of the fact that it would require collaborative nucleophilic attack of a potentially anionic component of the reagent. Addition of lithium bromide produced a rate decrease, attributed to the formation of $LiHgBr_3$ and the kinetic expression became

$$\text{Rate} = k_2[\text{R}_2\text{Hg}][\text{HgBr}_2.\text{LiBr}] \qquad (286)$$

The tribromide anion would, however, be expected to increase the rate if the reaction occurred *via* the $S_E i$ mechanism. The second-order kinetics and stereochemical retention of configuration were shown to apply to the one- and three-alkyl exchanges, and the retention of configuration for the two-alkyl exchange was confirmed by Jensen *et al.*[716].

Although the above work showed that for the symmetrization reaction (and its reverse reaction) the mechanism is $S_E 2$ with two steps of dissimilar rate, a symmetrical transition state (XLIII),

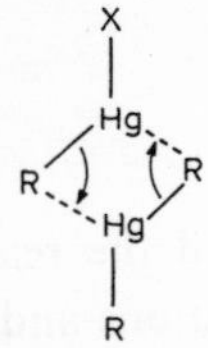

(XLIII)

was proposed by Reutov *et al.*[717] to explain the retention of configuration and distribution of radioactive label in reactions of alkylmercury compounds. Dessy and Lee[718] studied the reaction of the diarylmercuricals with mercuric salts in dioxan (Table 246). The reaction was cleanly second-order, and took place with

TABLE 246

RATE COEFFICIENTS ($k_2$) AND KINETIC PARAMETERS FOR THE REACTION OF $(XC_6H_4)_2Hg$ WITH $HgI_2$ IN DIOXAN[718]

| X | Temp. (°C) | | | | |
|---|---|---|---|---|---|
| | 25 | 35 | 45 | E | $\Delta S^{\ddagger}$ |
| 4-MeO | 71.5 | | | | |
| 4-Me | 13.1 | 22.7 | | 10.6 | −20 |
| 4-Ph | 2.25 | 4.20 | | 12.0 | −19 |
| H | 1.97 | 3.90 | 7.60 | 12.8 | −16 |
| 4-F | 0.42 | 0.729 | 1.40 | 12.9 | −19 |
| 4-Cl | 0.092 | 0.202 | 0.42 | 14.5 | −17 |

increasing ease along the series of mercuric salts: $HgI_2 < HgBr_2 < HgCl_2$. For the reaction of diphenylmercury with mercuric iodide, addition of up to 10 % of water caused less than a 4-fold rate increase (Table 247), and no appreciable effect followed from adding 5 mole % of lithium iodide to the reaction mixture; hence ionic iodide is not important in the rate-determining step. The activation entropies for the reaction are consistent with a transition state of fairly substantial cyclic character, and attack by an ion pair in a cyclic structure such as (XLIV) was favoured though it is apparent that the electrophilic attack on carbon must

TABLE 247

RATE COEFFICIENTS ($k_2$) FOR REACTION OF $Ph_2Hg$ WITH $HgI_2$ IN VARIOUS SOLVENTS[719]

| Solvent | Temp. (°C) | | | | | |
|---------|------|------|------|------|-------|------|
| | 15 | 25 | 35 | 45 | $E_a$ | $\Delta S^{\ddagger}$ |
| Dioxan | $-(29.2)^a$ | $2.0(58.0)^a$ | 3.90 | 7.60 | $12.8(12.5)^a$ | $-16(-11)^a$ |
| Dioxan(5 % $H_2O$) | | 3.5 | | | | |
| Dioxan (10 % $H_2O$) | | 6.8 | | | | |
| Ethanol | 32.1 | 62.8 | 113 | | 11.7 | $-13$ |
| Benzene | 19.0 | 29.2 | 43.0 | | 7.6 | $-28$ |
| Cyclohexane | | 15.9 | 23.2 | 34.0 | 7.6 | $-31$ |

$^a$ Data for PhEtHg.

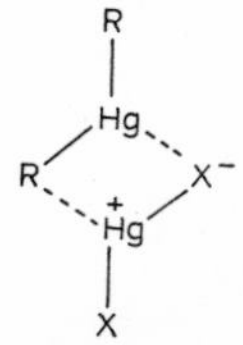

(XLIV)

be much more important than the nucleophilic attack on mercury. The substituent effects at 25 °C correlated with $\sigma$ values, $\rho = -5.87$; this is surprising since an electrophilic substitution with such a high $\rho$ factor should correlate with $\sigma^+$ rather than $\sigma$ values but this anomaly may partly derive from the presence of two aryl groups.

A subsequent study by Dessy et al.[719] of the rate of reaction of diphenylmercury with mercuric iodide in a variety of solvents produced the rate coefficients in Table 247. Significantly, the coordinating solvents produce higher activation energies and significantly less negative entropies. Coordination with mercury will reduce resonance between metal and aryl ring, so increasing the activation energy, as will increased solvation of the ground state by the coordinating solvents; this latter effect will also produce a less negative value of $\Delta S^{\ddagger}$ as observed. Alternative versions of the stereochemistry of the transition state, such as (XLV) were ruled out by the rate coefficients obtained with bis-o-phenylenedimercury (XLVI), viz.

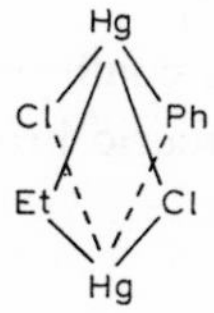

(XLV)

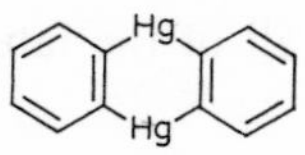

(XLVI)

of 25.3 (15 °C) and 44.8 (25 °C) for cleavage of the first aryl–mercury bond and 5.9 (15 °C) and 10.0 (25 °C) for cleavage of the second, leading to activation energies of 9.9 and 10.2, and entropies of −20 and −22, respectively, if the structure (XLV) was appropriate for the transition state, vastly different rate coefficients and kinetic parameters would have been expected in view of the steric hindrance to cleavage of the first aryl–mercury bond. These workers also reported that reaction of phenylethylmercury with labelled mercuric chloride gave a *statistical* distribution of mercury, but later workers report that this is not the case since cleavage occurs 95 % at the aryl–mercury bond so that most of the label ends up in the arylmercuric chloride, *i.e.* the transition state is predominantly (XLVII)[720], which is not appreciably different from (XLIV). (Other workers, however, have reported[721] that in ether at −20 °C, unsymmetrical mercuricals cleave predominantly at the alkyl–mercury bond, or, generally, at the R–Hg bond where R is the least electron-supplying group attached to mercury which is the opposite to that expected for an electrophilic substitution.)

(XLVII)

Subsequent kinetic work has amply confirmed the mechanistic picture described above. For example, the reaction of diphenylmercury with Ph(COOEt).CH.HgBr gives an almost instantaneous reaction with precipitation of phenylmercuric bromide, whereas reaction of the soluble product with a second molar equivalent of mercuric bromide gave a very slow (*ca.* 2 weeks) precipitation of phenyl-mercuric bromide[722], *i.e.* reaction involves (287) and (288)

$$Ph_2Hg + Ph(CO_2Et)CH.HgBr \xrightarrow[\text{(fast)}]{} Ph(CO_2Et)CH.HgPh + PhHgBr \quad (287)$$

$$Ph(CO_2Et)CH.HgPh + Ph(CO_2Et)CH.HgBr \xrightarrow[\text{(slow)}]{} [Ph(CO_2Et)CH]_2Hg + \\ PhHgBr \quad (288)$$

The fast reaction clearly involves the transition state (XLVIII) as confirmed by the observation of Reutov *et al.*[723] that all of the label in the mercuric bromide

Ph
|
Hg.
Ph ⟩ ⟨ Br
Hg*
|
PhCH·COOEt
(XLVIII)

PhCH·COOEt
|
Hg*
Ph ⟩ ⟨ Br
Hg
|
PhCH·COOEt
(XLIX)

Ph
|
H   Hg*
EtOOC–C ⟨ Br
Ph   Hg
|
PhCH·COOEt
(L)

remains in solution. However, this is not true of the second reaction which produces an even distribution of label. If the transition state were (XLIX) this would give no new products and an even distribution of label; on the other hand if it were (L) the correct product would have no label, consequently it was proposed that rapid equilibration *via* (XLIX) was followed by slow reaction *via* (L). The effects of substituents in the phenyl ring of the substituted benzyl moiety were considered to be consistent with such a mechanism.

Finally, the kinetics of the one-aryl exchange reaction, equilibrium (289)

$$PhHgCl + {}^{203}HgCl_2 \rightleftharpoons Ph{}^{203}HgCl + HgCl_2 \tag{289}$$

have been studied using labelled mercuric chloride and the reaction follows the established second-order pattern with an activation energy in anhydrous toluene of 6.09 (25.5–39.5 °C)[724]; the transition state for the reaction was vaguely interpreted as being analogous to (XLIII).

## 10.2 HALODEMERCURATION

The cleavage of phenylmercuric iodide by iodine in the presence of excess iodide ion (to suppress free-radical reactions) at 25 °C in aqueous dioxan was reported to be first-order in both aromatic and tri-iodide ion, and faster than the reaction of alkylmercuric iodides[724a]. A further study, together with bromodemercuration, both reactions being generally represented by

$$PhHgBr + Hal_2 = PhHal + HgHalBr \tag{290}$$

has been carried out using dimethylformamide, 80 % aqueous dioxan, and methanol as solvents, the reaction taking place very rapidly to the extent that it could not be studied under first-order conditions[725]. Hence it could not be established that the reaction, second-order overall, is first-order in each component, though it is highly unlikely to be otherwise. The progress of the reaction was studied photometrically and the rate coefficients for reaction with bromine (in the presence of excess ammonium bromide) and with iodine (in the presence of excess cadmium iodide) are given in Table 248 together with the kinetic parameters. For each reaction, the rate coefficients increase in the solvents dimethylformamide < 80 % aqueous dioxan < methanol, but the variation of the kinetic parameters with change of solvent shows no explicable pattern; in benzene as solvent, the

TABLE 248

RATE COEFFICIENTS AND KINETIC PARAMETERS FOR THE REACTION OF $Br_2$ OR $I_2$ WITH PhHgBr[725]

| Solvent | Temp. (°C) | $k_2$ ($Br_2$) | $k_2$ ($I_2$) | $E_a$ ($Br_2$) | $E_a$ ($I_2$) | $\Delta S^\dagger$ ($Br_2$) | $\Delta S^\ddagger$ ($I_2$) |
|---|---|---|---|---|---|---|---|
| Dimethylformamide | 5.9 | | 8.3 | 13.4 | 9.8 | −12.0 | −20.2 |
| | 10.0 | 1.88 | 11.1 | | | | |
| | 15.0 | 2.88 | 15.1 | | | | |
| | 20.0 | 4.25 | 20.2 | | | | |
| | 21.0 | 4.40 | | | | | |
| | 23.0 | | 23.6 | | | | |
| | 25.0 | 6.20 | | | | | |
| 80 % Aq. dioxan | 8.0 | 32.0 | | 14.0 | 11.2 | − 3.3 | −13.9 |
| | 9.9 | | 42.0 | | | | |
| | 12.0 | 45.9 | | | | | |
| | 15.0 | | 55.4 | | | | |
| | 16.0 | 65.8 | | | | | |
| | 20.0 | 89.4 | 78.0 | | | | |
| | 25.0 | | 115 | | | | |
| Methanol | 5.0 | 62.9 | | 8.2 | 11.1 | −21.7 | −12.7 |
| | 12.0 | 89.0 | | | | | |
| | 12.4 | | 88.5 | | | | |
| | 16.0 | | 114 | | | | |
| | 19.0 | 138 | | | | | |
| | 20.0 | | 148 | | | | |
| | 24.0 | | 190.5 | | | | |

reaction with iodine occurred *via* a radical mechanism. The transition state was proposed in involving a 4-centre reaction with assistance from halide ion but a more detailed description is inappropriate without further data. The transition state for halogen cleavage of alkylmercuric halides is likewise still not settled[610].

## 10.3 MERCURIDEBORONATION

Only one kinetic study of mercurideboronation has been carried out[726], this being the reaction between benzeneboronic acid and basic phenylmercuric perchlorate

$$PhB(OH)_2 + PhHgClO_4 + H_2O = Ph_2Hg + B(OH)_3 + HClO_4 \qquad (291)$$

In contrast to the analogous reaction with phenylmercuric chloride which gave complex kinetics, reaction (291) was second-order in 30–60 % aqueous ethanol (the progress of reaction being followed spectrophotometrically) and was first-order in each (organic) reagent. Rate coefficients were increased by a higher water content of the medium (Table 249) but this was not due to an increase in

TABLE 249

RATE COEFFICIENTS FOR REACTION OF $PhHgClO_4$ WITH $PhB(OH)_2$ IN AQUEOUS EtOH ($\mu = 0.01$) AT 25 °C[76]

| Water composition of solvent (%) | $k_2$ |
|---|---|
| 70 | 13.3 |
| 60 | 8.56, 8.22[a], 7.83[b] |
| 50 | 5.21 |
| 40 | 3.00, 3.05[b] |

[a] Ionic strength = 0.04.
[b] In presence of $\sim 10^{-4}$ $M$ $B(OH)_3$.

the ionic strength of the medium since a fourfold increase in this produced very little change in rate. The origin of this specific effect of water is not clear, though since the reaction is thought to occur between boronic acid and phenylmercuric hydroxide (produced initially and rapidly from the perchlorate by hydrolysis) the rate of this hydrolysis in solutions of different water content may be a significant factor.

The rate of the reaction in various buffer solutions, covering the pH range 4–8, was determined, and in hydrogen phosphate–dihydrogen phosphate buffers the rate at constant pH decreased as the concentration of dihydrogen phosphate increased. Similarly, with acetic acid–acetate and phosphoric acid–dihydrogen phosphate buffers the rate was inversely dependent upon the concentration of the molecular acid; in addition, with the latter buffer, the kinetic plots showed an unexplained departure from linearity after 50 % reaction.

The kinetic dependence of the reaction was explained in terms of a reaction between $PhB(OH)_3^-$ and $PhHg^+$. From analysis of the concentration of the species likely to be present in solution it was shown that reaction between these ions would yield an inverse dependence of rate upon molecular acid composition in buffer solutions, as observed for a tenfold change in molecular acid concentration, and that at high pH this dependence should disappear as found in carbonate buffers of pH $\sim$ 10. The form of the transition state could not be determined from the available data, and it would be useful to have kinetic parameters which might help to decide upon the likelihood of the 4-centre transition state, which was one suggested possibility.

*References pp. 388–406*

## 10.4 HYDROXYDEBORONATION

The kinetics of reaction (292), *viz.*

$$H_2O_2 + PhB(OH)_2 = PhOH + B(OH)_3 \tag{292}$$

are exceedingly complex and the investigations to date indicate the presence of at least four processes with different kinetic dependencies[727]. The first study, using aqueous buffer solutions of pH 2–6, showed that two reactions occurred, both were first-order in peroxide, and one was first-order in boronic acid, the other being second-order in boronic acid, *i.e.* the kinetic equation is

$$\text{Rate} = k_{obs}[H_2O_2][PhB(OH)_2] \tag{293}$$

(where $k_{obs} = k_2 + k_3[PhB(OH)_2]$) as indicated by the data in Table 250, which shows the experimentally observed rate coefficients and those calculated according to the kinetic equation (293). Both rates increase with pH, but are independent of the buffer concentration in the case of phthalate buffers (Table 251), and with acetate and chloracetate buffers the same specific rate coefficients were obtained

TABLE 250

RATE COEFFICIENTS FOR REACTION OF $PhB(OH)_2$ WITH $H_2O_2$ IN BUFFER SOLUTIONS AT 25 °C[727]

| [Acetate buffer] (M) | [Chloroacetate buffer] (M) | pH | $10^3$ [Boronic acid] (M) | $10^3 k_{obs}$ | $10^3 k_{calc}$ |
|---|---|---|---|---|---|
| 0.05 | | 4.88 | 20.0 | 9.92 | 9.9 |
| 0.05 | | 4.88 | 40.0 | 12.6 | 12.4 |
| 0.05 | | 5.24 | 10.0 | 19.8 | 20.2 |
| 0.05 | | 5.24 | 20.0 | 22.8 | 23.4 |
| | 0.22 | 3.75 | 40 | 1.73 | 1.66 |
| | 0.22 | 3.75 | 20 | 1.58 | 1.51 |
| | 0.055 | 3.67 | 40 | 1.61 | 1.54 |
| | 0.055 | 3.67 | 20 | 1.45 | 1.42 |

TABLE 251

NON-DEPENDENCE OF RATE COEFFICIENTS $k_2$ AND $k_3$ UPON BUFFER CONCENTRATION FOR REACTION OF $PhB(OH)_2$ WITH $H_2O_2$ AT 25 °C[727]

| [Phthalate buffer) (M) | $10^3 k_2$ | $10^3 k_3$ |
|---|---|---|
| 0.02 | 3.32 | 50.6 |
| 0.05 | 3.01 | 51.3 |
| 0.10 | 3.38 | 47.0 |

as with phthalate buffers. The existence of specific hydroxide ion catalysis was interpreted to mean that since the rate must be proportional to a species whose concentration depends upon the pH, then this must be $HOO^-$, and the mechanism of the second-order reaction was envisaged as

$$
\text{Ph}-\overset{\overset{\displaystyle OH}{|}}{\text{B}}-\text{OH} + \text{HOO}^- \underset{\text{fast}}{\rightleftharpoons}
\left[ \underset{(LI)}{\text{Ph}-\overset{\overset{\displaystyle OH}{|}}{\underset{\underset{\displaystyle O\cdots OH}{|}}{\text{B}}}-\text{OH}} \right]
\underset{\text{slow}}{\rightleftharpoons}
\left[ \underset{(LII)}{\text{Ph}-\overset{\overset{\displaystyle OH}{|}}{\underset{\underset{\displaystyle O^+}{|}}{\text{B}^-}}-\text{OH}} \right] + \text{OH}^-
\tag{294}
$$

$$
\text{PhOH} + \text{B(OH)}_3 \xleftarrow[\text{fast}]{\text{H}_2\text{O}} \text{PhOB(OH)}_2 \longleftarrow
$$

which is similar to that generally written for reactions of Caro's acid and percarboxylic acids with ketones to produce esters, and the mechanism of the third-order reaction as

$$
(LI) + \text{PhB(OH)}_2 \rightleftharpoons \text{H}_2\text{O} +
\left[ \underset{(LIII)}{\text{Ph}-\overset{\overset{\displaystyle OH}{|}}{\underset{\underset{\underset{\displaystyle OH}{|}}{\underset{\displaystyle O-B-Ph}{|}}}{\text{B}}}-\text{OH}} \right]
\longrightarrow (LII) + \text{Ph}\overset{}{\underset{\underset{\displaystyle O^-}{|}}{\text{B}}}\text{OH}
\tag{295}
$$

the peroxide bond to boron being considered as more easily broken in (LIII) than in (LI); hence the observed catalytic effect of the second molecule of boronic acid.

This study indicated that the slope of the log rate *versus* pH plot tends to become

TABLE 252

VARIATION IN RATE COEFFICIENT WITH CONCENTRATION OF ACID FOR REACTION OF PhB(OH)$_2$ WITH H$_2$O$_2$ AT 25 °C[727]

| $[HClO_4]$ (M) | 0.0025 | 0.005 | 0.0075 | 0.01 | 0.2 | 0.75 | 2.469 | 3.73 | 4.12 | 5.93 | 8.15 |
|---|---|---|---|---|---|---|---|---|---|---|---|
| $10^3 k_2$ | 0.781 | 0.771 | 0.773 | 0.793 | 1.56 | 3.94 | 16.0 | 31.9 | 44.0 | 137 | 605 |
| $10^3 k_2{}^{\text{a}}$ | | | | | 0.79 | 3.17 | 15.2 | 31.2 | 43.2 | 136 | 604 |

| $[H_3PO_4]$ (M) | 0.0725 | 2.267 | 5.19 | 6.87 | 9.67 |
|---|---|---|---|---|---|
| $10^2 k_2$ | 0.332 | 1.40 | 12.1 | 49.8 | 214 |
| $10^2 k_2{}^{\text{a}}$ | 0.255 | 1.32 | 12.0 | 49.7 | 214 |

| $[H_2SO_4]$ (M) | 0.193 | 0.386 | 1.13 | 2.26 | 3.64 | 6.38 | 8.99 |
|---|---|---|---|---|---|---|---|
| $10^2 k_2$ | 0.184 | 0.286 | 0.989 | 3.17 | 9.19 | 59.1 | 360 |
| $10^2 k_2{}^{\text{a}}$ | 0.107 | 0.209 | 0.912 | 3.09 | 9.11 | 59.0 | 360 |

[a] Corrected for the uncatalysed reaction of rate $\sim 7.7 \times 10^{-4}$.

zero at low pH so there is a possibility that a third mechanism involving molecular hydrogen peroxide becomes important in this range. A later investigation of the effect of phosphoric, perchloric and sulphuric acid catalysts on the reaction, *i.e.* in the low pH range, gave the data in Table 252[727]. The data for perchloric acid indicated the existence of this uncatalysed reaction which might be envisaged as

$$(LI) + H^+ \rightarrow H_2O + (LII) \tag{296}$$

Analysis of the rate coefficients, corrected for this uncatalysed reaction revealed a complex dependence of an acid-catalysed reaction, such that the logarithms of the rate depended on $H_0$ in phosphoric acid, upon $\log (a_{H_2O}.a_{HClO_4}^{\frac{1}{2}})$ in perchloric acid and upon $\log (a_{H_2O}.a_{H_2SO_4})^{\frac{1}{2}}$ in sulphuric acid; clearly, no satisfactory mechanistic picture is likely to ensue from these data for the acid-catalysed reaction without considerable further investigation.

Certain chelating agents produced a catalytic effect upon the reaction; among a range of 1,2 and 1,3-diols, only pinacol was effective, but a series of α-substituted

TABLE 253

RATE COEFFICIENTS FOR REACTION OF $H_2O_2$ WITH $XC_6H_4B(OH)_2$ CATALYSED BY $HClO_4$ (4.77 $M$) OR $H_3PO_4$ (5.29 $M$) AT 25°C[727]

| X | $10^2 k_2$ ($HClO_4$) | $10^2 k_2$ ($H_3PO_4$) |
|---|---|---|
| 4-MeO | 18.8 | 40.4 |
| 4-Me | 13.0 | 17.5 |
| H | 6.47 | 12.4 |
| 4-F | 4.64 | 8.75 |
| 3-MeO | 4.96 | 9.55 |
| 4-Cl | 4.23 | 8.09 |
| 3-F | 2.27 | 5.35 |
| 3-Cl | 2.53 | 6.10 |
| 3-Br | 2.69 | 6.31 |
| 3-NO$_2$ | 0.827 | 2.13 |

glycolic acids were effective in catalysing the second- and third-order reactions (substitution producing an increased catalytic effect), and this was suggested as arising from chelation with boronic acid so reducing the O–B–O angle such that subsequent substitution of $OOH^-$ occurs more easily. For citric acid, the catalytic effect appeared to reach a maximum at about pH 5, and α-hydroxyisobutyric acid produced a similar effect. The catalysis by these acids was assumed to arise from a reaction between free boronic acid and free hydroxy acid, but that in addition, a side reaction ties up these reagents in a complex such as (LIV), and volumetric analysis indicated that such complexes were formed.

(LIV)

Analysis of substituent effects in the reaction revealed random correlations of rate with $\sigma$ values of substituents for the uncatalysed reaction, but for the acid-catalysed reaction, reasonable Hammett plots with $\rho$ factors of $-1.27$ ($HClO_4$) and $-1.12$ ($H_3PO_4$) were obtained (Table 253).

### 10.5 HALODEBORONATION

#### 10.5.1 Bromodeboronation

Kinetic studies of bromodeboronation in aqueous acetic acid as solvent, *viz.*

$$ArB(OH)_2 + Br_2 \overset{H_2O}{=} ArBr + HBr + B(OH)_3 \qquad (297)$$

have shown that the reaction is second-order, first-order in each reagent for a 5-fold change in boronic acid concentration and a 17-fold change in initial bromine concentration[728]. In the absence of added sodium bromide departure from second-order behaviour occurs, since the bromide ion product reduces the concentration of free bromine; in the presence of sodium bromide good second-order kinetics result. At higher concentrations of reagents, *i.e.* of the order of 0.015 $M$ boronic acid, curvature of the second-order plots is also obtained in the presence of sodium bromide and thus was attributed to the proton liberated, since in the presence of added perchloric acid, the rate (which is reduced) is second-order throughout a run. Other acids also retard the rate and bases accelerate it; the latter effect is not merely a positive salt effect for although this was observed with sodium nitrate or perchlorate the rate being increased by 10–20 %, with bases the rate acceleration was up to eightfold. The rate was also accelerated by the addition of water and the respective second-order rate coefficients at 25 °C, obtained with 20 %, 30 %, and 50 % aqueous acid were 0.0738, 0.0374, and 0.0065.

The kinetics were interpreted in terms of mechanism (298)

$$ArB(OH)_2 + H_2O \rightleftharpoons \left[ \begin{array}{c} OH \\ | \\ Ar-B-OH \\ | \\ OH \end{array} \right]^- + H^+ \overset{Br_2}{\rightarrow} ArBr + HBr + B(OH)_3 \qquad (298)$$

which accommodates all the facts. An increase in water concentration will increase the concentration of the intermediate and acid will reduce it; formation of the intermediate will be promoted by salts. In the presence of bases the function of water is replaced by the base $(N^-)$ to give an intermediate of structure

$$\begin{bmatrix} \text{OH} \\ | \\ \text{Ar--B--OH} \\ | \\ \text{N} \end{bmatrix}^-$$

, and the concentration of this will be higher than the water-produced analogue; in each case the negatively charged intermediate is more reactive towards electrophiles than the uncharged boronic acid.

A more extensive examination of the kinetics of deboronation of 3-chloro-benzeneboronic acid, showed that for aqueous solutions of pH 2.10–4.74, second-order kinetics were again obtained for an eightfold variation in bromine concentration and a fivefold change in acid concentration. In the presence of various amounts of sodium bromide the reaction rate was also proportional to the concentration of free bromine, indicating that molecular bromine is the electrophile. Despite the acceleration previously noted with bases, the rate was reduced by an increasing acetate buffer concentration, both components producing retardation. The larger effect of acetic acid was attributed to its effect on the bromine–tribromide ion equilibrium, but the effect of acetate ion was not satisfactorily explained. The rate showed a linear dependence upon pH with almost unit slope, and this specific hydroxide ion catalysis obviously follows from reaction occurring on the boronate anion. As in the case of hydroxydeboronation, chelating agents increase the reaction rate, thus being attributed to the same cause, namely the reduction of the O–B–O angle from 120 °C to *ca.* 108 °C, thereby facilitating formation of the fourth covalency.

Bromodeboronation has acquired a particular significance in recent theories of electrophilic substitution and briefly this has arisen since it was supposed to have a very high $r$ factor relative to its $\rho$ factor in a Yukawa–Tsuno analysis. (For a fuller discussion see ref. 729). It had been suggested[729] that some of the rate coefficients determined for the reaction (Table 254) may be in error due to concurrent bromodeprotonation, and a reinvestigation[730] of this possibility has revealed a number of points:

(*i*) A gas chromatographic analysis of the reaction products showed that although some of the predicted side reactions do occur, their magnitude would not be sufficient to account for the large deviations of the Hammett plots.

(*ii*) The rate coefficient quoted by Kuivila *et al.*[728] for the 4-methoxy compound, when related to the concentration of reagents used (namely 0.01 $M$ of boronic acid and 0.008 $M$ of bromine), yields a half-life of 0.01 sec.

The titration method used is, in our experience, incapable of detecting a difference between a reaction rate of this speed or one a hundred times slower, since it takes a minimum of 45 sec to obtain the first reading. Since this is the major point on which the high $r$ factor was based, the arguments following from it are no longer tenable.

TABLE 254
RATE COEFFICIENTS FOR THE REACTION OF $Br_2$ WITH $XC_6H_4(OH)_3$ IN 80 % AQUEOUS HOAc AT 25 °C

| X | $10^5k_2$[a] | $k_{rel}$[a] | $10^5k_2$[b] | $k_{rel}$[b] |
|---|---|---|---|---|
| 4-MeO | $>7 \times 10^8$ | $1.45 \times 10^6$ | $\sim 6 \times 10^6$ | $1.1 \times 10^4$ |
| 4-Me | 38,000 | 78.7 | 27,500 | 51.5 |
| 4-Ph | 10,500 | 21.7 | 6,950 | 13.0 |
| 3-Me | 1,610 | 3.33 | 1,600[c] | 3.00[c] |
| 4-F | 1,360 | 2.81 | 1,370 | 2.56 |
| H | 484 | 1.0 | 534 | 1 |
| 4-Cl | 261 | 0.54 | | |
| 4-I | 241 | 0.497 | | |
| 4-Br | 200 | 0.413 | | |
| 3-I | 34.6 | 0.072 | | |
| 3-Br | 21.3 | 0.044 | | |
| 3-CO$_2$Et | 21.0 | 0.044 | | |
| 3-F | 11.7 | 0.039 | | |
| 3-Cl | 16.9 | 0.035 | | |
| 4-CO$_2$Et | 5.04 | 0.010 | | |
| 3-NO$_2$ | 1.46 | 0.003 | | |

[a] Ref. 727.
[b] Ref. 728.
[c] After correction for bromodeprotonation.

(*iii*) A redetermination of the rate coefficients for some of the substituents (Table 254) revealed considerably lower $k_{rel}$ values for the substituents which gave the major deviation from the Hammett plot. In addition, the 4-phenyl compound could not be dissolved in the quantities quoted by Kuivila *et al.* since this was over ten times the maximum solubility of the compound in 80 % aqueous acetic acid; rate coefficients for this compound were therefore obtained by dividing the first-order rates obtained with excess of boronic acid, by the maximum concentration of acid in 80 % aqueous acetic acid.

(*iv*) With the redetermined relative rates (and those of Kuivila for deactivated substituents) an almost linear correlation with $\sigma^+$ values is obtained. Obviously, it would be desirable to have a redetermination of the complete range of substituents, preferably in a less aqueous medium to give improved solubility and slower rates so that an accurate value of the rate coefficient for 4-methoxy compound can be obtained.

### 10.5.2 Iododeboronation

The kinetics of iododeboronation have been less extensively studied than for bromodeboronation, but the kinetic characteristics appear to be the same[731]. Thus in aqueous solution the reaction was found to be second-order for a fivefold variation in 4-methoxybenzeneboronic acid concentration and an eightfold variation in iodine concentration. If the hydrogen and iodide ion concentrations are not maintained constant, however, this order is not maintained, the rate showing an inverse dependence upon these concentrations. The inverse dependence upon hydrogen ion concentration was attributed to reaction occurring on the boronate anion, as also indicated by a linear dependence of rate upon pH (with unit slope), and the inverse dependence upon iodide ion concentration (Table 255) was at-

TABLE 255

DEPENDENCE OF SECOND-ORDER RATE COEFFICIENT ON $I^-$ FOR REACTION OF $PhB(OH)_2$ WITH $I_2$ AT 25 °C[731]

| $[I_2]$ (M) | $10^2 k_2$ | $10^2 k_2 [I^-]$ |
|---|---|---|
| 0.0999 | 4.99 | 0.499 |
| 0.152 | 3.20 | 0.487 |
| 0.100 | 2.48 | 0.497 |
| 0.350 | 1.38 | 0.497 |
| 0.400 | 1.21 | 0.485 |
| 0.499 | 0.955 | 0.477 |

tributed to reaction by molecular iodine since the concentration of this is inversely dependent upon the iodide concentration because of the large iodine–triodide ion equilibrium constant.

As in the case of bromodeboronation, an increase in acetate buffer concentration reduced the rate, the biggest effect arising from the acid component, and again the effect of increasing the concentration of acetate could not be explained. Chelates produced large rate accelerations as with bromodeboronation and sodium fluoride similarly produced a very large effect, which was considered to be related to the stability of the tetrafluoroborate anion.

The kinetics have also been examined by Brown et al.[732] using buffer solutions and with similar results; they pointed out that the reduction in rate caused by increasing the buffer concentration might conceivably have arisen from a slight change in pH since the rate coefficients are markedly dependent upon this; a change in pH of 0.03 units would have been sufficient to account for the results of Kuivila and Williams[731].

Rate coefficients and kinetic parameters for iododeboronation were determined for the benzene- and thiophene-boronic acids, and the results are given in Table 256. The relative reactivities derived from this work correlated well with those obtained in a number of other electrophilic substitutions[572], which is perhaps surprising in view of the large variation in the entropies of activation. These differences were explained by Brown et al.[732] in terms of the transition state for the phenyl compound occurring earlier along the reaction coordinate than for the

TABLE 256

RATE COEFFICIENTS ($10^2 k_2$) AND PARAMETERS FOR REACTION OF $ArB(OH)_2$ WITH $I_2$
IN BUFFER SOLUTION[732]

| | Temp. (°C) | | | | | | |
|---|---|---|---|---|---|---|---|
| Ar | 25 | 30 | 35 | 40 | 45 | $E_a$ | $\Delta S^\ddagger$ |
| Phenyl | 0.428 | 0.641 | 1.06 | 1.56 | 2.08 | 15.3 | − 1.7 |
| 2-Thienyl | 4,160 | 5,160 | 6,180 | 7,530 | 9,250 | 7.5 | − 9.8 |
| 3-Thienyl | 300 | 376 | 475 | 600 | 778 | 8.6 | −11 |

thienyl compounds, leading to less loss of freedom of solvent molecules as the
transition state is formed. This argument runs directly counter to the generally
accepted theory that the transition state for less reactive compounds occurs later,
not earlier; this explanation of the entropy differences is, therefore, unacceptable.

### 10.6 NITROSODECARBOXYLATION

Kinetic studies of the nitrosodecarboxylation reaction (299)

$$ArCOOH + HONO = ArNO + H_2O + CO_2 \qquad (299)$$

have been carried out with the object of comparing the results obtained in the
similar bromodecarboxylation and bromodesulphonation (see pp. 372 and 385),
and as in those studies the 3,5-dibromo-4-hydroxy (and also methoxy) derivative
was used[122a]. For the former compound, the reaction follows the kinetic equation

$$\text{Rate} = k_2[\text{HOArCOOH}][\text{NO}_2^-] \qquad (300)$$

and was not catalysed by base (acetate) unlike nitrosodeprotonation which takes
place 13 times slower at 30 °C; no spectroscopic evidence for the formation of
significant amounts of a quinonoid intermediate was obtained (cf. bromodesulphonation).

At pH < 5.0, the reaction rate was dependent upon the first power of the
hydrogen ion concentration and at pH > 5.0 upon the square of this concentration
as indicated by the data in Table 257. It was pointed out that the rate equation
(300) was equivalent to

$$\text{Rate} = k_3[\text{HOArCOO}^-][\text{NO}_2^-][\text{H}^+] \qquad (301)$$

and to

TABLE 257

RATE COEFFICIENTS FOR THE REACTION OF $NaNO_2$ WITH 4-HO-3,5-$Br_2C_6H_2$COOH IN
ACID ($\mu = 0.8$) AT 30 °C[122a]

| $[NaNO_2]$ (M) | pH | $10^6[H^+]$ (M) | $10^4k_1$ | $10^4k_1[H^+]^{-1}$ |
|---|---|---|---|---|
| 0.4 | 4.51 | 53.7 | 51.8 | 96.5 |
| 0.4 | 4.68 | 36.3 | 35.6 | 98.1 |
| 0.4 | 4.91 | 21.4 | 20.3 | 94.9 |
| 0.4 | 5.02 | 16.6 | 15.3 | 92.2 |
| 0.8 | 4.72 | 33.1 | 63.5 | 192 |
| 0.8 | 5.02 | 16.6 | 29.2 | 176 |
| 0.8 | 5.33 | 8.13 | 11.7 | 144 |
| 0.8 | 5.64 | 3.98 | 3.70 | 93 |

$$\text{Rate} = k_4[^-\text{OArCOO}^-][\text{NO}_2^-][\text{H}^+]^2 \tag{302}$$

and, so the argument runs, reaction occurs on the species indicated in these equations at pH < 5.0 and pH > 5.0 respectively; the role of the protons was thereafter completely ignored. In the writer's opinion, the kinetic evidence conclusively shows that reaction occurs on the neutral species. An argument which was supposed to show that reaction occurs on the phenol anion was the fact that the 4-methoxy compound was 300 times less reactive than the hydroxy compound and an erroneous calculation (which also used incorrect $\sigma^+$ values) was given to prove that a much smaller rate difference is predicted from theory. If, however, a $\rho$ factor of $-10$ is taken (it will certainly be greater than that, $-6.5$, for nitration), then the values of $\sigma_p^+$ for HO and MeO of $-1.0$ and $-0.778$ predict a reactivity difference of 150 i.e. close to that observed. If the phenol anion was involved, the observed rate difference would be expected to be very much greater.

The greater ease of reaction compared to nitrosodeprotonation may arise from the greater ease of carbon–carbon bond cleavage compared to carbon–hydrogen bond cleavage for in the latter reaction this is partly rate-determining and may so be in the decarboxylation as well.

## 10.7 HALODECARBOXYLATION

Only the bromodecarboxylation reaction has been subjected to kinetic studies[733], the reaction of bromine in 70–80 wt. % aqueous acetic acid with 3,5-dibromo-2(or 4)-hydroxybenzoic acids at 20 °C giving the titrimetically determined rate coefficients in Table 258. These results demonstrated that the reaction is first-order in aromatic and bromine, and that reaction rates are decreased by a decreasing water content of the solvent, by added acids, and by added bromide ion which is

TABLE 258

RATE COEFFICIENTS FOR THE REACTION OF $3,5\text{-}Br_2\text{-}2(OR\ 4)\text{-}OH\cdot C_6H_2COOH$ WITH $Br_2$
IN AQUEOUS HOAc AT 20 °C[733]

| Aq. HOAc (wt %) | [ArCOOH] (M) | [Br₂] (M) | [HClO₄] (M) | [LiBr] (M) | $10^2 k_2$ |
|---|---|---|---|---|---|
| 75.0 | 5.01[a] | 3.76 | 0.10 | | 5.8 |
| 75.0 | 5.01[a] | 7.50 | 0.10 | | 5.8 |
| 75.0 | 5.08[a] | 4.86 | 0.10 | 0.0040[b] | 5.6 |
| 75.0 | 5.01[a] | 7.44 | 0.10 | 0.0042[c] | 4.7 |
| 75.2 | 5.03[a] | 3.93 | 0.10 | 0.102 | 0.49 |
| 75.0 | 5.00 | 5.79 | 0.10 | | 200 |
| 80.0 | 5.01 | 5.81 | 0.10 | | 63 |
| 80.2 | 4.97 | 4.17 | 0.10 | 0.103 | 1.29 |
| 80.2 | 5.05 | 8.12 | 0.10 | 0.102 | 1.31 |
| 80.2 | 4.61 | 7.87 | 0.05 | 0.104 | 3.4 |
| 80.0 | 4.91 | 8.02 | 0.107[d] | | 0.72 |

[a] Data for the 2-hydroxy acid, other data is for the 4-hydroxyacid.
[b] Added lithium perchlorate.
[c] Added potassium bromide.
[d] Added hydrogen bromide.

not a salt effect; these observations were confirmed in media of constant ionic strength. The second-order rate coefficients were inversely proportional to the bromide ion concentration for the reaction of the 2-hydroxy compound, but for the 4 compound the dependence on bromide ion concentration was somewhat less. Rates decreased during a kinetic run due to production of bromide ion, but when the concentration of added bromide was greater than 0.3 $M$ no rate fall-off occurred. Rates depended upon the inverse square of the hydrogen ion concentration, so that the overall kinetic equation was approximately

$$\text{Rate} = k'[Br_2][ArH_2]_{\text{stoich}}[Br^-]^{-1}[H^+]^{-2} \tag{303}$$

which was interpreted to mean that the transition state contained two protons and one bromide ion less than the ground state. The reaction was, therefore, mechanistically different from nitrosodecarboxylation and was envisaged as equilibria (304)–(305), or (306) which is kinetically equivalent and indistinguishable, and (307)–(308)

$$\text{HOOCArOH} \rightleftharpoons \text{HOOCArO}^- + H^+ \quad \text{(fast)} \tag{304}$$

$$\text{HOOCArO}^- + Br_2 \rightleftharpoons \text{Br(HOOC)Ar}^+O^- + Br^- \quad \text{(slow)} \tag{305}$$

$$\text{HOOCArOH} + Br_2 \rightleftharpoons \text{Br(HOOC)Ar}^+O^- + H^+ + Br^- \quad \text{(slow)} \tag{306}$$

$$\text{Br(HOOC)Ar}^+O^- \rightleftharpoons \text{Br(O}^-OC)Ar^+O^- + H^+ \quad \text{(fast)} \tag{307}$$

$$Br(^-OOC)Ar^+O^- \rightarrow BrArO^- + CO_2 \tag{308}$$

The kinetic consequences of these reactions were examined in very great detail, and, briefly, the difference in the kinetic dependence of the two isomers appeared to arise from the extent to which equilibrium (306) was rate-determining, *i.e.* it would be slower for the 2 compound and this might reasonably be supposed to arise from greater steric hindrance to this step for this compound. Alternative mechanisms, such as involving rapid protodecarboxylation followed by slow bromination were eliminated by the greater reactivity of 2,6-dibromophenol over that of the corresponding 4-hydroxyacid (150–700 fold) and of 2,4-dibromophenol over that of the corresponding 2-hydroxyacid (100 fold), nor could it involve slow protodecarboxylation followed by rapid bromination since there would then no longer be a dependence on bromine concentration. It must be significant that this experimental result is exactly the reverse of that obtained in nitrosodecarboxylation of the same compound, which took place *faster* than the corresponding nitrosodeprotonation (see p. 372). Certainly one would expect the decarboxylation to occur more readily if reaction occurs on the carboxylate anion in view of the greater nucleophilicity of the aromatic, but perhaps the difference reflects the difference in steric hindrance between bromination and nitrosation. (See also chlorodesilylation, p. 379.)

These latter mechanisms were also eliminated by the observation of a $^{12}C:^{13}C$ kinetic isotope effect, which was dependent upon the bromide ion concentration. The isotope ratio of the evolved carbon dioxide after certain reaction times was compared with that in the carbon dioxide after total decarboxylation (and, therefore, in the original acid). At commencement of reaction a value of $1.002 \pm 0.003$ was obtained, whereas in the presence of 0.3 $M$ hydrogen bromide this increased to $1.045 \pm 0.001$, *i.e.* equilibrium (306) becomes reversed, thereby causing equilibria (307) and (308) to become partially rate-determining. In the absence of added bromide ion, but in the presence of 0.3 $M$ perchloric acid, an intermediate isotope effect (which increased with the extent of reaction) was obtained, and this was explained as due to bromide ion produced in the reaction bringing about reversal of equilibrium (306), though why this happens only in the presence of acid was not satisfactorily explained; more obvious is the effect on equilibrium (304) causing it to become partially rate-determining.

Doubt has been expressed as to the validity of the above mechanism by the observation that in the bromination of 2-hydroxy-4,6-methoxyacetophenone, bromine enters the 3 position and replaces the acyl group at a rate which is increased by acetylation of the hydroxy group, which should not be the case if a quinonoid intermediate is formed, as required above[734]. However, since the hydroxy group becomes acetylated during the course of the reaction, thereby partly changing the medium to bromine in acetic acid, this result is ambiguous and requires further investigation; in addition, of course, it is bromodeacylation and not bromodecarboxylation.

## 10.8 HALODEALKYLATION

### 10.8.1 Chlorodealkylation

The chlorination of $t$-butylbenzene in 99 % aqueous acetic acid has been studied kinetically in the context of chlorodeprotonation (see p. 99) and the overall kinetic form (for chlorodeprotonation and chlorodebutylation) is second-order, equation (124) (p. 98)[735]. Since, however, the extent of debutylation is very small it would be difficult to detect different kinetics for this reaction. A partial rate factor of 1.0 has been deduced for the process, and since the point of attachment of the $t$-butyl substituent is expected to have the highest charge density, this result is somewhat surprising, since it is rather unlikely that, in the absence of other factors, the second step of the reaction (carbon–carbon bond cleavage) is slower than carbon–hydrogen bond cleavage. Consequently, it would seem that the electrophile is hindered in its approach to the carbon atom bearing the $t$-butyl substituent and this is not as obvious as it might seem at first sight, for whereas solvation of this ring position is sterically hindered, formation of the $sp^3$-hybridised intermediate for substitution requires the $t$-butyl group to bend out of the plane of the aromatic ring thereby lessening steric hindrance between it and the incoming halogen.

### 10.8.2 Bromodealkylation

The bromination of $t$-butylbenzene by acidified hypobromous acid in 50 % aqueous dioxan at 25 °C follows the kinetic equation (89) (p. 84)[196], and the kinetic form of bromodebutylation is assumed to be the same since only 1.9 % of the total reaction ($k_3 = 7.25$) is debutylation, leading to a partial rate factor of 1.4; the same conclusions apply as outlined above.

## 10.9 MERCURIDESILYLATION

The kinetic order of mercuridesilylation

$$ArSiR_3 + Hg(OAc)_2 = ArHgOAc + R_3SiOAc \tag{309}$$

has not been very satisfactorily elucidated. The first study, by Benkeser et al.[736]

used acetic acid as solvent, the progress of the reaction being followed by dila-tometry. These workers stated that the reaction was second-order, and measured reaction rates under first-order conditions, with each reagent in excess in turn; rate coefficients obtained are given in Table 259. However, division of the first-

TABLE 259

RATE COEFFICIENTS FOR THE REACTION OF $XC_6H_4SiMe_3$ WITH $Hg(OAc)_2$ IN HOAc AT 25 °C[736]

| $X$ | $10^5k_1{}^a$ | $X$ | $10^5k_1{}^b$ |
|---|---|---|---|
| H | 1.83 | H | 1.03 |
| 2-Me | 19.8 | 2-Me | 11.6 |
| 3-Me | 3.75 | 3-Me | 2.67 |
| 4-Me | 21.2 | 4-Me | 11.0 |
| | | 2,3-Me$_2$ | 42.5 |
| | | 3,4-Me$_2$ | 28.0 |
| | | 2,4-Me$_2$ | 165 |
| | | 2,6-Me$_2$ | Too fast to measure |
| | | 3,5-Me$_2$ | 3.65 |
| | | 2,5-Me$_2$ | 25.0 |

$^a$ [ArSiMe$_3$) $= 0.017$ $M$, [Hg(OAc)$_2$] $= 0.1788$ $M$.
$^b$ [ArSiMe$_3$] $= 0.400$ $M$, [Hg(OAc)$_2$] $= 0.04$ $M$.

order rate coefficients by the concentration of the reagent in excess, does not yield the same rate for compounds studied under both conditions, the discrepancy being a factor of about four. It is, therefore, relevant that Eaborn and Webster[737] found that second-order kinetics could only be obtained in this reaction with 80 wt. % aqueous acid and that with 98.5 wt. % acid, the kinetic order appeared to be greater than three. Even with 80 wt. % acid, although good second-order kinetics could be obtained at a given (equal) concentration of aromatic and mercuric acetate, at different (equal) concentrations of reagents, slightly different second-order rate coefficients were obtained (Table 260)[670, 684, 737].

The results for the polymethyl compounds in Table 259, show departure from additivity for the 2,6 and 3,4-compounds (where steric acceleration is obviously responsible) and for the 3,5 compound, the result of which is inexplicable. Some additional relative rates were obtained by Benkeser et al.[738] as follows: 3-Me, 2.51; 3-i-Pr, 3.96; 3-t-Bu, 5.50; 4-Me, 10.6; 4-Et, 11.5; 4-i-Pr, 12.0; 4-t-Bu, 14.0. It is not clear which of the conditions used by Benkeser et al.[736] were employed for this study, especially since the relative rates quoted for 3-Me and 4-Me do not agree with either of the sets of data in Table 259.

The spread of rates in the mercuridesilylation reaction is close to that obtained in bromo- and proto-desilylation suggesting a similar transition state, but nothing

TABLE 260

RATE COEFFICIENTS ($10^4 k_2$) FOR REACTION OF ArSiMe$_3$ WITH Hg(OAc)$_2$ IN 80 WT. % AQUEOUS HOAc AT 25 °C[670, 684, 737]

| Ar | [ArSiMe$_3$] = [Hg(OAc)]$_2$ | | |
|---|---|---|---|
| | $\sim 0.05$ M | $\sim 0.01$ M | $\sim 0.005$ M |
| 4-MeC$_6$H$_4$ | | | 2,300 |
| 4-EtC$_6$H$_4$ | | | 2,050 |
| 4-i-PrC$_6$H$_4$ | | | 1,805 |
| 4-t-BuC$_6$H$_4$ | | | 1,975 |
| 2-MeC$_6$H$_4$ | | | 1,450 |
| 3-Me$_3$SiCH$_2$C$_6$H$_4$ | | 460 | 500 |
| 4-Ph.C$_6$H$_4$ | | | 435 |
| 4-Me$_3$SiC$_6$H$_4$ | | | 420[a] |
| 2-naphthyl | | | 382 |
| 2-Ph.C$_6$H$_4$ | | 310 | |
| 3-MeC$_6$H$_4$ | | 310 | |
| C$_6$H$_5$ | 102 | 122 | |
| 3-Ph.C$_6$H$_4$ | | 71 | |
| 4-FC$_6$H$_4$ | 42 | | |
| 4-ClC$_6$H$_4$ | 74 | | |
| 4-BrC$_6$H$_4$ | $\sim 6.65$ | | |
| 4-IC$_6$H$_4$ | $\sim 6.8$ | | |

[a] The $k_{rel}$ value would be obtained by dividing this value by 2.

is known as yet as to the extent of nucleophilic participation. Interestingly, the partial rate factor for 2-phenyl substitution (2.5) is considerably less than that for protodesilylation (6.0), which suggests that the expected steric acceleration is counterbalanced by steric hindrance to mercuration[739].

Some difficulty in obtaining good second-order rate coefficients was experienced by Moore[740], who studied the mercuridesilylation of 4-MeOC$_6$H$_4$SiR$_3$ compounds in 80 wt. % acetic acid at 25 °C, variation of up to 10 % in the rate coefficient within a particular run being observed. This was attributed to experimental error, but may be in fact more significant. Rate coefficients were not quoted, but relative reactivities (from the first 10 % of reaction) were: (R$_3$ = )Me$_3$, 190; Me$_2$CH$_2$Cl, 130; Me$_2$Ph, 69; MePh$_2$, 6.7; Me$_2$NHEt, 2.9; (4-MeC$_6$H$_4$)$_3$, 2.5; (3-MeC$_6$H$_4$)$_3$, 1.55; Ph$_3$, 1.0; (4-ClC$_6$H$_4$)$_3$, 0.50; (3-ClC$_6$H$_4$)$_3$, 0.15; the pattern of substituent effects here parallels that observed in protodesilylation. Significant differences are that the rate difference between the alkyl and aryl substituents are larger than in protodesilylation, whereas for substituted aryl substituents the rate differences are about the same. This is consistent with these differences arising partly through steric hindrance to the reagent (as proposed for protodesilylation p. 327), and consequently the differences are greater for mercuridesilylation.

## 10.10 NITRODESILYLATION

No kinetic studies, which have yielded rate coefficients of nitrodesilylation

$$ArSiR_3 + NO_2X = ArNO_2 + R_3SiX \tag{310}$$

have been carried out. The difficulty is that the reaction is complicated by con-current nitrodeprotonation, but in nitric acid in acetic anhydride, nitrodesilylation can be the predominant reaction. A determination of the percentage of reaction occurring in a given time, and the ratio $(R)$ of nitrodesilylation to nitration, has yielded the following results[741]:

(*i*) Since $R$ increased with increasing temperature, nitrodesilylation appears to have the higher activation energy. This cannot be stated with certainty, however, since the species responsible for the two reactions may be different, and the ratio of the species concentration may change with temperature.

(*ii*) A decrease in the nitric acid concentration caused $R$ to decrease as it did during a particular run (in which the acid concentration would decrease with time). Thus nitrodesilylation would seem to be of higher order than nitrode-protonation.

(*iii*) Addition of sulphuric acid increased the reaction rates and $R$, whereas added nitrate lowered both rates but did not affect $R$. Addition of acetic acid caused $R$ to pass through a maximum in a medium of composition *ca.* 20 % acetic anhydride and 80 % acetic acid.

(*iv*) In acetic acid–acetic anhydride media, dinitrogen pentoxide gave a much higher value of $R$ than nitric acid.

(*v*) With nitronium tetrafluoroborate in tetramethylene sulphone–acetonitrile $R$ was approximately zero, whereas this reagent usually gives similar nitration features to dinitrogen pentoxide, and nitric acid in acetic anhydride[73].

The mechanism of the reaction has been interpreted in terms of a 4- or 6-centre transition state (LV) or (LVI) in which nucleophilic assistance of C–Si bond cleavage occurs; this would not be possible for nitration by nitronium tetra-fluoroborate.

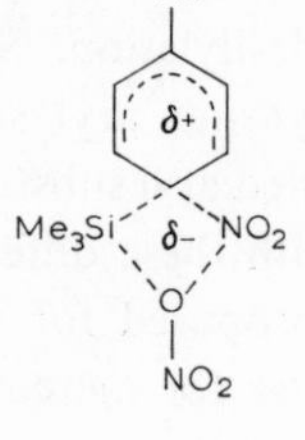

(LV) 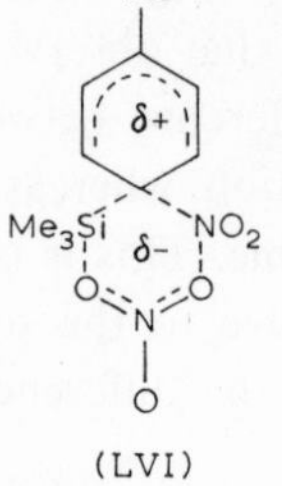                    (LVI)

## 10.11 HALODESILYLATION

### 10.11.1 Chlorodesilylation

A kinetic study of chlorodesilylation of phenyltrimethylsilane $(X = Cl)$

$$X_2 + PhSiMe_3 = PhX + Me_3X \tag{311}$$

in 98.5 wt. % aqueous acetic acid at 25 °C, and of chlorodeprotonation of 4-bromoanisole under the same conditions showed from a comparison of the extent of reaction at various times that the latter reaction proceeded at a constant value of 1.8 times faster than the former between 30 and 85 % of reaction. This indicated that the kinetic order was the same for both reactions, *i.e.* chlorodesilylation is a second-order reaction[742]. Since in chlorodeprotonation, 4-bromoanisole would be expected to react much faster than benzene, this indicates the greater ease of chlorodesilylation arising from the greater electron density at the carbon atom attached to silicon, *i.e.* other greater nucleophilicity of the arylsilicon compound. For both compounds, the second-order rate coefficients (which were not quoted) were stated to increase during a run, this being attributed to catalysis by the produced hydrogen chloride.

Chlorodesilylation of phenyltrimethylsilane in glacial acetic acid at 25 °C also gives second-order kinetics, a rate coefficient of $1.58 \times 10^{-2}$ being obtained, and this is 2–3 times less than that calculated from the data of Eaborn and Webster[742], the difference being attributed to the water content of the medium[743].

### 10.11.2 Bromodesilylation

Bromodesilylation (reaction (311), $X = Br$) has been studied under the same conditions as chlorodesilylation[670, 684, 742, 744]. The order of reaction was found to be one with respect to each reagent for $[Br_2] = 0.005$ $M$, but at higher bromine concentrations the order in bromine increased, the overall order being 3.5 at $[Br_2] = 0.025$–$0.050$ $M$. The apparent kinetic order increased during a run due to the production of bromide ion, and this increase was eliminated in the presence of added bromide ion. Lithium bromide caused a small rate increase ($\sim 10$–$20 \%$) as did other salts, lithium perchlorate being the most effective and giving a 76 % rate increase at 0.1 $M$ concentration. The lower order of reaction than that (*ca.* 3 see p. 114) found in bromodeprotonation under the same conditions is again consistent with the greater ease of desilylation relative to deprotonation. This is also manifest in the fact that bromodesilylation of phenyltrimethylsilane occurred only 2–3 times slower than chlorodesilylation compared to the much greater reactivity difference that applies in halogenodeprotonation (see Tables 61 and 62, pp. 104 and 105). Reaction rates were increased by an increasing water content of the medium, *e.g.* 4-chlorophenyltrimethylsilane reacted about 2.4 times faster in 95 than 98.5 wt. % acetic acid.

The method of determining reactivities was such that rate coefficients were not obtained, but a range of relative reactivities were (Table 261)[670, 684, 742]. The

TABLE 261

RELATIVE REACTIVITIES FOR REACTION OF $ArSiMe_3$ WITH $Br_2$ IN 98.5% AQUEOUS HOAc AT 25 °C[670, 684, 742]

| Ar | Relative reactivity | Ar | Relative reactivity | Ar | Relative reactivity | Ar | Relative reactivity |
|---|---|---|---|---|---|---|---|
| 1-naphthyl | 195 | $4\text{-}t\text{-}BuC_6H_4$ | 29.2 | $3\text{-}MeC_6H_4$ | 2.9 | $4\text{-}ClC_6H_4$ | 0.092 |
| $2\text{-}MeC_6H_4$ | 81.5 | $4\text{-}PhC_6H_4$ | 12.5 | $2\text{-}PhC_6H_4$ | 1.81 | $4\text{-}IC_8H_4$ | 0.088 |
| $4\text{-}MeC_6H_4$ | 48.8 | 2-Naphthyl | 11.5 | $C_6H_5$ | 1.00 | $4\text{-}BrC_6H_4$ | 0.071 |
| $4\text{-}EtC_6H_4$ | 45.4 | $3\text{-}Me_3SiCH_2C_6H_4$ | 8.5 | $4\text{-}FC_6H_4$ | 0.68 | $3\text{-}ClC_6H_4$ | 0.0030 |
| $4\text{-}i\text{-}PrC_6H_4$ | 32.5 | $4\text{-}Me_3SiC_6H_4$ | 3.05 | $3\text{-}PhC_6H_4$ | 0.41 | | |

spread of reactivities is much less than for molecular bromodeprotonation under the same conditions, which again accords with the greater reactivity of the aryl-silanes; the log $k_{rel}$ values plot very satisfactorily using the Yukawa–Tsuno equation[612] with $\rho = -6.8$ and $r = 0.79$, confirming that the transition state occurs earlier along the reaction coordinate than for bromodeprotonation.

Relative rates of bromodesilylation of $4\text{-}MeOC_6H_4SiR_3$ compounds have been measured under the same conditions (Table 262); some of these values are in-

TABLE 262

RELATIVE REACTIVITIES FOR REACTION OF $4\text{-}MeOC_6H_4SiR_3$ WITH $Br_2$ IN 98.5 % AQUEOUS HOAc AT 25 °C[670, 684, 745b]

| $R_3$ | $k_{rel}$ | $R_3$ | $k_{rel}$ | $R_3$ | $k_{rel}$ |
|---|---|---|---|---|---|
| $Me_3$ | 1,100 | $(EtO)_3$ | $8.5^a$ | $(3\text{-}MeC_6H_4)_3$ | 1.5 |
| $Me_2Ph$ | 355 | $Ph_3$ | 1.0 | $(4\text{-}ClC_6H_4)_3$ | 0.12 |
| $MePh_2$ | 15 | $(4\text{-}MeOC_6H_4)_3$ | $8.9^b$ | $(3\text{-}ClC_6H_4)_3$ | 0.052 |
| $(PhCH_2)_3$ | 30 | $(4\text{-}MeC_6H_4)_3$ | 3.4 | $(2\text{-}MeC_6H_4)_3$ | 0.023 |
| $ClCH_2Me_2$ | 65 | | | | |

[a] This rate probably refers to the hydrolysed and acetylated product[745a].
[b] After allowance for the availability of four Ar–Si bonds.

correctly quoted in ref. 745(a). The spread of rates is somewhat greater than in mercuridesilylation and protodesilylation of the same compounds and this accords with the larger $\rho$ factor for bromodesilylation. Although the low reactivity of the 2-methylphenyl compound (which was observed in protodesilylation) is apparent here too, the magnitude of the deactivation (relative to the phenyl compound) is much less than in that reaction, which is somewhat surprising since it is almost certainly a steric effect.

The nature of the transition state in bromodesilylation is problematical, since the reaction appears to take place in the non-polar solvents benzene and carbon tetrachloride with inversion of configuration at silicon, and, therefore, cannot proceed through a 4-centre intermediate (LVII) as this would lead to retention of configuration[746, 747]. The results are, however, consistent with a six-centre transition state (LVIII), which could follow from the high kinetic order in bromine

which is generally obtained with nonpolar solvents[748]. It would obviously be very useful if the configuration experiments could be carried out in acetic acid, under the condition of overall second-order kinetics; one would expect retention of configuration under these conditions.

### 10.11.3 Iododesilylation

Evidence for a cyclic transition state in iododesilylation of phenyltrimethylsilane by iodine monochloride (reaction (311), $X_2 = ICl$ giving PhI and $SiMe_3Cl$) has been obtained by comparing rates of chlorodesilylation and iododesilylation in acetic acid at 25 °C[743]. Good second-order kinetics for the latter reaction were obtained by application of equation (312)

$$Rate = k_2[ArSiMe_3][ICl]/(1 + K/[Cl^-]) \qquad (312)$$

where $K$ is the equilibrium constant for the reaction of iodine monochloride with the chloride ion produced. The average rate coefficient obtained was 0.133 compared with 0.0158 for chlorodesilylation. Since iododeprotonation takes place about 200 times less rapidly than chlorodeprotonation, a markedly different transition state was inferred and this was considered to be best represented by a four-centre structure, analogous to (LVII).

### 10.12 IODODESTANNYLATION

Iododestannylation has been the subject of two kinetic investigations using the spectrophotometric method, the results of which differ according to the solvent employed. The stoichiometry is

$$ArSnR_3 + I_2 = ArI + ISnR_3 \tag{313}$$

With a non-polar medium, carbon tetrachloride, third-order kinetics were obtained (in the initial reagent concentration range of $4–50 \times 10^{-4}$ $M$)[749], whereas with a polar medium, methanol, the order with respect to iodine was reduced to one[750].

With carbon tetrachloride as solvent, variation in the R groups produces a variation in rate different from that obtained in other cleavages (see pp. 327–342) in that the rates increase with increasing electron supply from the group and are not significantly affected by the size of the group. This is illustrated by the following rate coefficients $(10^2k_3)$ for the cleavage of $PhSnR_3$ at 25 °C: (R =) Ph, 0.057; Me, 3.11; Et, 15.8; cyclohexyl, 16.6. This agrees well with the reason advanced for the previous observation that the rate decreased with increasing size of the alkyl group, which was that the larger groups created steric hindrance to solvation at the reaction site; in the poor solvating medium in the present study, solvation and hence hindrance to it would be minimal.

Rate coefficients have been obtained for iododestannylation of a range of compounds $ArSnR_3$ and the rate coefficients are given in Table 263. No satis-

TABLE 263

RATE COEFFICIENTS $(10^2k_3)$ FOR REACTION OF $I_2$ WITH $XC_6H_4SnR_3$ IN $CCl_4$ AT 25 °C (R = CYCLOHEXYL EXCEPT WHERE INDICATED)[749]

| X | $10^2k_3$ | X | $10^2k_3$ | X | $10^2k_3$ | X | $10^2k_3$ |
|---|---|---|---|---|---|---|---|
| 4-MeO | 1,145, 1.19[a] | 4-Et | 167 | 3-MeO | 37.2 | 4-Cl | 1.67, 0.209[b] |
| 4-$(C_6H_{11})_3$Sn | 331 | 4-Me | 125, 23.4[b] | H | 16.6 | 4-Br | 1.325 |
| 4-Bu$^t$ | 232 | 3-Me | 70 | 2-Ph | 5.75 | 3-Cl | 0.642,0.0516[b] |
| 4-Pr$^i$ | 201 | 4-Ph | 47.7 | 4-F | 3.71 | 4-HOOC | 0.242 |

[a] R = Ph.
[b] R = Me.

factory correlation exists between the log $k_{rel}$ values and the Yukawa–Tsuno equation[612]. The discrepancy has been attributed to $\pi$ complex formation between iodine and the aromatic $\pi$ cloud (since $\pi$ complexes are found between aromatics and iodine in carbon tetrachloride), and which occurs significantly in the transition state of the rate-determining step; this would accord with the small spread of rates in the reaction which suggests that the transition state is not far displaced from the $\pi$ complex. In view of the subsequent substitution that occurs it was proposed that the iodine molecule lies over the $\pi$ bond adjacent to the reaction site as in (LIX), and this would account for the unusually small deactivation by normally deacti-

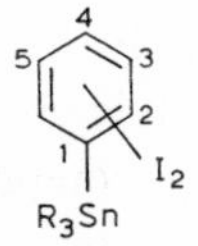

(LIX) 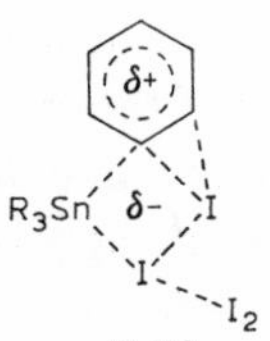    (LX)

vating substituents *meta* to the reaction site since a substituent in position 5 will have much the same effect as one in position 4, on the electron availability at the 1–2 bond. The transition to products was then envisaged as proceeding *via* (LX).

Inspection of the data in Table 263 shows that the effects of substituents are larger in the less reactive system and this is consistent with the general effect noted in electrophilic aromatic substitution.

Rate coefficients for iododestannylation in methanol have been determined over a range of temperatures (about 25–30 °C for each compound) with media of differing ionic strength. Generally, rate coefficients increased with an increase in the latter, but the variation was very dependent upon the reactivity of the arylstannane, being very large for the more reactive stannanes, and almost negligible for the least reactive. Rate coefficients, interpolated to 25 °C are given in Table 264.

TABLE 264

RATE COEFFICIENTS AND ARRHENIUS PARAMETERS FOR REACTION OF ArSnMe$_3$ WITH I$_2$ IN MeOH AT 25 °C[750]

| Ar | $\mu$ | $10^2 k_2$ | $E_a$ | log A |
|---|---|---|---|---|
| 4-MeOC$_6$H$_4$ | 0.1 | 324 | 3.9 | 7.4 |
| 4-MeC$_6$H$_4$ | 0.1 | 25.1 | 5.0 | 7.0 |
| 1-naphthyl | 0.01 | 11.0 | 6.3 | 7.7 |
| 2-naphthyl | 0.01 | 10.2 | 5.5 | 7.0 |
| 4-Me$_3$SnC$_6$H$_4$ | 0.1 | 8.70($\times$2) | | |
| 3-MeC$_6$H$_4$ | 0.1 | 7.59 | 5.5 | 6.9 |
| 9-phenanthryl | 0.01 | 7.40 | 6.0 | 7.2 |
| 4-Me$_3$SiC$_6$H$_4$ | 0.1 | 5.13 | | |
| C$_6$H$_5$ | 0.1 | 5.07[a] | 6.0 | 7.1 |
| 4-BrC$_6$H$_4$ | 0.1 | 1.21 | 6.7 | 7.0 |

[a] This rate coefficient was reported as 10.8 at 23 °C in a preliminary account[751]; no reason for this difference was given.

The log $k_{rel}$ values for the coefficients obtained at ionic strength 0.1 plotted satisfactorily using the Yukaw–Tsuno equation[612] with $\rho = -2.96$ and $r = 0.65$, and this demonstrates the difference between this reaction and that carried out in carbon tetrachloride; the point for the 2-naphthyl substituent also plots quite well though the data for the polycyclics cannot be compared directly since the variation in rate coefficient with ionic strength of the medium was not determined. The unexpectedly low reactivity of the 1-naphthyl and 9-phenanthryl compounds was attributed to steric hindrance, though surprisingly this is not apparent in the log $A$ values. The greater activation by a *para*-trimethylstannyl compared to a *para*-trimethylsilyl substituent accords with the conclusion previously reached by Eaborn et al.[749].

A fivefold variation in added iodide ion concentration produced no detectable trend in the rate coefficients, which had the mean value $99 \pm 10$ at 20 °C, and therefore nucleophilic attack of iodide ion on tin appears kinetically non-significant.

Contrary to the results obtained with carbon tetrachloride solvent and entirely in accord with the postulate that the effect arises from the steric hindrance to solvation, the rates of cleavage of $ArSnR_3$ compounds in methanol decrease on increasing the size of the group R. This is shown by the rate coefficients in Table 265, though it is difficult to draw any conclusion from the Arrhenius parameters

TABLE 265

RATE COEFFICIENTS AND ARRHENIUS PARAMETERS FOR REACTION OF $ArSnR_3$ WITH $I_2$ IN MeOH AT 25 °C ($\mu = 0.01$)[750]

| Ar | R | $k_2$ | $E_a$ | log $A$ |
|----|----|----|----|----|
| Phenyl | Me | 437 | 5.7 | 6.8 |
| | *n*-Bu | 107 | 5.9 | 6.3 |
| | *i*-Pr | 15.9 | 6.8 | 6.2 |
| 1-naphthyl | Me | 1,100 | 6.3 | 7.7 |
| | *n*-Bu | 398 | 7.5 | 8.1 |
| | *i*-Pr | 64.6 | 4.6 | 5.2 |
| 2-Naphthyl | Me | 1,020 | 5.5 | 7.0 |
| | *n*-Bu | 257 | 5.4 | 6.4 |
| | *i*-Pr | 28.8 | 10.3 | 8.6 |
| 3-Pyrenyl | *n*-Bu | 1,320 | | |

in the Table. Interestingly, the relative rates of substitution of phenyl, 1-naphthyl and 2-naphthyl, respectively, under the three conditions are: (R = Me$_3$) 1, 2.51, 2.34; (*n*-Bu) 1, 3.72, 2.40; (*i*-Pr) 1, 4.06, 1.81; this indicates either that the 2-naphthyl compound is the most so hindered or that steric acceleration is the important factor since the 1-naphthyl : 2-naphthyl ratio increases with the size of R. If the former effect is the most important it implies that steric hindrance to solvation must occur principally at the carbon atoms *ortho* to the reaction site, for in the case of 2-naphthyl compound only one of these sites lies between a peri-hydrogen and the hindered reaction site.

## 10.13 IODODEPLUMBYLATION

Only one kinetic measurement[751] has been carried out for this reaction, *viz.*

$$ArPbMe_3 + I_2 = ArI + IPbMe_3 \qquad (314)$$

The reaction is in methanol, like iododestannylation, first-order in each reagent. The second-order rate coefficient at 23 °C was reported as 20,900 and thus the reaction occurs very much more readily than cleavage of the corresponding tin compound, which repeats the pattern observed in acid-cleavage reactions (see p. 342). The magnitude of the rate coefficient may be subject to the same error that appears to be present in the measurement of the corresponding tin compound (see footnote to Table 264) since the rates were determined under the same conditions.

## 10.14 BROMODESULPHONATION

A kinetic study of bromodesulphonation,

$$ArSO_3H + Br_2 = ArBr + SO_3 + HBr \qquad (315)$$

has been carried out using iodometric and spectrophotometric methods to determine the reaction mechanism in aqueous solution[752].

For the reaction of bromine with sodium 3,5-dibromo-4-hydroxybenzene-sulphonate, in the presence of excess of either of these reagents, the reaction was first-order in the other. In the presence of excess sulphonate and 0.1 $M$ perchloric acid, the average rate coefficient was $9.2 \times 10^{-5}$ at 24.93 °C and $3.3 \times 10^{-6}$ at $-0.1$ °C leading to $\Delta H^{\ddagger} = 23$. In the absence of perchloric acid the former rate coefficient increased to $12.3 \times 10^{-5}$, whereas the presence of 0.15 $M$ sodium bromide caused a 30 % rate decrease and a change in the reaction colour from colourless to yellow orange, this latter being characteristic of the tribromide ion. Addition of potassium iodide to the reagents immediately after mixing gave a quantitative liberation of bromine, but after increasing time intervals the amount of liberated bromine decreased. First-order kinetics were followed to 90 % reaction if the sulphonate to bromine ratio was 15, but to only 56 % reaction when this ratio was only 2, which indicated a further reaction between bromine and the reaction product. On mixing the sulphonate and bromine, the ultraviolet spectrum of the former was immediately replaced by one characteristic of a 2,5-cyclo-hexadienone, which slowly turned into that of 2,4,6-tribromophenol.

*References pp. 388–406*

These observations indicated the following reaction mechanism, equilibria (316)–(318)

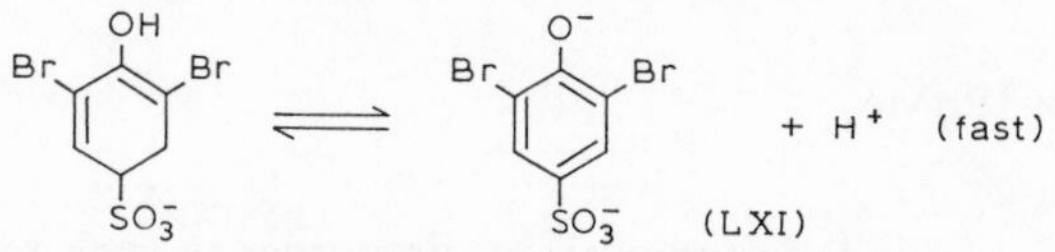

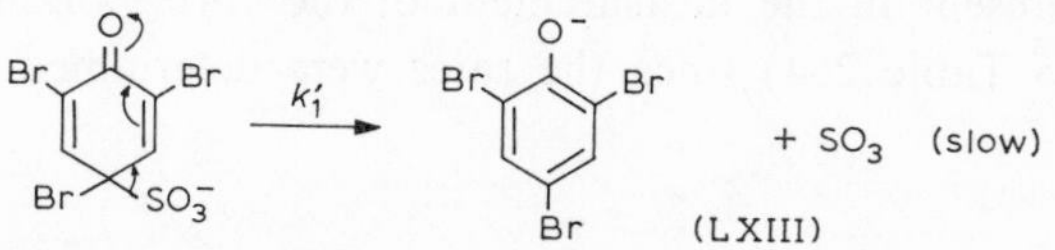

The rate reductions on adding protons, and bromide ion, follow from reversal of the first two equilibria (316) and (317), and the formation of tribromide ion from the excess bromine and bromide ion would then follow. The spectrophotometric data and the effect of added iodide ion follow from equilibria (316) and (317). The reaction may be compared to bromodecarboxylation, in which the first step of the reaction appears to be relatively fast and arises from the same cause, *viz.* the high negative charge at the reaction site due to the proximity of the anionic charge. The reaction between bromine and the reaction product is analogous to equilibrium (317) with Br instead of $SO_3^-$, and this was confirmed by the immediate disappearance of the bromine colour on mixing bromine with 2,4,6-tribromophenol, and the appearance of the characteristic 2,5-cyclobutadienone spectrum.

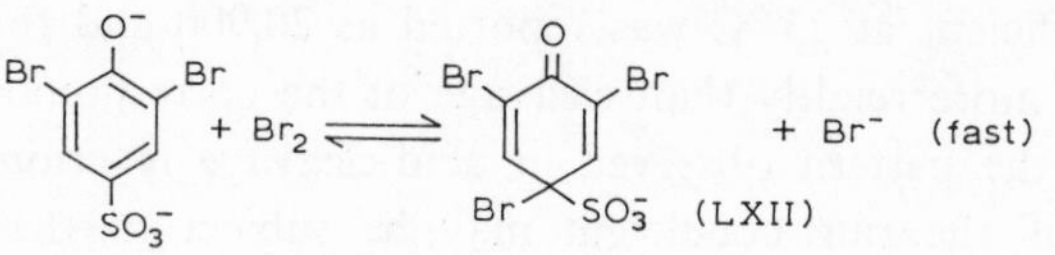

In the presence of excess bromine, the first-order rate coefficient was $10.3 \times 10^{-5}$, but kinetic studies here were complicated due to rapid reaction of bromine with the 2,4,6-tribromophenol to give the intermediate (LXII), with $SO_3^-$ replaced by Br) which slowly decomposed with a rate coefficient $k_2$ to give two products. By analysis in terms of two consecutive first-order reactions, the values of $k_1$ and $k_2$ were determined as $9.2 \times 10^{-5}$ and $3.75 \times 10^{-4}$ and the latter rate being the faster means that two moles of bromine were consumed for every mole of sulphonate undergoing substitution; in fact, more than two moles were consumed, the reason for this being undetermined.

Whereas the above study indicated that for the bromodesulphonation of the compound in question the stability of the first formed intermediate is sufficiently high for it to be formed immediately in high concentration, kinetic studies with other compounds indicated the formation of intermediates in concentrations ranging down to the "steady-state" region and in these cases, equilibrium (317) becomes the rate-determining step[752]. The compounds, sodium 3,5-dibromo-4-aminobenzenesulphonate(LXIV), sodium 4-methoxybenzenesulphonate(LXV), potassium 4-methylnaphthalenesulphonate(LXVI) and disodium 4-hydroxy-3,5-dinitrobenzenesulphonic acid (LXVII) were studied. All gave second-order kinetics at $-0.1$ °C but the dependence on bromide ion concentration varied. Since the second-order rate coefficient for a reaction in which equilibrium (317) is rate-determining is $k_2 = k_{obs}(1 + K[Br^-])$ (where $K$ is the equilibrium constant for tribromide ion formation) correction for this loss of bromine as tribromide ion should produce constant rate coefficients. In the present study this was not observed, the decrease in corrected rate for a 60-fold increase in bromide ion concentration was a factor of 1.3 for 2-methoxybenzoic acid, 1.4 for (LXVI), 4 for (LXV), and 57 for (LXIV) all at constant ionic strength; in the case of the latter compound, variation in the amount of added perchloric acid did not significantly alter the rate coefficients, indicating that the same species (almost certainly the free base) was undergoing reaction. The variations in rate dependence on bromide ion concentration were attributed to the steady-state concentration of the quinonoid intermediate being reduced by the added bromide ion to the extent that the subsequent step (318) becomes partially rate-determining. Obviously, the more stable the quinonoid intermediate the more likely is this subsequent step to be rate-determining, and the bigger the variation in rate coefficient with bromide ion concentration.

For compound (LXVII) reaction with bromine was very rapid, but it appeared that at low bromide concentration, formation of the intermediate was rate-determining, in contrast to the observation for rapid reaction of bromine with sodium 3,5-dibromo-4-hydroxybenzenesulphonate, and this was attributed to the stabilisation of the phenolate anion in the ground state of the molecule by the adjacent nitro groups. This stability is lost in the intermediate and regained on formation of the product. Consequently, the first energy barrier is higher than the second as it is for unreactive compounds. However, the stability of the quinonoid intermediate is still sufficiently high for the effect of added bromide ion in causing a return to starting materials to be comparable to that observed with a much less reactive compound such as (LXIV). Thus a 60-fold variation in added bromide ion concentration caused the rate (corrected for tribromide ion equilibrium) to vary by a factor of 65 for compound (LXVII) to 57 for (LXIV). The relative values for $k_2$ obtained in this study at $[Br^-] = 0.025$ $M$ were: (LXIV), 27.7; (LXV) 0.424; (LXVI), 0.141; (LXVII) 13.7; 2-methoxybenzoic acid, 19.5.

## REFERENCES

1 R. O. C. NORMAN AND R. TAYLOR, *Electrophilic Substitution in Benzenoid Compounds*, Elsevier, Amsterdam, 1965.
2 F. A. LONG AND M. A. PAUL, *Chem. Rev.*, 57 (1957) 935.
3 L. ZUCKER AND L. P. HAMMETT, *J. Am. Chem. Soc.*, 61 (1939) 2791.
4 E. GRUNWALD, A. HELLER AND F. S. KLEIN, *J. Chem. Soc.*, (1957) 2604.
5 J. KOSKIKALLIO AND E. WHALLEY, *Trans. Faraday Soc.*, 55 (1959) 815.
6 J. G. PRITCHARD AND F. A. LONG, *J. Am. Chem. Soc.*, 78 (1956) 2663, 2667, 6008; 80 (1958) 4162.
7 J. KOSKIKALLIO AND E. WHALLEY, *Can. J. Chem.*, 37 (1959) 788.
8 H. KWART AND A. L. GOODMAN, *J. Am. Chem. Soc.*, 82 (1960) 1947.
9 G. ARCHER AND R. P. BELL, *J. Chem. Soc.*, (1959) 3228.
10 R. H. BOYD, R. W. TAFT, A. P. WOLF AND D. R. CHRISTMAN, *J. Am. Chem. Soc.*, 82 (1960) 4729.
11 L. MELANDER AND P. C. MYHRE, *Arkiv Kemi*, 13 (1959) 507.
12 N. C. DENO, J. JARUZELSKI AND A. SHRIESHEIM, *J. Am. Chem. Soc.*, 77 (1955) 3044.
   N. C. DENO, P. T. GROVES AND G. STAINES, *J. Am. Chem. Soc.*, 81 (1959) 5790.
   N. C. DENO, P. T. GROVES, J. JARUZELSKI AND M. LUGASCH, *J. Am. Chem. Soc.*, 82 (1960) 4729.
13 R. L. HINMAN AND J. LANG, *Tetrahedron Letters*, (21) (1960) 12.
14 R. W. TAFT, *J. Am. Chem. Soc.*, 82 (1960) 2964.
15 A. J. KRESGE AND Y. CHIANG, *Proc. Chem. Soc.*, (1961) 81.
16 F. A. LONG AND J. SCHULZE, *J. Am. Chem. Soc.*, 83 (1961) 3340.
17 W. M. SCHUBERT AND R. H. QUACCHIA, *J. Am. Chem. Soc.*, 84 (1962) 3778.
18 A. J. KRESGE, G. W. BARRY, K. R. CHARLES AND Y. CHIANG, *J. Am. Chem. Soc.*, 84 (1962) 4343.
19 A. R. KATRITZKY, A. J. WARING AND K. YATES, *Tetrahedron*, 19 (1963) 465.
20 J. F. BUNNETT, *J. Am. Chem. Soc.*, 82 (1960) 499; 83 (1961) 4956, 4968, 4973, 4978.
21 P. B. D. DE LA MARE AND J. H. RIDD, *Aromatic Substitution*, Butterworths, London, 1959, p. 33.
22 E. BERLINER, *J. Am. Chem. Soc.*, 72 (1950) 4003.
23 P. GROSS, H. STEINER AND F. KRAUSS, *Trans. Faraday Soc.*, 32 (1936) 877.
   P. GROSS AND A. WISCHIN, *Trans. Faraday Soc.*, 32 (1936) 879.
   P. GROSS, H. STEINER AND H. SUESS, *Trans. Faraday Soc.*, 32 (1936) 883.
   J. C. HORNEL AND J. A. V. BUTLER, *J. Chem. Soc.*, (1936) 1361.
   W. J. C. ORR AND J. A. V. BUTLER, *J. Chem. Soc.*, (1937) 330.
   W. E. NELSON AND J. A. V. BUTLER, *J. Chem. Soc.*, (1938) 957.
24 E. A. HALEVI, F. A. LONG AND M. A. PAUL, *J. Am. Chem. Soc.*, 83 (1961) 305.
25 L. MELANDER, *Arkiv Kemi*, 2 (1950) 213.
26 E. BERLINER, *Progress in Physical Organic Chemistry*, Vol. 2, Wiley, New York. 1964, p. 258.
   R. O. C. NORMAN AND R. TAYLOR, *Electrophilic Substitution in benzenoid Compounds*, Elsevier, Amsterdam, 1965, p. 26.
27 H. ZOLLINGER, *Helv. Chim. Acta.*, 38 (1955) 1597, 1617.
28 R. P. BELL, *The Proton in Chemistry*, Methuen, London, 1959, p. 202; *Trans. Faraday Soc.*, 57 (1961) 961; *Discussions Faraday Soc.*, 39 (1965) 16.
   C. G. SWAIN, *J. Am. Chem. Soc.*, 83 (1961) 2154.
   F. H. WESTHEIMER, *Chem. Rev.*, 61 (1961) 265.
   L. MELANDER, *Isotope Effects on Reaction Rates*, The Ronald Press Co., New York, 1959.
   J. BIGELEISEN, *Pure Appl. Chem.*, 8 (1964) 217.
29 P. B. D. DE LA MARE AND J. H. RIDD, *Aromatic Substitution*, Butherworths, London, 1959, pp. 43–46.
30 J. E. LEFFLER, *J. Org. Chem.*, 20 (1955) 1202.
31 R. C. PETERSON, *J. Org. Chem.*, 29 (1964) 3133.
32 P. C. MYHRE AND M. BEUG, *J. Am. Chem. Soc.*, 88 (1966) 1569.
   P. C. MYHRE, M. BEUG AND L. L. JAMES, *J. Am. Chem. Soc.*, 90 (1968) 2105.

33 P. C. MYHRE, *Acta Chem. Scand.*, 14 (1960) 219.
34 See H. ZOLLINGER, *Advances in Physical Organic Chemistry*, V. GOLD, (Ed.), *Academic Press*, London, 1964, p. 168.
35 C. K. INGOLD, *Structure and Mechanism in Organic Chemistry*, Bell and Sons, London, 1953, p. 282.
36 E. S. HALBERSTADT, E. D. HUGHES AND C. K. INGOLD, *J. Chem. Soc.*, (1950) 2441.
37 C. A. BUNTON AND E. A. HALEVI, *J. Chem. Soc.*, (1952) 4917.
38 R. J. GILLESPIE AND D. G. NORTON, *J. Chem. Soc.*, (1953) 971.
   H. MARTINSEN, *Z. Phys. Chem.*, 50 (1904) 385; 59 (1907) 605.
   A. KLEMENC AND R. SCHÖLLER, *Z. Anorg. Chem.*, 141 (1924) 231.
   K. LAUER AND R. ODA, *J. Prakt. Chem.*, 144 (1936) 176.
   R. ODA AND U. VEDA, *Bull. Inst. Phys. Chem. Res. Japon*, 20 (1941) 335.
   F. H. WESTHEIMER AND M. S. KHARASCH, *J. Am. Chem. Soc.*, 68 (1946) 1871.
   G. M. BENNETT, J. C. D. BRAND, D. M. JAMES, T. G. SAUNDERS AND G. WILLIAMS, *J. Chem. Soc.*, (1947) 474.
   A. M. LOWEN AND G. WILLIAMS, *J. Chem. Soc.*, (1950) 3312.
   A. M. LOWEN, M. A. MURRAY AND G. WILLIAMS, *J. Chem. Soc.*, (1950) 3318.
   T. G. BONNER, F. BOWYER AND G. WILLIAMS, *J. Chem. Soc.*, (1952) 3274; (1953) 2650.
39 M. I. VINNICK, ZH. E. GRABOVSKAYA AND L. N. ARZAMASKOVA, *Zh. Fiz. Chem.*, 41 (1967) 580.
40 E. D. HUGHES, C. K. INGOLD AND R. I. REED, *J. Chem. Soc.*, (1950) 2400.
41 Ref. 21, pp. 57–59; ref. 35, pp. 269–274; ref. 1, pp. 62–3.
42 Ref. 21, pp. 63–64.
43 R. J. GILLESPIE AND D. G. NORTON, *J. Chem. Soc.*, (1953) 971.
44 P. A. H. WYATT, *J. Chem. Soc.*, (1954) 2647.
45 N. C. DENO AND R. STEIN, *J. Am. Chem. Soc.*, 78 (1956) 578.
46 J. G. TILLETT, *J. Chem. Soc.*, (1962) 5142.
47 R. G. COOMBES, R. B. MOODIE AND K. SCHOFIELD, *Chem. Commun.*, (1967) 352; *J. Chem. Soc.* (B), (1968) 800.
48 M. BRICKMAN, S. JOHNSON AND J. H. RIDD, *Proc. Chem. Soc.*, (1962) 228.
   M. BRICKMAN AND J. H. RIDD, *J. Chem. Soc.*, (1965) 6845.
   M. BRICKMAN, J. H. P. UTLEY AND J. H. RIDD, *J. Chem. Soc.*, (1965) 6851.
   S. R. HARTSHORN AND J. H. RIDD, *Chem. Commun.*, (1967) 133; *J. Chem. Soc.*, (B) (1968) 1063, 1068.
49 M. W. AUSTIN, M. BRICKMAN, J. H. RIDD AND B. V. SMITH, *Chem. Ind.* (*London*), (1962) 1057.
   M. W. AUSTIN AND J. H. RIDD, *J. Chem. Soc.*, (1963) 4204.
   M. W. AUSTIN, J. R. BLACKBOROW, J. H. RIDD AND B. V. SMITH, *J. Chem. Soc.*, (1965) 1051.
50 R. B. MOODIE, K. SCHOFIELD AND M. J. WILLIAMSON, *Chem. Ind.* (*London*), (1963) 1283; *Nitro Compounds, Proc. Intern. Symp., Warsaw*, 1962, Pergamon, London, 1964 p. 89.
   R. B. MOODIE, K. SCHOFIELD AND M. J. WILLIAMSON, *Chem. Ind.* (*London*), (1964) 1577.
   J. GLEGHORN, R. B. MOODIE, K. SCHOFIELD AND M. J. WILLIAMSON, *J. Chem. Soc.*, (B) (1966) 870.
   R. B. MOODIE, E. A. QUERESHI, K. SCHOFIELD AND J. T. GLEGHORN, *J. Chem. Soc.*, (B) (1968) 312, 316.
50a C. D. JOHNSON, A. R. KATRITZKY, B. J. RIDGEWELL AND M. VINEY, *J. Chem. Soc.*, (B) (1967), 1204.
   C. D. JOHNSON, A. R. KATRITZKY AND M. VINEY, *J. Chem. Soc.*, (B) (1967) 1211.
50b A. R. KATRITZKY AND M. KINGSLAND, *J. Chem. Soc.*, (B) (1968) 862.
50c C. D. JOHNSON, A. R. KATRITZKY, N. SHAKIR AND M. VINEY, *J. Chem. Soc.*, (B) (1967) 1213.
50d P. J. BRIGNELL, A. R. KATRITZKY AND H. O. TARHAN, *J. Chem. Soc.*, (B) (1968) 1477.
51 J. R. BLACKBOROW AND J. H. RIDD, *Chem. Commun.*, (1967) 132.
52 T. A. MODRO AND J. H. RIDD, *J. Chem. Soc.*, (B) (1968) 528.
53 A. GUSTAMINZA, T. A. MODRO, J. H. RIDD AND J. H. P. UTLEY, *J. Chem. Soc.*, (B) (1968) 534.
54 Ref. 1, p. 301.
55 J. H. P. UTLEY AND T. A. VAUGHAN, *J. Chem. Soc.*, (B) (1968) 196.

56  M. A. AKAND AND P. A. H. WYATT, *J. Chem. Soc.*, (B) (1967) 1327.
57  D. J. MILLEN, personal communication to P. B. D. DE LA MARE AND J. H. RIDD, quoted in ref. 21, p. 60.
    E. BRINER, B. SUZ AND P. FAVARGER, *Helv. Chim. Acta*, 18 (1935) 375.
    E. G. TAYLOR AND A. G. FOLLOWS, *Can. J. Chem.*, 29 (1951) 461.
    I. M. KOLTHOFF AND A. WILLMAN, *J. Am. Chem. Soc.*, 56 (1934) 1007.
58  G. A. BENFORD AND C. K. INGOLD, *J. Chem. Soc.*, (1938) 929.
59  D. W. COILETT AND S. D. HAMANN, *Trans. Faraday Soc.*, 57 (1961) 2231.
60  T. ASANO, R. GOTO AND A. SERA, *Bull. Chem. Soc. Japan*, 40 (1967) 2208.
61  G. A. OLAH, S. J. KUHN, S. H. FLOOD AND J. C. EVANS, *J. Am. Chem. Soc.*, 84 (1962) 3687.
62  K. L. NELSON AND H. C. BROWN, *J. Am. Chem. Soc.*, 73 (1951) 5605.
63  H. COHN, E. D. HUGHES, M. M. JONES AND M. G. PEELING, *Nature*, 169 (1952) 291.
64  L. M. STOCK, *J. Org. Chem.*, 26 (1961) 4120.
65  R. A. WIRKKALA, *Dissertation Abstr.*, 23 (1963) 2329.
66  J. D. ROBERTS, J. K. SANFORD, F. L. J. SIXMA, H. CERFONTAIN AND R. ZAGT, *J. Am. Chem. Soc.*, 76 (1954) 4525.
67  P. G. E. ALCORN AND P. R. WELLS, *Australian J. Chem.*, 18 (1965) 1377, 1391.
68  A. STREITWEISER AND R. C. FAHEY, *J. Org. Chem.*, 27 (1962) 2352.
69  T. G. BONNER, R. A. HANCOCK, R. L. WILLIAMS AND J. C. WRIGHT, *Chem. Commun.*, (1966) 109.
70  M. A. PAUL, *J. Am. Chem. Soc.*, 80 (1958) 5329.
71  F. G. BORDWELL AND E. W. GARBISCH, *J. Am. Chem. Soc.*, 82 (1960) 3588.
72  J. H. RIDD, *Studies on Chemical Structure and Reactivity*, Methuen, London, 1966, p. 143.
73  R. TAYLOR, *J. Chem. Soc.*, (B) (1966) 727; *Tetrahedron Letters*, (1966) 6093.
74  A. K. SPARKS, *J. Org. Chem.*, 31 (1966) 2299.
75  Ref. 21, p. 70.
76  A. FISCHER, A. J. READ, J. VAUGHAN AND G. J. WRIGHT, *Proc. Chem. Soc.*, (1961) 369; *J. Chem. Soc.*, (1964) 3687.
    A. FISCHER, A. J. READ AND J. VAUGHAN, *J. Chem. Soc.*, (1964) 3691.
77  Ref. 72, p. 145.
78  A. FISCHER, personal communication.
79  R. O. C. NORMAN AND G. K. RADDA, *J. Chem. Soc.*, (1961) 3030.
80  G. W. GRAY AND D. LEWIS, *J. Chem. Soc.*, (1961) 5156.
81  R. KETCHAM, R. CAVESTRI AND D. JAMBOTKAR, *J. Org. Chem.*, 28 (1963) 2139.
82  P. H. GRIFFITHS, W. A. WALKEY AND H. B. WATSON, *J. Chem. Soc.*, (1934) 631.
    K. HALVARSON AND L. MELANDER, *Arkiv Kemi*, 11 (1957) 77.
83  F. ARNALL AND T. LEWIS, *J. Chem. Soc. Ind.*, 48 (1929) 159T.
84  Ref. 72, p. 141.
85  C. K. INGOLD, A. LAPWORTH, E. ROTHSTEIN AND D. WARD, *J. Chem. Soc.*, (1931) 1959.
86  M. L. BIRD AND C. K. INGOLD, *J. Chem. Soc.*, (1938) 918.
87  J. R. KNOWLES, R. O. C. NORMAN AND G. K. RADDA, *J. Chem. Soc.*, (1960) 4885.
88  C. K. INGOLD AND M. S. SMITH, *J. Chem. Soc.*, (1938) 905.
89  F. G. BORDWELL AND K. ROHDE, *J. Am. Chem. Soc.*, 70 (1948) 1197.
90  J. R. KNOWLES AND R. O. C. NORMAN, *J. Chem. Soc.*, (1961) 2938.
91  C. K. INGOLD AND F. R. SHAW, *J. Chem. Soc.*, (1949) 575.
92  J. L. SPEIER, *J. Am. Chem. Soc.*, 75 (1953) 2930.
93  O. SIMAMURA AND Y. MIZUNO, *Bull. Chem. Soc. Japan*, 30 (1957) 196; *J. Chem. Soc.*, (1958) 3875.
94  M. J. S. DEWAR, T. MOLE, D. S. URCH AND E. W. T. WARFORD, *J. Chem. Soc.*, (1956) 3572.
    M. J. S. DEWAR, T. MOLE AND E. W. T. WARFORD, *J. Chem. Soc.*, (1956) 3576.
    M. J. S. DEWAR AND D. S. URCH, *J. Chem. Soc.*, (1958) 3079.
95  G. P. SHARNIN, I. E. MOISAK, E. E. GRYAZIN AND I. F. FALYAKOV, *J. Org. Chem. USSR*, 3 (1967) 1792.
96  M. J. S. DEWAR AND R. H. LOGAN, *J. Am. Chem. Soc.*, 90 (1968) 1924.
97  V. GOLD, E. D. HUGHES, C. K. INGOLD AND G. H. WILLIAMS, *J. Chem. Soc.*, (1950) 2452.
98  G. H. WILLIAMS, *Ph. D. Thesis*, University of London, (1948).

99 V. Gold, E. D. Hughes and C. K. Ingold, *J. Chem. Soc.*, (1950) 2467.
100 F. H. Cohen and J. P. Wibaut, *Rec. Trav. Chim.*, 54 (1935) 409.
101 Ref. 21, p. 73.
102 J. D. S. Goulden and D. J. Millen, *J. Chem. Soc.*, (1950) 2620.
   D. J. Millen, *J. Chem. Soc.*, (1950) 2600.
   R. J. Gillespie, J. Graham, E. D. Hughes, C. K. Ingold and E.R.A. Peeling, *J. Chem. Soc.*, (1950) 2504.
103 H. Martinsen, *Z. Phys. Chem.*, 50 (1904) 385.
   F. Arnall, *J. Chem. Soc.*, (1923) 3111.
   F. M. Lang, *Compt. Rend.*, 226 (1948) 1381; 227 (1948) 849.
104 J. Glazer, E. D. Hughes, C. K. Ingold, A. T. James, G. T. Jones and E. Roberts, *J. Chem. Soc.*, (1950) 2657.
105 C. A. Bunton, E. D. Hughes, C. K. Ingold, D. I. M. Jacobs, M. H. Jones, G. J. Minkoff and R. I. Reed, *J. Chem. Soc.*, (1950) 2628.
106 G. A. Olah, S. J. Kuhn and A. Mlinko, *J. Chem. Soc.*, (1956) 4527.
107 G. A. Olah and S. J. Kuhn, *J. Am. Chem. Soc.*, 84 (1962) 3684.
108 G. A. Olah, S. J. Kuhn and S. H. Flood, *J. Am. Chem. Soc.*, 83 (1961) 4571, 4581.
109 G. A. Olah and S. J. Kuhn, *Friedel Crafts and Related Reactions* Vol. III, Interscience, New York, 1964, p. 1393.
110 L. L. Ciaccio and R. A. Marcus, *J. Am. Chem. Soc.*, 84 (1962) 1838.
111 W. S. Tolgyesi, *Can. J. Chem.*, 43 (1965) 343.
112 G. A. Olah and N. A. Overchuk, *Can. J. Chem.*, 43 (1965) 3279.
113 Ref. 72, p. 152.
114 P. Kreienbühl and H. Zollinger, *Tetrahedron Letters*, (1965) 1739.
115 H. Cerfontain and A. Telder, *Rec. Trav. Chim.*, 86 (1967) 371.
116 P. Kovacic and J. J. Hiller, *J. Org. Chem.*, 30 (1965) 2871.
117 E. L. Blackall, E. D. Hughes and C. K. Ingold, *J. Chem. Soc.*, (1952) 28.
118 T. Suzawa, Z. Yssuoka, O. Manabe and H. Hiyama, *Chem. Abstr.*, 49 (1955) 13749.
   T. Suzawa and H. Hiyama, *Chem. Abstr.*, 50 (1956) 227.
119 D. A. Morrison and T. A. Turney, *J. Chem. Soc.*, (1960) 4827.
120 H. Schmid, G. Muhr and P. Riedl, *Monatsh. Chem.*, 97 (1966) 781.
121 B. C. Challis and A. J. Lawson, *Chem. Commun.*, (1968) 818.
122 H. Ladenheim and M. L. Bender, *J. Am. Chem. Soc.*, 82 (1960) 1895.
122a K. M. Ibne-Rasa, *J. Am. Chem. Soc.*, 84 (1962) 4962.
123 J. B. Conant and W. D. Peterson, *J. Am. Chem. Soc.*, 52 (1930) 1220.
124 R. Wister and P. D. Bartlett, *J. Am. Chem. Soc.*, 63 (1941) 413.
125 J. H. Binks and J. H. Ridd, *J. Chem. Soc.*, (1957) 2398.
126 K. H. Meyer and H. Tochtermann, *Chem. Ber.*, 54 (1921) 2283.
127 H. Zollinger, *Helv. Chim. Acta*, 38 (1955) 1597.
128 A. Grimison and J. H. Ridd, *J. Chem. Soc.*, (1959) 3019.
129 H. Zollinger, *Helv. Chim. Acta*, 38 (1955) 1617, 1623.
130 R. Ernst, O. A. Stamm and H. Zollinger, *Helv. Chim. Acta*, 41 (1958) 2274.
131 E. Helgstrand and B. Lamm, *Arkiv Kemi*, 20 (1962) 193.
131a I. Dobas, V. Sterba and M. Vecera, *Chem. Ind. (London)*, (1968) 1814.
132 Ref. 1, pp. 110–116.
133 S. L. Friess, A. H. Soloway, B. K. Morse and W. C. Ingersoll, *J. Am. Chem. Soc.*, 74 (1952) 1305.
134 R. D. Chambers, P. Goggin and W. K. R. Musgrove, *J. Chem. Soc.*, (1959) 1804.
135 P. D. Bartlett and K. Nozaki, *J. Am. Chem. Soc.*, 69 (1947) 2299.
136 C. Walling and R. B. Hodgdon, *J. Am. Chem. Soc.*, 80 (1958) 228.
137 P. C. Myhre, G. S. Owen and L. L. James, *J. Am. Chem. Soc.*, 90 (1968) 2115.
138 A. F. Holleman, *Chem. Rev.*, 1 (1925) 187.
139 D. R. Vicary and C. N. Hinshelwood, *J. Chem. Soc.*, (1939) 1372.
   K. D. Wadsworth and C. N. Hinshelwood, *J. Chem. Soc.*, (1944) 469.
   E. Dresel and C. N. Hinshelwood, *J. Chem. Soc.*, (1944) 649.
140 J. C. D. Brand and A. Rutherford, *J. Chem. Soc.*, (1952) 3916.

141 H. CERFONTAIN AND A. TELDER, *Rec. Trav. Chim.*, 84 (1965) 1613.
    H. CERFONTAIN, H. J. HOFMAN AND A. TELDER, *Rec. Trav. Chim.*, 83 (1964) 493.
142 H. CERFONTAIN, A. TELDER AND L. VOLLBRACHT, *Rec. Trav. Chim.*, 83 (1964) 1103.
143 J. K. BOSSCHER AND H. CERFONTAIN, *Rec. Trav. Chim.*, 57 (1968) 873; *Tetrahedron*, 24 (1968)
    6543: *J. Chem. Soc.*, (B) (1968) 1524.
144 J. A. WALSH, *Dissertation Abstr.*, 24 (1964) 5013.
145 F. J. STUBBS, C. D. WILLIAMS AND C. N. HINSHELWOOD, *J. Chem. Soc.*, (1948) 1065.
146 U. BERGLAND-LARSSON AND L. MELANDER, *Arkiv Kemi*, 6 (1953) 219.
    U. BERGLAND-LARSSON, *Arkiv Kemi*, 10 (1957) 549.
147 C. EABORN AND R. TAYLOR (a) *J. Chem. Soc.*, (1960) 1480; (b) *J. Chem. Soc.*, (1961) 1012;
    (c) *J. Chem. Soc.*, (1961) 2388.
148 H. MARTINSON, *Z. Physik. Chem.*, 62 (1908) 713.
149 I. S. IOFFE, *J. Gen. Chem. USSR*, 3 (1933) 437.
150 J. PINNOW, *Z. Elektrochem.*, 21 (1915) 380; 23 (1917) 243.
151 K. LAUER AND R. ODA, (a) *J. Prakt. Chem.*, 142 (1935) 258; 144 (1935) 32; (b) *Chem.
    Ber.*, 70 (1937) 333.
152 W. A. COWDREY AND D. S. DAVIES, *J. Chem. Soc.*, (1949) 1871.
153 K. LAUER AND Y. HIRATA, *J. Prakt. Chem.*, 145 (1936) 287.
154 K. LAUER AND K. IRIE, *J. Prakt. Chem.*, 145 (1936) 281.
    K. LAUER, *Chem. Ber.*, 70 (1937) 1707.
155 R. LANZ, *Bull. Soc. Chim. France*, 2 (1935) 2092.
156 T. F. YOUNG AND G. E. WALRAFEN, *Trans. Faraday Soc.*, 57 (1961) 34.
157 A. W. KAANDORP, H. CERFONTAIN AND F. L. J. SIXMA, *Rec. Trav. Chim.*, 81 (1962) 969.
158 J. C. D. BRAND, *J. Chem. Soc.*, (1950) 1004.
    J. C. D. BRAND AND W. C. HORNING, *J. Chem. Soc.*, (1952) 3922.
159 J. C. D. BRAND, A. W. P. JARVIE AND W. C. HORNING, *J. Chem. Soc.*, (1959) 3844.
160 V. GOLD AND D. P. N. SATCHELL, *J. Chem. Soc.*, (1956) 1635.
161 H. CERFONTAIN, *Rec. Trav. Chim.*, 80 (1961) 296; 84 (1965) 551.
162 A. J. PRINSEN, *Ph. D. Thesis*, University of Amsterdam, (1968).
163 M. KILPATRICK AND M. W. MEYER, *J. Phys. Chem.*, 65 (1961) 530.
164 H. CERFONTAIN, *Rec. Trav. Chim.*, 79 (1960) 935.
165 M. KILPATRICK, M. W. MEYER AND M. L. KILPATRICK, *J. Phys. Chem.*, 64 (1960) 1433.
166 C. EABORN AND R. TAYLOR, *J. Chem. Soc.*, (1960) 3301.
167 A. W. KAANDORP, H. CERFONTAIN AND F. L. J. SIXMA, *Rec. Trav. Chim.*, 82 (1963) 113, 923.
168 H. CERFONTAIN, A. W. KAANDORP AND F. L. J. SIXMA, *Rec. Trav. Chim.*, 82 (1963) 565.
169 J. M. ARENDS AND H. CERFONTAIN, *Rec. Trav. Chim.*, 85 (1966) 93.
170 H. DE VRIES AND H. CERFONTAIN, *Rec. Trav. Chim.*, 86 (1967) 873.
171 M. KILPATRICK, M. W. MEYER AND M. L. KILPATRICK, *J. Phys. Chem.*, 65 (1961) 1189.
172 A. W. KAANDORP, Ph. D. Thesis, University of Amsterdam (1963).
173 YA. L. LEITMAN AND M. S. PEVZNER, *J. Pract. Chem. USSR*, 32 (1959) 1842.
    YA. L. LEITMAN AND I. N. DIYAROV, *J. Pract. Chem. USSR*, 34 (1961) 376.
174 H. CERFONTAIN AND A. TELDER, *Rec. Trav. Chim.*, 86 (1967) 527.
175 R. TAYLOR AND G. G. SMITH, *Tetrahedron*, 19 (1963) 923.
176 C. W. F. KORT AND H. CERFONTAIN, *Rec. Trav. Chim.*, 86 (1967) 865.
177 C. W. F. KORT AND H. CERFONTAIN, *Rec. Trav. Chim.*, 87 (1968) 24.
178 Ref. 1. p. 305.
179 P. A. H. WYATT, *Discussions Faraday Soc.*, 24 (1957) 162; *Trans. Faraday Soc.*, 56 (1960)
    490.
180 S. C. J. OLIVIER, *Rec. Trav. Chim.*, 33 (1914) 91.
181 S. C. J. OLIVIER, *Rec. Trav. Chim.*, 33 (1914) 244.
182 S. C. J. OLIVIER, *Rec. Trav. Chim.*, 35 (1915) 109.
183 S. C. J. OLIVIER, *Rec. Trav. Chim.*, 35 (1915) 166.
184 F. R. JENSEN AND H. C. BROWN, *J. Am. Chem. Soc.*, 80 (1958) 4038, 4042, 4046.
185 R. E. VAN DYKE AND H. E. CRAWFORD, *J. Am. Chem. Soc.*, 73 (1951) 2018.
186 M. KOBAYASHI, K. HONDA AND A. YAMAGUCHI, *Tetrahedron Letters*, (1968) 487.
187 F. R. JENSEN AND G. GOLDMAN, in *Friedel-Crafts and Related Reactions*, G. A. OLAH (Ed.),

Interscience, New York, 1964, p. 1319.

188  H. C. Brown and G. Marino, *J. Am. Chem. Soc.*, 81 (1959) 3308.

189  E. Shilov and N. Kaniaev, *Compt. Rend. Acad. Sci. USSR*, 24 (1939) 890.

190  C. F. Prutton and S. H. Maron, *J. Am. Chem. Soc.*, 57 (1935) 1652.

191  W. J. Wilson and F. G. Soper, *J. Chem. Soc.*, (1949) 3376.

192  D. H. Derbyshire and W. A. Waters, *J. Chem. Soc.*, (1950) 564.

193  A. E. Bradfield, G. I. Davies and E. Long, *J. Chem. Soc.*, (1949) 1389.

194  S. J. Branch and B. Jones, *J. Chem. Soc.*, (1954) 2317.

195  P. B. D. de la Mare and J. T. Harvey, *J. Chem. Soc.*, (1956) 36.

196  P. B. D. de la Mare and J. T. Harvey, *J. Chem. Soc.*, (1957) 131.

197  P. B. D. de la Mare and M. Hassan, *J. Chem. Soc.*, (1957) 3004.

198  P. B. D. de la Mare and I. C. Hilton, *J. Chem. Soc.*, (1962) 997.

199  P. B. D. de la Mare and J. L. Maxwell, *J. Chem. Soc.*, (1962) 4829; *Chem Ind. (London)*, (1961) 553.

200  Ref. 1, pp. 147–8.

200a Y. Furuya. A. Morita and I. Urasaki, *Bull. Chem. Soc. Japan*, 41 (1968) 997.

201  P. B. D. de la Mare, T. M. Dunn and J. T. Harvey, *J. Chem. Soc.*, (1957) 923.

202  D. H. Derbyshire and W. A. Waters, *J. Chem. Soc.*, (1951) 73.

203  P. B. D. de la Mare, A. D. Ketley and C. A. Vernon, *Research (London)*, 6 (1953) 125; *J. Chem. Soc.*, (1954) 1290.

204  M. Anbar and I. Dostrovsky, *J. Chem. Soc.*, (1954) 1094.

204a P. B. D. de la Mare, J. T. Harvey, M. Hassan and S. Varma, *J. Chem. Soc.*, (1958) 2756.

205  C. G. Swain and A. D. Ketley, *J. Am. Chem. Soc.*, 77 (1955) 3410.

206  Ref. 1, p. 121.

207  G. Stanley and J. Shorter, *J. Chem. Soc.*, (1958) 246, 256.

208  P. B. D. de la Mare, I. C. Hilton and C. A. Vernon, *J. Chem. Soc.*, (1960) 4039.

209  P. B. D. de la Mare, I. C. Hilton and S. Varma, *J. Chem. Soc.*, (1960) 4044.

209a M. Hassan and G. Yousif, *J. Chem. Soc.*, (B) (1968) 459.

210  L. O. Brown and F. G. Soper, *J. Chem. Soc.*, (1953) 3576.

211  M. D. Carr and B. D. England, *Proc. Chem. Soc.*, (1958) 350.

212  V. Cofman, *J. Chem. Soc.*, (1919) 1040.

213  F. G. Soper and G. F. Smith, *J. Chem. Soc.*, (1927) 2757.

214  B. S. Painter and F. G. Soper, *J. Chem. Soc.*, (1947) 342.

215  E. Berliner, *J. Am. Chem. Soc.*, 72 (1950) 4003.

216  E. Berliner, *J. Am. Chem. Soc.*, 73 (1951) 4307.

217  W. C. Buss and J. E. Taylor, *J. Am. Chem. Soc.*, 92 (1960) 5991.

218  E. Grovenstein and D. C. Kilby, *J. Am. Chem. Soc.*, 79 (1957) 2972.

219  E. Grovenstein and N. S. Aprahamian, *J. Am. Chem. Soc.*, 84 (1962) 212.

220  E. Berliner, *J. Am. Chem. Soc.*, 82 (1960) 5435.

221  E. Berliner, *J. Am. Chem. Soc.*, 78 (1956) 3632.

222  E. Berliner, *J. Am. Chem. Soc.*, 80 (1958) 856.

223  E. A. Shilov and F. Weinstein, *Nature*, 182 (1958) 1300; *Dokl. Akad. Nauk SSSR*, 123 (1958) 93.

224  F. M. Vainshtein and E. A. Shilov, *Proc. Acad. Sci. USSR*, 133 (1960) 821.

225  A. Grimison and J. H. Ridd, *J. Chem. Soc.*, (1959) 3019.

226  K. J. P. Orton and H. King, *J. Chem. Soc.*, (1911) 1369.

227  K. J. P. Orton and A. E. Bradfield, *J. Chem. Soc.*, (1927) 986.

228  K. J. P. Orton, F. G. Soper and G. Williams, *J. Chem. Soc.*, (1928) 998.

229  A. E. Bradfield and B. Jones, *J. Chem. Soc.*, (1928) 1006.

230  A. E. Bradfield and B. Jones, *J. Chem. Soc.*, (1928) 3073.

231  P. B. D. de la Mare and P. W. Robertson, *J. Chem. Soc.*, (1943) 279.

232  K. Lauer and R. Oda, *Chem. Ber.*, 69 (1936) 1061.

233  P. W. Robertson, R. M. Dixon, W. G. M. Goodwin, I. R. McDonald and J. F. Scaife, *J. Chem. Soc.*, (1949) 294.

234  P. W. Robertson, *J. Chem. Soc.*, (1954) 1267.

235  R. E. Roberts and F. G. Soper, *Proc. Roy. Soc. (London), Ser. A*, 140 (1933) 71.

236  M. J. S. DEWAR AND T. MOLE, *J. Chem. Soc.*, (1957) 342.
237  L. J. ANDREWS AND R. M. KEEFER, *J. Am. Chem. Soc.*, 81 (1959) 1063; 79 (1957) 5169.
238  R. M. KEEFER AND L. J. ANDREWS, *J. Am. Chem. Soc.*, 84 (1962) 3635.
239  L. M. STOCK AND A. HIMOE, *J. Am. Chem. Soc.*, 83 (1961) 4605.
240  E. BACIOCCHI AND L. MANDOLINI, *J. Chem. Soc.*, (B) (1967) 1361.
241  A. E. BRADFIELD AND B. JONES, *J. Chem. Soc.*, (1931) 2903.
242  B. JONES, *J. Chem. Soc.*, (1938) 1414.
243  B. JONES, *J. Chem. Soc.*, (1941) 267; (1943) 445.
244  A. E. BRADFIELD, W. O. JONES AND F. SPENCER, *J. Chem. Soc.* (1931) 2907.
245  B. JONES, *J. Chem. Soc.*, (1934) 210.
246  B. JONES, *J. Chem. Soc.*, (1936) 1231.
247  B. JONES, *J. Chem. Soc.*, (1935) 1831, 1835.
248  B. JONES, *J. Chem. Soc.*, (1936) 1854.
249  B. JONES, *J. Chem. Soc.*, (1943) 430.
250  B. JONES, *J. Chem. Soc.*, (1941) 358.
251  A. E. BRADFIELD AND B. JONES, *Trans. Faraday Soc.*, 37 (1941) 726.
     B. JONES, *J. Chem. Soc.*, (1942) 418, 676.
252  H. C. BROWN AND L. M. STOCK, *J. Am. Chem. Soc.*, 79 (1957) 5175, 5615.
253  R. M. KEEFER AND L. J. ANDREWS, *J. Am. Chem. Soc.*, 79 (1957) 4348.
254  P. B. D. DE LA MARE AND M. HASSAN, *J. Chem. Soc.*, (1958) 1519.
255  E. BACIOCCHI AND G. ILLUMINATI, (a) *Chem. Ind.* (*London*), (1958) 917; (b) *J. Am. Chem. Soc.* 86 (1964) 2677.
256  S. F. MASON, *J. Chem. Soc.*, (1959) 1233.
257  Ref. 1, p. 145.
258  O. M. H. EL DUSUOQUI AND M. HASSAN, *J. Chem. Soc.* (B) (1966) 374.
259  L. M. STOCK AND F. W. BAKER, *J. Am. Chem. Soc.*, 84 (1962) 1660.
260  P. B. D. DE LA MARE, D. M. HALL, M. M. HARRIS, M. HASSAN, E. A. JOHNSON AND N. V. KLASSEN, *J. Chem. Soc.*, (1962) 3784.
261  P. W. ROBERTSON, P. B. D. DE LA MARE AND B. E. SWEDLAND, *J. Chem. Soc.*, (1953) 782.
262  P. B. D. DE LA MARE AND E. A. JOHNSON, *J. Chem. Soc.*, (1963) 4076.
263  P. B. D. DE LA MARE, E. A. JOHNSON AND J. S. LOMAS, *J. Chem. Soc.*, (1964) 5317.
264  P. B. D. DE LA MARE, O. M. H. EL DUSUOQUI AND E. A. JOHNSON, *J. Chem. Soc.*, (B) (1966) 521.
265  G. MARINO, *Tetrahedron*, 21 (1965) 843.
266  L. M. STOCK AND A. HIMOE, *J. Am. Chem. Soc.*, 83 (1961) 1957.
267  L. J. ANDREWS AND R. M. KEEFER, *J. Am. Chem. Soc.*, 82 (1960) 5823.
268  E. BACIOCCHI, G. ILLUMINATI AND G. SLEITER, *Tetrahedron Letters*, (23) (1960) 30.
269  P. B. D. DE LA MARE AND J. S. LOMAS, *Rec. Trav. Chim.*, 86 (1967) 1082.
270  M. HASSAN AND S. A. OSMAN, *J. Chem. Soc.*, (1965) 2194.
271  P. SCHONKEN, J. LE PAGE AND J. C. JUNGERS, *Bull. Soc. Chim. France*, (1957) 1394.
272  R. M. KEEFER AND L. J. ANDREWS, *J. Am. Chem. Soc.*, 82 (1960) 4547.
273  L. J. ANDREWS AND R. M. KEEFER, *J. Am. Chem. Soc.*, 79 (1957) 5169.
274  L. LE PAGE AND J. C. JUNGERS, *Bull. Soc. Chim. France*, (1960) 525.
275  G. A. OLAH, S. J. KUHN AND B. A. HARDIE, *J. Am. Chem. Soc.*, 86 (1964) 1055.
276  S. Y. CAILLE AND R. J. P. CORRIU, *Chem. Commun.* (1967) 1251.
277  R. BOLTON AND P. B. D. DE LA MARE, *J. Chem. Soc.*, (B) (1967) 1044.
277a R. BOLTON, *J. Chem. Soc.*, (B) (1968) 712, 714.
278  A. E. BRADFIELD, B. JONES AND K. J. P. ORTON, *J. Chem. Soc.*, (1929) 2810.
279  P. W. ROBERTSON, P. B. D. DE LA MARE AND W. T. G. JOHNSTON, *J. Chem. Soc.*, (1943) 276.
280  K. LAUER AND R. ODA, *Chem. Ber.*, 69 (1936) 978.
281  R. M. KEEFER, A. OTTENBERG AND L. J. ANDREWS, *J. Am. Chem. Soc.*, 78 (1956) 255.
281a J. RAJARAM AND J. C. KURIACOSE, *Australian. J. Chem.*, 21 (1968) 3069.
282  R. M. KEEFER AND L. J. ANDREWS, *J. Am. Chem. Soc.*, 78 (1956) 3637.
283  S. F. MASON, *J. Chem. Soc.* (1958) 4329.
284  E. BERLINER AND M. C. BECKETT, *J. Am. Chem. Soc.*, 79 (1957) 1425.
285  E. BERLINER AND J. C. POWERS, *J. Am. Chem. Soc.*, 83 (1961) 905.

286 E. BERLINER AND B. J. LANDRY, *J. Org. Chem.*, 27 (1962) 1083.
287 U.-J. P. ZIMMERMAN AND E. BERLINER, *J. Am. Chem. Soc.*, 84 (1962) 3953.
288 E. BERLINER, D. M. FALCIONE AND J. L. RIEMENSCHNEIDER, *J. Org. Chem.*, 30 (1965) 1812.
289 I. K. LEWIS, R. D. TOPSOM, J. VAUGHAN AND G. J. WRIGHT, *J. Org. Chem.*, 33 (1968) 1497.
290 L. M. YEDDANAPALLI AND N. S. GNANAPRAGASAM, *J. Chem. Soc.*, (1956) 4934; *J. Ind. Chem. Soc.*, 36 (1959) 745.
291 P. LINDA AND G. MARINO, *Chem. Commun.*, (1967) 499; *J. Chem. Soc.*, (B) (1968) 392.
292 R. P. BELL AND E. N. RAMSDEN, *J. Chem. Soc.*, (1958) 161.
293 J. E. DUBOIS, P. ALCAIO AND G. BARBIER, *Compt. Rend.*, 254 (1962) 3000.
294 R. P. BELL AND T. SPENCER, *J. Chem. Soc.*, (1959) 1156.
295 R. P. BELL AND D. J. RAWLINSON, *J. Chem. Soc.*, (1961) 63.
296 E. BERLINER AND F. GASKIN, *J. Chem. Soc.*, 32 (1967) 1660.
297 G. ILLUMINATI AND G. MARINO, *J. Am. Chem. Soc.*, 78 (1956) 4975.
298 R. M. KEEFER, J. H. BLAKE AND L. J. ANDREWS, *J. Am. Chem. Soc.*, 76 (1954) 3062.
299 H. C. BROWN AND R. A. WIRKKALA, *J. Am. Chem. Soc.*, 88 (1966) 1447.
300 H. V. ANSELL AND R. TAYLOR, *J. Chem. Soc.*, (B) (1968) 526.
301 P. G. FARRELL AND S. F. MASON, *Nature*, 183 (1959) 250.
302 P. G. FARRELL AND S. F. MASON, *Nature*, 197 (1963) 590.
303 M. CHRISTEN AND H. ZOLLINGER, *Helv. Chim. Acta*, 45 (1962) 2057, 2066.
304 E. BACIOCCHI, G. ILLUMINATI, G. SLEITER AND F. STEGEL, *J. Am. Chem. Soc.*, 89 (1967) 125.
305 E. BERLINER AND K. SCHUELLER, *Chem. Ind.* (*London*), (1960) 1444.
306 E. HELGSTRAND, *Acta Chem. Scand.*, 19 (1965) 1583.
307 E. HELGSTRAND AND A. NILSSON, *Acta Chem. Scand.*, 20 (1966) 1465.
308 A. N. BOURNS, quoted by H. ZOLLINGER in *Advances in Physical Organic Chemistry*, Vol. 2, V. GOLD, (Ed.), 1964, p. 163.
309 P. B. D. DE LA MARE AND O. M. H. EL DUSOUQI, *J. Chem. Soc.*, (B) (1967) 251.
310 G. C. ISRAEL, A. W. N. TUCK and F. G. SOPER, *J. Chem. Soc.*, (1945) 547.
311 V. S. KARPINSKII AND V. D. LYASHENKO, *J. Gen. Chem. USSR*, 30 (1960) 164; 32 (1962) 3922; 33 (1963) 599.
312 L. BRUNER, *Z. Phys. Chem.*, 41 (1902) 514.
313 C. C. PRICE, *J. Am. Chem. Soc.*, 58 (1936) 2101.
C. C. PRICE AND C. E. ARNTZEN, *J. Am. Chem. Soc.*, 60 (1938) 2835.
C. C. PRICE, *Chem. Rev.*, 29 (1941) 37.
314 P. W. ROBERTSON, J. E. ALLAN, K. N. HALDANE AND M. G. SIMMERS, *J. Chem. Soc.*, (1949) 933.
315 N. S. GNANAPRAGASAM, N. V. RAO AND L. M. YEDDANAPALLI, *J. Indian Chem. Soc.*, 36 (1959) 777.
316 T. TSURATA, K. SASAKI AND J. FURUKAWA, *J. Am. Chem. Soc.*, 74 (1952) 5995; 76 (1954) 994.
317 J. H. BLAKE AND R. M. KEEFER, *J. Am. Chem. Soc.*, 77 (1955) 3707.
318 R. JOSEPHSON, R. M. KEEFER AND L. J. ANDREWS, *J. Am. Chem. Soc.*, 83 (1961) 2128.
319 R. PAJEAU, *Compt. Rend.*, 207 (1938) 1420; *Bull. Soc. Chim. France*, 6 (1939) 1187.
320 L. J. ANDREWS AND R. M. KEEFER, *J. Am. Chem. Soc.*, 78 (1956) 4549.
321 R. JOSEPHSON, R. M. KEEFER AND L. J. ANDREWS, *J. Am. Chem. Soc.*, 83 (1961) 3562.
322 V. HATANAKA, R. M. KEEFER AND L. J. ANDREWS, *J. Am. Chem. Soc.*, 87 (1965) 4280.
323 G. A. OLAH, S. J. KUHN, S. H. FLOOD AND B. A. HARDIE, *J. Am. Chem. Soc.*, 86 (1964) 1039, 1044.
324 Y. OGATA, Y. FURUYA AND K. OKAMO, *Bull. Chem. Soc. Japan*, 37 (1964) 960.
325 C. H. LI, *J. Am. Chem. Soc.*, 64 (1942) 1147.
326 K. W. DOAK AND A. H. CORWIN, *J. Am. Chem. Soc.*, 71 (1949) 159.
327 E. GROVENSTEIN AND F. C. SCHMALSTIEG, *J. Am. Chem. Soc.*, 89 (1967) 5084.
328 L. J. LAMBOURNE AND P. W. ROBERTSON, *J. Chem. Soc.*, (1947) 1167.
329 R. M. KEEFER AND L. J. ANDREWS, *J. Am. Chem. Soc.*, 78 (1956) 5623.
330 L. J. ANDREWS AND R. M. KEEFER, *J. Am. Chem. Soc.*, 79 (1957) 1412.
331 Y. OGATA AND K. NAKAJIMA, *Tetrahedron*, 20 (1964) 43.
332 E. M. CHEN, R. M. KEEFER AND L. J. ANDREWS, *J. Am. Chem. Soc.*, 289 (1967) 428.
333a D. B. STEELE, *J. Chem. Soc.*, (1903) 1470.

333bL. F. Martin, P. Pizzolato and L. S. McWaters, *J. Am. Chem. Soc.*, 57 (1935) 2584.
333cH. Clement, *Ann. Chim.*, 13 (1940) 243.
334  H. Ulich and G. Heyne, *Z. Electrochem.*, 41 (1935) 509.
335  S. J. C. Olivier and G. Berger, *Rec. Trav. Chim.*, 45 (1926) 710.
336  N. O. Calloway, *J. Am. Chem. Soc.*, 59 (1937) 1474.
337  F. E. Condon, *J. Am. Chem. Soc.*, 70 (1948) 2265.
338  F. E. Condon, *J. Am. Chem. Soc.*, 71 (1949) 3544.
339  H. C. Brown and K. L. Nelson, *J. Am. Chem. Soc.*, 75 (1953) 6292.
340  H. C. Brown and M. Grayson, *J. Am. Chem. Soc.*, 75 (1953) 6285.
341  H. C. Brown and H. Jungk, *J. Am. Chem. Soc.*, 77 (1955) 5584.
342  H. C. Brown and H. Jungk, *J. Am. Chem. Soc.*, 78 (1956) 2182.
343  H. Jungk, H. C. Brown and C. R. Smoot, *J. Am. Chem. Soc.*, 78 (1956) 2185.
344  C. R. Smoot and H. C. Brown, *J. Am. Chem. Soc.*, 78 (1956) 6245, 6249.
345  S. U. Choi and H. C. Brown, *J. Am. Chem. Soc.*, 81 (1959) 3315.
346  H. C. Brown and A. H. Neyens, *J. Am. Chem. Soc.*, 84 (1962) 1233, 1655.
347  H. C. Brown and B. A. Bolto, *J. Am. Chem. Soc.*, 81 (1959) 3320.
348  R. H. Allen and L. D. Yats, *J. Am. Chem. Soc.*, 83 (1961) 2799.
349  N. N. Lebedev, *J. Gen. Chem. USSR*, 24 (1954) 673.
350  N. N. Lebedev, *J. Gen. Chem. USSR*, 28 (1958) 1211.
351  N. N. Lebedev, *J. Gen. Chem. USSR*, 27 (1957) 2520.
352  N. N. Lebedev, *Chem. Abstr.*, 53 (1959) 9106.
353  H. Hart and F. A. Cassis, *J. Am. Chem. Soc.*, 76 (1954) 1634.
354  H. Hart, F. A. Cassis and J. J. Bordeaux, *J. Am. Chem. Soc.*, 76 (1954) 1639.
355  G. Chuchani, H. Diaz and J. Zabicky, *J. Org. Chem.*, 31 (1966) 1573.
356  G. A. Olah, S. J. Kuhn and S. H. Flood, *J. Am. Chem. Soc.*, 84 (1962) 1688, 1695.
357  G. A. Olah, S. H. Flood, S. J. Kuhn, M. E. Moffatt and N. A. Overchuck, *J. Am. Chem. Soc.*, 86 (1964) 1046.
358  G. A. Olah, S. H. Flood and M. E. Moffatt, *J. Am. Chem. Soc.*, 86 (1964) 1060.
359  G. A. Olah and N. A. Overchuck, *J. Am. Chem. Soc.*, 87 (1965) 5786.
360  G. A. Olah, N. A. Overchuck and J. C. Lapiere, *J. Am. Chem. Soc.*, 87 (1965) 5785.
361  C. D. Nenitzescu, S. Titeica and V. Ioan, *Bull. Soc. Chim., France*, 20 (1955) 1272, 1279.
     V. Ioan, D. Sandulescu, S. Titeica and C. S. Nenitzescu, *Tetrahedron*, 19 (1963) 323, 335.
362  C. D. Nenitzescu, S. Titeica and V. Ioan, *Acta Chim. Acad. Sci, Hung.*, 12 (1957) 195.
363  N. N. Lebedev, *J. Gen. Chem. USSR*, 24 (1954) 1751.
364  R. N. Volkov and S. V. Zavgorodnii, *Proc. Acad. Sci. USSR*, 133 (1960) 869.
365  D. Bethell and V. Gold, *J. Chem. Soc.*, (1958) 1905.
366  D. Bethell, V. Gold and T. Riley, *J. Chem. Soc.*, (1959) 3134.
367  D. Bethell and V. Gold, *J. Chem. Soc.*, (1958) 1930.
368  T. G. Bonner, J. M. Clayton and G. Williams, *J. Chem. Soc.*, (1957) 2867.
369  E. Berliner, *J. Am. Chem. Soc.*, 66 (1944) 533.
370  L. K. Brice and R. D. Katstra, *J. Am. Chem. Soc.*, 82 (1960) 2669.
371  C. K. Bradsher and F. A. Vingiello, *J. Am. Chem. Soc.*, 71 (1949) 1434.
372  F. A. Vingiello and J. G. van Oot, *J. Am. Chem. Soc.*, 73 (1951) 5070.
373  F. A. Vinigiello, J. G. van Oot and H. H. Hannabass, *J. Am. Chem. Soc.*, 74 (1952) 4546.
374  F. A. Vingiello, M. O. L. Spangler and J. E. Bondurant, *J. Org. Chem.*, 25 (1960) 2091.
375  F. A. Vingiello and M. M. Schlechter, *J. Org. Chem.*, 28 (1963) 2448.
376  R. Taylor, *J. Chem. Soc.*, (1962) 4881.
377  T. G. Bonner, M. P. Thorpe and T. M. Williams, *J. Chem. Soc.*, (1955) 2351.
378  T. G. Bonner and M. Barnard, *J. Chem. Soc.*, (1958) 4181.
379  T. G. Bonnor and M. Barnard, *J. Chem. Soc.*, (1958) 4176.
380  T. G. Bonner and J. M. Watkins, *J. Chem. Soc.*, (1955) 2358.
381  G. Vavon, J. Bolle and J. Calin, *Bull. Soc. Chim. France*, 6 (1939) 1025.
382  H. H. Szmant and J. Dudek, *J. Am. Chem. Soc.*, 71 (1949) 3763.
383  G. S. Mironov, M. I. Farberov, V. D. Sheiv and I. I. Bespalova, *J. Org. Chem. USSR*, 2 (1966) 1615.

384 Y. OGATA AND M. OKANO, *J. Am. Chem. Soc.*, 78 (1956) 5423.

385 Y. ISHII AND Y. YAMASHITA, *J. Chem. Soc. Japan*, (*Ind. Chem. Sect.*) 56 (1953) 104.

386 I. N. NAZAREV AND A. V. SEMENOSKY, *Bull. Acad. Sci.*, (1957) 997.

387 Ref. 1, pp. 174–177.

388 G. A. OLAH, S. J. KUHN, W. J. TOLGYESI AND E. B. BAKER, *J. Am. Chem. Soc.*, 84 (1962) 2733.

389 *Friedel-Crafts and Related Reactions*, G. A. OLAH (Ed.), Vol. III, Interscience, New York, 1964.

390 B. MENSCHUTKIN, *J. Russ. Phys. Chem. Soc.*, 45 (1913) 1710; 46 (1914) 259; *J. Chem. Soc.* (*Abstr.*), 106 (1914) I, 189, 673.

391 O. C. DERMER, D. M. WILSON, F. M. JOHNSON AND V. H. DERMER, *J. Am. Chem. Soc.*, 63 (1941) 2881.

392 O. C. DERMER AND R. A. BILLMEIER, *J. Am. Chem. Soc.*, 64 (1942) 464.

393 O. C. DERMER, P. J. MORI AND S. SUGUITAN, *Chem. Abstr.*, 46 (1952) 7538.

394 F. R. JENSEN AND H. C. BROWN, *J. Am. Chem. Soc.*, 80 (1958) 3039.

395 H. C. BROWN AND F. R. JENSEN, *J. Am. Chem. Soc.*, 80 (1958) 2291.

396 S. C. J. OLIVIER, *Rec. Trav. Chim.*, 37 (1918) 205.

397 H. ULICH AND P. V. FRAGSTEIN, *Chem. Ber.*, 72 (1939) 620.

398 N. N. LEBEDEV, *J. Gen. Chem.*, 21 (1951) 1788.

399 N. N. GREENWOOD AND K. WADE, *J. Chem. Soc.*, (1956) 1527.

400 J. M. TEDDER, *Chem. Ind.*, (*London*), (1954) 630.

401 G. BADDELEY AND D. VOSS, *J. Chem. Soc.*, (1954) 418.

402 G. BADDELEY, *J. Chem. Soc.*, (1949) S99.

403 E. H. MAN AND C. R. HAUSER, *J. Org. Chem.*, 17 (1952) 397.

404 P. H. GORE, *Chem Ind.* (*London*), (1954) 1385; *Chem. Rev.*, 55 (1955) 229.

405 Y. YAMASE, *Bull. Chem. Soc. Japan*, 34 (1961) 480, 484.

406 Y. YAMASE AND R. GOTO, *Nippon Kagaku Zasshi*, 81 (1960) 1906.

407 H. C. BROWN AND F. R. JENSEN, *J. Am. Chem. Soc.*, 80 (1958) 2296.

408 P. A. GOODMAN AND P. H. GORE, *J. Chem. Soc.*, (C) (1968) 966.

409 E. ROTHSTEIN AND R. W. SAVILLE, *J. Chem. Soc.*, (1949) 1950, 1959.
M. E. GRUNDY, W. H. HSÜ AND E. ROTHSTEIN, *J. Chem. Soc.*, (1958) 581.

410 H. C. BROWN AND H. L. YOUNG, *J. Org. Chem.*, 22 (1957) 724.

411 H. C. BROWN AND H. L. YOUNG, *J. Org. Chem.*, 22 (1957) 719.

412 H. C. BROWN, B. A. BOLTO AND F. R. JENSEN, *J. Org. Chem.*, 23 (1958) 414, 417.

413 F. SMEETS AND J. VERHULST, *Bull. Soc. Chim. Belges*, 63 (1954) 439.

414 F. R. JENSEN, *J. Am. Chem. Soc.*, 79 (1957) 1226.

415 F. R. JENSEN, G. MARINO AND H. C. BROWN, *J. Am. Chem. Soc.*, 81 (1959) 3303.

416 R. CORRIU, M. DORE AND R. THOMASSIN, *Tetrahedron Letters*, (1968) 2759.

417 H. C. BROWN AND G. MARINO, *J. Am. Chem. Soc.*, 81 (1959) 3308.

418 H. C. BROWN, G. MARINO AND L. M. STOCK, *J. Am. Chem. Soc.*, 81 (1959) 3310.

419 H. F. MCDUFFIE AND G. DOUGHERTY, *J. Am. Chem. Soc.*, 64 (1942) 297.

420 Y. OGATA AND R. ODA, *Bull. Inst. Phys. Chem. Res.*, 71 (1942) 728.

421 H. C. BROWN AND G. MARINO, *J. Am. Chem. Soc.*, 81 (1959) 5611; 84 (1962) 1236.

422 H. C. BROWN AND G. MARINO, *J. Am. Chem. Soc.*, 81 (1959) 5929; 84 (1962) 1658.

422a P. H. GORE, C. K. THADANI AND S. THORBURN, *J. Chem. Soc.*, (C) (1968) 2502.

423 D. J. CRAM, W. H. WECHTER AND R. W. KIERSTEAD, *J. Am. Chem. Soc.*, 80 (1958) 3126.

424 Y. L. GOLDFARB, V. P. LITVINOV AND V. I. SHNEDOV, *Zh. Obshch. Khim.*, 30 (1960) 534.

425 E. S. NOVIKOVA, *Chem. Abstr.*, 54 (1960) 400.

425a P. A. GOODMAN AND P. H. GORE, *J. Chem. Soc.*, (C) (1968) 2452.

426 R. B. GIRDLER, P. H. GORE AND J. A. HOSKINS, *J. Chem. Soc.*, (C) (1966) 181.

427 R. B. GIRDLER, H. GORE AND C. K. THADANI, *J. Chem. Soc.*, (C) (1967) 2619.

428 M. ROSENBLUM AND W. G. HOWELLS, *J. Am. Chem. Soc.*, 84 (1962) 1167.

429 M. ROSENBLUM, J. O. SANTER AND W. G. HOWELLS, *J. Am. Chem. Soc.*, 85 (1963) 1450.

430 M. ROSENBLUM AND F. W. ABBATE, *J. Am. Chem. Soc.*, 88 (1966) 4178.

431 R. A. BENKESER, Y. NAGAI AND J. HOOZ, *J. Am. Chem. Soc.*, 86 (1964) 3742.

432 P. LINDA AND G. MARINO, *Tetrahedron*, 33 (1967) 1739.

433 S. CLEMENTI, G. GEMEL AND G. MARINO, *Chem. Commun.*, (1967) 498.
434 S. KIMOTO, *J. Pharm. Soc. Japan*, 75 (1955) 727; *Chem. Abstr.*, 50 (1956) 3293.
434a D. S. NOYCE, P. A. KETTLE AND E. H. BARRITT, *J. Org. Chem.*, 33 (1968) 1500.
434b D. B. DENNEY AND P. P. KLEMCHUK, *J. Am. Chem. Soc.*, 80 (1958) 3285, 6014.
435 F. H. WESTHEIMER, E. SEGEL AND R. M. SCHRAMM, *J. Am. Chem. Soc.*, 69 (1947) 773.
436 R. M. SCHRAMM, W. KLAPPROTH AND F. H. WESTHEIMER, *J. Phys. Chem.*, 55 (1951) 843.
437 C. PERRIN AND F. H. WESTHEIMER, *J. Am. Chem. Soc.*, 85 (1963) 2773.
438 R. P. BELL, *J. Phys. Chem.*, 55 (1951) 859.
439 A. J. KRESGE, M. DUBECK AND H. C. BROWN, *J. Org. Chem.*, 32 (1967) 745.
440 H. C. BROWN AND C. W. MCGARY, *J. Am. Chem. Soc.*, 77 (1955) 2306.
441 H. C. BROWN AND C. W. MCGARY, *J. Am. Chem. Soc.*, 77 (1955) 2300.
442 A. J. KRESGE AND J. F. BRENNAN, *Proc. Chem. Soc.*, (1963) 215; *J. Org. Chem.*, 32 (1967) 752.
443 C. W. MCGARY AND G. GOLDMAN, *Ph. D. Theses*, Purdue University, 1955 and 1961.
444 H. C. BROWN AND C. W. MCGARY, *J. Am. Chem. Soc.*, 77 (1955) 2310.
445 H. C. BROWN AND M. DUBECK, *J. Am. Chem. Soc.*, 81 (1959) 5608; 82 (1960) 1939.
446 H. C. BROWN, M. DUBECK AND G. GOLDMAN, *J. Am. Chem. Soc.*, 84 (1962) 1229.
447 H. C. BROWN AND G. GOLDMAN, *J. Am. Chem. Soc.*, 84 (1962) 1650.
448 A. J. KRESGE AND H. C. BROWN, *J. Org. Chem.*, 32 (1967) 756.
449 C. G. SWAIN, T. E. C. KNEE AND A. J. KRESGE, *J. Am. Chem. Soc.*, 79 (1957) 505.
449a R. MOTOYAMA, S. NISHIMURA, E. IMOTO, Y. MURAKAMA, K. HARRI AND J. OGAWA, *Chem. Abstr.*, 54 (1966) 14224.
450 H. C. BROWN AND R. A. WIRKKALA, *J. Am. Chem. Soc.*, 88 (1966) 1447, 1453, 1456.
451 C. K. INGOLD, C. G. RAISIN AND C. L. WILSON, *Nature*, 134 (1934) 734; *J. Chem. Soc.* (1936) 915, 1637.
    A. P. BEST AND C. L. WILSON, *J. Chem. Soc.*, (1938) 28.
452 M. KOIZUMI AND T. TITANI, *Bull. Chem. Soc. Japan*, 13 (1938) 681.
    M. KOIZUMI, *Bull. Chem. Soc. Japan*, 14 (1939) 353.
453 M. KOIZUMI AND T. TITANI, *Bull. Chem. Soc. Japan*, 13 (1938) 318, 595, 631; 14 (1939) 40.
454 M. KOIZUMI, *Bull. Chem. Soc. Japan*, 14 (1939) 531; 15 (1940) 8, 37.
455 M. KOIZUMI AND T. TITANI, *Bull. Chem. Soc. Japan*, 13 (1938) 95.
456 M. KOIZUMI AND T. TITANI, *Bull. Chem. Soc. Japan*, 12 (1937) 107; 13 (1938) 85, 298, 307; 14 (1939) 453.
457 V. GOLD AND F. A. LONG, *J. Am. Chem. Soc.*, 75 (1953) 4543.
458 V. GOLD AND D. P. N. SATCHELL, *J. Chem. Soc.*, (1955) 3609.
459 V. GOLD AND D. P. N. SATCHELL, *J. Chem. Soc.*, (1955) 3622.
460 V. GOLD, R. W. LAMBERT AND D. P. N. SATCHELL, *Chem. Ind.* (*London*), (1959) 1312; *J. Chem. Soc.*, (1960) 2461.
461 V. GOLD AND D. P. N. SATCHELL, *J. Chem. Soc.*, (1955) 3619.
462 F. A. LONG AND M. A. PAUL, *Chem. Rev.*, 57 (1957) 1.
463 V. GOLD AND D. P. N. SATCHELL, *J. Chem. Soc.*, (1956) 2743.
464 C. EABORN AND R. TAYLOR, *Chem. Ind.* (*London*), (1959) 949.
465 R. BAKER, C. EABORN AND R. TAYLOR, *J. Chem. Soc.*, (1961) 4927.
466 C. EABORN AND R. TAYLOR, *J. Chem. Soc.*, (1961) 247.
467 S. OLSSON AND L. MELANDER, *Acta Chem. Scand.*, 8 (1954) 523.
    S. OLSSON, *Arkiv. Kemi*, 14 (1959) 85; 16 (1960) 489.
468 R. TAYLOR, unpublished work.
469 S. OLSSON, *Arkiv. Kemi*, 15 (1960) 259.
470 L. MELANDER AND S. OLSSON, *Acta Chem. Scand.*, 10 (1956) 879.
471 C. EABORN, P. M. JACKSON AND R. TAYLOR, *J. Chem. Soc.*, (B) (1966) 613.
472 B. D. BATTS AND V. GOLD, *J. Chem. Soc.*, (1964) 4284.
473 A. J. KRESGE AND Y. CHIANG, *J. Am. Chem. Soc.*, 84 (1962) 3976.
474 D. P. N. SATCHELL, *J. Chem. Soc.*, (1956) 3911.
475 D. P. N. SATCHELL, *J. Chem. Soc.*, (1959) 463.
476 D. P. N. SATCHELL, *J. Chem. Soc.*, (1958) 1927, 3910.
477 E. L. MACKOR, P. J. SMIT AND J. H. VAN DER WAALS, *Trans. Faraday Soc.*, 53 (1957) 1309.
478 D. P. N. SATCHELL, *J. Chem. Soc.*, (1958) 3904.

479  K. YATES AND B. STEVENS, *Can. J. Chem.*, 42 (1964) 1967.
480  A. J. KRESGE AND Y. CHIANG, *J. Am. Chem. Soc.*, 81 (1959) 5509; 83 (1961) 2877.
481  A. J. KRESGE AND Y. CHIANG, *J. Am. Chem. Soc.*, 89 (1967) 4411.
482  C. G. SWAIN, E. C. STIVERS, J. F. RENVER AND L. J. SCHAAD, *J. Am. Chem. Soc.*, 80 (1958) 5885.
483  A. STREITWEISER, R. H. JAGOW, R. L. FAHEY AND S. SUZUKI, *J. Am. Chem. Soc.*, 80 (1958) 2236.
     E. A. HALEVI, *Progr. Phys. Org. Chem.*, 1 (1963) 109.
484  S. SELTZER, *J. Am. Chem. Soc.*, 83 (1961) 2625.
485  A. J. KRESGE, Y. CHIANG AND Y. SATO, *J. Am. Chem. Soc.*, 89 (1967) 4418.
486  F. A. LONG AND J. SCHULZE, *J. Am. Chem. Soc.*, 86 (1964) 327.
487  R. J. THOMAS AND F. A. LONG, *J. Am. Chem. Soc.*, 86 (1964) 4770.
488  R. J. THOMAS AND F. A. LONG, *J. Org. Chem.*, 29 (1964) 3411.
489  J. SCHULZE AND F. A. LONG, *J. Am. Chem. Soc.*, 86 (1964) 331.
490  J. COLAPIETRO AND F. A. LONG, *Chem. Ind. (London)*, (1960) 1056.
491  L. C. GRUEN AND F. A. LONG, *J. Am. Chem. Soc.*, 89 (1967) 1287.
492  C. G. SWAIN AND A. S. ROSENBERG, *J. Am. Chem. Soc.*, 83 (1961) 2154.
     J. BIGELEISEN, *Pure Appl. Chem.*, 8 (1964) 217.
493  J. A. LONGRIDGE AND F. A. LONG, *J. Am. Chem. Soc.*, 89 (1967) 1292.
494  A. J. KRESGE, *Discussions Faraday Soc.*, 39 (1965) 49.
494a A. J. KRESGE, D. P. ONWOOD AND S. SLAE, *J. Am. Chem. Soc.*, 90 (1968) 6982.
495  B. C. CHALLIS AND F. A. LONG, *J. Am. Chem. Soc.*, 85 (1963) 2524.
496  R. L. HINMAN AND E. B. WHIPPLE, *J. Am. Chem. Soc.*, 84 (1962) 2534.
497  B. C. CHALLIS AND F. A. LONG, *J. Am. Chem. Soc.*, 87 (1965) 1196.
498  A. J. KRESGE, R. A. MORE O'FERRALL, L. E. HAKKA AND V. P. VITULLO, *Chem. Commun.*, (1965) 46.
499  L. MELANDER, *Arkiv Kemi*, 17 (1961) 291.
500  V. GOLD AND F. L. TYE, *J. Chem. Soc.*, (1952) 2181.
501  V. GOLD, *Proc. Chem. Soc.*, (1961) 453.
502  B. B. P. TICE, I. LEE AND F. H. KENDALL, *J. Am. Chem. Soc.*, 85 (1963) 329.
     I. LEE AND F. H. KENDALL, *J. Am. Chem. Soc.*, 88 (1966) 3813.
503  A. C. LING AND F. H. KENDALL, *J. Chem. Soc.*, (B) (1967) 440.
504  Ref. 1, Chapt. 8.
505  A. C. LING AND F. H. KENDALL, *J. Chem. Soc.*, (B) (1967) 445.
506  G. P. BEAN AND A. R. KATRITZKY, *J. Chem. Soc.*, (B) (1968) 864.
507  A. R. KATRITZKY AND B. J. RIDGEWELL, *Proc. Chem. Soc.*, (1962) 114; *J. Chem. Soc.*, (1963) 3753.
508  G. P. BEAN, C. E. JOHNSON, A. R. KATRITZKY, B. J. RIDGEWELL AND A. W. WHITE, *J. Chem. Soc.*, (B) (1967) 1219.
509  G. P. BEAN, P. J. BRIGNELL, C. D. JOHNSON, A. R. KATRITZKY, B. J. RIDGEWELL, H. O. TARHAN AND A. W. WHITE, *J. Chem. Soc.*, (B) (1967) 1222.
510  P. BELLINGHAM, C. D. JOHNSON AND A. R. KATRITZKY, *J. Chem. Soc.*, (B) (1967) 1226.
511  P. BELLINGHAM, C. D. JOHNSON AND A. R. KATRITZKY, *Chem. Commun.*, (1967) 1047; *J. Chem. Soc.*, (B) (1968) 866.
512  Ref. 1, p. 298.
513  A. R. KATRITZKY AND I. POJARLIEFF, *J. Chem. Soc.*, (B) (1968) 873.
514  A. R. KATRITZKY, M. KINGSLAND AND O. S. TEE, *J. Chem. Soc.*, (B) (1968) 1484.
515  W. W. PAUDLER AND L. S. HELMICK, *J. Org. Chem.*, 33 (1968) 1087.
516  A. E. COMYNS, R. A. HOWALD AND J. E. WILLARD, *J. Am. Chem. Soc.*, 78 (1956) 3989.
517  D. P. N. SATCHELL, *J. Chem. Soc.*, (1960) 4388.
518  A. I. SHATENSHTEIN, A. P. SANNIKOV AND P. P. ALIKHANOV, *J. Gen. Chem. USSR*, 75 (1965) 418.
519  A. P. SANNIKOV, E. Z. UTYANSKAYA, P. P. ALIKHANOV AND A. I. SHATENSHTEIN, *J. Gen. Chem. USSR*, (1966) 2027.
520  C. EABORN AND D. R. KILLPACK, personal communication.
521  A. I. SHATENSHTEIN, K. I. ZHDANOV, L. N. VINOGRADOV AND V. R. KALINACHENKO, *Dokl.*

*Akad. Nauk SSSR*, 102 (1955) 779.

A. I. SHATENSHTEIN, K. I. ZHDANOV AND V. M. BASMANOVA, *J. Gen. Chem. USSR*, 33 (1961) 232.

522 R. TAYLOR, *Chimia (Aarau)*, 22 (1968) 1.

523 V. GOLD, *Friedel Crafts and Related Reactions*, Vol. II, Interscience, New York, 1964, p. 1253.

524 R. BAKER, *Ph. D. Thesis*, University of Leicester, 1962.

525 R. BAKER, C. EABORN AND J. A. SPERRY, *J. Chem. Soc.*, (1962) 2382.

526 K. C. C. BANCROFT, *Ph. D. Thesis*, University of Leicester, 1963.

527 R. TAYLOR, G. J. WRIGHT AND A. J. HOMES, *J. Chem. Soc.*, (B) (1967) 780.

528 J. VAUGHAN AND G. J. WRIGHT, *J. Org. Chem.*, 33 (1968) 2580.

529 H. V. ANSELL AND R. TAYLOR, *J. Chem. Soc.*, (B) (1968) 526.

530 H. V. ANSELL AND R. TAYLOR, unpublished work.

531 R. E. SPILLETT, *Ph. D. Thesis*, University of Leicester, (1963).

532 A. I. SEREBRYANSKAYA, A. V. ELTSOV AND A. I. SHATENSHTEIN, *J. Org. Chem. USSR*, 3 (1967) 343.

533 K. C. C. BANCROFT, R. W. BOTT AND C. EABORN, *J. Chem. Soc.*, (1964) 4806.

534 R. BAKER AND C. EABORN, *J. Chem. Soc.*, (1961) 5077.

535 A. I. SHATENSHTEIN, YU. I. RANNEVA AND T. T. KOVALENKO, *J. Gen. Chem., USSR*, 32 (1962) 954.

536 R. BAKER, R. W. BOTT AND C. EABORN, *J. Chem. Soc.*, (1963) 2136.

537 R. BAKER, R. W. BOTT, C. EABORN AND P. M. GREASLEY, *J. Chem. Soc.*, (1964) 627.

538 J. M. BLATCHLY AND R. TAYLOR, *J. Chem. Soc.*, (1964) 4641.

539 A. STREITWEISER, G. R. ZIEGLER, P. C. MOWERY, A. LEWIS AND R. G. LAWLER, *J. Am. Chem. Soc.*, 90 (1968) 1357.

540 D. R. KILLPACK, *D. Phil. Thesis*, University of Sussex, 1969.

541 C. EABORN, P. GOLBORN, R. E. SPILLETT AND R. TAYLOR, *J. Chem. Soc.*, (B) (1968) 1112.
R. W. BOTT, C. EABORN AND R. E. SPILLETT, *Chem. Commun.* ,(1965) 147.
C. EABORN, P. GOLBORN AND R. TAYLOR, *J. Organometal Chem.*, 10 (1967) 171.

542 R. TAYLOR, *J. Chem. Soc.*, (B) (1968) 1559.

543 P. P. ALIKHANOV AND A. I. SHATENSHTEIN, *J. Gen. Chem. USSR*, 38 (1968) 222.

544 W. HANSTEIN AND T. G. TRAYLOR, *Tetrahedron Letters*, (1967) 4451.

545 A. N. NESMEYANOV, D. N. KURSANOV, V. N. SETKINA, N. V. KISLYAKOVA AND N. S. KO-CHETKOVA, *Tetrahedron*, (1961) 41; *Bull. Acad. Sci. USSR* (1962) 1845.

546 M. N. NEFEDOVA, D. N. KURSANOV, V. N. SETKINA, E. N. PEREVALOVA AND A. N. NES-MEYANOV, *Proc. Acad. Sci. USSR*, 166 (1966) 84.

547 W. M. LAUER, G. W. MATSON AND G. STEDMAN, *J. Am. Chem. Soc.*, 80 (1958) 6433, 6437, 6439.

548 W. D. BLACKLEY, *Dissertation Abstr.*, 22 (1961) 1755.

549 Ref. 1, p. 211.

550 D. E. RICE, *Dissertation Abstr.*, 21 (1961) 3961.

551 T. J. EVENSON, *Dissertation Abstr.*, 20 (1960) 3953.

552 W. M. LAUER AND C. B. KOONS, *J. Org. Chem.*, 24 (1959) 1169.

553 A. J. KRESGE AND D. P. N. SATCHELL, *Tetrahedron Letters*, (13) (1959) 20.

554 A. R. BUTLER AND C. EABORN, *J. Chem. Soc.*, (B) (1968) 370.

555 A. I. SHATENSHTEIN, A. G. KAMRAD, I. O. SHAPIRO, YA. I. RANNEVA AND E. N. ZVYAGIN-TSEVA, *Proc. Acad. Sci. USSR*, 168 (1966) 502.
E. N. ZVYAGINTSEVA, T. A. YUKUSHINA AND A. I. SHATENSHTEIN, *J. Gen. Chem. USSR*, (1968) 1933.

556 D. N. KURSANOV, V. N. SETKINA, M. N. NEFEDOVA AND A. N. NESMEYANOV, *Bull. Acad. Sci. USSR*, (1965) 2187.

557 E. V. BYKOVA, V. N. SETKINA AND D. N. KURSANOV, *Proc. Acad. Sci. USSR*, 178 (1968) 31.

558 F. S. YAKUSHIN, V. N. SETKINA, E. A. YAKOVLEVA, A. I. SHATENSHTEIN AND D. N. KURSANOV, *Bull. Acad. Sci. USSR*, (1967) 202.

559 R. TAYLOR, *Ph. D. Thesis*, University of Leicester, 1959.

560 C. PARKYANI, Z. DOLEJSEK AND R. ZAHRADNIK, *Tetrahedron Letters*, (1963) 1897; *Collection*

*Czech. Chem. Comm.*, 33 (1968) 1211.

561 G. DALLINGA, A. VERRIJN STUART, P. J. SMIT AND E. L. MACKOR, *Z. Elektrochem.*, 61 (1957) 1019.

562 G. DALLINGA, P. J. SMIT AND E. L. MACKOR, in *Steric Effects in Conjugated Systems*, G. W. GRAY (Ed.), Butterworths, London, 1958, p. 150.

563 G. DALLINGA, P. J. SMIT AND E. L. MACKOR, *Mol. Phys.*, 3 (1960) 130.
C. MACLEAN AND E. L. MACKOR, *Mol. Phys.*, 3 (1960) 233.

564 A. N. NESMEYANOV, D. N. KURSANOV, V. N. SETKINA, N. V. KISLYAKOVA, N. S. KOCHETKOVA AND R. B. MATERIKOVA, *Proc. Acad. Sci. USSR*, 143 (1962) 199.

565 S. OAE, A. OHNO AND W. TAGAKI, *Bull. Chem. Soc. Japan*, 35 (1962) 681.

566 W. M. LAUER AND J. T. DAY, *J. Am. Chem. Soc.*, 77 (1955) 1904.

567 A. I. SHATENSHTEIN, E. A. RABINOVICH AND V. A. PAVLOV, *J. Gen. Chem. USSR*, 34 (1964) 4050.

568 M. G. KUZMIN, B. M. UZHINOV, G. S. GYORGY AND I. V. BEREZIN, *J. Phys. Chem. USSR*, 41 (1967) 400.

569 S. OLSSON, *Arkiv Kemi*, 15 (1960) 275.

570 G. VAN DYKE TIERS, *J. Am. Chem. Soc.*, 78 (1956) 4165.

571 K. SCHWETLIK, K. UNVERFERTH AND R. MAYER, *Z. Chemie.*, 7 (1967) 58.

572 R. TAYLOR, *J. Chem. Soc.*, (B) (1968) 1397; (1970) 1364.

573a A. I. SHATENSHTEIN AND YU. I. RANNEVA, *J. Gen. Chem. USSR*, 31 (1961) 1317.

573b A. I. SHATENSHTEIN, E. N. ZYAGINTSEVA AND Z. N. OVCHINIKOVA, *J. Gen. Chem., USSR*, 31 (1961) 1324.

574 YA. M. VARSHAVSKII AND A. I. SHATENSHTEIN, *Dokl. Akad. Nauk SSSR*, 95 (1954) 297.

575 YA. M. VARSHARVSKII, M. G. LOZHKINA AND A. I. SHATENSHTEIN, *Z. Fiz. Khim.*, 31 (1957) 1377.

576 A. I. SHATENSHTEIN, A. V. VEDENEEV AND P. P. ALIKHANOV, *J. Gen. Chem. USSR*, 28 (1958) 2666.

577 YA. M. VARSHARVSKII, V. R. KALINACHENKO AND A. I. SHATENSHTEIN, (a) *Z. Fiz. Khim.*, 30 (1956) 2093; (b) *Z. Fiz. Khim.*, 30 (1956) 2098.

578 A. KOROLEV, A. I. SHATENSHTEIN, E. YURIGINA, V. KALINACHENKO AND P. P. ALIKHANOV, *J. Gen. Chem. USSR*, 26 (1956) 1869.

579 E. N. YURIGINA, P. ALIKHANOV, E. A. IZRAILEVICH, P. N. MANOCHKINA AND A. I. SHATEN-SHTEIN, *J. Phys. Chem. USSR*, 34 (1960) 277.

580 A. I. SHATENSHTEIN, V. R. KALINACHENKO, E. N. YURYGINA AND V. M. BASMANOVA, *Zh. Obshch. Khim.*, 29 (1959) 849.

581 A. I. SHATENSHTEIN AND P. P. ALIKHANOV, *J. Gen. Chem., USSR*, 30 (1960) 1007.

582 A. I. SHATENSHTEIN AND E. A. IZRAILEVICH, *Dokl. Akad. Nauk SSSR*, 94 (1954) 923; *Z. Fiz. Khim.*, 32 (1958) 2711.
N. M. DYKHNO AND A. I. SHATENSHTEIN, *Zh. Fiz. Khim.*, 28 (1954) 14.

583 A. I. SHATENSHTEIN AND E. A. IZRAILEVICH, *Zh. Fiz. Khim.*, 28 (1954) 3.

584 A. I. SHATENSHTEIN, A. P. TALANOV AND YU. I. RANNEVA, *Zh. Obshch. Kim.*, 30 (1960) 583.

585 A. I. SHATENSHTEIN AND A. V. VEDENEEV, *Zh. Obshch. Khim.*, (1958) 2644.

586 A. I. SHATENSHTEIN AND E. A. IZRAILEVICH, *J. Gen. Chem. USSR*, (1962) 1912.

587 G. G. ISAEVA, E. A. YAKOVLEVA AND A. I. SHATENSHTEIN, *J. Gen. Chem. USSR*, (1968) 1399.

588 G. E. HALL, R. PICCOLINI AND J. D. ROBERTS, *J. Am. Chem. Soc.*, 77 (1955) 4540.

589 A. I. SHATENSHTEIN, *Isotopic Exchange and the Replacement of Hydrogen in Organic Compounds*, Consultants Bureau, New York, 1962, p. 105.

590 A. I. SHATENSHTEIN, *Tetrahedron*, 18 (1962) 95; *Advances in Physical Organic Chemistry*, Vol. 1, Academic Press, 1963, p. 156.

591 YU. I. SHAPIRO, L. I. BELENKII, I. A. ROMANSKII, F. M. STOYANOVICH, YA. L. GOLDFARB AND A. I. SHATENSHTEIN, *J. Gen. Chem., USSR*, (1968) 1938.

591a D. N. KURSANOV, V. N. SETKINA, N. K. BARANETSKAYA, E. I. FEDIN, K. N. ANISIMOV AND V. M. URINYUK, *Proc. Acad. Sci. USSR*, (1968) 1118.

592 A. STREITWEISER AND R. G. LAWLER, *J. Am. Chem. Soc.*, 85 (1963) 2852; 87 (1965) 5388.
A. STREITWEISER, R. G. LAWLER AND C. PERRIN, *J. Am. Chem. Soc.*, 87 (1965) 5383.

593 A. STREITWEISER AND R. A. CALDWELL, *J. Am. Chem. Soc.*, 87 (1965) 5394.

594 A. Streitweiser, R. A. Caldwell, R. G. Lawler and G. R. Ziegler, J. Am. Chem. Soc. 87 (1965) 5399.

595 A. Streitweiser and F. Mares, J. Am. Chem. Soc., 90 (1968) 644.

596 A. Streitweiser and J. S. Humphrey, J. Am. Chem. Soc., 89 (1967) 3769.

597 A. Streitweiser, J. A. Hudson and F. Mares, J. Am. Chem. Soc., 90 (1968) 648.

598 I. F. Tupitsyn and N. K. Semenova, Chem. Abstr., 60 (1964) 6721.

599 Y. Kawazoe, M. Ohnishi and Y. Yoshioka, Chem. Pharm. Bull. (Tokyo), 12 (1964) 1384.

600 J. A. Zoltewicz and C. L. Smith, J. Am. Chem. Soc., 89 (1967) 3358.

601 J. A. Zoltewicz and J. D. Meyer, Tetrahedron Letters (1968) 421.

602 R. E. Dessy and R. M. Salinger, J. Org. Chem., 26 (1961) 3519.

603 L. V. Guild, C. A. Hollingworth, D. H. McDaniel, S. K. Podder and J. H. Wotz, J. Org. Chem., 27 (1962) 762.

604 M. S. Kharasch and L. Chalkley, J. Am. Chem. Soc., 46 (1924) 1211.
   M. S. Kharasch and M. N. Grafflin, J. Am. Chem. Soc., 47 (1925) 1948.
   M. S. Kharasch and R. Marker, J. Am. Chem. Soc., 48 (1926) 3130.
   M. S. Kharasch and A. L. Flenner, J. Am. Chem. Soc., 54 (1932) 674.
   M. S. Kharasch, H. Pines and J. M. Levine, J. Org. Chem., 3 (1938–9) 347.
   M. S. Kharasch, R. R. Legault and W. R. Sprowls, J. Org. Chem., 3 (1938–9) 409.

605 A. H. Corwin and M. A. Naylor, J. Am. Chem. Soc., 69 (1947) 1004.

606 F. Kaufman and A. H. Corwin, J. Am. Chem. Soc., 77 (1955) 6280.

607 E. A. Braude, J. Chem. Soc., (1948) 1971.
   E. A. Braude and E. S. Stern, J. Chem. Soc., (1948) 1976.

608 S. Winstein and T. G. Traylor, J. Am. Chem. Soc., 77 (1955) 3747.

609 R. E. Dessy, G. F. Reynolds and J. Y. Kim, J. Am. Chem. Soc., 81 (1959) 2683.

610 F. R. Jenson and B. Rickborn, Electrophilic Substitution in Organomercurials, McGraw Hill, New York, 1968.

611 R. E. Dessy and J-Y. Kim, J. Am. Chem. Soc., 82 (1960) 686.

612 Y. Yukawa and Y. Tsuno, Bull. Chem. Soc. Japan, 32 (1959) 971.

613 R. E. Dessy and J-Y. Kim, J. Am. Chem. Soc., 83 (1961) 1167.

613a V. S. Petrosyan and O. A. Reutov, Proc. Acad. Sci., USSR, 180 (1968) 514.

614 A. N. Nesmeyanov, A. E. Borisov and I. S. Saveleva, Proc. Acad. Sci., USSR, 155 (1964) 280.

615 A. N. Nesmeyanov, E. G. Perevalova, S. P. Gubin and A. G. Kozlovskii, J. Organometal Chem., 11 (1968) 577.

616 F. Nerdel and S. Makover, Naturwissenschaften, 45 (1958) 490.

617 I. P. Beletskaya, A. E. Myshkin and O. A. Reutov, Bull. Acad. Sci. USSR, (1965) 226.

618 I. P. Beletskaya, A. E. Myshkin and O. A. Reutov, Bull. Acad. Sci. USSR, (1967) 232, 239.

619 M. M. Kreevoy and R. L. Hansen, J. Phys. Chem., 65 (1961) 1055.

620 F. R. Jensen and D. Heyman, J. Am. Chem. Soc., 88 (1966) 3438.

621 B. F. Hegarty, W. Kitching and P. R. Wells, J. Am. Chem. Soc., 89 (1967) 4816.

622 I. P. Beletskaya, L. A. Federov and O. A. Reutov, Proc. Acad. Sci. USSR, 163 (1965) 1381.

622a I. P. Beletskaya, A. L. Kurts and O. A. Reutov, J. Org. Chem. USSR, 3 (1967) 1884.

623 R. D. Brown, A. S. Buchanan and A. A. Humffray, Australian J. Chem., 18 (1965) 1507.

624 R. D. Brown, A. S. Buchanan and A. A. Humffray, Australian J. Chem., 18 (1965) 1513.

625 H. G. Kuivila and K. V. Nahebedian, Chem. Ind., (London), (1959) 1120; J. Am. Chem. Soc., 83 (1961) 2159, 2164, 2167.

626 V. Gold, Trans. Faraday Soc., 56 (1960) 255.

627 H. G. Kuivila, J. F. Reuwer and J. A. Mangravite, Can. J. Chem., 41 (1963) 3081.

628 H. G. Kuivila, J. F. Reuwer and J. A. Mangravite, J. Am. Chem. Soc., 86 (1964) 2666.

629 A. D. Ainley and F. Challenger, J. Chem. Soc., (1930) 2171.

630 R. D. Brown, A. S. Buchanan and A. A. Humffray, Australian J. Chem., 18 (1965) 1521.

631 A. Klages and G. Lickroth, Chem. Ber., 32 (1899) 1549.

632 A. Schenkel, Helv. Chim. Acta, 29 (1946) 436.

633 H. Schenkel and M. Schenkel-Rudin, Helv. Chim. Acta, 31 (1948) 514.

634 W. M. Schubert, *J. Am. Chem. Soc.*, 71 (1949) 2639.
635 W. M. Schubert, J. Donohue and J. D. Gardner, *J. Am. Chem. Soc.*, 76 (1954) 9.
636 F. M. Beringer and S. Sands, *J. Am. Chem. Soc.*, 75 (1953) 3319.
637 C. A. A. Bothner-By and J. Bigeleisen, *J. Chem. Phys.*, 19 (1951) 755.
638 W. H. Stevens, J. M. Pepper and M. Lonnsbury, *J. Chem. Phys.*, 20 (1952) 192.
639 B. R. Brown, D. L. Hammick and J. B. Scholefield, *J. Chem. Soc.*, (1950) 778.
640 B. R. Brown, W. W. Elliott and D. W. Hammick, *J. Chem. Soc.*, (1951) 1384.
641 W. M. Schubert and J. D. Gardner, *J. Am. Chem. Soc.*, 75 (1953) 1401.
642 W. M. Schubert, R. E. Zahler and J. Robins, *J. Am. Chem. Soc.*, 77 (1955) 2293.
643 R. W. Hay and M. J. Taylor, *Chem. Commun.*, (1966) 525.
643a K. R. Lynn and A. N. Bourns, *Chem. Ind. (London)*, (1963) 782.
643b A. V. Willi, *Zeit. Naturforsch.*, 132 (1958) 997.
644 A. V. Willi and J. F. Stocker, *Helv. Chim. Acta*, 37 (1954) 1113.
645 A. V. Willi, *Helv. Chim. Acta*, 40 (1957) 1053; *Trans. Faraday Soc.*, 55 (1959) 433.
646 W. H. Stevens, J. M. Pepper and M. Lonnsbury, *Can. J. Chem.*, 30 (1952) 529.
646a G. E. Dunn, P. Leggate and I. E. Scheffler, *Can. J. Chem.*, 43 (1965) 3080.
646b J. M. Los, R. F. Rekker and C. H. T. Tonsbeek, *Rec. Trav. Chim.*, 86 (1967) 622.
647 A. V. Willi, C. M. Won and P. Vilk, *J. Phys. Chem.*, 72 (1968) 3142.
648 E. Louise, *Ann. Chim. Phys.*, 6 (1885) 206.
649 R. T. Arnold and E. Rondestvedt, *J. Am. Chem. Soc.*, 68 (1946) 2176.
650 W. M. Schubert and H. K. Latourette, *J. Am. Chem. Soc.*, 74 (1952) 1829.
651 W. M. Schubert and R. E. Zahler, *J. Am. Chem. Soc.*, 76 (1954) 1.
652 W. M. Schubert and H. Burkett, *J. Am. Chem. Soc.*, 76 (1956) 64.
653 W. M. Schubert and P. C. Myhre, *J. Am. Chem. Soc.*, 80 (1958) 1755.
654 H. Burkett, W. M. Schubert, F. Schultz, R. D. Murphy and R. Talbot, *J. Am. Chem. Soc.*, 81 (1959) 3923.
655 F. S. Kipping and N. W. Cusa, *J. Chem. Soc.*, (1935) 1088.
    H. Gilman and F. J. Marshall, *J. Am. Chem. Soc.*, 71 (1949) 2066.
656 C. Eaborn, *J. Chem. Soc.*, (1953) 3148.
657 J. E. Baines and C. Eaborn, *J. Chem. Soc.*, (1956) 1436.
658 R. A. Benkeser and H. Krysiak, *J. Am. Chem. Soc.*, 76 (1954) 6353.
659 R. W. Bott, C. Eaborn and P. M. Greasley, *J. Chem. Soc.*, (1964) 4804.
660 C. Eaborn and K. C. Pande, *J. Chem. Soc.*, (1960) 1566.
661 R. W. Bott, C. Eaborn, K. C. Pande and T. W. Swaddle, *J. Chem. Soc.*, (1962) 1217.
662 R. W. Bott, C. Eaborn and D. R. M. Walton, *J. Organometallic Chem.*, 2 (1964) 154.
663 D. R. M. Walton, *J. Organometallic Chem.*, 3 (1965) 438.
664 R. A. Benkeser, R. A. Hickner and D. I. Hoke, *J. Am. Chem. Soc.*, 80 (1958) 2279.
    R. A. Benkeser and F. S. Clark, *J. Am. Chem. Soc.*, 82 (1960) 4881.
665 R. A. Benkeser, W. Schroeder and O. H. Thomas, *J. Am. Chem. Soc.*, 80 (1958) 2283.
666 C. Eaborn and K. C. Pande, *J. Chem. Soc.*, (1961) 297.
666a C. Eaborn and K. C. Pande, *J. Chem. Soc.*, (1961) 5082.
667 R. W. Bott, C. Eaborn and P. M. Jackson, *J. Organometallic Chem.*, 7 (1967) 79.
    R. C. Moore, *Ph. D. Thesis*, University of Leicester, 1961.
668 R. W. Bott and C. Eaborn, in *Organometallic Compounds of the Group IV Elements*, A. G. MacDiarmid (Ed.), Marcel Dekker, New York, 1968, p. 414.
669 C. Eaborn, *J. Chem. Soc.*, (1956) 4858.
670 C. Eaborn, Z. Lasocki and D. E. Webster, *J. Chem. Soc.*, (1959) 3034.
671 C. Eaborn and D. R. M. Walton, *J. Organometallic Chem.*, 3 (1965) 169.
672 A. R. Bassindale, C. Eaborn and D. R. M. Walton, *J. Chem. Soc.*, (B) (1969) 12.
673 C. Eaborn and P. M. Jackson, *J. Chem. Soc.*, (B) (1969) 21.
674 C. Eaborn and J. A. Sperry, *J. Chem. Soc.*, (1961) 4921.
675 C. Eaborn and R. C. Moore, *J. Chem. Soc.*, (1959) 3640.
676 K. C. Pande, *Ph. D. Thesis*, University of Leicester, 1960.
677 J. A. Sperry, *Ph. D. Thesis*, University of Leicester, 1960.
678 R. W. Bott, C. Eaborn and K. Leyshon, *J. Chem. Soc.*, (1964) 1971.
679 F. B. Deans and C. Eaborn, *J. Chem. Soc.*, (1959) 2299.

680  F. B. Deans and C. Eaborn, *J. Chem. Soc.*, (1959) 2303.

681  C. Eaborn, D. R. M. Walton and D. J. Young, *J. Chem. Soc.*, (B) (1969) 15.

682  R. W. Bott, B. F. Dowden and C. Eaborn, *J. Chem. Soc.*, (1965) 6306.

683  C. Eaborn, J. A. Treverton and D. R. M. Walton, *J. Organometallic Chem.*, 9 (1967) 259.

684  F. B. Deans, C. Eaborn and D. E. Webster, *J. Chem. Soc.*, (1959) 3031.

685  C. Eaborn and J. F. R. Jaggard, *J. Chem. Soc.* (B) (1969) 892.

686  R. A. Benkeser, R. A. Hickner, D. I. Hoke and O. H. Thomas, *J. Am. Chem. Soc.*, 80 (1958) 5289.

687  J. Nasielski and M. Planchon, *Bull. Soc. Chim. Belges*, 69 (1960) 123.

688  G. Fritz and D. Kummer, *Z. Anorg. Allgem. Chem.*, 309 (1961) 105.

689  G. Marr and D. E. Webster, *J. Chem. Soc.*, (B) (1968) 202.

690  G. Marr and D. E. Webster, *J. Organometallic Chem.*, 2 (1964) 99.

691  C. Eaborn and J. A. Waters, *J. Chem. Soc.*, (1961) 542.

692  C. Eaborn, A. R. Thompson and D. R. M. Walton, *J. Chem. Soc.* (B) (1969) 859.

693  O. Buchman, M. Grosjean and J. Nasielski, *Bull. Soc. Chim. Belges*, 72 (1963) 286; *Helv. Chim. Acta.*, 47 (1964) 1695.

694  A. Delhaye, J. Nasielski and M. Planchon, *Bull. Soc. Chim. Belges*, 69 (1960) 134.

695  C. Eaborn, H. L. Hornfeld and D. R. M. Walton, *J. Chem. Soc.*, (B) (1967) 1036.

696  H. Cerfontain and J. M. Arends, *Rec. Trav. Chim.*, 85 (1966) 358.

697  Ya. K. Syrkin, V. I. Yakerson and S. E. Shnol, *J. Gen. Chem. USSR*, 29 (1959) 189.

698  J. Crafts, *Chem. Ber.*, 34 (1901) 1350; *Bull. Soc. Chim. France*, 1 (1907) 917.

699  R. Lantz, *Bull. Soc. Chim. France*, 2 (1935) 2092.

700  R. Lantz, *Bull. Soc. Chim. France*, 12 (1945) 1004.

701  G. Baddeley, G. Holt and J. Kenner, *Nature*, 154 (1944) 361.

702  I. R. Bellabono, *Chem. Abstr.*, 55 (1961) 27225.

703  M. Kilpatrick and M. W. Meyer, *J. Phys. Chem.*, 65 (1961) 1312.

704  Y. Muramoto, *Chem. Abstr.*, 54 (1960) 16416.

705  A. A. Spryskov and N. A. Ovsyankina, *J. Gen. Chem.*, 20 (1950) 1083; 21 (1951) 1649; 24 (1954) 1777; *Sb. Statei Obshch. Khim.*, 2 (1953) 882.

706  Y. I. Leitman and M. S. Pevsner, *J. Appl. Chem. USSR*, 32 (1956) 2830.

707  H. Feilchenfeld, *Ind. Eng. Chem.*, 48 (1956) 1935.

708  A. C. M. Wanders and H. Cerfontain, *Rec. Trav. Chem.*, 86 (1967) 1199.

709  A. J. Shoesmith, A. C. Hetherington and R. H. Slater, *J. Chem. Soc.*, 125 (1924) 1312, 2278.

710  B. H. Nicolet and J. R. Sampey, *J. Am. Chem. Soc.*, 49 (1927) 1796.
     B. H. Nicolet and W. L. Ray, *J. Am. Chem. Soc.*, 49 (1927) 1801.

711  D. Gold and M. Whittaker, *J. Chem. Soc.*, (1951) 1184.

712  H. S. Choguill and J. H. Ridd, *J. Chem. Soc.*, (1961) 822.

713a H. B. Charman and C. K. Ingold, *J. Chem. Soc.*, (1959) 2523.

713b H. B. Charman, E. D. Hughes and C. K. Ingold, *J. Chem. Soc.*, (1959) 2530.

714c H. B. Charman, E. D. Hughes, C. K. Ingold and F. G. Thorpe, *J. Chem. Soc.*, (1961) 1121.
     E. D. Hughes, C. K. Ingold, F. G. Thorpe and H. C. Volger, *J. Chem. Soc.*, (1961) 1133.

714d G. F. Wright, *Can. J. Chem.*, 30 (1952) 268.

715a S. Winstein, T. G. Traylor and C. S. Garner, *J. Am. Chem. Soc.*, 77 (1955) 3741.

715b S. Winstein and T. G. Traylor, *J. Am. Chem. Soc.*, 78 (1956) 2597.

716  F. R. Jensen, L. H. Gaile, L. D. Whipple and D. K. Wedergaertner, *J. Am. Chem. Soc.*, 81 (1959) 1262; 82 (1960) 2469.

717  O. A. Reutov, T. P. Karpov, E. V. Uglova and V. A. Malyanov, *Proc. Acad. Sci. USSR*, 134 (1960) 1017; *Tetrahedron Letters*, 19, (1960) 6; (these papers appear to be identical reports).

718  R. E. Dessy and Y. K. Lee, *J. Am. Chem. Soc.*, 82 (1960) 689.

719  R. E. Dessy, Y. K. Lee and J.-Y. Kim, *J. Am. Chem. Soc.*, 83 (1961) 1163.

720  A. N. Nesmeyanov and O. A. Reutov, *Proc. Acad. Sci. USSR*, 144 (1962) 405; *Tetrahedron*, 20 (1964) 2803.

721 K. Broderson and V. Schlenker, *Chem. Ber.*, 94 (1961) 3304.

722 F. R. Jensen and J. Miller, *J. Am. Chem. Soc.*, 86 (1964) 4735.

723 I. P. Beletskaya, G. A. Artamkina and O. A. Reutov, *Proc. Acad. Sci. USSR*, 166 (1966) 242.

724 T. A. Smolina, M. Chan and O. A. Reutov, *Bull. Acad. Sci. USSR*, (1966) 390.

724a J. Keller, *Ph. D. Thesis*, University of California, Los Angeles, 1948.

725 I. P. Beletskaya, A. V. Ermanson and O. A. Reutov, *Bull. Acad. Sci. USSR*, (1965) 218.

726 H. G. Kuivila and T. C. Muller, *J. Am. Chem. Soc.*, 84 (1962) 377.

727 H. G. Kuivila, *J. Am. Chem. Soc.*, 76 (1954) 870; 77 (1955) 4014.
   H. G. Kuivila and R. A. Wiles, *J. Am. Chem. Soc.*, 77 (1955) 4830.
   H. G. Kuivila and A. G. Armour, *J. Am. Chem. Soc.*, 79 (1957) 5659.

728 H. G. Kuivila and E. K. Easterbrook, *J. Am. Chem. Soc.*, 73 (1951) 4629;
   H. G. Kuivila and A. R. Hendrickson, *J. Am. Chem. Soc.*, 74 (1952) 5068;
   H. G. Kuivila and E. J. Soboczenski, *J. Am. Chem. Soc.*, 76 (1954) 2675;
   H. G. Kuivila and L. E. Benjamin, *J. Am. Chem. Soc.*, 77 (1955) 4834.

729 Ref. 1, pp. 256–7.

730 C. A. Holder and R. Taylor, unpublished work.

731 H. G. Kuivila and R. M. Williams, *J. Am. Chem. Soc.*, 76 (1954) 2679.

732 R. D. Brown, A. S. Buchanan and A. A. Humffray, *Australian J. Chem.*, 18 (1965) 1527.

733 E. Grovenstein and U. V. Henderson, *J. Am. Chem. Soc.*, 78 (1956) 569;
   E. Grovenstein and G. A. Ropp, *J. Am. Chem. Soc.*, 78 (1956) 2560.

734 J. A. Donnelly and F. Policky, *Chem. Ind. (London)*, (1965) 1338.

735 P. B. D. de la Mare, J. T. Harvey, M. Hassan and S. Varma, *J. Chem. Soc.*, (1958) 2756.

736 R. A. Benkeser, R. A. Hickner and D. I. Hoke, *J. Am. Chem. Soc.*, 80 (1958) 5294.

737 C. Eaborn and D. E. Webster, personal communication from C. Eaborn.

738 R. A. Benkeser, T. V. Listen and G. Stanton, *Tetrahedron Letters*, (15) (1960) 1.

739 R. Taylor and G. G. Smith, *Tetrahedron*, 19 (1963) 937; ref. 1, p. 266.

740 Ref. 668, pp. 426–8.
   R. C. Moore, *Ph. D. Thesis*, University of Leicester, 1961.

741 Ref. 668 pp. 421–3.
   K. Leyshon, *Ph. D. Thesis*, University of Leicester, 1962.

742 C. Eaborn and D. E. Webster, *J. Chem. Soc.*, (1960) 179.

743 L. M. Stock and A. R. Spector, *J. Org. Chem.*, 28 (1963) 3272.

744 C. Eaborn and D. E. Webster, *J. Chem. Soc.*, (1957) 4449.

745a Ref. 668, p. 419.

745b R. C. Moore, *Ph. D. Thesis*, University of Leicester, 1962.

746 C. Eaborn and O. W. Steward, *Proc. Chem. Soc.*, (1963) 59; *J. Chem. Soc.*, (1965) 521.

747 L. H. Sommer, K. W. Michael and W. D. Korte, *J. Am. Chem. Soc.*, 89 (1967) 868.

748 Ref. 668, p. 418.

749 R. W. Bott, C. Eaborn and J. A. Waters, *J. Chem. Soc.*, (1963) 681.

750 O. Buchman, M. Grosjean and J. Nasielski, *Helv. Chim. Acta.*, 47 (1964) 1679, 1688, 2037.

751 A. Delhaye, J. Nasielski and M. Planchon, *Bull. Soc. Chim. Belges*, 69 (1960) 134.

752 L. G. Cannell, *J. Am. Chem. Soc.*, 79 (1957) 2927, 2932.

753 In contrast to these studies, bromination of 1,5-dimethylnaphthalene in 90 % aqueous acetic acid at 25°C did not give a linear correlation of $k_{obs}$. *versus* $K/(K+[\text{Br}^-])$, thereby indicating a change in mechanism, so that rate-determining incursion of C–H bond-breaking through steric hindrance was postulated (E. Berliner, J. B. Kim and M. Link, *J. Org. Chem.*, 33 (1968) 1160).

   The relative rates of bromination from the above and additional studies yielded the following relative rates with partial rate factors for the indicated position in parentheses. These rates were mostly obtained by extrapolation of observed rates to those expected in 50 % aqueous acetic acid and therefore probably contain small errors: benzene, 1.0 (1.0); biphenyl, $1.54 \times 10^3 (4-, 4.34 \times 10^3)$; naphthalene, $1.24 \times 10^5 (1-, 1.84 \times 10^5; 2-, 1.86 \times 10^3)$; phenanthrene, $7.43 \times 10^5 (9-, 2.23 \times 10^6)$; fluoranthrene, $2.30 \times 10^6 (3-, 6.90 \times 10^6)$; chrysene, $1.25 \times 10^7 (6-, 3.75 \times 10^7)$; pyrene, $2.84 \times 10^{10} (1-, 4.26 \times 10^{10})$; 1,2-benzanthracene, $2.44 \times 10^{10} (7-, 1.46 \times 10^{11})$; anthracene, $7.87 \times 10^{11} (9-, 2.36 \times 10^{12})$; 2,6-dimethylnaphthalene, $9.0 \times 10^7$; acenaphthene, $5.49 \times 10^{10}(5-, 1.65 \times 10^{11})$; fluorene, $3.53 \times 10^6(2-, 1.03 \times 10^7)$ (L. ALTSCHULER AND E. BERLINER, *J. Am. Chem. Soc.*, 88 (1966) 5837).

754 Rates $(10^7k)$ of dedeuteration by 0.6 $M$ $KNH_2$ in liquid ammonia at 0° were also determined by Hall *et al.* as follows: Benzene, 1000; toluene, 410 (*ortho*), 360 (*meta*), 440 (*para*); *t*-butylbenzene, 32 (*ortho*), 210 (*para*). The toluene values are in fair agreement with those in Table 177, and the low reactivity at the *ortho*-position of *t*-butylbenzene was considered to be of steric origin. (G. E. HALL, E. M. LIBBY AND E. L. JAMES, *J. Org. Chem.*, 28 (1963) 311).

# Chapter 2

# Nucleophilic Aromatic Substitution: the $S_N2$ Mechanism

S. D. ROSS

## 1. Introduction

Nucleophilic aromatic substitution has been the subject of frequent and extensive reviews[1-10]. The data on reaction rates, reaction products, substituent effects, salt effects, etc. are all readily available and need not be reassembled here. In spite of this abundance of both data and discussion, some questions of mechanism remain incompletely resolved.

Our knowledge of mechanism is most sophisticated for the $S_N2$ nucleophilic displacements, where the reactive agent is an electron donor, where the leaving group is a halogen ion or some other group capable of some stability as an anion, and where the activation is due to electron-withdrawing substituents, suitably positioned in the aromatic substrate. The most widely accepted mechanism for this category of reactions is one involving an intermediate complex, formed by addition of the nucleophile to the carbon atom undergoing substitution and converting that carbon atom to one with its substituents arranged in a tetrahedral configuration. The best evidence for this interpretation comes from studies of reaction rates, and, in particular, from the observation of the effects of basic catalysts on the rates of reactions with selected substrates and selected nucleophiles.

It is regrettable that the evidence afforded by reaction kinetics is rarely, if ever, uniquely consistent with a single mechanism or a single explanation. The results for nucleophilic aromatic substitution reactions are no exception. Legitimate questions can be raised with respect to the extent to which observations made on a particular system permit generalization to other systems. Even for the specific systems studied points of detail arise, and choices have to be made where alternatives are possible. Every such choice introduces an element of uncertainty and imposes a limitation on the extent to which the reaction mechanism is, in fact, known.

A treatise on kinetics is a logical and fitting medium in which to analyze and discuss just such limitations and uncertainties of mechanism. The present chapter will attempt such a treatment for the $S_N2$ mechanism in nucleophilic aromatic substitution. An effort will be made to pinpoint every assumption and highlight every instance where alternate choices are possible. The end result hoped for is a clearer delineation of the known and the probable from the uncertain and the unknown.

## 2. The intermediate complex mechanism

### 2.1 HISTORICAL

Early work on the mechanism of nucleophilic aromatic substitution reactions was marked by a strong predisposition in favor of a two-stage mechanism, with a discrete intermediate intervening between reactants and products. The major influence promoting this bias was the demonstration by Jackson and Boos[11] and by Meisenheimer[12] that stable, isolable $\sigma$-complexes are formed by the interaction of alkali alkoxides with picryl chloride[11] or picryl ethers[12]. These observations made it both logical and necessary to consider the possibility that similar intermediates are involved in other nucleophilic aromatic substitutions.

These early results have since been confirmed and extended by a vast and still growing body of research. All of the contemporary spectroscopic techniques (IR, UV, visible, NMR, ESR) have been brought to bear on the problem, and further confirmation has come from cryoscopic and conductometric studies. The early confusion that resulted from the coexistence of both $\sigma$-complexes and $\pi$-complexes (donor–acceptor or non-covalently-bonded complexes) has been clarified. This research has been extensively reviewed[10, 13–15] and will not be detailed here.

An equally important influence favoring the intermediate complex mechanism was the parallel study of electrophilic aromatic substitution. Progress in this field was more rapid, largely because Melander[16, 17] had both brilliantly formulated the problem and devised incisive experiments to test his hypotheses. This work generated wide acceptace for the two-stage mechanism in electrophilic aromatic substitution. It was, therefore, natural to apply similar reasoning to nucleophilic substitutions, particularly since Wheland's calculations[18] had indicated that aromatic substitutions by electrophilic, nucleophilic and radical reagents can all be treated from a unified point of view.

Finally, the most frequently observed order of halogen mobility in nucleophilic aromatic substitution ($F \gg Cl \approx Br > I$) was the reverse of that for $S_N2$ displacements at saturated carbon ($I > Br > Cl > F$)[19]. This suggested that displacements at saturated carbon and in aromatic systems involved different mechanisms and created a presumption in favor of a two-stage mechanism for the latter reactions. The order of halogen mobility proved to be an uncertain and misleading criterion of mechanism, and, in fact, reactions showing the order, $I > Br > Cl > F$, were found[20, 21]. Nevertheless, this consideration had a significant influence on the direction of the research in this field.

In early rate studies the nucleophile was frequently an anion, and the intermediate complex mechanism was formulated as

$$Y^- + \underset{A}{\overset{X}{\bigcirc}} \quad \underset{N}{} \underset{k_R}{\overset{k_F}{\rightleftharpoons}} \quad \underset{C}{\overset{Y\ X}{\bigcirc}} \quad \overset{k_p}{\longrightarrow} \quad \underset{D}{\overset{Y}{\bigcirc}} \quad \underset{E}{+ X^-}$$

Applying the steady-state approximation to the concentration of the intermediate, C, one obtains the rate equation

$$\frac{d[D]}{dt} = \frac{k_F k_p [A][N]}{k_R + k_p} \tag{1}$$

It was immediately apparent that two limiting situations are possible. With $k_p \gg k_R$, eq. (1) reduces to

$$\frac{d[D]}{dt} = k_F[A][N], \tag{2}$$

and with $k_R \gg k_p$, eq. (1) reduces to

$$\frac{d[D]}{dt} = \frac{k_F k_p}{k_R}[A][N] = K k_p[A][N], \tag{3}$$

where $K$ is the equilibrium constant for formation of the intermediate, C. The experimental rate coefficient would, in any case, refer to a second-order process, first-order in A and first-order in N, but the kinetic measurements afford no way of determining whether the value obtained corresponds to the composite of rate coefficients, $k_F k_p/(k_R + k_p)$, according to eqn. (1) or to $k_F$, the rate coefficient of intermediate formation, according to (2) or to $K k_p$, the product of the equilibrium constant for intermediate formation and the rate coefficient of decomposition of intermediate to product, according to eqn. (3). Consequently, the data permit no distinction between a synchronous $S_N2$ mechanism and an intermediate complex mechanism.

As a result the research emphasis in this field focused on efforts to design experiments in which it might be possible to determine to which one of the foregoing three rate equations the observed second-order rate coefficient actually corresponded. More specifically, the objective was to observe one and the same system first under conditions in which complex decomposition $(k_p)$ was rate-determining and then under conditions in which complex formation $(k_F)$ was rate-determining. A system in which either formation or decomposition was subject to some form of catalysis was thus indicated. In displacements with primary and secondary amines the transformation of reactants to products necessarily involves the transfer of a proton at some stage of the reaction. Such reactions are potential-

ly capable of showing base catalysis, and displacements with amines as nucleophiles, therefore, became a subject of intensive investigation in many laboratories.

## 2.2 BASE CATALYSIS

In discussing base catalysis it will prove convenient to adopt, at the outset, a
distinction first proposed by Bunnett and Garst[22], who noted that the observed
cases of catalysis in nucleophilic aromatic substitution could be broadly divided
into two categories. The classification was in terms of the relative rates of the
catalyzed and uncatalyzed reactions. Since all of the systems could be accommodated empirically by eqn. (4),

$$k_2^* = k' + k''[B] \tag{4}$$

in which $k_2^*$ is the experimental second-order rate coefficient, $k'$ is the rate coefficient for the uncatalyzed reaction, $k''$ is that for the catalyzed reaction and B is
the base, the distinction was made in terms of the ratio, $k''/k'$, which has units of
l.mole$^{-1}$. The first category contained systems showing strong catalysis with
$k''/k'$ frequently greater than 50 l.mole$^{-1}$. The second category included those
reactions the rates of which could still be fitted by eqn. (4) but for which the ratios
$k''/k'$ were significantly lower than 50 l.mole$^{-1}$ and more commonly 5 l.mole$^{-1}$
or lower.

The number of reactions that fit into this second class are manifold. Some
typical examples are the amine-catalyzed reactions of 2,4-dinitrochlorobenzene
with $n$-butylamine in chloroform $(k''/k' = 2.59$ l.mole$^{-1})$[23], with allylamine in
chloroform $(k''/k' = 4.60$ l.mole$^{-1})$[24] and with allylamine in ethanol $(k''/k' = 0.36$
l.mole$^{-1})$[25] and the amine-catalyzed reaction of $p$-nitrofluorobenzene with piperidine in polar solvents $(k''/k' < 3.2$ l.mole$^{-1})$[26]. A typical example of a strongly
catalyzed system is the reaction of 2,4-dinitrophenyl phenyl ether with piperidine
in 60 % dioxan–40 % water[27].

For this reaction, catalyzed by piperidine, $k''/k'$ is equal to 53 l.mole$^{-1}$, and
with hydroxide ion as the base $k''/k'$ is equal to 370 l.mole$^{-1}$.

These two categories of base catalysis may also be differentiated in terms of the
character of the leaving group. In reactions strongly accelerated by base the
leaving group is generally a "poor" one, whereas reactions only mildly catalyzed
by base usually involve a "good" leaving group. The character of a leaving group
may be classified semi-quantitatively in terms of the p$K_A$ of its conjugate acid in
water, with the larger p$K_A$ values corresponding to the poorer leaving groups.
For the present purpose it is convenient to classify all leaving groups whose
conjugate acids have p$K_A$'s significantly greater than 3 as poor leaving groups and
all leaving groups whose conjugate acids have p$K_A$'s appreciably lower than 3 as

good leaving groups. The basis for this somewhat arbitrary separation is the fact that fluorides occupy a pivotal position and can show either strong or weak base catalysis depending on the nature of the nucleophile and the remaining substituents on the substrate.     Thus 2,4-dinitrofluorobenzene shows weak base catalysis in its reaction with aniline and very strong base catalysis in the reaction with N-methylaniline. For the reaction with aniline $k''/k'$ is 2 l.mole$^{-1}$ for the hydroxide ion-catalyzed reaction in 60 % dioxan–40 % water, 5–6 l.mole$^{-1}$ for the acetate ion-catalyzed reaction in methanol and 2.6 l.mole$^{-1}$ for the aniline catalyzed reaction in $t$-butanol[22]. For the reaction with N-methylaniline, $k''/k'$ is 150 l.mole$^{-1}$ for the acetate ion-catalyzed reaction in ethanol and 350 l.mole$^{-1}$ for the hydroxide ion-catalyzed reaction in 60 % dioxan–40 % water[28].

The strongly base-catalyzed reactions have two additional, highly significant, distinguishing characteristics. At a high concentration of the basic catalyst the experimental second-order rate coefficient, $k_2^*$, approaches a limiting value. Plots of $k_2^*$ versus $[B]_0$ are curvilinear, rising rapidly at low $[B]_0$ and flattening out to a nearly constant value of $k_2^*$ at high $[B]_0$. Equally important, there is at least a qualitative relationship between the basic strength of the catalyst and its effectiveness in accelerating the reaction. Thus hydroxide ion is a more effective catalyst than acetate ion which is, in turn, a more effective catalyst than an amine such as aniline.

In contrast, the mildly catalyzed reactions show no signs of approaching a limiting rate with increasing concentration of the base. Plots of $k_2^*$ versus $[B]_0$ are linear and show no signs of curvature at high $[B]_0$. Moreover, there is little or no relationship between the basic strength of the catalyst and its ability to increase the reaction rate. Amines, hydroxide ion, acetate ion and even nitro compounds all appear to be almost equally effective in serving this purpose.

### 3. The mechanism of strongly base-catalyzed reactions

From the standpoint of mechanism the most significant characteristic of strongly base-catalyzed, nucleophilic aromatic substitution reactions is the fact that they approach or attain a limiting rate at high concentrations of the catalyst. The essential experimental observation is that as the concentration of the catalyst is increased, the second-order rate coefficient first increases rapidly and then levels off and becomes almost independent of the catalyst concentration. This behaviour pattern can be interpreted as involving a change in the rate-determining step, which in turn requires that there be an intermediate in the reaction pathway. Such observations, therefore, constitute convincing evidence for an addition intermediate in nucleophilic aromatic substitution.

This characteristic dependence of rate on catalyst concentration was first observed by Bunnett and Randall[28] for the reaction of 2,4-dinitrofluorobenzene with N-methylaniline catalyzed by hydroxide ion. More recently it has been observed, in a more convincing manner, by Kirby and Jencks[29] for the base-catalyzed reactions of secondary amines with p-nitrophenyl phosphate and by Bunnett and Garst[22, 27] and by Bunnett and Bernasconi[30] for the reactions of ethers of 2,4-dinitrophenol with piperidine.

All of these reactions were discussed in terms of the following mechanism,

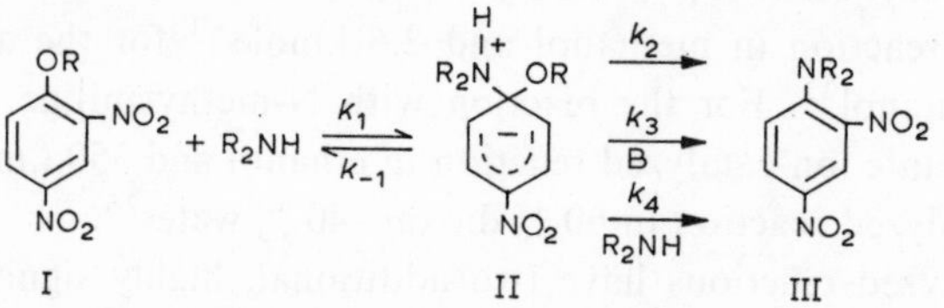

where $R_2NH$ is the nucleophile (e.g. piperidine in the reaction with 2,4-dini-trophenyl phenyl ether) and B is the basic catalyst (e.g. hydroxide ion in the above reaction). Applying the steady-state approximation with respect to the inter-mediate, II, one obtains the rate equation,

$$\frac{d[III]}{dt} = \frac{k_1[I][R_2NH](k_2+k_3[B]+k_4[R_2NH])}{k_{-1}+k_2+k_3[B]+k_4[R_2NH]} \tag{5}$$

For experimental purposes it is customary to define a second-order rate coeffi-cient, $k_2^*$, usually determined in the limit of zero time, as follows,

$$k_2^* = \frac{1}{[I][R_2NH]}\frac{d[III]}{dt} = \frac{k_1(k_2+k_3[B]+k_4[R_2NH])}{k_{-1}+k_2+k_3[B]+k_4[R_2NH]} \tag{6}$$

Equation (6) is then discussed in terms of two limiting conditions. The first is

$$k_2+k_3[B]+k_4[R_2NH] \gg k_{-1} \tag{7}$$

in which case it reduces to

$$k_2^* = k_1 \tag{8}$$

and this defines the limiting rate in the presence of a large concentration of the catalyst B. The second limiting condition is

$$k_{-1} \gg k_2+k_3[B]+k_4[R_2NH] \tag{9}$$

in which case eqn. (6) becomes

$$k_2^* = \frac{k_1}{k_{-1}} \left( k_2 + k_3[B] + k_4[R_2NH] \right) \tag{10}$$

By determining $k_2^*$ at constant $[I]_0$ and $[R_2NH]_0$, and varying $[B]_0$, and plotting $k_2^*$ versus $[B]_0$, one obtains a curve which is linear at low $[B]_0$, with slope equal to $k_1 k_3/k_{-1}$ and intercept equal to $(k_1/k_{-1})(k_2 + k_4[R_2NH]_0)$, and which at high $[B]_0$ flattens out to $k_1$. It is also possible to make measurements at low and constant $[B]_0$, constant $[I]_0$ and varying $[R_2NH]_0$. If catalysis by the nucleophile, $R_2NH$, is significant, a plot of $k_2^*$ versus $[R_2NH]_0$ will have an initial linear portion with slope equal to $k_1 k_4/k_{-1}$ and intercept equal to $(k_1/k_{-1})(k_2 + k_3[B]_0)$.

The available experimental results are completely in accord with this formulation. Both of these limiting conditions have been observed experimentally, and plots of both $k_2^*$ versus $[B]_0$ and $k_2^*$ versus $[R_2NH]_0$ have been shown to have characteristics consistent with this proposed mechanism. These observations thus constitute very convincing evidence for the intermediate complex mechanism in nucleophilic aromatic substitution.

The transition state diagram ($\Delta G^{\ddagger}$ versus reaction co-ordinate) for this mechanism consists of two free energy barriers separated by a minimum corresponding to the intermediate, II. At low base concentration the second barrier, which must be traversed to convert intermediate to product, is the larger one, and this step is rate-limiting. With increasing concentration of base this barrier decreases, until at higher concentrations of base the first barrier becomes the larger one and intermediate formation becomes rate-limiting.

The conversion of the intermediate, II, to product, III, involves the removal of a proton from nitrogen and the breaking of the bond to the leaving group. These two events may occur in a single, concerted, synchronous process, and a variety of transition states, consistent with such a process, are possible. These warrant discussion.

The simplest and most obvious possibility is that depicted by F

F

in which hydrogen bonding by the catalyst to the ammonium hydrogen increases the mobility of the electron pair in the N–H bond and facilitates the changes indicated by the arrows. The major attraction of this transition state is its similarity to the much better established transition state for the E-2 elimination reaction. There is, however, a significant difference in the present case in that the proton is being removed from nitrogen rather than carbon, and proton transfers from nitrogen have been shown to be exceedingly fast and generally diffusion-controlled[31]. The merits[6] and limitations[29] of this transition state have been discussed and will not be repeated here.

Also worthy of consideration is the more highly organized, doubly hydrogen-bonded transition state suggested by Capon and Rees[32]. This has been represented in G as involving $R_2NH$ as the catalyst, but it can be

$$O_2N\!-\!\overset{R_2}{\underset{NO_2}{\fbox{$\;-\;$}}}\overset{N-H\cdots}{\underset{O\cdots H}{\big\langle}}\overset{}{\underset{R}{\big\rangle}}N\!-\!R_2$$

G

depicted equally well with $OH^-$ as the catalyst. This may be looked upon as a special case of bifunctional catalysis with the catalyst supplying both nucleophilic assistance to remove the proton and electrophilic aid in separating the leaving group. The striking efficacy of such catalysis, in suitably organized systems, was first demonstrated by Swain and Brown[33] in their study of the mutarotation of tetramethylglucose in benzene solution. α-Pyridone is a ten thousand-fold weaker base than pyridine and a hundred-fold weaker acid than phenol, yet with α-pyridone in 0.05 $M$ concentration the observed rate of mutarotation was more than fifty times the total rate with both 0.05 $M$ pyridine and 0.05 $M$ phenol. This striking effect was attributed to the fact that α-pyridone contains both a nucleophilic and an electrophilic function suitably positioned in the molecule to supply both forms of assistance simultaneously to the reaction.

Bifunctional catalysis in nucleophilic aromatic substitution was first observed by Bitter and Zollinger[34], who studied the reaction of cyanuric chloride with aniline in benzene. This reaction was not accelerated by phenols or γ-pyridone but was catalyzed by triethylamine and pyridine and by bifunctional catalysts such as α-pyridone and carboxylic acids. The carboxylic acids did not function as purely electrophilic reagents, since there was no relationship between catalytic efficiency and acid strength, acetic acid being more effective than chloracetic acid, which in turn was a more efficient catalyst than trichloroacetic acid. For catalysis by the carboxylic acids Bitter and Zollinger proposed the transition state depicted by H.

$$Cl\!-\!\overset{N}{\underset{N\cdots N}{\fbox{$\;-\;$}}}\overset{Cl\cdots H-O}{\underset{\underset{Ar}{N-H\cdots O}}{\big\langle}}C\!-\!R$$

H

Bifunctional catalysis has also been observed by Pietra and Vitali[35] for a more typical nucleophilic aromatic substitution reaction, that of 2,4-dinitrofluorobenzene and piperidine in benzene. For this reaction triethylamine does not have an accelerating effect, while piperidine and α-pyridone are both efficient catalysts, with the latter being more than five times as effective as the former. For catalysis by α-pyridone transition state I was suggested.

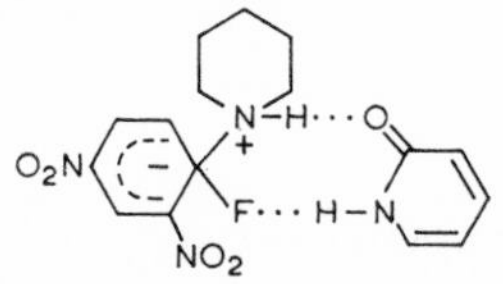

I

These observations lend some support to the suggestion of Capon and Rees[32]. It is, however, questionable that an amine or hydroxide ion would supply significant electrophilic assistance for separation of the leaving group, and acetate ion would be effective only in its enolic form, J

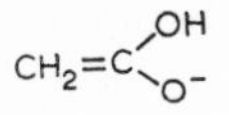

J

Moreover, in the transition state, G, the N–H ... O and the N–H ... N hydrogen bonds would almost certainly be linear, and G would have geometrical characteristics which might better be described by a four-membered ring than a six-membered ring. This is a possible disadvantage, but it is conceivable that it might be overcome by incorporating a solvent molecule, in this case water, as a bridge in the transition state as shown in K

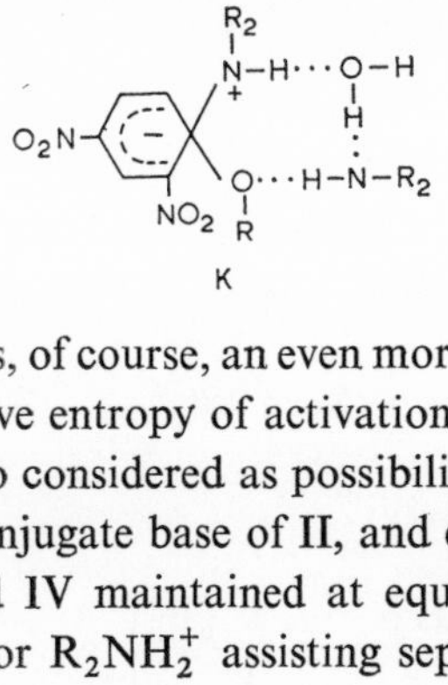

K

This is, of course, an even more highly organized transition state and entails a more negative entropy of activation, but this need not necessarily be prohibitive.

Also considered as possibilities have been transition states involving IV, below, the conjugate base of II, and either $R_2NH_2^+$ or $BH^+$ [29, 30]. This mechanism, with II and IV maintained at equilibrium by rapid, reversible proton transfers and $BH^+$ or $R_2NH_2^+$ assisting separation of the leaving group from intermediate IV in the rate-limiting step, may be formulated as

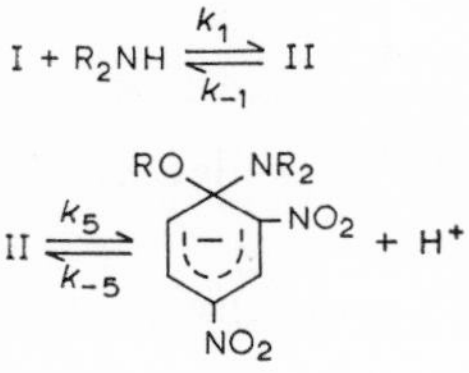

$$I + R_2NH \underset{k_{-1}}{\overset{k_1}{\rightleftarrows}} II$$

IV

$$IV \begin{cases} \xrightarrow{k_6} III + OR^- \\ \xrightarrow[R_2NH_2^+]{k_7} III + ROH + R_2NH \\ \xrightarrow[BH^+]{k_8} III + ROH + B \end{cases}$$

Applying the steady-state approximation to the sum, ([II] + [IV]), one obtains

$$\frac{d([II]+[IV])}{dt} = 0 = k_1[I][R_2NH] - k_{-1}[II] - k_6[IV] - k_7[IV][R_2NH_2^+]$$
$$- k_8[IV][BH^+] \tag{11}$$

which with introduction of the equilibrium relationship,

$$K_1 = \frac{k_5}{k_{-5}} = \frac{[IV][H^+]}{[II]} \tag{12}$$

leads to the rate equation

$$\frac{d[III]}{dt} = \frac{k_1[I][R_2NH](k_6 + k_7[R_2NH^+] + k_8[BH^+])}{(k_{-1}[H^+]/K_1) + k_6 + k_7[R_2NH_2^+] + k_8[BH^+]} \tag{13}$$

With the aid of two additional equilibrium relationships

$$K_2 = \frac{[R_2NH][H^+]}{[R_2NH_2^+]} \tag{14}$$

$$K_3 = \frac{[B][H^+]}{[BH^+]} \tag{15}$$

eqn. (13) can be transformed into

$$\frac{d[III]}{dt} = \frac{k_1[I][R_2NH]\left(\dfrac{k_6 K_1}{[H^+]} + \dfrac{k_7 K_1[R_2NH]}{K_2} + \dfrac{k_8 K_1[B]}{K_3}\right)}{k_{-1} + \dfrac{k_6 K_1}{[H^+]} + \dfrac{k_7 K_1[R_2NH]}{K_2} + \dfrac{k_8 K_1[B]}{K_3}} \tag{16}$$

a rate expression which, for an experiment at constant [H$^+$], is identical in form with eqn. (5).

It follows from the result obtained by differentiating eqn. (12) with respect to time at constant [H$^+$] that

$$\frac{d([II]+[IV])}{dt} = \frac{d[II]}{dt} = \frac{d[IV]}{dt} = 0 \tag{17}$$

The individual expressions for $d[II]/dt$, (18), and $d[IV]/dt$, (19).

$$\frac{d[II]}{dt} = 0 = k_1[I][R_2NH] - k_{-1}[II] - k_5[II] + k_{-5}[IV][H^+] \tag{18}$$

$$\frac{d[IV]}{dt} = 0 = k_5[II] - k_{-5}[IV][H^+] - k_6[IV] - k_7[IV][R_2NH_2^+]$$

$$- k_8[IV][BH^+] \tag{19}$$

may be used to define the conditions, in terms of rate coefficients and of concentrations, required to maintain II in equilibrium with IV and $H^+$. Equation (18) may be transformed to

$$1 - \frac{k_1[I][R_2NH] - k_{-1}[II]}{k_5[II]} = \frac{k_{-5}[IV][H^+]}{k_5[II]} \tag{20}$$

Since, with the equilibrium maintained, the right hand side of eqn. (20) is equal to one, the required condition or restriction is given by

$$1 \gg \frac{k_1[I][R_2NH] - k_{-1}[II]}{k_5[II]} \tag{21}$$

Equation (19) can be transformed to

$$\frac{k_5}{k_{-5}} = \frac{[IV][H^+]}{[II]} \left(1 + \frac{k_6 + k_7[R_2NH_2^+] + k_8[BH^+]}{k_{-5}[H^+]}\right) \tag{22}$$

whence it follows that (23) must obtain if (16) is to be valid.

$$1 \gg \frac{k_6 + k_7[R_2NH_2^+] + k_8[BH^+]}{k_{-5}[H^+]} \tag{23}$$

It is desirable and perhaps necessary to make the conditions, (21) and (23), explicit, since substitution of the equilibrium condition (12) into either (18) or (19), separately, gives results that makes this derivation appear suspect. The derivation is, however, valid, and it corresponds to a transition-state diagram containing three peaks. The last and highest one, which is lowered as the concentration of base increases, corresponds to the conversion of IV to product; the first corresponds to conversion of starting materials to II; separating these two peaks is a lower-lying, small hillock corresponding to the reversible interconversion of II and IV.

It is also possible to derive a rate equation for a reaction sequence which does not differ essentially from that which led to (16) without introducing the equilibrium assumption (12)[36]. For convenience the mechanism is now rewritten as

$$I + R_2NH \underset{k_{-1}}{\overset{k_1}{\rightleftharpoons}} II$$

$$II + R_2NH \underset{k_{-9}}{\overset{k_9}{\rightleftharpoons}} IV + R_2NH_2^+$$

$$II + B \underset{k_{-10}}{\overset{k_{10}}{\rightleftharpoons}} IV + BH^+$$

$$IV \begin{cases} \xrightarrow{k_6} III + OR^- \\ \xrightarrow[R_2NH_2^+]{k_7} III + ROH + R_2NH \\ \xrightarrow[BH^+]{k_8} III + ROH + B \end{cases}$$

Applying the steady-state approximation to both [II] and [IV] individually leads to

$$[IV] = \frac{k_1[I][R_2NH](k_9[R_2NH] + k_{10}[B])}{k_{-1}(k_{-9}[R_2NH_2^+] + k_{-10}[BH^+]) + (k_{-1} + k_9[R_2NH] \atop + k_{10}[B])(k_6 + k_7[R_2NH_2^+] + k_8[BH^+])} \qquad (23)$$

for the steady-state concentration of IV and, for the rate equation, to

$$\frac{d[III]}{dt} = \frac{k_1[I][R_2NH](k_9[R_2NH] + k_{10}[B])(k_6 + k_7[R_2NH_2^+] + k_8[BH^+])}{k_{-1}(k_{-9}[R_2NH_2^+] + k_{-10}[BH^+]) + (k_{-1} + k_9[R_2NH] \atop + k_{10}[B])(k_6 + k_7[R_2NH_2^+] + k_8[BH^+])} \qquad (24)$$

For the special case where

$$(k_{-1} + k_9[R_2NH] + k_{10}[B])(k_6 + k_7[R_2NH_2^+] + k_8[BH^+]) \gg k_{-1}(k_{-9}[R_2NH^+] \atop + k_{-10}[BH^+]) \qquad (25)$$

eqn. (24) reduces to

$$\frac{d[III]}{dt} = \frac{k_1[I][R_2NH](k_9[R_2NH] + k_{10}[B])}{k_{-1} + k_9[R_2NH] + k_{10}[B]} \qquad (26)$$

and once again it is possible to define an experimental second-order rate coefficient, $k_2^*$, as

$$k_2^* = \frac{1}{[I][R_2NH]} \frac{d[III]}{dt} = \frac{k_1(k_9[R_2NH]+k_{10}[B])}{k_{-1}+k_9[R_2NH]+k_{10}[B]} \tag{27}$$

Experimentally $k_2^*$ is normally determined in the limit of zero time, at which time the special case defined by (25) might be expected to hold, since $[R_2NH_2^+] \approx 0$ and $[BH^+] \approx 0$. Moreover, $k_9 > k_{-9}$ and $k_{10} > k_{-10}$, since usually B and $R_2NH$ are probably stronger bases than IV.

It is apparent that when condition (25) is satisfied this treatment affords a fit for the experimental observations that is completely satisfactory. Equations (26) and (27) are identical in from with eqns. (5) and (6). A plot of $k_2^*$ *versus* $[B]_0$ will give exactly the characteristics observed. The curve will rise rapidly at low $[B]_0$ and flatten out to a constant value at high $[B]_0$.

The transition-state diagram for this mechanism also contains three peaks. At low $[B]_0$ the middle peak, corresponding to the proton transfer from II to give IV, is the highest, and this step is rate-limiting. With increasing $[B]_0$ the height of this barrier is lowered, until at high $[B]_0$, the first peak, corresponding to formation of II, becomes rate-limiting. The third peak' for conversion of IV to product is lower-lying than either of the foregoing peaks, and the observed rates are independent of the rates of these product-forming steps.

There are, thus, at least three distinguishably different mechanisms which adequately account for the measured reaction rates in systems which show strong catalysis at low $[B]_0$ and attain or approach a limiting rate at high $[B]_0$. The number of possible product-forming transition states are even larger in number. It should be noted that systems, which have been found to show these characteristics, are, as of this moment, extremely limited in number. It is surely conceivable that there are reactions for which none of these formulations, all of which contain some approximation or assumption, will prove adequate. Some, perhaps, will require a more complex rate equation such as (24), with its additional parameters; for still others, depending on the relative magnitudes of the rate coefficients involved, even the steady-state approximation may be improper, and only extensive, computer-aided curve fitting will suffice.

Nevertheless, it is pertinent to ask if there is any basis for choice or preference among the three mechanisms which have been considered. They differ in the timing of the proton transfer relative to the separation of the leaving group and the degree to which these two processes may or may not be concerted. Where the proton transfer is part of the rate-determining step a kinetic hydrogen–deuterium isotope effect would be anticipated. If one inspects kinetic hydrogen–deuterium isotope effects in nucleophilic aromatic substitution generally, the effects observed are invariably smaller than had been predicted for a rate-determining step which involves cleavage of an N–H bond[37]. The overall isotope effects are frequently unity or very close to unity, and even inverse effects have been observed[38], suggesting the possibility that secondary rather than primary isotope effects may be involved[39].

If one limits the consideration to only that limited number of reactions which clearly belong to the category of nucleophilic aromatic substitutions presently under discussion, only a few experimental observations are pertinent. Bunnett and Bernasconi[30] and Hart and Bourns[40] have studied the deuterium solvent isotope effect and its dependence on hydroxide ion concentration for the reaction of 2,4-dinitrophenyl phenyl ether with piperidine in dioxan–water. In both studies it was found that the solvent isotope effect decreased with increasing concentration of hydroxide ion, and Hart and Bourns were able to estimate that $k^H/k^D$ for conversion of intermediate to product was approximately 1.8. Also, Pietra and Vitali[41] have reported that in the reaction of piperidine with cyclohexyl 2,4-dinitrophenyl ether in benzene, the reaction becomes 1.5 times slower on substitution of the N-deuteriated amine at the highest amine concentration studied.

Both of these reported kinetic hydrogen–deuterium isotope effects are disturbingly small, yet they are probably too large to be considered secondary isotope effects. These results lend support to the intermediate complex hypothesis, but they can be accommodated equally well by all three of the mechanisms that have been considered. These results, therefore, afford no basis for discrimination among the possible mechanisms.

The third mechanism discussed, rate equation (26), differs from the first two in that the bond to the leaving group is not broken in the rate-determining step. For the reaction of 2,4-dinitrophenyl phenyl ether with piperidine this mechanism would, therefore, predict that a primary kinetic oxygen isotope effect would not be observed. Hart and Bourns[40] have, in fact, observed a primary oxygen isotope effect for this reaction, with $k^{16}/k^{18}$ for conversion of intermediate to product having an estimated value of 1.012–1.013. It must, therefore, be concluded that this mechanism is not appropriate for this specific reaction and, furthermore, that this mechanism cannot be general. However, it does not follow that the mechanism is eliminated from all consideration. Barnett and Jencks[42] have demonstrated that the rate-determining step of the intramolecular S to N acetyl transfer of S-acetyl-mercaptoethylamine above pH 2.3 is a diffusion-controlled proton transfer and have noted that this system (in reverse) is analogous to the diffusion-controlled proton removal by hydroxide ion from the (potential) carbonyl oxygen atom in the decomposition of hemithioacetals to aldehydes. Comparable situations are conceivable for nucleophilic aromatic substitution.

It has also been argued[10, 40] that the second mechanism (rapid, reversible interconversion of II and IV) cannot be general. The basis for this contention is the fact that electrophilic catalysis is rare in nucleophilic aromatic substitution of non-heterocyclic substrates, an exception being the 2000-fold acceleration by thorium ion of the rate of reaction of 2,4-dinitrofluorobenzene with thiocyanate ion in methanol[43]. More directly pertinent is the observation that there is no electrophilic catalysis by added methanol in the reaction of 2,4-dinitrophenyl phenyl ether with piperidine in benzene[44]. This argument has force but is not compelling.

The first mechanism (rate equation (5)) can have a product-forming transition state, F, analogous to the E2 elimination reaction, or multicenter transition states, such as G or K. Hart and Bourns[40] have suggested that the multicenter process cannot be general, since Bernasconi and Zollinger[45] have observed catalysis by tertiary amines in the reaction of 2,4-dinitrofluorobenzene with piperidine in benzene. It is questionable that this reaction properly belongs in the class of nucleophilic aromatic substitutions showing both strong catalysis and a limiting rate at high catalyst concentration. The observed rate accelerations were, in fact, modest, and, as has been noted, fluoride is a borderline leaving group.

It is probably unrealistic, at this time, to demand that a single mechanism be general for all of the reactions under consideration. What should be stressed is the fact that the number of pertinent examples is at present extremely limited and the further fact that a variety of mechanisms, which can account for all the present experimental observations, is available. It is even possible that the advent of new data will expand rather than constrict the mechanistic possibilities. One promising approach is the direct measurement of the rates of some of the individual steps, and such work is now in progress[46].

## 4. The mechanism of weakly base-catalyzed reactions

The reactions discussed to this point are of great significance because of their implications with regard to the mechanism of nucleophilic aromatic substitution generally and because they demonstrate unequivocally the correctness of the intermediate complex hypothesis, at least for those reactions which show base catalysis and reach a limiting rate at high catalyst concentration. The most useful substitutions on an aromatic nucleus (and those most frequently used for synthetic purposes) commonly involve substrates with good leaving groups, chlorides, bromides and iodides. Although it has been demonstrated that these reactions are subject to catalysis by bases[6], the rate accelerations observed are relatively modest in magnitude, a limiting rate at high catalyst concentration has not been observed, and the catalysis, if it is indeed catalysis, appears to be different in kind from that already discussed.

The observed second-order rate coefficients can, in all cases, be treated empirically as consisting of the sum of a second-order rate coefficient (for the uncatalyzed reaction) and the product of a third-order rate coefficient and the catalyst concentration (for the catalyzed reaction), as indicated by eqn. (4). If the catalyst is a base this constitutes base catalysis in a formal sense, at least. Nevertheless, this catalysis shows atypical features, which should be stressed. Is most striking characteristic is the fact that the efficiency of a catalyst is very little dependent on its strength as a base. Table 1 presents some catalytic coefficients, $k''$, for various compounds acting as bases in the reaction of 2,4-dinitrochlorobenzene and al-

TABLE 1

CATALYTIC COEFFICIENTS IN THE REACTION OF 2,4-DINITROCHLOROBENZENE AND ALLYLAMINE IN 2-PHENYLETHANOL AT $24.8 \pm 0.1$ °C

| Catalyst | $k'' \times 10^4$ $(l^2.mole^{-2}.sec^{-1})$ |
|---|---|
| m-Dinitrobenzene | 0.68 |
| Benzil | 0.77 |
| Dimethylsulfoxide | 2.3 |
| Dimethylsulfone | 1.5 |
| Allylamine | 1.2 |
| N-Allyl-2,4-dinitroaniline | 1.3 |
| 2,3-Dinitrochlorobenzene | 0.70 |

lylamine in 2-phenylethanol at $24.8 \pm 0.1$ °C[25]. The uncatalyzed rate coefficient, $k'$, in this system is $1.35 \times 10^{-4}$ l.mole$^{-1}$.sec$^{-1}$ and the ratio, $k''/k'$, is, in every case, small. Of the catalysts in Table 1 only allylamine and N-allyl-2,4-dinitroaniline are bases in the most conventional sense, and the latter amine which is a much weaker base than allylamine is a slightly more effective catalyst. In another system, the reaction of n-butylamine with 2,4-dinitrochlorobenzene in 50 % dioxan–50 % water, the amine is an effective catalyst, but the much stronger base, hydroxide ion, is relatively ineffective[47]. All of the catalysts in Table 1 are, however, bases in a Lewis sense; all of them are electron-donors and all of them can function as acceptors in hydrogen bond formation. It may, indeed, be this last characteristic which determines their suitability to serve as catalysts in these reactions.

The solvent effects also suggest that hydrogen bonding may play an important role in these reactions. The most comprehensive studies of solvent effects are those reported by Suhr[26, 48], and some of his results for the reaction of p-nitrofluorobenzene and piperidine at 50 °C are summarized in Table 2, where $k'$ is the rate coefficient for the uncatalyzed reaction and $k''$ is the rate coefficient for the reaction catalyzed by a second molecule of piperidine. The solvents have been arranged in the order of increasing polarity as indicated by the dielectric constant at 25 °C. It is immediately apparent that polarity is the most important solvent parameter determining rate, with the more polar solvents giving faster reactions, since this is a reaction in which neutral molecules give a transition state with extensive charge separation and eventually an ionic product, the amine hydrofluoride. The least effective solvents are the two ethers, butyl ether and anisole, where the values of the ratio $k''/k'$ are relatively large. The most effective solvents

TABLE 2

SOLVENT EFFECTS ON THE RATE OF REACTION OF $p$-NITROFLUOROBENZENE AND PIPERIDINE AT 50 °C

| Solvent | $\varepsilon$ (25 °C, Debye) | $k' \times 10^6$ ($l.mole^{-1}.sec^{-1}$) | $k'' \times 10^6$ ($l^2.mole^{-2}.sec^{-1}$) | $k''/k'$ ($l.mole^{-1}$) |
|---|---|---|---|---|
| Butylether | 3.06 | $2 \pm 1$ | $97 \pm 3$ | 48.5 |
| Anisole | 4.33 | $9 \pm 2$ | $102 \pm 1$ | 11.3 |
| Ethyl acetate | 5.96 | $43 \pm 1$ | $44 \pm 2$ | 1.0 |
| Methyl acetate | 6.82 | $65 \pm 2$ | $55 \pm 3$ | 0.85 |
| Tetrahydrofuran | 7.50 | $69 \pm 1$ | $69 \pm 1$ | 1.0 |
| Dimethoxyethane | 9.50 | $52 \pm 1$ | $60 \pm 2$ | 1.2 |
| Acetone | 20.9 | $360 \pm 1$ | $140 \pm 10$ | 0.39 |
| Ethanol | 24.3 | $140 \pm 5$ | 44 | 0.32 |
| Methanol | 32.6 | $140 \pm 2$ | $30 \pm 1$ | 0.21 |
| Nitromethane | 34.3 | $820 \pm 20$ | $780 \pm 40$ | 0.95 |
| Acetonitrile | 36.5 | $1150 \pm 30$ | | |
| Dimethylformamide | 37.7 | $8100 \pm 400$ | | |
| Dimethylsulfoxide | 48.5 | $27400 \pm 900$ | | |

are the dipolar, aprotic solvents, dimethylformamide and dimethylsulfoxide, where catalysis by piperidine is not experimentally observable. These two solvents are the most polar solvents in Table 2, but they are also, at the same time, the most efficient acceptors for hydrogen bond formation. Ethanol and methanol are more polar solvents than acetone, but the rate of reaction is faster in acetone, probably because the carbonyl group is the better acceptor for hydrogen bonding.

Similar solvent effects are observed in the reaction of 2,4-dinitrochlorobenzene and $n$-butylamine at 24.8 °C[6]. These results are summarized in Table 3. The reac-

TABLE 3

SOLVENT EFFECTS IN THE REACTION OF 2,4-DINITROCHLOROBENZENE AND $n$-BUTYL-AMINE AT 24.8 °C

| Solvent | $k' \times 10^4$ ($l.mole^{-1}.sec^{-1}$) | $k'' \times 10^4$ ($l^2.mole^{-2}.sec^{-1}$) | $k''/k'$ ($l.mole^{-1}$) |
|---|---|---|---|
| Chloroform | 2.00 | 5.17 | 2.59 |
| Ethanol | 9.0 | 6.7 | 0.74 |
| 50 % dioxan–50 % water | 22 | 6.3 | 0.29 |

tion is slowest in chloroform, which is a good donor in hydrogen bonding but relatively ineffective as an acceptor, and in this solvent catalysis by $n$-butylamine contributes significantly to product formation. In the more polar solvents the rates increase, but catalysis becomes less important since $k''$ stays essentially constant while $k'$ increases more than ten-fold. Data for this reaction in dimethylformamide are not available, but the reaction of 2,4-dinitroiodobenzene and $n$-butylamine in dimethylformamide at 24.8 °C has a $k'$ of $92 \times 10^{-4}$ l.mole$^{-1}$.sec$^{-1}$, and catalysis by the amine is not observed[49]. It is reasonably certain that the chloride would react two to five times faster and would not show amine catalysis.

The available data on the effect of added neutral salts on the rates of these reactions, although not extensive, are suggestive and may be significant. The most directly pertinent data available is on the reaction of 2,4-dinitrochlorobenzene and *n*-butylamine in chloroform, which was chosen as the solvent, since it is a poor acceptor for hydrogen bonding although a suitable donor[23, 50]. The studies include three sets of data. The first was with no salt added, the chloride constant at 0.05 $M$ and the initial amine concentrations varied from 0.1 to 1.5 $M$. In the second set, the initial chloride and amine concentrations were constant at 0.05 $M$ and 0.2 $M$ respectively, and a neutral salt, benzyltriethylammonium nitrate, was added at concentrations ranging from 0 to 0.34 $M$. In the third set, the neutral salt was constant at 0.05 $M$, the chloride was again constant at 0.05 $M$, and the amine was varied from 0.2 to 0.8 $M$.

Although individual runs for the first set of experiments follow the second-order rate law, the observed second-order rate coefficients, $k_2^*$, are strongly dependent on the initial amine concentrations, with the rate increasing regularly as the amine concentration increases. Nevertheless, for all of the measurements, a plot of $k_2^*$ *versus* the initial amine concentration is linear, and the data can be fitted with eqn. (4), with $k'$ equal to $1.87 \times 10^{-4}$ l.mole$^{-1}$. sec$^{-1}$ and $k''$ equal to $5.63 \times 10^{-4}$ l$^2$ . mole$^{-2}$ . sec$^{-1}$.

The addition of the neutral salt (the second set of experiments) accelerates the reaction rate far more than does an equivalent concentration of excess amine. In the absence of the salt, a fifteen-fold increase in the initial amine concentration (from 0.1 $M$ to 1.5 $M$) raises the measured rate coefficient by a factor of four, but at a salt concentration of 0.34 $M$ the measured second-order rate coefficient is nine times larger than with the salt absent.

These large increases in rate might be attributed to the operation of a neutral salt effect, and, in fact, a plot of log $k_2^*$ *versus* the square root of the ionic strength, $\mu^{\frac{1}{2}}$, is linear. However, the reactants, in this case, are neutral molecules, not ions; in the low dielectric constant solvent, chloroform, ionic species would be largely associated, and the Brönsted–Bjerrum theory of salt effects[51, 52], which is valid only for dilute-solution reactions between ions at small $\mu$ (below 0.01 $M$ for 1 : 1 electrolytes), does not properly apply.

This is a reaction in which neutral molecules react to give a dipolar or ionic transition state, and some rate acceleration from the added neutral salt is to be expected[53], since the added salt will increase the polarity or effective dielectric constant of the medium. Some of the rate increases due to added neutral salts are attributable to this cause, but it is doubtful that they are all thus explained. The set of data for constant initial chloride and initial salt concentrations and variable initial amine concentrations affords some insight into this aspect of the problem.

A plot of $k_2^*$ *versus* the initial amine concentrations, for these data, is again linear, giving values of $5.30 \times 10^{-4}$ l. mole$^{-1}$. sec$^{-1}$ for $k'$ and $8.27 \times 10^{-4}$ l$^2$. mole$^{-2}$. sec$^{-1}$ for $k''$, with these new values corresponding to a salt concentration or $\mu$ of 0.05 $M$. In the absence of the salt, $k'$ was $1.87 \times 10^{-4}$ l.mole$^{-1}$.sec$^{-1}$ and $k''$ was $5.63 \times 10^{-4}$ l$^2$.mole$^{-2}$.sec$^{-1}$. Thus the addition of 0.05 $M$ benzyltriethylammonium nitrate increases $k''$ by 47 % and $k'$ by more than 180 %.

One possible explanation for the above results is that the transition state for the uncatalyzed reaction is either more ionic or has its charges more highly separated than does the transition state for the catalyzed reaction. A consideration of possible transition state structures makes this explanation improbable, since the transition state for the catalyzed reaction would, in fact, be expected to show the greater charge separation, and this would be equally the case for both the transition state for intermediate formation and the transition state for conversion of intermediate to product.

An alternate possibility is that one of the component ions of the salt is actually involved in the rate-determining transition state and can catalyze the reaction as does the second molecule of amine. This possibility is supported by the fact that at constant initial concentrations of the substrate and the amine and variable salt concentrations, the experimental data also give a linear plot of $k_2^*$ *versus* the initial concentration of the salt.

For the reactions under consideration, the intermediate has carbon, at the seat of substitution, bonded to nitrogen from the amine nucleophile and to the leaving group—fluoride or chloride or iodide etc. Since only substrates with good leaving groups are under consideration and since the amine is a very poor leaving group, it is most certain that the transformation of intermediate to product will be very much favored over its reversion to starting materials. If there is to be base catalysis in these circumstances it is more plausible to have it occur in the step in which the intermediate is formed, as first suggested by Kirby and Jencks[29], rather than in the step in which the intermediate is converted to product.

This possibility warrants more detailed consideration. For the reaction of 2,4-dinitrochlorobenzene with an amine it is highly probable that the rate-determining step is intermediate formation. One possible mode of base catalysis in this system would involve transfer of the proton in the step in which the intermediate is formed.

One can conceive of a transition state such as L collapsing to the intermediate, M, and $R_2NH_2^+$, *viz.*

This need not necessarily involve a termolecular collision, since it may be preceded by the equilibrium

$$2\,R_2NH \rightleftharpoons R_2NH \ldots \overset{\overset{\displaystyle H}{\displaystyle |}}{N}-R_2$$

$$O$$

to form the hydrogen-bonded complex, O, which would be a better nucleophile than the amine itself. The second amine molecule is not unique in being able to perform this function. A solvent molecule, *e.g.* dimethylformamide, would be effective, *viz.*

$$R_2NH + H-\overset{\overset{\displaystyle O}{\displaystyle \|}}{C}-N(CH_3)_2 \rightleftharpoons R_2N-H \ldots O=\overset{\overset{\displaystyle H}{\displaystyle |}}{C}-N(CH_3)_2$$

as would an anion such as nitrate ion, *viz.*

$$R_2NH + NO_3^- \rightleftharpoons NO_3^- \ldots H-N-R_2$$

In fact any species capable of serving as an acceptor for hydrogen bonding could exert a catalytic effect. The intermediate formed in the catalyzed reaction would be the conjugate base of the intermediate formed in the uncatalyzed reaction. Since its proton has already been transferred, its conversion to product is no longer subject to catalysis by a base. Electrophilic catalysis to assist the removal of the leaving group is possible but probably unnecessary and, in any event, would not manifest itself in the measured rates, if intermediate formation is rate-determining.

If the catalysis in these systems has its source in the presence of a hydrogen bond from the amino hydrogen of the nucleophile to a suitable acceptor in the transition state for intermediate formation, as in L, it is not surprising that the efficiency of a catalyst is very little dependent on its strength as a base. Hydrogen bonds are, in general, weak bonds, and in the gamut of hydrogen bonds the N–H . . . O bonds and the N–H . . . N bonds are amongst the weakest, with bond energies[54, 55] as low as 1.3–2.0 kcal.mole$^{-1}$.

The preceding discussion is not offered as proof that base catalysis in reactions on substrates having good leaving groups represents catalysis of intermediate formation. It is presented only to indicate that this hypothesis is both possible and plausible. The mechanism under consideration may be formulated as follows.

V                         VI                        VII

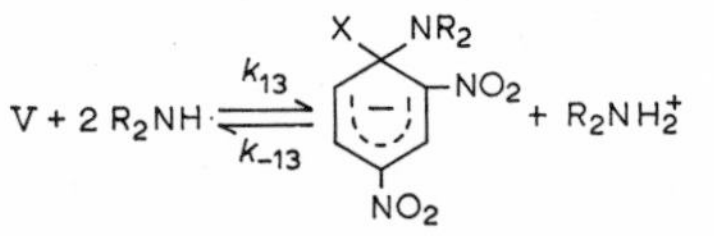

$$V + 2 R_2NH \underset{k_{-13}}{\overset{k_{13}}{\rightleftharpoons}} \text{VIII} + R_2NH_2^+$$

VIII

$$V + R_2NH + B \underset{k_{-14}}{\overset{k_{14}}{\rightleftharpoons}} \text{VIII} + BH^+$$

$$\text{VIII} \overset{k_{15}}{\to} \text{VII} + X^-$$

The desired rate is given by

$$\frac{d[\text{VII}]}{dt} = k_{12}[\text{VI}] + k_{15}[\text{VIII}] \tag{28}$$

Applying the steady-state approximation to both [VI] and [VIII] one obtains (29) for the steady-state concentration of VI and (30) for the steady-state concentration of VIII, *viz.*

$$[\text{VI}] = \frac{k_{11}[\text{V}][\text{R}_2\text{H}]}{k_{12} + k_{-11}} \tag{29}$$

$$[\text{VIII}] = \frac{k_{13}[\text{V}][\text{R}_2\text{NH}]^2 + k_{14}[\text{V}][\text{R}_2\text{NH}][\text{B}]}{k_{15} + k_{-13}[\text{R}_2\text{NH}_2^+] + k_{-14}[\text{BH}^+]} \tag{30}$$

The general rate equation is given by

$$\frac{d[\text{VII}]}{dt} = \frac{k_{11}k_{12}[\text{V}][\text{R}_2\text{NH}]}{k_{12} + k_{-11}} + k_{15}\left\{\frac{k_{13}[\text{V}][\text{R}_2\text{NH}]^2 + k_{14}[\text{V}][\text{R}_2\text{NH}][\text{B}]}{k_{15} + k_{-13}[\text{R}_2\text{NH}_2^+] + k_{-14}[\text{BH}^+]}\right\} \tag{31}$$

The present mechanism is being applied only to reactions involving good leaving groups. In these reactions the intermediates, VI and VIII, collapse to products much more rapidly than they return to starting materials. These are, therefore, reactions in which conditions (32) and (33)

$$k_{12} \gg k_{-11} \tag{32}$$

$$k_{15} \gg k_{-13}[\text{R}_2\text{NH}_2^+] + k_{-14}[\text{BH}^+] \tag{33}$$

are valid and eqn. (31) simplifies to

$$\frac{d[\text{VII}]}{dt} = k_{11}[\text{V}][\text{R}_2\text{NH}] + k_{13}[\text{V}][\text{R}_2\text{NH}]^2 + k_{14}[\text{V}][\text{R}_2\text{NH}][\text{B}] \tag{34}$$

Equation (34) can be expressed in more general form as

$$\frac{d[VII]}{dt} = [V][R_2NH](k_{11}+\Sigma_i k_i[B]_i) \tag{35}$$

where B is now any species capable of serving as an acceptor in hydrogen bond formation. The experimental second-order rate coefficient, $k_2^*$, will now be given by

$$k_2^* = \frac{1}{[V][R_2NH]} \frac{d[VII]}{dt} = k_{11}+k_{13}[R_2NH]+k_{14}[B] \tag{36}$$

and again (36) can be expressed more generally as

$$k_2^* = \frac{1}{[V][R_2NH]} \frac{d[VII]}{dt} = k_{11}+\Sigma_i k_i[B]_i \tag{37}$$

It should be immediately apparent that this formulation is in complete accord with all of the rate measurements for these systems. Equation (37) demands that $k_2^*$ increase linearly with increasing catalyst concentration, and this is, in fact, what is observed.

In the above formulation the proton is transferred in the step in which the intermediate is formed. Such proton transfer is not essential for base catalysis. An alternate mode of catalysis is one in which the transition state for intermediate formation is a hydrogen-bonded complex, e.g. L, but in which this complex collapses to VI and the catalyst rather than to VIII. For such a formulation the only significant intermediate determining the rates would be VI, which would now be formed by the additional steps

$$V+2 R_2NH \rightleftharpoons VI+R_2NH$$

and

$$V+R_2NH+B \rightleftharpoons VI+B$$

It is obvious that this formulation would also lead to eqn. (37) and would be consistent with the experimental data.

This hypothesis of base catalysis of intermediate formation does no violence to and very conveniently accommodates all other experimental findings on these reactions. The "element effect" of Bunnett et al.[56, 57] follows logically as the result of selecting all the reactions or carrying them out under conditions such that intermediate formation is the rate-determining step.

The "ortho effect" has its origin in the ability of an ortho substituent to form a hydrogen bond with a hydrogen on the amine nucleophile[58-61]. The present hypothesis requires that this hydrogen bonding occur in the transition state for intermediate formation rather than in the transition state for conversion of intermediate to product. The appropriate transition state can be represented as in P.

P

The basis for the faster rates with *ortho* substituents thus resides in the ability of the *ortho* group to participate in this intramolecular hydrogen bonding, a function for which a *para* substituent is not suitably situated. The strong solvent dependence of the *ortho* effect for amine reactions also follows logically from the present hypothesis. The *ortho* : *para* ratio, $k_o/k_p$, is largest in non-polar solvents, where it varies from 50–80, and smallest in polar solvents, where it is, nevertheless, greater than unity. The polar solvents are invariably efficient acceptors for hydrogen bond formation and compete with the *ortho* substituent in this role.

Finally and most importantly, this mechanism affords a satisfactory explanation for the small deuterium isotope effects ($k^H/k^D$ very close to one) observed in these reactions. The results are completely in accord with the "solvation rule" of Swain *et al.*[62], which in the present context might have been given the more limited name, "hydrogen bond rule". This rule states that a proton being transferred from one oxygen to another or from nitrogen to oxygen or from nitrogen to nitrogen in an organic reaction, in which there are bond changes on carbon in the rate-determining step, should lie in an entirely stable potential at the transition state and not form reacting bonds nor give rise to primary hydrogen isotope effects. As Swain *et al.* state, "this rule is a restatement of thermodynamic laws governing the structures of all stable aggregates and so should apply to reactants as well as to the stable parts of transition states". It is, thus, directly pertinent to the mechanism proposed and correctly predicts the isotope effect for a rate-determining reaction having a transition state such as L.

## 5. Conclusion

Two types of base catalysis in nucleophilic aromatic substitution reactions have been discussed. Strong catalysis, which occurs in reactions with substrates having poor leaving groups, has been identified as catalysis of the product-forming step. In these systems it is possible to observe a change in the rate-limiting step, with product formation being rate-limiting at low catalyst concentration and intermediate formation being rate-determining at high catalyst concentration. This constitutes unequivocal proof of the intermediate complex mechanism in these reactions. Reactions with substrates having good leaving groups show weak base catalysis, and these reactions do not reach a limiting rate at high catalyst concentration. This catalysis has been identified as catalysis of intermediate formation. In these systems the rates increase linearly with increasing catalyst concentration, and it has not been possible to observe a change in the rate-limiting step of the reaction. One is not, therefore, compelled to postulate the presence of an intermediate or intermediates along the reaction pathway. Nevertheless, it is highly probable that intermediates are involved and certainly convenient to treat all nucleophilic aromatic substitutions in a unified manner in terms of the intermediate complex mechanism.

## REFERENCES

1 J. F. BUNNETT AND R. E. ZAHLER, Chem. Rev., 49 (1951) 273.
2 J. MILLER, Rev. Pure Appl. Chem., 1 (1951) 171.
3 J. F. BUNNETT, Quart. Rev., 12 (1958) 1.
4 J. F. BUNNETT, Theoretical Organic Chemistry, Butterworths, London, 1959, p. 144.
5 J. SAUER AND R. HUISGEN, Angew. Chem., 72 (1960) 294.
6 S. D. ROSS, Progr. Phys. Org. Chem., 1 (1963) 31.
7 G. ILLUMINATI, Advan. Heterocyclic Chem., 3 (1964) 285.
8 B. CAPON, M. J. PERKINS AND C. W. REES, Organic Reaction Mechanisms-1965, Interscience, London, 1966, p. 50.
9 B. CAPON, M. J. PERKINS AND C. W. REES, Organic Reaction Mechanisms-1966, Interscience, London, 1967, p. 44.
10 E. BUNCEL, A. R. NORRIS AND K. E. RUSSELL, Quart. Rev., 22 (1968) 123.
11 C. L. JACKSON AND W. F. BOOS, Am. Chem. J., 20 (1898) 444.
12 J. MEISENHEIMER, Ann., 323 (1902) 205.
13 G. BRIEGLEB, Electronen–Donator–Acceptor Komplexe, Springer-Verlag, Berlin, 1961.
14 L. J. ANDREWS AND R. M. KEEFER, Molecular Complexes in Organic Chemistry, Holden-Day, San Francisco, 1964.
15 E. M. KOSOWER, Progr. Phys. Org. Chem., 3 (1965) 81.
16 L. MELANDER, Arkiv Kemi, 2 (1950) 211.
17 U. BERGLUND-LARSSON AND L. MELANDER, Arkiv. Kemi, 6 (1953) 219.
18 G. W. WHELAND, J. Am. Chem. Soc., 64 (1942) 900.
19 C. K. INGOLD, Structure and Mechanism in Organic Chemistry, Cornell Univ. Press, Ithaca, New York, 1953, p. 338.
20 P. J. C. FIERENS AND A. HALLEUX, Bull. Soc. Chim. Belges, 64 (1955) 717.
21 G. S. HAMMOND AND L. R. PARKS, J. Am. Chem. Soc., 77 (1955) 340.
22 J. F. BUNNETT AND R. H. GARST, J. Am. Chem. Soc., 87 (1965) 3875.
23 S. D. ROSS AND M. FINKELSTEIN, J. Am. Chem. Soc., 79 (1957) 6547.
24 S. D. ROSS, M. FINKELSTEIN AND R. C. PETERSEN, J. Am. Chem. Soc., 81 (1959) 5336.
25 S. D. ROSS, J. E. BARRY AND R. C. PETERSEN, J. Am. Chem. Soc., 83 (1961) 2133.
26 H. SUHR, Ber. Bunsenges. Physik. Chem., 67 (1963) 893.
27 J. F. BUNNETT AND R. H. GARST, J. Am. Chem. Soc., 87 (1965) 3879.
28 J. F. BUNNETT AND J. J. RANDALL, J. Am. Chem. Soc., 80 (1958) 6020.
29 A. J. KIRBY AND W. P. JENCKS, J. Am. Chem. Soc., 87 (1965) 3217.
30 J. F. BUNNETT AND C. BERNASCONI, J. Am. Chem. Soc., 87 (1965) 5209.
31 M. EIGEN, Pure Appl. Chem., 6 (1963) 97.
32 B. CAPON AND C. W. REES, Ann. Rept. Chem. Soc. (London), 60 (1963) 279.
33 C. G. SWAIN AND J. F. BROWN, JR., J. Am. Chem. Soc., 74 (1952) 2538.

34  B. BITTER AND H. ZOLLINGER, *Angew. Chem.*, 70 (1958) 246; *Helv. Chim. Acta*, 44 (1961) 812.
35  F. PIETRA AND D. VITALI, *Tetrahedron Letters*, (1966) 5701.
36  S. D. ROSS AND R. C. PETERSEN, *Tetrahedron Letters*, (1968) 4699.
37  K. WIBERG, *Chem. Rev.*, 55 (1955) 713.
38  H. ZOLLINGER, *Advan. Phys. Org. Chem.*, 2 (1964) 163.
39  E. A. HALEVI, *Progr. Phys. Org. Chem.*, 1 (1963) 180.
40  C. R. HART AND A. N. BOURNS, *Tetrahedron Letters*, (1966) 2995.
41  F. PETRA AND D. VITALI, *Chem. Commun.*, (1968) 692.
42  R. BARNETT AND W. P. JENCKS, *J. Am. Chem. Soc.*, 90 (1968) 4199.
43  K. B. LAM AND J. MILLER, *Chem. Commun.*, (1966) 642.
44  F. PIETRA, *Tetrahedron Letters*, (1965) 2405.
45  C. BERNASCONI AND H. ZOLLINGER, *Tetrahedron Letters*, (1965) 1083.
46  C. F. BERNASCONI, *J. Am. Chem. Soc.*, 90 (1968) 4982.
47  S. D. ROSS, *J. Am. Chem. Soc.*, 80 (1958) 5319.
48  H. SUHR, *Chem. Ber.*, 97 (1964) 3277.
49  S. D. ROSS, *J. Am. Chem. Soc.*, 81 (1959) 2113.
50  S. D. ROSS, *Tetrahedron*, 25 (1969) 4427.
51  J. N. BRÖNSTED, *Z. Physik Chem.*, 102 (1922) 169; 115 (1925) 337.
52  N. BJERRUM, *Z. Physik. Chem.*, 108 (1924) 82; 118 (1925) 251.
53  K. J. LAIDLER, *Chemical Kinetics*, McGraw-Hill, New York, 1950, p. 132; See also, L. C. BATEMAN, M. G. CHURCH, E. D. HUGHES, C. K. INGOLD AND N. A. TAHER, *J. Chem. Soc.*, (1940) 979.
54  M. M. DAVIES, *Ann. Rept. Chem. Soc.*, (*London*), 43 (1946) 5.
55  L. N. FERGUSON, *Electron Structures of Organic Molecules*, Prentice-Hall, New York, 1952, p. 57.
56  J. F. BUNNETT, E. W. GARBISCH, JR. AND K. M. PRUITT, *J. Am. Chem. Soc.*, 79 (1957) 385.
57  J. F. BUNNETT AND W. D. MERRITT, JR., *J. Am. Chem. Soc.*, 79 (1957) 5967.
58  R. R. BISHOP, E. A. S. CAVELL AND N. B. CHAPMAN, *J. Chem. Soc.*, (1952) 437.
59  J. F. BUNNETT AND R. J. MORATH, *J. Am. Chem. Soc.*, 77 (1955) 5051.
60  W. GREIZERSTEIN AND J. A. BRIEUX, *J. Am. Chem. Soc.*, 84 (1962) 1032.
61  S. D. ROSS AND M. FINKELSTEIN, *J. Am. Chem. Soc.*, 85 (1963) 2603.
62  C. G. SWAIN, D. A. KUHN AND R. L. SCHOWEN, *J. Am. Chem. Soc.*, 87 (1965) 1553.

*Chapter 3*

# Aromatic Rearrangements

### D. L. H. WILLIAMS

## 1. Introduction

There are a large number of chemical reactions which can be, and have been, classed as aromatic rearrangements which represent a wide variety of reaction types. Considerable effort has gone into attempts to determine the detailed reaction mechanism in many cases. The techniques that have been most fruitful in this respect are the application of kinetic methods and the use of isotopic labelling in product analyses. Often both these methods have been used to complement each other and it would be unwise to attempt to separate the contributions from each, nevertheless examples have been chosen for this chapter where kinetic methods have been of particular value in determining reaction mechanisms. The results of product analyses, particularly the use of isotopes in establishing the intra- or intermolecularity of a reaction, are, of course, discussed where relevant. The choice of reactions is not meant to be completely comprehensive but rather, illustrative. A number of excellent books[1-3] covering aspects of aromatic rearrangements are available together with a number of review articles dealing with individual reactions, to which reference will be made in the appropriate section. The literature has been covered until the end of 1969.

## 2. Rearrangements of N-substituted aromatic amines

### 2.1 THE ORTON REARRANGEMENT (N-HALOANILIDES)

The rearrangement of N-chloroanilides to give *para* (mainly) and *ortho* chloroanilides has been known since 1886 when Bender[4] converted N-chloroacetanilide to the *para* chloro isomer by treatment with concentrated hydrochloric acid or by heating with water or simply by melting N-chloroacetanilide. The reaction was much studied by Orton and co-workers between 1909 and 1928 after whom it is now generally known. In all cases the reaction is specifically catalysed by hydrogen halides (most of the work being concerned with hydrogen chloride) and can be represented overall by

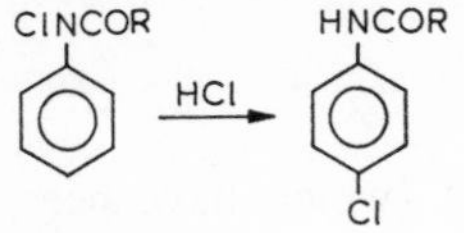

$$\tag{1}$$

*References pp. 481–486*

Slosson[5] showed that the N-bromoanilides I and III also underwent rearrangement to give the corresponding *para* bromo compounds II and IV, *viz.*

In general the *para* isomer is the main product but about 5 % *ortho* isomer was detected[6] when the solvent was aqueous acetic acid. All the evidence suggests that the rearrangement is intermolecular involving the reversible formation of free chlorine and the anilide, reaction (2), followed by C-chlorination of the anilide, reaction (3)

$$ClArNCOR + HCl \rightleftharpoons Cl_2 + ArNHCOR \qquad (2)$$

$$ArNHCOR + Cl_2 \rightarrow p\text{-}ClArNHCOR \qquad (3)$$

The evidence for this is based on the facts that (*a*) chlorine can be aspirated from the reaction mixture, (*b*) in the presence of more reactive aromatic compounds cross-chlorination occurs as in the reaction

and (*c*) the ratio of isomers formed in the rearrangement is the same as that obtained for the direct chlorination of the anilide by chlorine under the same experimental conditions[7]. Tracer experiments using [36]Cl have shown[8] that N-chloroacetanilide equilibrates with chloride ion in solution, an observation that is consistent with the formation of free chlorine during the rearrangement.

The other halogen acids also bring about the rearrangement[9,10], N-chloro-acetanilide with hydrobromic acid forming *ortho* and *para* bromoacetanilide and hydriodic acid yielding the corresponding iodo derivatives. This has been interpreted as involving the formation of bromine chloride, reaction (5), and iodine monochloride which are well-known brominating and iodinating agents, respectively.

The results of kinetic studies on the rearrangement (which have been summarised by Hughes and Ingold[11]) completely support the intermolecular mechanism suggested above on the basis of product analyses. The rate law for the disappearance of N-chloroacetanilide is[12]

$$\text{Rate} = k[\text{PhNClCOCH}_3][\text{HCl}]^2.$$

This third-order rate equation is interpreted as meaning that the process is first-order in each reactant, *viz.* N-chloroacetanilide, chloride ion and hydrogen ion. This has been confirmed[10] for the reaction of N-chloroacetanilide with hydrogen bromide in a variety of aqueous media, under conditions where the dechlorination is rate-determining. The rate equation is

$$\text{Rate} = k[\text{C}_6\text{H}_5\text{NClCOCH}_3][\text{H}^+][\text{Br}^-]$$

which cannot be re-written in a molecular form since protons are derived from both acids.

Orton and Jones[13] showed clearly that an equilibrium existed between the anilide and free chlorine on one side and the N-chloroanilide and hydrogen chloride on the other as in

$$\text{ArNHCOCH}_3 + \text{Cl}_2 \rightleftharpoons \text{ArNClCOCH}_3 + \text{HCl} \tag{6}$$

This was achieved by working with an anilide where the rate of C-chlorination was much reduced by the introduction into the ring of electron-attracting substituents as in V and VI (below). Both the amounts of chlorine and N-chloroanilide

were determined as a function of time and it was found that the equilibrium was set up very rapidly, to be disturbed as the amount of C-chlorination became significant. The equilibrium constant which was determined using aqueous acetic acid as solvent was dependent on the solvent composition, being very small, *i.e.* over to the side of free chlorine in glacial acetic acid. All the evidence points to a scheme such as

$$\tag{7}$$

Orton *et al.*[14] demonstrated that N- and C-chlorination, steps (ii) and (iii), occurred simultaneously by starting with chlorine and acetanilide and showing that the ratio of N- to C-chlorinated products was constant during the course of the reaction thus obeying the requirement for two simultaneous reactions of the same kinetic order. Reaction (i) in (7) was conveniently studied by Soper[15], who carried out the reaction in the presence of *p*-cresol and phenol to trap out the chlorine as soon as it was formed thus preventing reactions (ii) and (iii) from occurring. The reaction was found to be first-order in N-chloroanilide and second-order in hydrochloric acid thus confirming the earlier work. The most reasonable mechanism for the de-chlorination step of the rearrangement involves the protonation of the N-chloroanilide followed by attack by chloride ion, *viz.*

$$(8)$$

Similar mechanisms of course can be written for the reaction with the other hydrogen halide acids, and indeed for the relatively slow decomposition of N-chloroacetanilide in solutions of other strong acids[15], *viz.*

$$(9)$$

Hypochlorous acid is a poorer chlorinating agent than chlorine itself so that C-chlorination does not generally take place, although hypochlorous acid can be removed with a more reactive aromatic substrate such as *p*-cresol.

In this, as in many catalysed reactions, the protonated substrate is postulated as an intermediate, and although the proposed reaction scheme in fact accords with all the known experimental facts it perhaps would be instructive to determine the dependence of the rate coefficient on the Hammett acidity function at high acid concentration and also to investigate the solvent isotope effect $k_{D_2O}/k_{H_2O}$. Both these criteria have been used successfully (see Sections 2.2–2.4) to confirm the intermediacy of the protonated substrate in other acid-catalysed aromatic rearrangements.

The Orton rearrangement also takes place in aprotic solvents such as chlorobenzene and the reaction has been much studied under these conditions particularly by Bell *et al.*[16]. The findings present nothing like as clear a picture of the mechanism as do the data concerning the reaction in hydroxylic solvents. Some N-haloacetanilides undergo rearrangement in chlorobenzene at reasonable rates at temperatures around room temperature. Examples are N-bromoacetanilide and N-iodoformanilide, catalysis being effected by acetic, halogenoacetic, benzoic and picric acids. N-chloroacetanilide was found[17] to be relatively unreactive, rate measurements being carried out at 100 °C. Under these conditions the reaction was found to be subject to general acid catalysis as expected, there being a direct relationship involving the Brönsted catalysis law relating the catalytic coefficients for the different acids with their dissociation constants. Unfortunately the reaction is not strictly first-order in N-halogenoanilide[16,18], although the reaction has been regarded as first-order for mechanistic purposes and the results have usually been calculated from initial rate studies. The dependence on acidity also is not straightforward, in some cases the orders in acid appear less than one (which may be attributable to dimerisation of the carboxylic acid) but in other cases it appears to be greater than one. For the purposes of analysing the acid catalysis, Bell and Danckwerts[17] have reduced all the data to zero acid concentration. It is not settled whether the rearrangement is intra- or intermolecular and the mechanism of the acid-catalysis is not clear cut. Trans-halogenation has been observed[19] under these conditions but this neither confirms nor denies the inter-molecularity of the mechanism as it is quite conceivable that the N-halogen-oanilide or its protonated form acts as a halogenating agent without the formation of free chlorine. More recently it has been shown[20] that the chlorination of p-cresol by chloramine-T does not involve the formation of free hypochlorous acid as was earlier thought[21], but rather a dichloroamide which is the effective chlorinating agent. For an analogous situation where the protonated substrate is thought to be an effective nitrosating agent in the rearrangement of N-nitrosami-nes see Section 2.4.

It has been suggested[19] that the Orton rearrangement catalysed by carboxylic acids in aprotic solvents is an intermolecular process, the halogenating agent being the acyl hypohalite, whereas Scott and Martin[22] have preferred a termolecular mechanism for the trans-halogenation. The situation is obviously more complex here than with hydroxylic solvents; it cannot be claimed that the mechanism is completely understood. It is conceivable as Bell suggested that there is an intra-molecular pathway open to the Orton rearrangement under certain circumstances.

## 2.2 THE BENZIDINE REARRANGEMENT (HYDROAZOAROMATICS)

It has been long known[23] that hydrazobenzene (VII) is converted in acid solution into benzidine (VIII), which is the main product, and diphenyline (IX), *viz.*

$$\tag{10}$$

Later it was shown[24] that *ortho* and *para* semidines (X and XI) could also be formed in this reaction together with oxidation and disproportionation products (azo-compounds and amines respectively) with certain substituted substrates. More work has been done on this reaction than on any other aromatic rearrangement of N-substituted amines and the subject has been dealt with comprehensively in a recent review by Banthorpe[25], who has done so much to develop the present-day accepted mechanism of the benzidine rearrangement.

Early theories as to the mechanism of the benzidine rearrangement involved in one case[26] homolysis of the N–N bond leading to free radical formation and in another case[27] heterolysis of the N–N bond forming a positive and a negative species. All mechanisms involving N–N bond breaking forming two fragments (whatever their nature) have been shown to be invalid by a large number of "cross-over" experiments which demonstrate conclusively the intramolecular nature of the rearrangement. For example for a large number of hydrazo-compounds which have different substituents in each ring[24], e.g. XII

the products were always of the form of XV and never of XIII or XIV. Further, it has been demonstrated that when two different symmetrical hydrazo compounds, which react at comparable rates, are subjected to rearrangement together, no unsymmetrical cross-products are detected. This was first shown by Ingold and Kidd[28], who carried out the concurrent rearrangement of 2,2′-dimethoxy- and 2,2′-diethoxy-hydrazobenzenes (XVI and XVII) and showed by melting point curves of the product that less than 6 %, if any, of the cross-over product XVIII was formed. This has been elegantly confirmed more recently by a $^{14}$C labelling experiment[29], in which 2,2'-dimethyl- (XIX), and 2-[14]C-methyl-hydrazobenzene (XX) were rearranged together. Analysis of the 3,3'-dimethylbenzidine formed set the upper limit of 0.03 % on the yield of the labelled product XXI.

Similar experiments on the analysis of the products from the three isomeric hydrazonaphthalenes[30] and substituted hydrazobenzenes[31] confirm the intramolecularity of the reaction. In addition, products resulting from attack on the solvent of fragments from a hydrazo-compound during rearrangement have never been detected[30].

A turning point in the history of the benzidine rearrangement was the analysis of kinetic forms of the reaction particularly with regard to the dependence on the acidity. Early workers used a tedious gravimetric method for following the reaction whereas most of the recent workers have used the redox dye Bindschedler's Green to determine the amount of unreacted hydrazo compound at any time. This water-soluble dye oxidizes hydrazoaromatics almost instantaneously[32], and the excess can be conveniently titrated with titanous chloride. Carlin et al.[33] were able to show by a spectrophotometric method that the rate of disappearance of hydrazobenzene was also the rate of formation of both products benzidine and diphenyline. More recently several electroanalytical methods have been successfully developed[34] for following the rate of the rearrangement. All results show that the reaction is first-order with respect to the hydrazobenzene, but there was considerable uncertainty for a time regarding the dependence on acidity. The very early work of van Loon[35] indicated that the rearrangement of hydrazobenzene itself was second-order in acid. This work has subsequently been confirmed by Hammond and Shine[36]. These results prompted the suggestion that one of two mechanisms, (11) and (12),

$$
\left.
\begin{aligned}
& B + H^+ \rightleftarrows BH^+ \\
& BH^+ + H^+ \rightleftarrows BH_2^{++} \xrightarrow{slow} \text{Products}
\end{aligned}
\right\} \tag{11}
$$

$$
\left.
\begin{aligned}
& B + H^+ \rightleftarrows BH^+ \\
& BH^+ + H^+ \xrightarrow{slow} \text{Products}
\end{aligned}
\right\} \tag{12}
$$

B = hydrazoaromatic

were involved. However Carlin and Odioso[37] showed clearly that the second-order dependence in acid was not applicable to 2,2'-dimethylhydrazobenzene where the only product was 3,3'-dimethylbenzidine when the rearrangement was carried out in 95 % aqueous ethanol over the range 0.03 to 0.1 $M$ hydrochloric acid. Rather, a non-integral order of 1.6 in acidity was observed, *viz.*

$$-\frac{d[B]}{dt} = k[B][H^+]^{1.6} \tag{13}$$

Subsequently many other substituted hydrazobenzenes were shown to have a non-integral acidity dependence between 1 and 2, for example with 4-chloro 4'-methylhydrazobenzene (XXII) the order was 1.6. Further, the order in acid was found to vary for any one substituted hydrazobenzene substrate, increasing

(*e.g.* for XXIII)[38] from 1.1 at $8 \times 10^{-4}$ $M$ acid to 2.0 at 0.6 $M$ acid. The explanation[39] which is now generally accepted is that the rearrangement proceeds by two concurrent mechanisms, one of which involves one proton transfer to the base and the other two protons. The rate equation resulting from such processes is

$$-\frac{d[B]}{dt} = k_2[B][H^+] + k_3[B][H^+]^2 \tag{14}$$

and would of course account for an overall dependence on acidity of between 1 and 2 which would approach the value of 1 at low acidities and 2 at high acidities as the first, and then the second term of the rate equation becomes dominant. With certain substrates the values of $k_2$ and $k_3$ are such that only either first-order or second-order dependence on acidity is observed, whatever the acid concentration. Banthorpe *et al.*[40], using 2,2'-dimethylhydrazobenzene in 60 % dioxan–water solvent showed that the plot of $-(d[B]/dt)/[B][H^+]$ *versus* $[H^+]$ was strictly linear as demanded by equation (14) and obtained values of $k_2$ and $k_3$ from the intercept and slope of the line.

It is important for acid-catalysed reactions to determine whether the reaction is specifically catalysed by hydrogen ions or whether general acid catalysis takes place. Specific acid catalysis has been conclusively demonstrated for the benzidine rearrangement by three different sorts of kinetic experiments. In the first, it has been shown[41] by the standard test for general acid catalysis (by measuring the rate of reaction in a buffered solution at constant pH over a range of concentration of organic acids) that for 4-methoxyhydrazobenzene the reaction is specifically catalysed by the hydrogen ion, *i.e.* the rate coefficient did not vary at constant pH as the general acid concentration was changed.

Secondly, it has been found that the benzidine rearrangement is subject to a solvent isotope effect $k_{D_2O}/k_{H_2O} > 1$. If a proton is transferred from the solvent to the substrate in a rate-determining step the substitution of protium by deuterium will lead to a retardation in the rate of reaction (primary isotope effect) whereas if a proton is transferred in a fast equilibrium step preceeding the rate-determining step as in

$$B + H^+ \overset{fast}{\rightleftharpoons} BH^+ \qquad (15)$$

$$BH^+ \overset{slow}{\rightarrow} products$$

the substitution of $H_2O$ by $D_2O$ produces a rate enhancement of about 2–3 fold. The reason for this isotope effect arises from the fact that $D_3O^+$ in $D_2O$ is a stronger acid than is $H_3O^+$ in $H_2O$ and since the rate is governed by the concentration of $BH^+$ a scheme like (15) should produce $k_{D_2O}/k_{H_2O}$ of about 2–3 as, in $D_2O$, the concentration of the conjugate acid will be greater than in $H_2O$. This solvent isotope effect has been observed for many acid-catalysed reactions of this type, *e.g.* ester hydrolyses. Banthorpe *et al.*[40,42,38,] have determined the solvent isotope effect for the benzidine rearrangement under conditions where the kinetic order in acid is 1, transitional and 2. The results in Table 1 show conclusively that

TABLE 1

SOLVENT ISOTOPE EFFECTS IN THE REARRANGEMENTS OF RNHNHR′ IN DIOXAN–WATER

| R | R′ | $[H^+]$ or $[D^+]$ | Order in $H^+$ | $k_{D_2O}/k_{D_2O}$ | $f_m^{\ddagger}$ |
|---|---|---|---|---|---|
| $1\text{-}C_{10}H_7$ | $1\text{-}C_{10}H_7$ | 0.01 | 1.0 | 2.3 | 2.3 |
| $2\text{-}C_{10}H_7$ | $C_6H_5$ | 0.02 | 1.2 | 2.6 | 2.2 |
| | | 0.31 | 1.8 | 3.8 | 2.1 |
| $2\text{-}CH_3C_6H_4$ | $2\text{-}CH_3C_6H_4$ | 0.01 | 1.3 | 2.1 | 1.8 |
| | | 0.29 | 1.9 | 3.5 | 1.9 |
| $C_6H_5$ | $C_6H_5$ | 0.19 | 2.0 | 4.8 | 2.2 |

$f_m^{\ddagger}$ is the rate enhancement per proton transferred.

*all* proton transfers (*i.e.* the *one* in the one-proton mechanism and *both* in the two-proton mechanism) occur in fast equilibrium steps before the rate-determining steps.

The third approach is based on the correlation of rate coefficients with the Hammett acidity function, and is known as the Hammett–Zucker hypothesis. Although this test is not always reliable[43] it has often been used to support other evidence. The basis of the test is that if log (rate coefficient) is proportional to $H_0$ (an acidity function defined in terms of the equilibrium $B + H^+ \rightleftarrows BH^+$, where B is a neutral base) then the reaction is specific acid catalysed, whereas if the correlation is with pH then the reaction is subject to general acid catalysis. In each case the order with respect to the acid is given by the slope of the line. Below about 0.25 $M$ acid the test cannot be applied as $H_0 = \text{pH}$ (or $h_0 = [H^+]$) up to this sort of acidity; this has limited the test to the two-proton mechanism for the benzidine rearrangement as the one-proton mechanism is too rapid at acidities where there are significant differences between $h_0$ and $[H^+]$. An excellent linear $H_0$ correlation has been obtained[44] for the rearrangement of hydrazobenzene over the range 0.2–1.0 $M$ hydrochloric acid in dioxan–water solvent. The order in acid varied from 2.1 to 2.7 depending on whether the ionic strength was maintained at a constant value or not. Similar plots, with slopes of 2.0 to 2.3 have been found for N-1-naphthyl-N'-phenylhydrazine[45] and other hydroazobenzenes. Strictly the theory predicts a slope of 2.0 for a dependence on $h_0 h_+$ rather than on $h_0^2$, since $h_0$ is defined for the transfer of a proton to a neutral base whereas $h_+$ would be the appropriate function for the transfer of a proton to a positively charged base, as in the case of the second proton-transfer in the benzidine rearrangement. Values of $h_+$ have not been determined but it is not too unreasonable to expect that they would parallel $h_0$ values at these acidities.

The kinetic data based on the demonstration of specific acid catalysis in buffers, solvent isotope effects and acidity functions all support mechanisms where the proton-transfers are fast. It is possible to write equations which accommodate these facts together with the first-order dependence on hydrazo-compound and the concurrent first and second-order dependence on acidity. These are

$$RNHNHR' + H_3O^+ \overset{fast}{\rightleftarrows} R\overset{+}{N}H_2NHR' + H_2O$$

$$R\overset{+}{N}H_2NHR' \rightarrow Products + H^+$$

$$R\overset{+}{N}H_2NHR' + H_3O^+ \overset{fast}{\rightleftarrows} R\overset{+}{N}H_2\overset{+}{N}H_2R' + H_2O$$

$$R\overset{+}{N}H_2\overset{+}{N}H_2R' \rightarrow Products + 2\ H^+$$

A number of investigations concerning the effect of added electrolytes and change in the solvent polarity have been reported (see ref. 25 pp. 17–19 for a more detailed account). The addition of neutral salts causes a large increase in the rate of reaction in both the one and two-proton mechanisms. In addition, the effect of increasing the water content in a dioxan–water solvent, provided that the concentration of water is above a certain threshold value, also produces a large increase in the rate of reaction, *e.g.* for 2,2'-hydrazonaphthalene[46] the rate coefficient is increased about 80 fold on increasing the water concentration from 30 to 52 %. Both the salt and solvent effects would tend to support a mechanism involving a highly polar transition state for the one and two-proton reactions.

There is one further piece of kinetic evidence which throws light on an aspect of the benzidine rearrangement mechanism, and this is comparison of the rates of reaction of ring-deuterated substrates with the normal $^1$H compounds. If the final proton-loss from the benzene rings is in any way rate-determining then substitution of D for H would result in a primary isotope effect with $k_D < k_H$. This aspect has been examined in detail[42] for two substrates, hydrazobenzene itself where second-order acid dependence is found and 1,1'-hydrazonaphthalene where the acid dependence is first-order. The results are given in Tables 2 and 3.

TABLE 2

THE EFFECT OF RING-DEUTERIUM SUBSTITUTION ON THE RATE AND PRODUCTS OF THE REARRANGEMENT OF HYDRAZOBENZENE IN 90 % ETHANOL AT 0 °C

| Position of D | $10^3 k_3$ $(l^2.mole^{-2}.sec^{-1})$ | Product distribution (%) | |
|---|---|---|---|
| | | Benzidine | Diphenyline |
| None | 2.80 | 71.8 | 27.2 |
| All except 4,4' | 2.87 | 71.4 | 27.9 |
| Only 4,4' | 2.85 | 72.8 | 26.2 |

TABLE 3

THE EFFECT OF RING-DEUTERIUM SUBSTITUTION ON THE RATE AND PRODUCTS OF THE REARRANGEMENT OF 1,1'-HYDRAZONAPHTHALENE IN 60 % DIOXAN AT 0 °C

| Position of D | $k_2$ $(l.mole^{-1}.sec^{-1})$ | Products distribution (%) | | |
|---|---|---|---|---|
| | | 4,4' Diamine (XXV) | 2,2' Diamine (XXVII) | Carbazole (XXVIII) |
| None | 1.65 | 63.6 | 17.0 | 16.7 |
| 2,2' | 1.64 | 63.1 | 6.5 | 29.5 |
| 4,4' | 1.63 | 62.3 | 18.1 | 18.5 |

It is clear from the results that there is no kinetic isotope effect when deuterium is substituted for hydrogen in various positions in hydrazobenzene and 1,1'-hydrazonaphthalene. This means that the final removal of hydrogen ions from the aromatic rings (which is assisted either by the solvent or anionic base) in a positively charged intermediate or in a concerted process, is not rate-determining (*cf.* most electrophilic aromatic substitution reactions[47]). The product distribution in the case of hydrazobenzene is also unaffected by deuterium substitution and so must be decided before the final proton-removal from the rings. There is, however, a product isotope effect in the reaction of 1,1'-hydrazonaphthalene (XXIV) where the relative amounts of 2,2'-diamine (XXVII) and carbazole (XXVIII) are changed when 2,2'-hydrogen atoms are replaced by deuterium. This demonstrates that there is competition between the formation of the 2,2'-linked product (XXVII) and the carbazole (XXVIII) *after* the overall rate-determining step of the reaction, during the final proton-loss from XXVI. As the rate of formation of XXVII is decreased by deuterium substitution (primary isotope effect) more of the carbazole is correspondingly formed.

The conclusion from the experiments with ring-deuterated substrates is that the final proton-loss from the aromatic rings is not rate determining although in some cases it may effect the product distribution.

Much of the kinetics and products work already described has been due to Banthorpe *et al.* who have produced a mechanism for the benzidine rearrangement[42] which adequately explains the known facts. This has been called the Polar-Transition-State Mechanism and is currently accepted as being the most satisfactory description of the rearrangement. Other mechanisms have been proposed over the years and their limitations discussed (for detailed account see ref. 48).

The Polar-Transition-State theory based on earlier ideas by Hughes and Ingold[49], has as its main feature the heterolysis of the N–N bond in the mono- or di-protonated hydrazo molecule as the transition state is approached, with a de-localisation of the positive charge in the mono-protonated case and of one of the positive charges in the di-protonated case, *viz.* (16) and (17), respectively

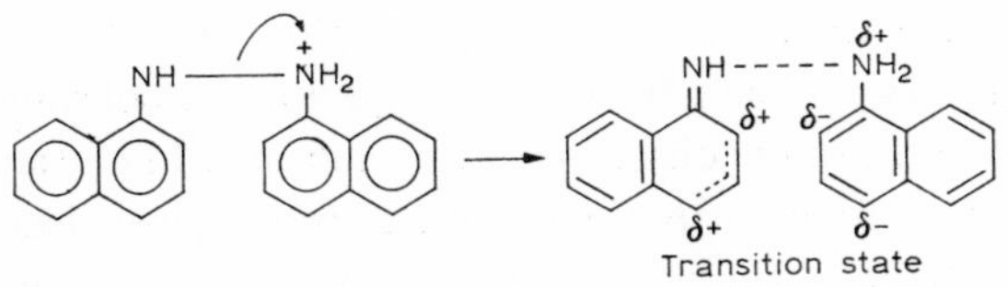

Transition state

(16)

Transition state

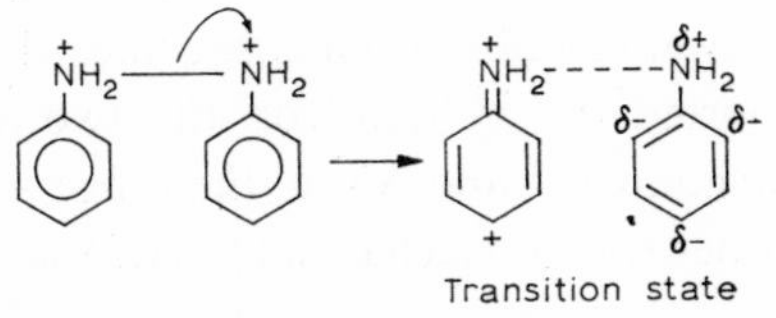

(17)

New bonds, carbon–carbon or carbon–nitrogen (depending on which product is formed) are formed synchronously with N–N bond fission. It can be seen from the diagrams of the transition states in (16) and (17) that all the possible products from the rearrangement can be derived when bonding is effected between the appropriate charged sites. The quantitative distribution of these charges in the transition state will decide the relative rates of formation of each of the products which will be greatly affected by the nature of the ring (and N) substituents in the hydrazo compound; it has been possible to correlate the nature of the substituents with the overall product distribution. Further it has been observed that electron-releasing substituents increase the rates of rearrangement *via* the one and two-proton mechanism, for example[50] the third-order rate coefficients for the reaction of hydrazobenzene, 2,2′-dimethyl and 4,4′-dimethyl derivatives in aqueous dioxan are 1.7, 8.5 and $14000 \times 10^{-3}$ $l^2.mole^{-2}.sec^{-1}$, respectively. This is consistent with what is known about substituent effects with regard to the ability of substituents to promote heterolysis and to stabilise the charges generated in the transition state. This is shown qualitatively by the fact that the unsymmetrically substituted hydrazo compound XXIX forms the *o*-semidine XXX rather than its isomer XXXI, *viz.*

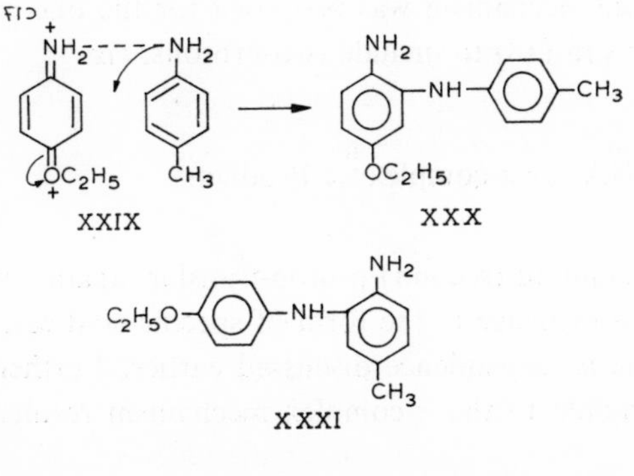

XXIX                    XXX

XXXI

consistent with the better charge-stabilising power of the $-OC_2H_5$ group rather than the $-CH_3$ group. The qualitative correlations of relative rates of reaction and substituents are discussed more fully elsewhere[51]. Quantitative correlations of rates with Hammett $\sigma$ values are difficult because the equilibrium constants for the protonation steps are included in the observed rate coefficients. In one study[41] an approximate allowance for the contributions from N-protonation has been made resulting in a reasonably good Hammett relationship which gave a $\rho$ value of $-6.1$. Similarly information from heats and entropies of activation are not easily correlated with substituents and a given reaction mechanism since they are also composite quantities, but it is reported[37, 52, 53] that for the two-proton mechanism $\Delta H^{\ddagger}$ lies in the range 19–22 kcal.mole$^{-1}$ and $\Delta S^{\ddagger}$ in the range $-10$ to $+3$ cal.deg$^{-1}$.mole$^{-1}$ whereas for the one-proton mechanism[53] $\Delta H^{\ddagger}$ is about 5 kcal.mole$^{-1}$ lower.

It has been claimed[54] that the polar-transition-state theory does not adequately explain the formation of $p$-semidine products because of the larger bonding distances that are involved, and it has been suggested that N–N bond fission preceeds C–C or C–N bond formation. This claim has been refuted[55], and it has been argued that the N–4' distance can be reduced to about 3A when the two rings are relatively displaced. Both $o$- and $p$-semidine should be formed in higher yields when the rings can thus be displaced parallel to each other in the transition state. This would be more likely at higher temperatures and when the reaction is carried out in a solvent of low polarity. There is some evidence[56] that under these conditions (and also under heterogeneous reaction conditions[57]) $o$- and $p$-semidine formation is more favoured.

Two other theories as to the mechanism of the benzidine rearrangement have been advocated at various times. The first is the $\pi$-complex mechanism first put forward and subsequently argued by Dewar (see ref. 1 pp 333-343). The theory is based on the heterolysis of the mono-protonated hydrazo compound to form a $\pi$-complex, $i.e.$ the formation of a delocalised covalent $\pi$ bond between the two rings which are held parallel to each other. The rings are free to rotate and product formation is thought of as occurring by formation of a localised $\sigma$-bond between appropriate centres. Originally the mechanism was proposed for the one-proton catalysis but was later modified as in (18) to include two-protons, $viz.$

$$\text{ArNHNHAr} \underset{\text{H+}}{\rightleftarrows} \overset{+}{\text{ArNH}_2}\text{NHAr} \rightleftarrows \pi\text{-complex} \overset{\text{H+}}{\rightarrow} \text{Products} \tag{18}$$

This scheme requires a rate-determining (second) proton-transfer, against which there is considerable experimental evidence in the form of specific-acid catalysis, the solvent isotope effect and the $h_0$ dependence discussed earlier. Further, application of the steady-state principle to the $\pi$-complex mechanism results in a rate equation of the form

$$\text{Rate} = \frac{a[B][H^+]^2}{1 + b[H^+]} \tag{19}$$

which is fundamentally different from equation (14) for which there is experimental support. Equation (19) falls down particularly in its prediction of second-order acid dependence at low acid concentration and first-order dependence at high acid, whereas the reverse is true. The theory has also been criticised on other kinetic grounds, particularly since it predicts that where the rate of formation of 4,4'-benzidine is second-order in acid that of semidine formation should be first-order, whereas in practice many examples of second-order acid dependence are known. In addition this theory does not account satisfactorily for the products of the rearrangement so that generally it cannot be accepted as an accurate account of the reaction mechanism.

The other mechanism which has been advocated[58] is that known as the "radical-pair mechanism", in which two cation radicals are thought of as intermediates held in a solvent cage so preserving the intramolecularity of the reaction, *viz.*

$$\overset{+}{Ar}\overset{+}{NH_2}NH_2Ar \rightarrow [Ar\overset{+}{\dot{N}}H_2\overset{+}{\dot{N}}H_2Ar] \rightarrow \text{products} \tag{20}$$

Recombination of the ion radicals within the cage is thought of as forming the path to rearrangement whilst escape of the radicals and subsequent reaction with the hydrazo compound leads to the formation of disproportionation products often observed. The theory is mainly directed at the two-proton mechanism and does not accommodate well the one-proton mechanism, since this requires the formation of a cation and a neutral radical, *viz.*

$$\overset{+}{Ar}NH_2NHAr \rightarrow [Ar\overset{+}{\dot{N}}H_2NHAr] \rightarrow \text{products} \tag{21}$$

It seems unlikely, especially in view of the known reactions of radical pairs, that both these intermediates would remain completely within the solvent cage with no detectable products from attack on the solvent (no such products have been observed) and that both should also give the same rearrangement products. Further this sort of bond fission would hardly account for the large solvent and salt effects that have been observed for the benzidine rearrangement, these are better rationalised in terms of a more polar transition state. No radicals were detected[59] by ESR spectroscopy during the rearrangement of hydrazobenzene (one-proton mechanism) and 1,1'-hydrazo naphthalene (two-proton mechanism), neither could polymerisation of added monomers be brought about. Radicals have been detected in some cases[60], but have been attributed to side-oxidations of products. Finally it is difficult to rationalise the distribution of products by this mechanism. Clearly, the large quantity of experimental facts concerning the benzidine rearrangement, which have been derived from a variety of kinetic techniques as well as product analyses, are at present best accommodated by the polar-transition-state theory.

A so-called "thermal rearrangement" takes place when hydrazo compounds are heated in neutral solvents. The products are much the same as for the acid-catalysed reaction[61]. The kinetics are often complex, but a first-order dependence on the substrate was shown[62] for the rearrangement of 2,2'-hydrazonaphthalene at 80 °C in a number of solvents. Reaction is faster in hydroxylic solvents and there is something of a correlation between the rate of reaction and the polarity of the solvent. This suggests that there is a substantial separation of charges in the transition state and a mechanism similar to the polar-transition-state theory (but involving the uncharged species) has been put forward[63].

## 2.3 THE NITRAMINE REARRANGEMENT (N-NITROAMINES)

N-Nitro substituted aromatic amines (XXXII) readily undergo rearrangement in the presence of mineral acids or Lewis acids in a variety of solvents and at temperatures around room temperature to give mainly *ortho* (XXXIII) but also in many cases significant amounts of the *para* (XXXIV) nitroamine. The reaction

was first discovered by Bamberger[64], who established its generality for aniline and naphthylamine derivatives. When the *ortho* and *para* positions are blocked, a diazonium ion, which can undergo the usual coupling reactions, is found[65], and small quantities of nitrous acid (but never nitric acid) have often been detected in the reaction solution. The nitro-group can also be transferred to another different molecule, *i.e.* transfer-nitration can occur, as in the rearrangement of N-methyl N-2,4-trinitroaniline (XXXV) in 80 % sulphuric acid in the presence of phenol[66].

The reaction products were shown to contain the expected product of rearrangement (XXXVI) together with the denitration product (XXXVII) and nitrophenol (XXXVIII).

As will be seen, most of the available evidence points to an intramolecular reaction, although in one case[67] it has been suggested that there exists an inter-molecular component to the rearrangement which takes place simultaneously with the intramolecular part. Evidence against de-nitration followed by C-nitration of the amine comes from a comparison[66] of the isomer proportions from the re-arrangement of N-nitroaniline (XXXII, X=R=H) in 84 % sulphuric acid, with the isomer proportions from the nitration of aniline under the same reaction conditions. In the rearrangement 93 % *ortho* and 7 % *para* nitroaniline are formed whereas for the direct nitration the isomer proportions are 6 % *ortho*, 34 % *meta* and 59 % *para*. These figures indicate that the two processes are entirely different particularly with reference to the *meta* isomer, since this has never been detected in the products from rearrangement. More convincing evidence for the intramolecularity of the reaction, however, comes from labelling experiments using the $^{15}$N isotope[68-70]. No detectable $^{15}$N was found in the rearrangement products of N-nitroaniline, N-methyl-N-nitro-1-naphthylamine and N-nitro-1-naphthylamine when the reac-tions were carried out in the presence of the $^{15}$N label in the form of nitrate or nitrite over a wide range of acid concentration. It is certain that any fragment of disproportionation ($NO_2^+$, $NO_2^-$, $NO_2$, $NO^+$ or $NO$) would exchange very rapidly with either nitrate or nitrite in aqueous solution. Two other sets of experiments based on labelled cross-over products gave conflicting results: (*a*) The concurrent rearrangement of [$^{15}NO_2$] N-nitroaniline and N-nitro-*para*-toluidine[71] gave the rearrangement product of the latter (3-nitro-*para*-toluidine) free from excess $^{15}$N whereas (*b*) the concurrent rearrangement of [$^{15}NO_2$] *para*-fluoro-N-methyl-N-nitroaniline and N-methyl-N-nitroaniline gave *para*-nitro-N-methylaniline enriched in $^{15}$N[67]. If it is assumed that in each experiment the pairs of substrates rearrange at comparable rates then it is difficult to reconcile the results of the two experiments although it is perhaps conceivable that the N-methyl substituent in (*b*) could promote a degree of (reversible) disproportionation which would account for the $^{15}$N exchange.

Rate measurements on the rearrangement of N-nitroaniline were carried out spectrophotometrically[69]. Samples were withdrawn from the reaction mixture, neutralised and the absorption of the free nitroaniline products measured at 400 m$\mu$. The kinetics were strictly first-order in the nitroamine as evidenced by the constancy of the first-order rate coefficients during all runs and over a ten-fold range of initial nitroamine concentration. This rules out most chain processes and a mechanism whereby one molecule of N-nitroaniline reacts with another. The dependence of the rate coefficient on acidity is shown for perchloric acid in Figs. 1 and 2. The log $k$ *versus* $-$pH plot (Fig. 1) is clearly curved whereas the log $k$ *versus* $-H_0$ plot (Fig. 2) is a good straight line of slope 1.3. Similar results were obtained for sulphuric acid where the slope of the $H_0$ plot was 1.25. The solvent isotope effect $k_{D_2O}/k_{H_2O}$ was determined as 3.3 and the rate coefficient remained unchanged when deuterium was substituted in the 2, 4 and 6 positions in the benzene ring. These kinetic results are in accord with a fast proton-transfer, analogous to

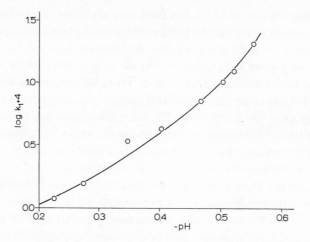

Fig. 1. Variation of log $k$ with pH for the rearrangement of N-nitroaniline in perchloric acid at 25 °C.

that proposed for the benzidine rearrangement (see section 2.2), followed by a slow unimolecular reaction of the protonated amine, *viz.*

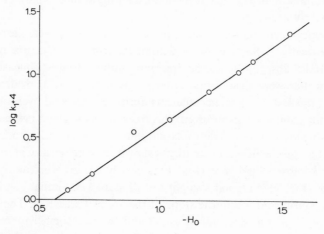

$$\text{(21)}$$

These data provide no further information regarding the nature and course of the slow step or series of steps apart from revealing that the final proton transfer

Fig. 2. Variation of log $k$ with $H_0$ for the rearrangement of N-nitroaniline in perchloric Acid at 25 °C.

from the benzene ring is not rate-limiting. The same conclusions were obtained from rate measurements for N-nitro-1-naphthylamine and its N-methyl derivative in perchloric acid in 60 % dioxan–water solvent[72].

Further evidence concerning the mechanism of the rate determining step(s) was obtained from product analyses using ring deuterium-substituted materials. The *ortho : para* ratios of nitroanilines derived from N-nitroaniline were found to be somewhat dependent on the presence of deuterium in the *ortho* and *para* positions. For the *ortho* deuterated compound (it was experimentally convenient to work with the 2,3,5,6 tetradeuterated compound) the *ortho : para* ratio was the same as for the undeuterated compound at high acidity, whereas at lower acidity it changed from 2.6 to 2.1. When both *ortho* and *para* positions were deuterated the *ortho : para* ratio increased on deuteration from 3.0 to 4.5 in 50 % perchloric acid and from 2.3 to 4.9 in 26 % perchloric acid. This suggested that there was direct competition in the post-rate-determining step between proton-loss from the C-protonat-

XXXIXa    XXXIX b

ed precursor of *ortho*-nitroaniline (**XXXIXa**) and its conversion to the C-protonated precursor of the *para* isomer (**XXXIXb**), and that such steps were reversible. On the basis of the $^{15}N$ experiments, the rate measurements and the deuterium product isotope effect, a mechanism involving cyclic transition states, *viz.*

has been proposed. The mechanism is analogous to the accepted mechanism for the Claisen rearrangement of allyl aryl ethers (see Section 3.1) and the thermal rearrangement of N-allylamines[73]. This scheme is able to account for the kinetic forms observed and isotope effects. It can also accommodate the formation of the many by-products that have been observed on many occasions (nitrophenols, tar, products derived from diazonium ions, etc.) by disproportionation of intermediate XL, as shown in (22), to form a dication and nitrite ion from which

$$
\overset{+}{N}H_2NO_2 \qquad\longrightarrow\qquad \overset{+}{N}H_2 \;+\; NO_2^- \tag{22}
$$

XL

these by-products could be formed. The direction of N–N bond fission is of course the same as in the path leading to rearrangement.

White *et al.*[74] have obtained similar kinetic results for the acid-catalysed re-arrangement of N-nitro-N-methylaniline, *i.e.* a first-order dependence on the nitroamine with a linear $H_0$ plot of slope 1.19 for phosphoric acid, and a deuterium solvent isotope effect of about three, although the results have only been presented in preliminary form*. Further, an excellent Hammett $\sigma^+$ correlation was claimed for thirteen *para* substituted nitroamines which gave a $\rho$ value of $-3.9$. Since it is expected that the rate coefficients would correlate with $\sigma$ (rather than $\sigma^+$) if sub-stituents were merely reflecting the different basicities of the amines, the $\sigma^+$ cor-relation implies that the amino nitrogen is electron-deficient in the transition state, *i.e.* that N–N bond breaking takes place in the sense $N - - - \overset{\delta-}{N}O_2$ rather than $N \overset{\delta+}{- - -} NO_2$ sense. Support for this comes from the fact that nitrous acid (rather than nitric acid) or products derived from nitrous acid have often been detected as side products, together with small quantities of organic products such as quinones and imines[65,75] which could have formed from the dication, see (22). On the basis of the partial $^{15}N$ cross-over experiment, of the kinetic results and from the fact that the yield of rearranged products is decreased on addition of excess hydroquinone (which in itself proves nothing) White proposed a tentative mech-anism (similar to the radical-pair mechanism proposed for the benzidine re-arrangement, Section 2.2) in which the rate-determining step is the homolytic fission of the N–N bond to form a radical and a radical ion (XLI) which are held together in a solvent cage, *viz.*

$$
\overset{+}{R}NH\cdot \quad \cdot NO_2
$$

XLI

These fragments either combine intramolecularly to form the *ortho* and *para* nitro compounds or dissociate completely and then undergo an intermolecular reaction to form the same products. The theory was not developed to include a detailed transition state and no mention was made of how the *para* isomer was formed. Reduction of the cation-radical could give the amine (which was observed ex-perimentally[76]), but one would expect the concurrent formation of nitrogen dioxide and hence nitrite and nitrate ions; however, the latter has never been

---

* The full papers have recently appeared[183].

detected in the products of rearrangement. Further, the mechanism does not account for the observed product isotope-effects and the absence of the *meta* product isomer. Tests using ESR have failed to show the presence of radicals[69], and no initiation of polymerisation was observed. Such a mechanism which does not adequately explain all the experimental facts cannot be accepted. It is conceivable that, under the conditions employed (low acid concentration and the relatively high temperature of 40 °C) White *et al.* have observed a thermal rearrangement as has been established for the rearrangement of N-methyl-N-nitro-1-naphthyl-amine[77].

An earlier theory[78] as to the mechanism of the nitramine and other rearrangements was based on the de-nitration of the protonated nitroamine to form the nitronium ion $NO_2^+$ which migrates to the *ortho* and *para* positions of the aromatic ring method without becoming kinetically free, *via* a $\pi$ complex. This seems to be ruled out by the Hammett $\sigma^+$–$\rho$ correlation described earlier which requires that the amino-nitrogen and not the nitro-nitrogen becomes electron-deficient in the transition state, and also by the complete absence of any *meta*-isomer products. Further, it might have been expected that disproportionation would lead to nitrate ion which has never been detected.

It is not possible to rule out an intramolecular rearrangement *via* a transition

XLII

state of the type XLII where the nitrite ion does not become free but rather is held to the aromatic nucleus by electrostatic forces as has been proposed in the polar-transition state theory for the benzidine rearrangement. The positive charges are written in the dication on the amino-nitrogen and the *para* carbon atoms but it is thought that appreciable charge would also be present at the *ortho* positions to allow the formation of the *ortho* rearranged product.

It is not easy to envisage further experiments (kinetic or otherwise) which would throw more light on the mechanism of the rearrangement. As far as the author is aware no activation parameters have ever been determined. Conceivably information regarding the magnitude of the entropy of activation $\Delta S^{\ddagger}$ might prove useful in confirming a cyclic transition state, but there remains the difficulty of estimating the contribution to the experimentally determined value of $\Delta S^{\ddagger}$ of the entropy charge $\Delta S^0$ associated with the protonation of the nitroamine, since the determination of the equilibrium constant for the protonation step is made difficult by the subsequent rearrangement of the protonated form.

## 2.4 THE FISCHER-HEPP REARRANGEMENT (N-NITROSOAMINES)

It is well known that N-nitroso aromatic amines (XLIII) rearrange in acid solution to form the C-nitroso isomer. In most cases that have been recorded the *para* isomer is formed (XLIV), *viz.*

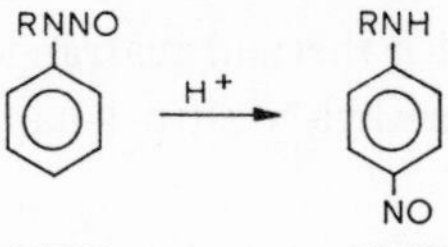

RNNO    $\xrightarrow{\text{H}^+}$    RNH, NO

XLIII      XLIV

although in the naphthylamine series the 2-naphthylamine derivative (XLV) yields the 1-nitroso product (XLVI), *viz.*

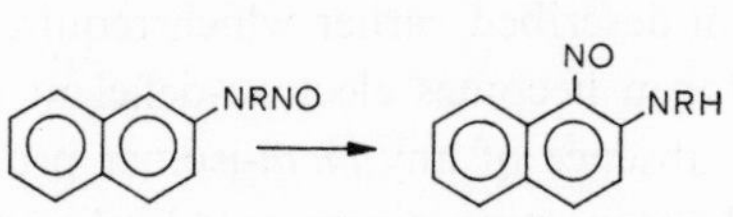

NRNO → NO, NRH

XLV      XLVI

There is no reference to the formation of the *ortho* nitroso amine in the benzene series, indeed the formation and characterisation of *ortho* nitroso amines from any reaction has only been reported once or twice[79]. When the *para* position is blocked by a substituent, de-nitrosation so the secondary amine can occur[80]; a certain amount (which depends on the conditions) of denitrosation occurs also, concurrently with the rearrangement[81], so that N-nitroso-N-methylaniline (XLVII) yields N-methylaniline (XLVIII) as well as the rearrangement product *p*-nitroso-N-methylaniline XLIX, *viz.*

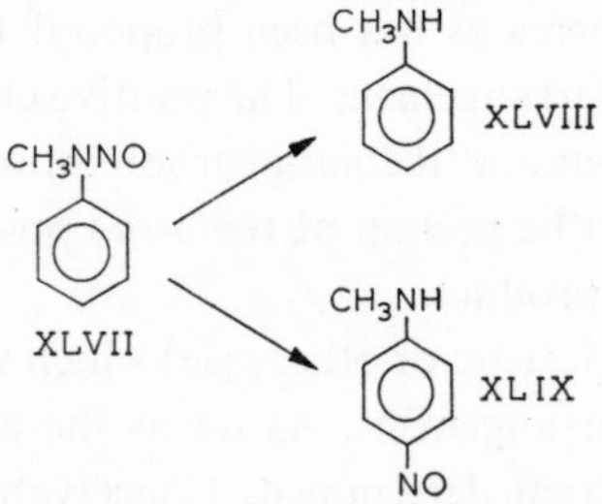

CH$_3$NNO    →    CH$_3$NH (XLVIII)

XLVII

CH$_3$NH, NO (XLIX)

On the preparative scale the best conditions for a high yield of rearrangement product appear to be those used by the original discoverers of the reaction Fischer and Hepp[82], who treated a solution of the N-nitrosoamine in dry ether or ethanol with a solution of hydrogen chloride in ethanol. In aqueous acid solution the yield of rearranged product is generally lower[83], and more de-nitrosated product is formed. Within the range of acids studied it seems that the best yields of rearranged product are obtained with hydrochloric acid[84]. Trans-nitrosation reactions have been observed; N-methyl-N-nitroso-*p*-toluidine will nitrosate added diphenylamine[84], and N-methyl-N-nitrosoaniline with hydrogen chloride gave the nitrosyl chloride adduct (L) when the reaction was carried out in the presence of the olefin

LI. On the basis of the evidence of trans-nitrosation and on the apparent specific catalysis by hydrochloric acid, a mechanism first proposed by Houben[85] has become generally accepted [84,86,87]. This involves a reversible de-nitrosation forming the secondary amine and nitrosyl chloride followed by C-nitrosation in the *para* position, *viz.*

Actually the evidence by no means requires this mechanism since there is no reason why the nitrosamine itself should not act as a primary nitrosating agent (thus allowing cross-nitrosation) and there was no rate data available to support the idea of specific catalysis by hydrochloric acid.

It may seem surprising that until very recently no kinetic experiments had been reported on this reaction. Now two investigations have been carried out; the results from one are not yet published in full[81,185], but will be discussed later. Russian workers[88] have found that rearrangement takes place in aqueous solution of sulphuric acid even in the presence of vast excess of either urea or sulphamic acid which are known to react with nitrous acid at rates much greater than that of the rearrangement. It was noted that the yield of *para*-nitroso-N-methylaniline actually increased as the urea concentration was increased. The rate of reaction was measured spectrophotometrically at 25 °C by following the disappearance of the absorption band of the nitrosamine. The reaction was found to be first order in the nitrosamine as evidenced by the constancy of the first-order rate coefficient as the initial concentration of the nitrosamine was varied from $1 \times 10^{-4}$ *M* to $4 \times 10^{-4}$ *M*. The rates correlated approximately with $H_0$, but the slope of the line varied from 1.58 for N-nitroso-N-methylaniline to 0.75 for the *meta*-OH substrate and to 1.95 for *meta*-CH$_3$. The rearranged C-nitroso product was in all cases formed together with the product of de-nitrosation, *viz.* the secondary amine, and from the ratio of the two products the authors calculated the rate coefficients for both rearrangement and de-nitrosation assuming (without any evidence) that the latter step leads to intermolecular rearrangement in the absence of urea or sulphamic acid. At high acid concentrations the de-nitrosation step was favoured whereas in the presence of electron-releasing *meta*-substituents more of the rearranged product was found. No conclusions were drawn regarding the mechanism beyond the statement that the rearrangement was, in part, intramolecular.

In another investigation[81,185], the rates of formation of *para*-nitroso-N-methyl-aniline from N-nitroso-N-methylaniline in hydrochloric acid solution at 31 °C were measured by noting the appearance of the absorption at 340 m$\mu$ due to the protonated from of the *para*-nitroso-N-methylaniline. A typical run is shown in Fig. 3. The decreasing absorption at about 260 m$\mu$ represents the disappearance

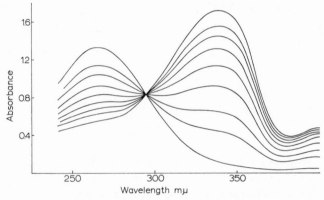

Fig. 3. Visible–*UV* absorbance curves for the rearrangement of N-nitroso-N-methylaniline in hydrochloric acid at 31 °C.

of the nitrosamine. There is an excellent isosbestic point at 295 m$\mu$ confirming that a simple A → B process is taking place. Unfortunately the trace at $t = \infty$ did not pass through this point, and so the optical density at $\infty$ at 340 m$\mu$ could not be reliably used in the first-order rate equation; consequently the Guggenheim method[89] was used where log $[(O.D.)_t - (O.D.)_{t+T}]$ was plotted against the time $t$ and the rate coefficient calculated from the slope. Good first-order plots were obtained and there was no variation in the first-order rate coefficient as the initial concentration of the nitrosamine was varied. There was a good correlation (see Fig. 4) between log $k$ and the Hammett acidity function $H_0$ in the acid range 3–6 $M$, the slope of the line being 1.2. The line becomes curved at high acidity at a point corresponding to the change from the free base N-nitroso-N-methyl-aniline to its protonated form as observed by a change in its ultraviolet spectrum. A similar independence of rate on acidity at high acidity was observed[97] in the phenylhydroxylamine rearrangement (see Section 2.5) and was attributed to complete protonation of the starting material.

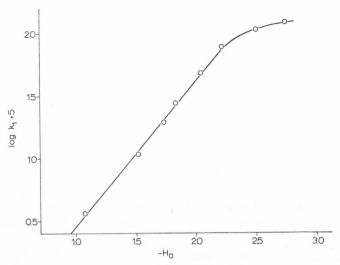

Fig. 4. Variation of log $k$ with $H_0$ for the rearrrangement of N-nitroso-N-methylaniline in hydrochloric acid at 31 °C.

These kinetic results are consistent with a fast reversible protonation of the nitrosamine followed by a unimolecular decomposition, *viz.*

The observed solvent isotope effects of $k_{D_2O}/k_{H_2O}$ of 1.6 at 5.5 $M$ acid and 2.1 at 4 $M$ acid confirm that reaction takes place by some rate-determining reaction of the protonated substrate. The kinetic pattern is thus similar to that observed in the benzidine and nitramine rearrangements. It appears that that chloride ion is not involved here as was previously thought, and the evidence is against the formation of nitrosyl chloride and its involvement as an effective agent. Rearrangement was found to take place in the presence of a large excess of urea and a preliminary experiment[90] using [15N] labelling showed that no pick up of label occurred when the reaction was carried out in the presence of sodium [15N] nitrite in ethanol–water solvent containing either sulphuric or hydrochloric acid. In aqueous acid, however, exchange of [15N] took place between N-nitroso-N-methylaniline and sodium [15N] nitrite very rapidly before any rearrangement occurred. This exchange problem remains to be cleared up. Experiments[185] with ring deuterium-labelled materials showed that cross-nitrosation occurred, again before any significant amount of rearrangement took place. 2,4,6-Trideuterated-N-nitroso-N-methylaniline (LII) transferred the –NO group to N-methylaniline in the presence of a large excess of urea in aqueous hydrochloric acid, *viz.*

AROMATIC REARRANGEMENTS

MeNNO $\underset{\text{LII}}{\overset{\text{D}\quad\text{D}}{\bigcirc}}$ D $\xrightarrow{\text{H}^+}$ $\underset{\text{LIII}}{\overset{\text{MeN}^+\text{HNO}}{\bigcirc}}$ $\xrightarrow{\text{C}_6\text{H}_5\text{NHMe}}$ $\overset{\text{MeNH}}{\bigcirc}$ $+$ $\overset{\text{MeNNO}}{\bigcirc}$

No accurate rate measurements have yet been carried out, but qualitative experiments showed that the exchange was acid-catalysed so that it is likely that the protonated form of the nitrosamine (LIII) acts itself as a nitrosating agent. This would account for all the earlier cross-nitrosations that have been observed without necessitating the prior formation of nitrous acid, nitrosyl chloride or any other carrier of $NO^+$.

One further piece of kinetic evidence has been obtained pertinent to the mechanism of the rearrangement, that is the observation[185] that the deuterated compound LII formed the rearranged product at a slower rate than the protium compound, i.e. there was an isotope effect $k_H/k_D$ of 1.7 at 6 $M$ HCl. It is considered that this is too large to be a secondary effect and so means that the final proton loss to form para-nitroso-N-methylaniline must be at least in part, rate-determining. This is rather unusual in the field of electrophilic aromatic substitution generally and more specifically is not observed in the benzidine or nitramine rearrangements.

The product of de-nitrosation, the secondary amine is often (if not always) formed along with the rearranged product in aqueous acid solution, particularly at high acid concentration and also in strong acids. The kinetics of the reverse reaction, the N-nitrosation of N-methylaniline, have been studied by Kalatzis and Ridd[91] who found that below 1 $M$ perchloric acid nitrosation of the amine by nitrous anhydride ($N_2O_3$) took place, whereas above this acidity a strongly acid-catalysed reaction became important which was interpreted as attack by a positive nitrosating agent (e.g. $NO^+$) on the protonated amine. Russian workers[92] have attempted to look at the denitrosation reaction in isolation by using para-substituted materials (to remove the possibility of rearrangement) and by working in the presence of a trap for $NO^+$ carriers (urea). A correlation of rate coefficients with $H_0$ was claimed, but the slope varied from 0.78 to 1.56 as the para substituent was changed and most para substituents caused an increase in the rate of reaction over that with the unsubstituted material. Reactions in hydrochloric acid were correspondingly faster than those in sulphuric acid. The facts have been interpreted as involving two concurrent steps in one of which the anion ($Cl^-$) is considered to play an important part in the removal of the nitroso group.

So far as the rearrangement is concerned, the scheme consistent with all the known facts, for the reactions of N-nitrosoaromatic amines in acid solution is set out below. It incorporates the rearrangement and the de-nitrosation, viz.

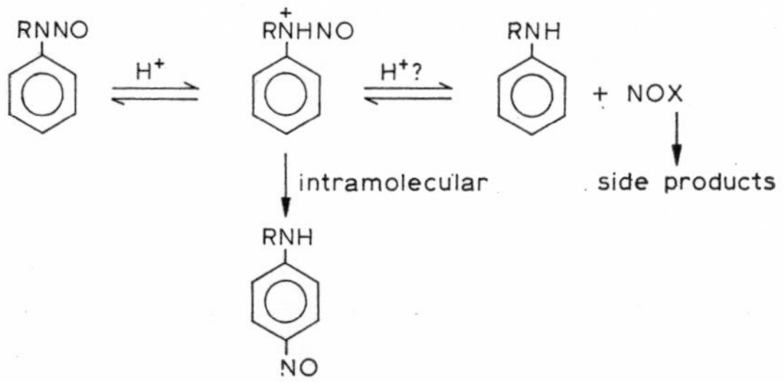

RNNO $\quad$ $\overset{+}{\text{RNHNO}}$ $\quad$ RNH

$\xrightarrow{\text{H}^+}$ $\qquad$ $\xleftrightharpoons{\text{H}^+?}$ $\qquad$ + NOX

intramolecular $\qquad$ side products

RNH

NO

This scheme has been confirmed recently[185] by noting the variation of the observed first-order rate coefficient for the reaction of N-nitroso-N-methylaniline in hydrochloric acid as the concentration of added N-methylaniline is changed. The measured first-order rate coefficient is, of course, a function of the individual rate coefficients; the above scheme would predict that if sufficient N-methylaniline were added, then the de-nitrosation reaction might be suppressed completely leaving only the rearrangement reaction. In terms of the rate coefficients, one would expect the observed rate coefficient to decrease to a limiting value as the concentration of added N-methylaniline is increased. This limiting value would then be the rate coefficient for rearrangement. Simultaneously, the yield of rearranged product should increase towards 100 %. The results are set out below.

| [Added PhNHMe] $(10^3\ M)$ | $10^4\ k_{obs}$ $(sec^{-1})$ | Yield rearrangement $(\%)$ |
|---|---|---|
| 0 | 5.07 | 28 |
| 0.44 | 3.74 | 54 |
| 0.94 | 3.61 | 60 |
| 3.08 | 2.80 | 76 |
| 3.96 | 2.78 | 78 |
| 5.71 | 2.84 | 80 |

It is claimed that the limiting value of $k_{obs}$, $2.81 \times 10^{-4}$ sec$^{-1}$, represents the rate coefficient for the rearrangement reaction above $(k_r)$. The ring deuterium isotope effect $k_H : k_D$ was re-determined for this individual rate coefficient for rearrangement by finding the limiting value in the presence of added N-methylaniline and was found to be 2.4 at two different acidities, as compared with 1.7 for the ratio of the observed composite rate coefficients, as expected, since no isotope effect would be predicted for the de-nitrosation step.

Rate measurements were also carried out[81] in the presence of added urea in an attempt to remove the free nitrite species formed and so prevent N-nitrosation (the reverse of de-nitrosation). The observed rate coefficient was now found to be the sum of the rate coefficients for rearrangement $(k_r)$ and de-nitrosation $(k_d)$

together with a contribution $k_u$[urea] which represents the reaction between the starting material (or more probably its protonated form) and urea, *i.e.*

$$k_{obs} = k_r + k_d + k_u[urea]$$

This is another example (*cf.* the deuterium exchange experiments earlier) where the protonated N-nitroso amine itself reacts as a nitrosating agent transferring the $NO^+$ group to a suitable acceptor, without the formation of a free nitrite species such as $NO^+$, $HNO_2$, $N_2O_3$ or NOCl.

As to the mechanism of the intramolecular rearrangement there is no evidence available which permits its formulation. There appear to be three possibilities by analogy with schemes put forward for other aromatic rearrangements. (*a*) A caged radical-ion mechanism (*b*) a polar cyclic transition state theory as in the nitramine rearrangement and (*c*) a $\pi$-complex mechanism. There is no conclusive evidence for (*a*) or (*c*) in other rearrangements and so there is no reason to invoke similar mechanisms here. The distance involved in the bond making step seems rather large for a mechanism like (*b*), although recently a mechanism has been proposed[93] for the rearrangement of sulphanilic acid involving bonding to positive charge anchored on the *para* carbon atom. The analogous intermediate for this reaction would be LIV, *viz.*

LIV

An attractive feature of a mechanism of this type would be its ability to explain the observed ring-hydrogen isotope effect. A difficulty is the absence of the *ortho* isomer product, although conceivably as yet unidentified products derived from the (unstable) *ortho* isomer may be formed. An intermediate analogous to $\pi$-complex has been suggested[94] during the diazotisation of aniline at high acidities, *viz.*

where B is a base and $X^+$ the effective nitrosating agent ($NO^+$ or its hydrated form).

As has been indicated, since there is a ring isotope effect there must be a degree of C–H bond breaking in the transition state of the rate-determining stage. Clearly further work is required in this system before a definitive mechanism can be established for the intramolecular rearrangement.

Very recently the kinetics of the rearrangement of N-nitrosodiphenylamine in methanol and methanol–toluene mixtures containing hydrogen chloride have been reported[184]. The reaction was first-order in both the nitroso compound and hydrogen chloride and there was no dependence (apart from a salt effect) on added chloride ion. The author suggests a mechanism involving the slow formation of nitrosyl chloride *via* a cyclic four-centred transition state. This does not accord with the fact that rearrangement does take place in the presence of traps for free NOX compounds nor with the kinetic acidity-dependence on $H_0$, the solvent-isotope effect, or the observed ring-isotope effect[81].

## 2.5 THE PHENYLHYDROXYLAMINE REARRANGEMENT

Phenylhydroxylamine (LV) is converted in 2 $N$ aqueous sulphuric acid at room temperature to *para*-aminophenol (LVI), *viz.*

together with small quantities of by-products. This reaction was extensively examined by Bamberger[95] after whom the arrangement is sometimes known, and has received very little attention in recent years. All the product analyses (mainly by Bamberger) point to the fact that the reaction is intermolecular. When the reaction is carried out in ethanolic acid solution the products are *ortho*- and *para*-phenetidines (LVII and LVIII), *viz.*

and reaction in methanol yields anisidines. If hydrochloric acid is used the *ortho*- and *para*-chloroanilines are formed and reaction in the presence of phenol yields some $(p)OHC_6H_4C_6H_4NH_2(p)$. There are many other examples which support the view that the reaction is intermolecular, the OH group being replaceable by any suitable nucleophile Y, and further that the reaction is a *nucleophilic* inter-molecular rearrangement (as opposed to an electrophilic reaction as in the Orton rearrangement), *i.e.* with the OH group departing in the OH⁻ sense rather than OH⁺. Direct confirmation of the intermolecularity of the reaction was provided by the observation[96] that full $^{18}O$ uptake took place when the reaction was carried out in $H_2{}^{18}O$.

Very little kinetic work has been carried out on this reaction, but what there is confirms the conclusion above that the rearrangement is of the nucleophilic intermolecular type. Heller *et al.*[97], showed that at low acidity the rate was proportional to the acid concentration. At higher concentrations of acid this proportionality disappeared consistent with a scheme whereby a protonated intermediate is formed through which reaction takes place, and at high acid the base becomes completely protonated. There was no sign of a dependence on $[H^+]^2$ as in the benzidine rearrangement nor on $[Cl^-]$ when the reaction was carried out in hydrochloric acid. There are two possible sites for protonation, on the nitrogen atom (LIX) and the oxygen atom (LX). Product formation is best accounted for by assuming protonation on the oxygen atom. Nucleophilic attack at the *para* and

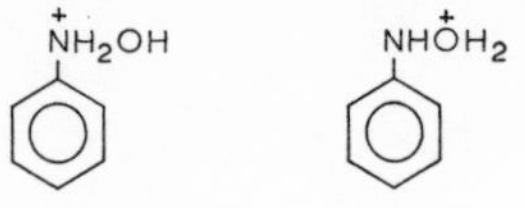

LIX          LX

*ortho* positions of LX is not thought to occur directly, as, where chloride ions are present, their concentration does not appear in the rate equation. The mechanism is best thought of as a rate-determining heterolysis of LX to form ion LXI followed by a rapid reaction with any nucleophile present at the *ortho* and *para* positions.

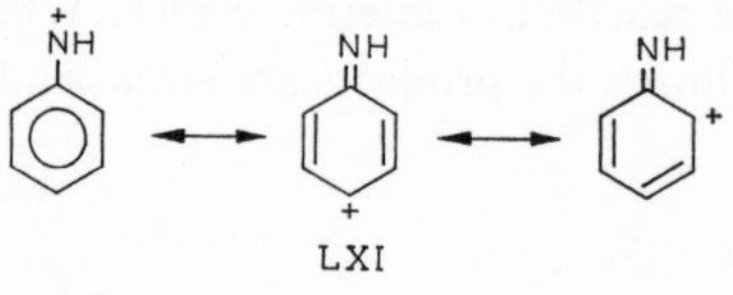

LXI

Yukawa[98] has written a reaction scheme involving reaction *via* LIX, but this mechanism does not account so readily for the products.

N-substituted phenylhydroxylamine derivatives, *e.g.* N-acetyl and N-sulphonic acid, also form the *para*-aminophenol[99], but more bulky groups prevent reaction, it is thought by steric hindrance to the approach of the hydronium ion. O-substituted phenylhydroxylamines, on the other hand form only the *ortho* product, it is thought *via* an intramolecular rearrangement, *e.g.*

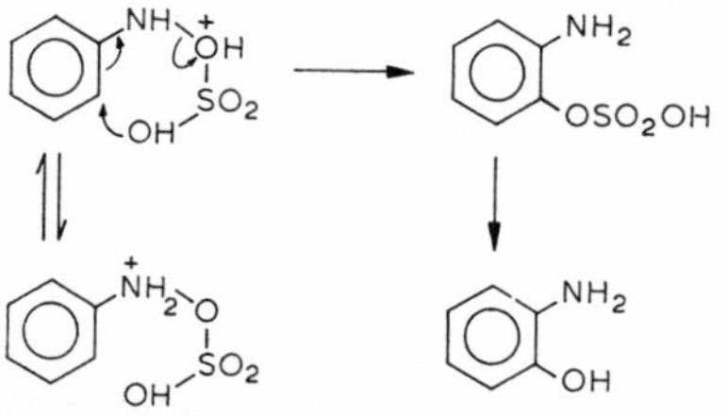

(23)

This reaction has been shown[100] to be involved in the degradation *in vivo* of carcinogenic arylamines by N-hydroxylation and subsequent rearrangement.

## 2.6 REARRANGEMENTS OF OTHER N-SUBSTITUTED AROMATIC AMINES

A large number of rearrangement reactions of N-substituted aromatic amines are known apart from the five examples already discussed. In LXII

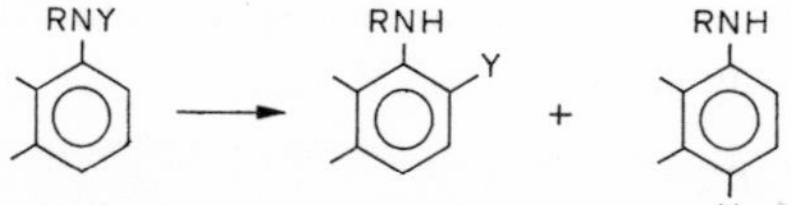

LXII

Y can be any of the following groups, halogen, –NHAr, –NO$_2$, –NO, –OH, alkyl, –SO$_3$H, –N=NAr, cyclohexadienone, –CH$_2$SAr, –SO$_2$Ar. Much work has been done on the first three and relatively little kinetic work on the last six where the mechanism has very largely been deduced from the products and by analogy with other reactions. There follows a short section indicating where kinetic results have proved useful in establishing or confirming a mechanism for the less-well-studied reactions.

Diazoamino compounds, (LXIII) produced by coupling a diazonium ion with a primary aromatic amine, *viz.*

$$ArN_2^+ + ArNH_2 \rightarrow ArN=N–NHAr + H^+$$
LXIII

undergo rearrangement in acid solution to form *para*-amino azocompounds[101] (LXIV), *viz.*

$$ArN=N–NHAr \xrightarrow{H^+} ArN=NAr(p\text{-}NH_2)$$
LXIV

If the *para* position is blocked then the amino group will enter the *ortho* position. Friswell and Green[102] suggested an acid-catalysed fission of LXIII to form the diazonium ion and the primary aromatic amine (the reversal of its formation) followed by an electrophilic aromatic substitution at the *para* position of the amine (as in aromatic diazo-coupling). This was supported by much product work including the fact that added amines[103] and phenols[104] underwent diazo coupling reactions. Kinetic measurements by Goldschmidt *et al.*[105], who used aniline as a solvent, showed that the disappearance of reactant was first order and that the reaction was subject to general acid catalysis. In strong acids such as hydrochloric acid or nitric acid the reaction was approximately first order in the acid but appeared to contain a second-order component, whereas, when weaker acid catalysts such as substituted benzoic acids were used, the order in acid approached two. This has been interpreted by Hughes and Ingold[106] as involving a third-order reaction as in the Orton rearrangement, first-order in each of the diazo-amino compound, hydrogen ion and a nucleophile. In strong acid solutions in aniline the nucleophile is aniline itself which is present in vast excess and so does not show up in the rate equation, but, when weak acids are used, their more strongly nucleophilic anions take over the role of nucleophile from the solvent. The mechanism is then seen as a protonation of the –NH– site in the substrate followed by attack by a nucleophile B: at the central nitrogen atom which is synchronous with the N–N bond breaking shown in LXV, *viz.*

The rearrangement of quinamines as in (24)

(24)

is formally similar to the benzidine rearrangement. Reaction takes place in alcohol or acetic acid solvents containing, for example, hydrogen chloride[107]. By varying substituents A, B, X and Y it was shown as for the benzidine rearrangement that the reaction was intramolecular by the absence of cross-over products. The reaction was observed to be first order in quinamine and also in acid.

The rate of reaction increased as A and B were made electron donors whilst electron-withdrawing groups in these positions retarded reaction. The effect of changing substituents X and Y was less marked. These results have been taken[107] to indicate reaction by protonation on nitrogen rather than oxygen, and a mechanism involving a π-complex has been put forward, although this seems a matter for speculation only on the evidence available. Similar benzidine-like rearrangements have been reported for imidazole[108], pyridine[109] and thiazole hydrazo compounds[110].

A rearrangement reaction which has been much studied, and in many ways resembles the reaction of phenylhydroxylamines, is the Wallach rearrangement[111] of azoxybenzene (LXVI) to give *para* hydroxyazobenzene (LXVII), *viz.*

This reaction does not strictly fall into the class of N-substituted aromatic amines, but is included here because of its similarity with many reactions in this section. A recent comprehensive review of the reaction and related reactions is available[112]. The reaction takes place in strongly acid solutions (usually sulphuric acid). The intermolecularity was established by $^{18}O$ labelling experiments. Reaction in $^{18}O$-labelled sulphuric acid gave LXVII which contained the $^{18}O$ content of the solvent, whilst $^{18}O$-labelled LXVI gave unlabelled LXVII[113]. It has been claimed that the formation of small quantities of the *ortho* isomer is by an intramolecular process. Some evidence concerning the intermediate has been obtained from $^{15}N$ work[114]. When LXVI labelled in one nitrogen only with $^{15}N$ was rearranged in 83 % sulphuric acid the $^{15}N$ became symmetrically distributed in the product LXVII, and LXVI recovered from the reaction mixture still had the same unsymmetrical degree of labelling. This requires a symmetrical intermediate, which has been formulated by one group of workers[114] as the cyclic oxide LXVIII and by another as the dication LXIX, *viz.*

The reaction is clearly acid-catalysed and analysis[115] of the variation of the rate-coefficient with acidity ($H_0$) in the range 75–96 % sulphuric acid has indicated that a second proton transfer is taking place. This is because, at these acidities, spectroscopic measurements show that the azoxybenzene is completely protonated, yet the observed first-order rate coefficient increases along the range from $0.016 \times 10^{-5}$ to $26.1 \times 10^{-5}$ sec$^{-1}$ at 25 °C. For a scheme like

$$S + HA \rightleftharpoons SH^+ + A^-$$

$$SH^+ + HA \rightleftharpoons SH_2^{2+} + A^- \tag{25}$$

$$SH_2^{2+} \xrightarrow{slow} products$$

it is possible to derive (26) and (27) for the variation of the observed first-order rate-coefficient $k_{obs}$ with acidity, *viz.*

$$\log k_{obs} - \log \frac{[SH^+]}{[S]+[SH^+]} = -H_+ + \text{constant} \qquad (26)$$

$$\log k_{obs} - \log \frac{[S]}{[S]+[SH^+]} = -H_0 + \log [HA] + \text{constant} \qquad (27)$$

making the usual assumptions regarding the ratios of activity coefficients. The left-hand sides of (26) and (27) plotted against $-H_0$ and $-H_0 + \log [H_2SO_4]$, respectively, ($H_+$ is assumed to parallel $H_0$) gave straight lines for the range 75–87 % sulphuric acid with slopes 1.3 and 2.3 respectively. Above 87 % sulphuric acid the plots curved which suggests that there is not a single mechanism operative over the whole acid range. It is difficult to obtain further information regarding the possibility of specific or general acid catalysis at these high acidities. The site for the first protonation has been shown[116] to be the azoxy oxygen atom. The following scheme has been proposed for the complete mechanism, viz.

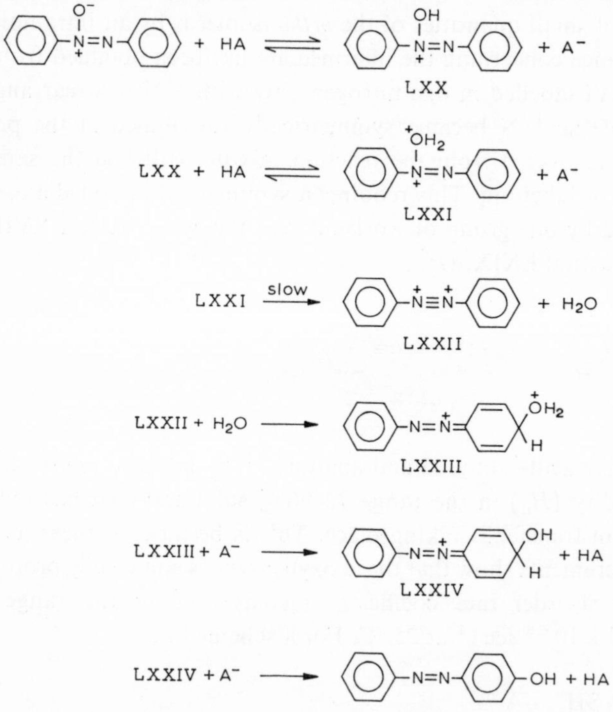

but it is not yet generally accepted and a different mechanism has been put forward by Hahn et al.[117] following work with azoxybenzene itself and several substituted azoxybenzenes. It is interesting to note that para-methylazoxybenzenes show rather different reaction characteristics and it appears that the mechanism of rearrangement here is quite markedly changed. Further work is needed to establish the mechanism unambiguously.

## 3. Rearrangements of ethers and esters

### 3.1 THE CLAISEN REARRANGEMENT

The rearrangement of an O-allyl ether (LXXV) to form the isomeric *ortho* allyl phenol (LXXVI)

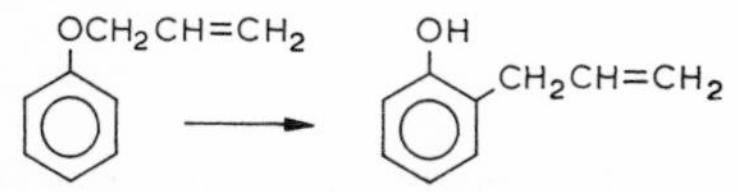

LXXV         LXXVI

was discovered in 1912 by Claisen[118] and this type of reaction has been extensively studied since. The reaction is not confined to aromatic systems but is a general one for the rearrangement of O-allyl enol ethers to form unsaturated carbonyl compounds or the corresponding enols, but since much of the work done has been concerned with rearrangement within an aromatic system it is thought pertinent to include that section in this chapter. This and related unsaturated rearrangements have been thoroughly reviewed recently[119, 120].

Reaction is brought about thermally usually at around 200 °C either in an inert solvent (diphenyl ether, cyclohexane and diethylaniline have been used) or in the absence of a solvent. If the *ortho* positions in the aromatic ether are blocked (as in LXXVII) then rearrangement to the *para* position LXXVIII occurs, *viz.*

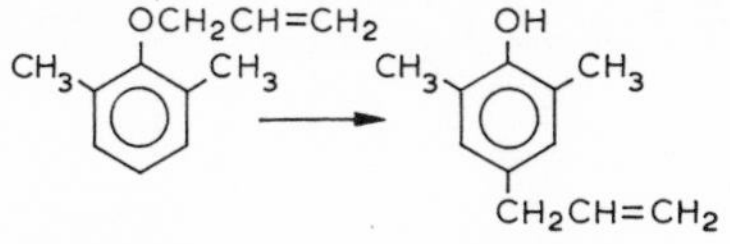

LXXVII         LXXVIII

The reaction is undoubtedly intramolecular. Evidence for this comes from the absence of crossed-over products when two different ethers were rearranged together[121]. Further, optically active *ortho*-1,3-dimethylallyl phenol is formed from optically active 1,3-dimethylallyl phenyl ether[122] and the presence of dienone intermediates has been demonstrated. Claisen and Tietze[123] first proposed intermediate dienones such as LXXIX and LXXX. This was established[124] when

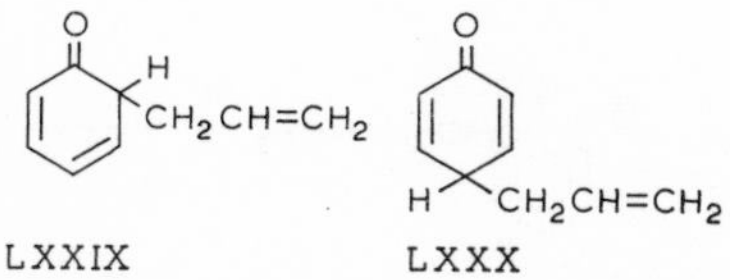

LXXIX         LXXX

the Diels-Alder adducts were isolated from the reaction carried out in the presence of maleic anhydride. In one case[125] the dienone was synthesized and it was possible to complete the rearrangement to the phenol, regenerate the allyl ether and also to obtain the maleic anhydride adduct. Starting with the dienone it was possible to measure the rate coefficients for the reverse step (to form the ether) and for the formation of the product phenol, $k_{-1}$ and $k_2$ respectively, $viz$.

$$\text{Ether} \underset{k_{-1}}{\overset{k_1}{\rightleftarrows}} \text{Dienone} \overset{k_2}{\rightarrow} \text{Phenol}$$

from the relative amounts of ether and phenol formed since under the chosen conditions $k_1$ was negligible. Elegant work by Schmid and Schmid and by Ryan and O'Connor[126] who used $^{14}$C-labelling techniques, showed that the $\gamma$-carbon atom in LXXXI became the $\alpha$-carbon atom in the $ortho$ rearranged product LXXXII, $viz$.

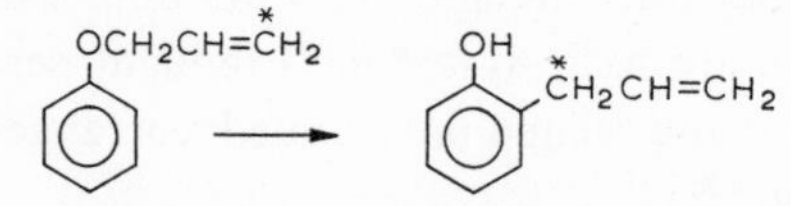

LXXXI          LXXXII

whereas for the $para$ rearrangement (LXXXIII to LXXXIV) the position of the $^{14}$C remained in the $\gamma$-position in the allyl group upon rearrangement, $viz$.

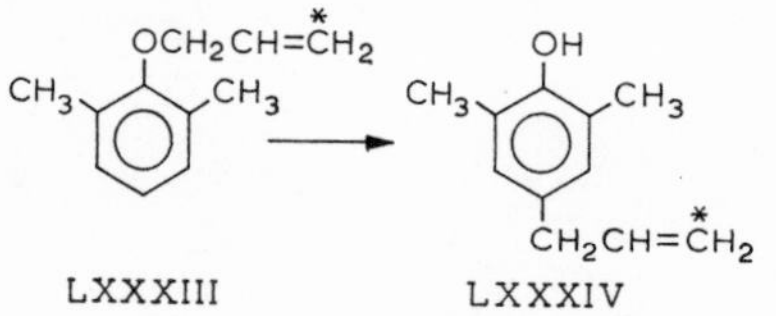

LXXXIII          LXXXIV

Experiments with $^{14}$C-labelled substrates also demonstrated conclusively the intramolecularity of the rearrangement. The generally accepted scheme involves the formation of the dienone LXXXV which can lose a hydrogen atom if R = H, to form the $ortho$ product or if R $\neq$ H further rearrangement to LXXXVI occurs with subsequent formation of the $para$ product, $viz$.

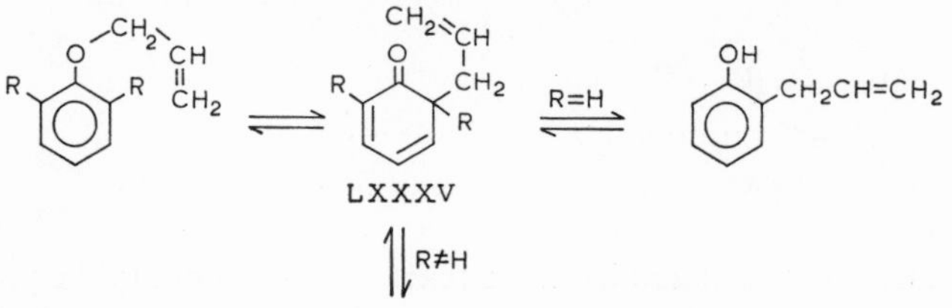

LXXXV

LXXXVI

This scheme requires a cyclic transition state for the formation of **LXXXV** where a carbon–carbon bond is being formed synchronously with the breaking of a carbon–oxygen bond together with an analogous cyclic transition state for the conversion of **LXXXV** to **LXXXVI**.

Early kinetic work[127] showed that the formation of both *ortho* and *para* products was a first-order process and that the rates of reaction were insensitive to added acid or base and to change of solvent. The activation parameters were of the same order of magnitude for both reactions and the suggestion was made that both had a similar rate-determining step. Schmid *et al.*[128] showed that the formation of a dienone intermediate in the *para* rearrangement was also reversible since the radioactivity from allyl 2,6-dimethyl-4-allyl-$\gamma$-[14]C phenyl ether **LXXXVII** became uniformly distributed in the $\gamma$ carbon atoms of the O- and C-allyl groups

LXXXVII                    LXXXVIII

as in **LXXXVIII** when the material was subjected to the conditions for rearrangement. Further, Haegele and Schmid[129] again using [14]C-labelled materials carried out a full kinetic analysis by determining the radioactivity in the O- and C-allyl groups as a function of time, and were able to calculate the rate coefficients for the individual steps of the reaction scheme.

The cyclic nature of the transition states for both *ortho* and *para* rearrangements is confirmed by the sign and magnitude of the entropy of activation $\Delta S^{\ddagger}$. In general $\Delta S^{\ddagger}$ is large and negative. Typical values are $-12$ cal.deg$^{-1}$.mole$^{-1}$ and $-11.5$ cal.deg$^{-1}$.mole$^{-1}$ for the rearrangement of O-allyl phenyl ether in diphenyl ether[130] and carbitol[131] solvents respectively. Similarly for the *para* rearrangement $\Delta S^{\ddagger}$ is $-11.7$ cal.deg$^{-1}$.mole$^{-1}$ for the allyl ether of methyl *ortho*-cresotinate[132] and in the non-aromatic case $\Delta S^{\ddagger}$ is $-7.7$ cal.deg$^{-1}$.mole$^{-1}$ for the rearrangement of allyl vinyl ether in the gas phase[133]. The volume of activation $\Delta V^{\ddagger}$ obtained[134] from the variation of the rate coefficient with the pressure, is also large and negative $(\Delta V^{\ddagger} \sim -18$ cm$^3$.mole$^{-1})$. These are all taken to mean that in the transition state there is a restriction on the freedom of movement (*i.e.* a loss of entropy) as compared with the initial state as expected for the cyclic transition state proposed for the Claisen rearrangement.

From the study with the synthesised dienone[127] and the [14]C work[129] it was possible not only to evaluate the individual rate coefficients for the separate steps involving the dienone intermediates, *viz.*

$$\text{Ether} \underset{k_{-1}}{\overset{k_1}{\rightleftarrows}} o\text{-Dienone} \overset{k_2}{\rightarrow} p\text{-Dienone}$$

$$\downarrow \qquad\qquad\qquad \downarrow$$

$$o\text{-Allyl phenol} \quad p\text{-Allyl phenol}$$

but also, from their variation with temperature, the energy and entropy of activation for each step. The results are given in Table 4; $k$ refers to the overall observed

### TABLE 4

RATE COEFFICIENTS AND ACTIVATION PARAMETERS FOR THE INDIVIDUAL STEPS OF
THE CLAISEN REARRANGEMENT AT 170 °C

| Rate coefficients | LXXXIX | XC | XCI |
|---|---|---|---|
| $10^5 k$ (sec$^{-1}$) | 6.5 | 7.0 | 0.76 |
| $k_2/k_{-1}$ | 2.7 | 3.0 | |
| $k_{-1}/k_1$ | 330 | | |
| $10^5 k_1$ (sec$^{-1}$) | 4.4 | 4.7 | 0.38 |
| $10^5 k_{-1}$ (sec$^{-1}$) | $1.5 \times 10^3$ | | |
| $10^5 k_2$ (sec$^{-1}$) | $4 \times 10^3$ | | |

| Activation parameters | LXXXIX | | | XC | XCI |
|---|---|---|---|---|---|
| | $k_1$ | $k_{-1}$ | $k_2$ | $k_1$ | $k_1$ |
| $E_a$ (kcal.mole$^{-1}$) | 31.0 | 26.8 | 26.8 | 30.5 | 33.1 |
| $\Delta S^{\ddagger}$ (cal.deg$^{-1}$.mole$^{-1}$) | $-11.3$ | $-9.3$ | $-7.3$ | $-12.2$ | $-11.5$ |

composite first-order rate coefficient. Not all the data refer to the same solvent but the reaction is generally not very sensitive to solvent change (see later), and in some cases temperature extrapolations have been carried out. It appears that for the *para* rearrangement (LXXXIX and XC) there is little difference in the polar or steric effects of the methyl and allyl groups in this reaction, as reflected in the

TABLE 5

RATE COEFFICIENTS FOR REARRANGEMENT OF ALLYL $p$-CRESYL ETHER AT 185 °C[130]

| Solvent | $10^5k$ $(sec^{-1})$ |
|---------|----------------------|
| Phenol | 45 |
| Ethylene glycol | 18 |
| Benzyl alcohol | 9.7 |
| 1-Octanol | 9 |
| Carbitol | 3.6 |
| Methyl salicylate | 2.45 |
| Benzonitrile | 2.49 |
| N,N-Dimethylaniline | 2.46 |
| Acetophenone | 2.41 |
| Diphenyl ether | 2.08 |
| Diphenyl methane | 2.12 |
| Decalin | 1.56 |

TABLE 6

RATE COEFFICIENTS FOR THE REARRANGEMENT OF ALLYL $p$-X-PHENYL ETHER IN CARBITOL AT 181 °C

| X | $10^5k$ $(sec^{-1})$ | X | $10^5k$ $(sec^{-1})$ |
|------|----------------------|------|----------------------|
| NO$_2$ | 1.03 | Br | 2.77 |
| CN | 1.13 | Ph | 3.68 |
| MeSO$_2$ | 1.34 | Me | 4.42 |
| Ac | 1.58 | NHAc | 5.79 |
| Bz | 1.60 | MeO | 9.16 |
| H | 2.56 | NH$_2$ | 21.3 |
| Cl | 2.63 | | |

agreement in the rate coefficients and also the activation parameters.

The relatively small dependence of the rate on solvent is demonstrated in Table 5. The rate coefficients are a little greater in the more polar solvents but the overall effects are small. This is taken to indicate that there is only little polar character to the transition state. Another probe to determine the polarity of the transition state is the effect of substituents on the rate. The results obtained for the *ortho* rearrangement by varying the *para* substituents are shown in Table 6[131] for the conversion

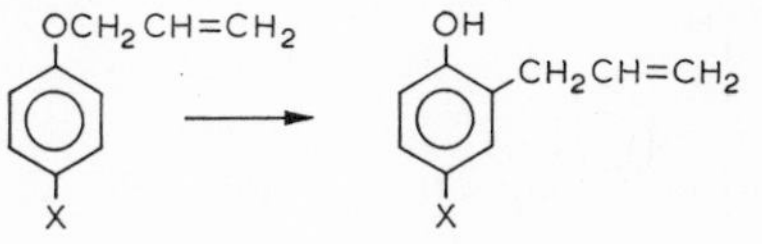

XCII          XCIII

In general, electron-releasing substituents increase $k$ whereas electron-attracting substituents decrease $k$ when compared with X = H. These results have been tested using linear free energy relationships. The best fit is obtained using the values $\sigma^+$ for *para* substituents and a $\rho$ value of $-0.6$ is obtained. It is not clear why the correlation is with $\sigma^+$ rather than any of the many other $\sigma$ scales but this result has been interpreted as demonstrating that there is some polar character to the transition state with a depletion of electrons at the oxygen atom. There is an indication also from these results that the bond-breaking process (oxygen–carbon) is more important than the bond-making process (carbon–carbon). Similar results were obtained by Goering and Jacobson[130] who carried out the rearrangement in diphenyl ether as solvent. Again a correlation with $\sigma^+$ was found with $\rho = -0.5$. This suggests the electronic movements within the molecule indicated by XCIV.

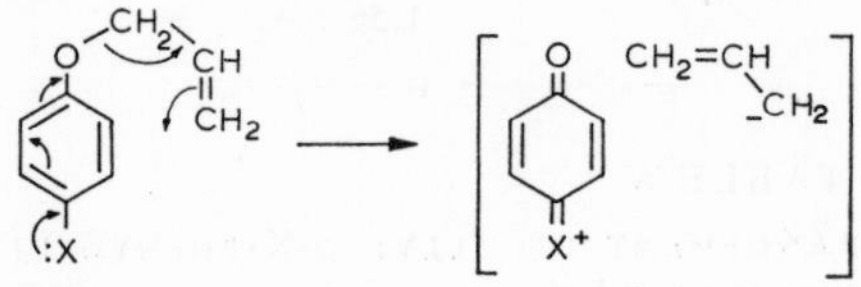

XCIV

However, variation of the substituent in the allyl group, for the rearrangement of XCV to XCVI, which are given in Table 7[135], shows that again there is a correla-

TABLE 7

RATE COEFFICIENTS FOR THE REARRANGEMENT OF X-CINNAMYL *p*-TOLYL ETHERS IN CARBITOL AT 180 °C

| $X$ | $10^5k$ $(sec^{-1})$ | $X$ | $10^5k$ $(sec^{-1})$ |
|---|---|---|---|
| $m$-NO$_2$ | 1.90 | $m$-OMe | 3.62 |
| $m$-CN | 1.99 | $m$-Me | 4.04 |
| $m$-Cl | 3.00 | $p$-Me | 6.08 |
| $p$-Cl | 3.39 | $p$-OMe | 9.61 |
| H | 3.41 | | |

tion of $\sigma^+$ with a small negative value for $\rho(-0.4)$, so that the rate is increased by

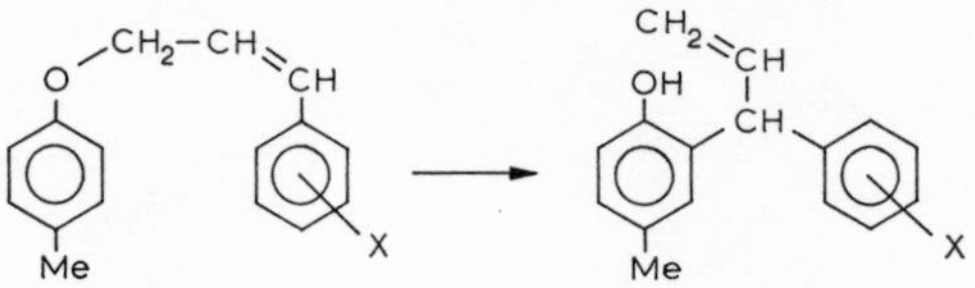

XCV        XCVI

electron-release from both groups of the ether. This makes it difficult to describe the nature of the transition state with any certainty; it has been suggested[135,136] that it would be better represented by homolytic fission and the formation of quasi-radical structures, although there is no compelling evidence to support this claim.

Much attention has been directed recently at the stereochemistry of the Claisen rearrangement following the claim by Alexander and Kluiber[122] that the rearrangement was stereospecific and proceeded with retention of configuration. The stereochemistry is discussed fully in the recent review articles[120] and will not be documented here beyond the statement that a chair-like transition state is predicted by molecular orbital symmetry relationships[137] to have a lower energy than the boat-like form and that this is borne out by experiment[138] since the *trans* isomers of XCVII and XCVIII and others react twice as fast as do the corresponding *cis* isomers.

A so-called abnormal Claisen rearrangement has been observed on a number of occasions usually concurrently with the normal reaction. Here the $\beta$-carbon atom in the ether becomes bonded to the *ortho* position in the phenyl group as in the formation of C from XCIX[139], *viz.*

It is now thought[140] that the abnormal product is formed reversibly from the normal product *via* a spiran-type intermediate.

Many rearrangements analogous to the Claisen rearrangement but with other atoms or groups instead of the oxygen atom are known. For example, N-allyl-1-naphthylamine rearranges at 260 °C to 2-allyl-1-naphthylamine[141] and the carbon analogue has been recognised[142] in the equilibration of butenylbenzenes and *ortho* propenyltoluenes. In the sulphur analogue (thio-Claisen), phenyl allyl thioether CI forms the cyclised products CII and CIII *via*, it is thought[143] the thiirane intermediate CIV, *viz.*

The Claisen rearrangement has also been observed in heterocyclic systems, for example see ref. 144.

### 3.2 THE FRIES REARRANGEMENT

A reaction which has proved to be of much use in synthetic organic chemistry is the formation of the *ortho* and/or the *para* isomers of a hydroxyketone (CVI and CVII) by treatment of a phenolic ester (CV) with an acid catalyst, *viz.*

The acid is usually a Lewis acid, *e.g.* aluminium chloride[145]. Although it has often been used for synthetic purposes, very little mechanistic work has been carried out and the detailed mechanism of the reaction is far from settled. The reaction has been carried out heterogeneously and homogeneously in a variety of inert solvents such as nitrobenzene, ethylene dichloride and light petroleum and also in the absence of a solvent. Other catalysts have included the tetrachlorides of tin, zirconium and titanium and boron trifluoride. The *ortho* : *para* ratio of the products varies enormously with the temperature, the solvent and the substrate used[146]. Generally the *ortho* isomer is favoured at higher temperatures and the *para* at lower temperatures but this is by no means always the case. The situation is further complicated by the conversion of the *para* to the *ortho* when heated, for example with aluminium chloride at 170 °C[146,147].

As to the question of intra- or intermolecularity of the rearrangement there are three opinions, one which states that the reaction is completely intermolecular[148], another supports a concurrent intra- and intermolecular mechanism[149], whilst a third claims that the reaction is completely intramolecular[150]. Evidence for the intermolecular mechanism is based on trapping experiments such as the reaction of *meta*-tolyl acetate (CVIII) in the presence of *meta*-chlorobenzoyl chloride (CIX) when the acylation product (CX) is formed rather than the products of rearrangement of CVIII, *viz.*

CVIII     CIX     CX

implying the intermediate formation of *meta*-cresol. Acetyl chloride has also been recovered from reaction mixtures[151] as have the phenolic products of de-acyl-ation[149]. Further, the four *ortho*-hydroxy ketones CXIII–CXVI were isolated[152] when the acetate ester CXI and the benzoate ester CXII were rearranged in the presence of aluminium chloride, *viz.*

CXI     CXII     CXIII     CXIV     CXV     CXVI

These results, although suggestive of an intermolecular mechanism do not necessarily demand it.

Simultaneous inter- and intramolecular reaction have been suggested on the basis[153] of the varying yields of the *ortho* and *para* isomers and also of the acetylated product of added diphenyl ether as the dilution was changed 5-fold in the reaction of *meta*-tolyl acetate in the presence of diphenyl ether. In particular it was noted that as the dilution increased the amounts of acetylated diphenyl ether and the *para* product both decreased whilst that of the *ortho* product rose very slightly.

The result of a $^{14}$C-labelling experiment[150] is indicative that the rearrangement is completely intramolecular. Phenyl acetate in solution containing 1-$^{14}$C acetic anhydride and aluminium chloride picked up $^{14}$C from the acetic anhydride. The label was also found in the *ortho* and *para* products. The significant part was that the extent of labelling in both products was the same but was smaller than that found in the recovered ester. This was interpreted as demonstrating that intramolecular rearrangement (to both products) was taking place concurrently with the exchange of acyl group between the ester and the acetic anhydride.

On the basis of all these experiments various mechanisms have at some stage been advanced for the Fries rearrangement involving the free acylium ion or as a tightly bound ion pair, $\pi$-complexes and cyclic intermediates. It is clearly impossible to reconcile all the experimental data by one reaction mechanism. It is probable that many such mechanisms are possible, each one operative under a certain set of conditions.

Approximate rate measurements have been carried out by Cullinane *et al.*[149]

who stopped the reaction at appropriate times and by a series of extraction methods actually isolated the unchanged ester, both (in some cases) *ortho* and *para* products as well as the phenolic product of deacylation. The reaction appeared to be first-order in the ester and the rate increased with the catalyst : ester ratio up to two and the remained constant. Reaction, however, took place to completion even when this ratio was one. This was taken to mean that reaction could take place *via* a 1 : 1 or 2 : 1 complex, the latter being the energetically most favourable route. It was claimed that the non-zero extrapolated rate at time zero both for the formation of rearranged products and product of de-acylation argued against an intermolecular mechanism on the grounds that there should be an induction period if reaction proceeds *via* an acylium ion or acetyl chloride. This spurious argument has been criticised[154]. The kinetic data obtained prove nothing regarding the inter- or intramolecularity of the reaction and assist only little in the quest for a mechanism beyond that one molecule of the ester and two of the catalyst are required for the most favourable reaction path. Clearly there remains much to be done. A thorough kinetic study of the reaction would be of great value in an attempt to establish firmly the reaction mechanism.

## 3.3 REARRANGEMENT OF ALKYL ARYL ETHERS

Alkyl aryl ethers (CXVII) undergo rearrangement in the presence of Lewis acid and strong acid catalysts to give the corresponding alkyl phenols (CXVIII and CXIX), *viz.*

For example *t*-butyl phenyl ether with aluminium chloride forms *para-t*-butyl phenol[155]. Often the de-alkylated phenol is also formed in considerable quantity. The reaction formally resembles the Fries and Claisen rearrangements. Like the Fries rearrangement the question of inter- or intramolecularity has not been settled, although may experiments based on cross-over studies[156], the use of optically active ethers[157] and comparison with product distribution from Friedel-Crafts alkylation of phenols[158] have been carried out with this purpose in view.

Tarbell and Petropoulos[159] measured the rate of formation of phenol from benzyl phenyl ether catalysed by aluminium bromide and found it to be a first-order process. Moreover, the rate coefficient was the same as that for the formation of phenol from *ortho* benzyl phenol. Since the ratio of products (phenol : *ortho* benzyl phenol) was unaffected by a large change in the temperature it was argued that they were formed from a common intermediate. A case was made out for a fast intramolecular rearrangement. In one other instance[160] the disappearance of *para* tolyl benzyl ether was shown to be a first-order process and an argument was presented for concurrent inter- and intramolecular processes.

Very recently[161] it has been shown that a π-complex is not an intermediate in the rearrangement of alkyl aryl ethers. The π-complex mechanism[158] requires complete retention of configuration of the migrating alkyl group. However, the rearrangement of *sec*-butyl phenyl ether gave the *ortho* and *para* alkylated phenols with respectively 14 % and 52 % net inversion of configuration. The results were claimed to be consistent with two parallel processes, an $S_N2$ type displacement by the ether on its aluminium bromide complex and also an ion-pair mechanism.

## 4. Rearrangement of alkylaromatics

The migration of alkyl groups from one position to another in an aromatic ring can be brought about by acids, usually Lewis acids together with a hydrogen halide. Thus the radioactivity of the 1 position in $[1\text{-}^{14}C]$ toluene decreases in the presence of aluminium bromide and hydrogen bromide[162]. The level of activity in the other ring positions (relative to the methyl group) increases in such a way as to suggest that the change arises by successive 1,2 shifts of the methyl group, either, it is thought, *via* a three-membered ring bridged ion CXX or a π-complex CXXI, *viz.*

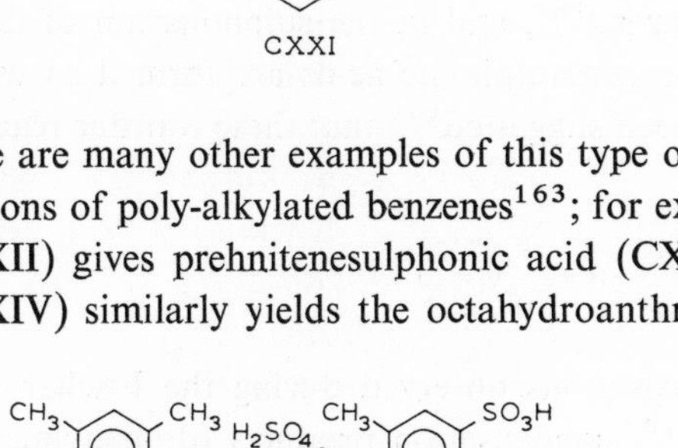

CXX

CXXI

There are many other examples of this type of rearrangement particularly in the reactions of poly-alkylated benzenes[163]; for example, the sulphonation of durene (CXXII) gives prehnitenesulphonic acid (CXXIII) and octahydrophenanthrene (CXXIV) similarly yields the octahydroanthracenesulphonic acid (CXXV), *viz.*

CXXII

CXXIII

The trimethylbenzenes are isomerised to mesitylene in the presence of boron trifluoride and anhydrous hydrogen fluoride[164]. Rearrangement of alkyl- (and halogeno-) aromatics during sulphonation is known as the Jacobsen rearrangement. Differences have been observed in the isomers formed in sulphonation as compared with rearrangement in $BF_3$–HF[163,165]; for example, in sulphonation durene is rearranged to prehnitene (after hydrolysis) whereas in $BF_3$–HF isodurene is formed. The results have been taken to show that in sulphonation rearrangement takes place in the sulphonic acid derivative rather than in the parent hydrocarbon.

There is no evidence which absolutely confirms the intermediacy of a π-complex (CXXI) or a bridged structure (CXX) in these rearrangements, or, in fact, that reaction proceeds via an intermediate at all. These are merely convenient ways of representing the reaction. It is generally agreed that the rearrangement is intramolecular and it is thought that the first step involves protonation of the aromatic hydrocarbon by $SO_3H^+$ or $BF_3$–HF, etc. to form a σ complex CXXVI which in turn forms the π-complex or bridged ion intermediate, viz.

from which the rearranged product is formed. Support for a σ complex comes from the fact that salts of benzene and alkylbenzenes have been prepared[166] which appear to be in the benzeneonium ion form and the observed link between the stability of the conjugate acids of alkylbenzenes and the relative positions and numbers of substituted methyl groups[167].

Cross-alkylations have been reported on a number of occasions. Thus, ethylbenzene when treated with aluminium bromide and hydrogen bromide at 0 °C forms some benzene and diethylbenzene[168], and in the sulphonation of durene some trimethyl- and pentamethyl-benzenesulphonic acids are formed as well as the tetra-methyl compound. It has been suggested[169] that these transfer reactions involve an $S_N2$ type process

and are analogous to the cross-nitrosations observed during the Fischer–Hepp rearrangement (See 2.4). Recently[170], condensation products (diphenylmethane derivatives) have been described from the reaction of monochlorotetramethylbenzenes with sulphuric acid.

Relative rate coefficients for alkyl group migrations in alkyltoluenes were obtained by Allen[171] for the various possible rearrangements *ortho → para*, *para → ortho*, *meta → para*, *para → meta*, *ortho → meta* and *meta → ortho*, and are shown in Table 8. These results were obtained by using small amounts of

TABLE 8

RELATIVE RATE COEFFICIENTS FOR ALKYL GROUP MIGRATIONS IN ALKYLTOLUENES IN TOLUENE SOLVENT[171]

| | $o \to p$ | $p \to o$ | $m \to p$ | $p \to m$ | $o \to m$ | $m \to o$ |
|---|---|---|---|---|---|---|
| Xylenes[a] | 0.0 | 0.0 | 2.1 | 6.0 | 3.6 | 1.0 |
| Ethyltoluenes[b] | 2.9 | 1.0 | 24.6 | 60.8 | 38.2 | 5.3 |
| Cymenes[c] | 19.9 | 1.0 | 2.5 | 5.5 | 45.8 | 1.0 |

[a] At 50 °C.
[b] At room temperature.
[c] At 0° C.

catalyst ($Al_2Cl_6$–HCl) so that little substrate became irreversibly complexed with the catalyst and intermolecular transfer to the solvent gave an alkyltoluene. From the results it can be seen that no direct transfer takes place between the *ortho* and *para* positions in the xylenes but does occur for the ethyltoluenes and the cymenes. The figures are taken to indicate the relative ease of formation of the $\sigma$-complexes (*e.g.* CXXVI) and also the migration of the alkyl group in the $\sigma$-complex. The formation of *meta*-xylene from the *para* isomer is faster by a factor of 1.7 than is its formation from *ortho*-xylene. Comparison[172] of the respective enthalpies and entropies of formation of the isomers suggest that protonation of *ortho* xylene is more favourable than that of *para* xylene. Other calculations[173] support this. It is not at all obvious why the protonated *para* isomer should rearrange faster than the *ortho* isomer. Other workers[164,174] have shown that this is the case when sufficient catalyst was present to prevent further reaction of *meta*-xylene; here it appeared that the rate of rearrangement of the protonated *para* isomer must be more than five times that of the protonated *ortho* isomer.

Both prehnitene (CXXVII) and durene (CXXVIII) rearrange to isodurene (CXXIX), *viz.*

in the presence of excess $BF_3$–HF. The rate of reaction is about the same for both isomers[175] although statistically one would expect CXXVIII to be the more reactive since rearrangement of any one of the four methyl groups leads to the formation

*References pp. 481–486*

of CXXIX whereas for CXXVII only two methyl groups can undergo rearrangement. It is found[173], however, that the basicity of the carbon atoms bearing
these two methyl groups is about twice that of the four positions in CXXVIII thus
accounting for the observed rate difference.

## 5. Rearrangement of halogenoaromatics

In the presence of an acid catalyst certain halogen-substituted aromatic compounds undergo rearrangement, halogen appearing to migrate to another position
in the benzene ring. For example[176], each of the dichlorobenzenes when treated
with aluminium chloride at 200 °C results in the formation of the equilibrium
mixture 8 % *ortho*, 60 % *meta*, and 33 % *para*-dichlorobenzene. Another example,
often quoted, is the Reverdin rearrangement[177] where migration of the iodine
atom takes place during nitration of *para* iodoanisole, the main product being
2-iodo 4-nitroanisole (CXXX), *viz.*

Diiodoanisole is also formed in this reaction, which has been taken[178] as evidence
that the reaction is intermolecular. Halogen migration during sulphonation
(Jacobsen rearrangement) is also well-known[163].

Among the halogens, fluorine does not undergo rearrangement, and what
evidence there is suggests that the rearrangement of chlorine is an intramolecular
process (1,2 shift) whereas that of bromine appears to take place by both inter-
and intramolecular routes. Less is known about iodine migration.

Reaction is again, as in the case of alkyl migration, thought of as occurring
through a σ-complex CXXXI which can either undergo the rearrangement
intramolecularly to form CXXXII or dissociate forming the halogen cation (or
a species derived therefrom), from which the products are formed by the usual
electrophilic substitution process, *viz.*

Complications arise due to halogenation of the starting material; thus, bromo-benzene gives[179] initially 15 % *ortho* and 85 % *para*-dibromobenzene, whereas if the reaction is allowed to proceed further the *meta* isomer is formed up to its equilibrium concentration of 62 %. So far as the migration of chlorine is concerned, in chlorofluorobenzenes and dichlorobenzenes[176] the formation of the *para* isomer from the *ortho* compound and vice versa occurs only after some of the *meta* isomer has been formed, thus suggesting strongly that rearrangement occurs only intramolecularly by 1,2 shifts.

Relative rate coefficients have been obtained for the rearrangements of the various bromotoluenes[180] and are shown in Table 9. The rates of the *ortho* → *para*

TABLE 9

RELATIVE RATE COEFFICIENTS FOR BROMINE MIGRATION IN THE BROMOTOLUENES[180]

| Migration | Relative rate coeff. | Migration | Relative rate coeff. |
|---|---|---|---|
| m → o | 1.0 | o → p | 6.0 |
| m → p | 1.5 | p → o | 13.4 |
| o → m | 1.2 | p → m | 4.2 |

and *para* → *ortho* reactions are the greatest, at first sight possibly confirming the intermolecular mechanism suggested earlier[181]. However, it was found[180] that, although there was a gradual increase in the amount of toluene formed (by dis-proportionation) during the reaction, there was no increase in the rate of formation of the *ortho*-bromotoluene suggesting that the latter is formed by some process other than that involving disproportionation. An intramolecular mechanism in-volving two consecutive 1,2 shifts was proposed.

The equilibrium concentrations of many disubstituted benzenes (containing alkyl and halogen substituents) show that the *meta* isomer is in nearly all cases the most thermodynamically stable. It is not obvious why this should be so. Shine[182] had discussed this problem in terms of the relative sizes of the standard enthalpy and entropy changes between any pair of isomers.

REFERENCES

1 P. DE MAYO (Ed.), *Molecular Rearrangements*, Part I, Interscience, New York, 1963.
2 H. J. SHINE, *Aromatic Rearrangements*, Elsevier, Amsterdam, 1967.
3 B. S. THYAGARAJAN (Ed.), *Mechanisms of Molecular Migrations*, Vols. I and II, Interscience, New York, 1968 and 1969.
4 G. BENDER, *Chem. Ber.*, 19 (1886) 2272.
5 E. SLOSSON, *Chem. Ber.*, 28 (1895) 3265.

6   F. D. Chattaway and K. J. P. Orton, *J. Chem. Soc.*, 77 (1900) 797.
7   K. J. P. Orton and A. E. Bradfield, *J. Chem. Soc.*, (1927) 986.
    C. Beard and W. J. Hickinbottom, *J. Chem. Soc.*, (1958) 2982.
8   A. R. Olsen, G. W. Porter, F. A. Long and R. S. Halford, *J. Am. Chem. Soc.*, 58 (1936) 2467.
    A. R. Olsen, R. S. Halford and J. C. Hornel, *J. Am. Chem. Soc.*, 59 (1937) 1613.
9   A. E. Bradfield, K. J. P. Orton and I. C. Roberts, *J. Chem. Soc.*, (1928) 782.
10  M. Richardson and F. G. Soper, *J. Chem. Soc.*, (1929) 1873.
11  E. D. Hughes and C. K. Ingold, *Quart. Rev.*, 6 (1952) 34.
12  H. S. Harned and H. Seltz, *J. Am. Chem. Soc.*, 44 (1922) 1475.
13  K. J. P. Orton and W. J. Jones, *J. Chem. Soc.*, 95 (1909) 1456.
14  K. J. P. Orton, F. G. Soper and G. Williams, *J. Chem. Soc.*, (1928) 998.
15  F. G. Soper, *J. Phys. Chem.*, 31 (1927) 1192.
    F. G. Soper and D. R. Pryde, *J. Chem. Soc.*, (1927) 2761.
16  R. P. Bell, *Proc. Roy. Soc. (London)*, *Ser. A*, 143 (1934) 377; R. P. Bell and R. V. H. Levinge, *Proc. Roy. Soc. (London)*, *Ser. A*, 151 (1935) 211.
    R. P. Bell, *J. Chem. Soc.*, (1936) 1154.
    R. P. Bell and J. F. Brown, *J. Chem. Soc.*, (1936) 1520.
17  R. P. Bell and P. V. Danckwerts, *J. Chem. Soc.*, (1939) 1774.
18  J. M. W. Scott, *Can. J. Chem.*, 38 (1960) 2441.
19  G. C. Israel, A. W. N. Tuck and F. G. Soper, *J. Chem. Soc.*, (1945) 547.
20  T. Higuchi and A. Hussain, *J. Chem. Soc.*, B, (1967) 549.
21  L. R. Pryde and F. G. Soper, *J. Chem. Soc.*, (1931) 1510.
22  J. M. W. Scott and J. G. Martin, *Can. J. Chem.*, 43 (1965) 732; 44 (1966) 2901.
23  A. W. Hoffmann, *Proc. Roy. Soc. (London)*, 12 (1863) 576.
    H. Schmidt and G. Shultz, *Chem. Ber.*, 11 (1878) 1754; *Ann. Chem.*, 207 (1881) 320, 348.
24  P. Jacobson, *Ann. Chem.*, 367 (1909) 304; 428 (1922) 76.
25  D. V. Banthorpe, *Topics Carbocyclic Chem.*, 1 (1969) 1.
26  M. Tichwinsky, *J. Russ. Phys. Chem. Soc.*, 35 (1903) 667.
27  J. Stieglitz, *Am. Chem. J.*, 29 (1903) 62.
28  C. K. Ingold and H. V. Kidd, *J. Chem. Soc.*, (1933) 984.
29  J. M. Smith, G. W. Wheland and W. M. Schwartz, *J. Am. Chem. Soc.*, 74 (1952) 2282.
30  D. V. Banthorpe and E. D. Hughes, *J. Chem. Soc.*, (1962) 3308.
31  M. Vecera, J. Gasparic and J. Petranek, *Chem. Ind. (London)*, (1957) 299; *Collection Czech. Chem. Commun.*, 22 (1957) 1603.
32  M. J. S. Dewar, *J. Chem. Soc.*, (1946) 777.
33  R. B. Carlin, R. G. Nelb and R. C. Odioso, *J. Am. Chem. Soc.*, 73 (1951) 1002.
34  D. M. Oglesby, J. D. Johnson and C. N. Reilley, *Anal. Chem.*, 38 (1966) 385.
    W. M. Schwartz and I. Shain, *J. Phys. Chem.*, 69 (1965) 30.
35  J. P. van Loon, *Rec. Trav. Chim.*, 23 (1904) 62.
36  G. S. Hammond and H. J. Shine, *J. Am. Chem. Soc.*, 72 (1950) 220.
37  R. B. Carlin and R. C. Odioso, *J. Am. Chem. Soc.*, 76 (1954) 100.
38  D. V. Banthorpe, *J. Chem. Soc.*, (1962) 2429.
39  D. A. Blackadder and C. Hinshelwood, *J. Chem. Soc.*, (1957) 2898.
40  D. V. Banthorpe, C. K. Ingold, J. Roy and S. M. Somerville, *J. Chem. Soc.*, (1962) 2436.
41  A. Cooper, Mechanism of the benzidine rearrangement, *Ph. D. Thesis*, London, 1966.
42  D. V. Banthorpe, E. D. Hughes and C. K. Ingold, *J. Chem. Soc.*, (1964) 2864.
43  J. Hine, *Physical Organic Chemistry*, 2nd edn., McGraw-Hill, New York, 1962, pp. 71, 121.
44  C. A. Bunton, C. K. Ingold and M. Mhala, *J. Chem. Soc.*, (1957) 1906.
    D. V. Banthorpe, E. D. Hughes, C. K. Ingold and J. Roy, *J. Chem. Soc.*, (1962) 3294.
45  D. V. Banthorpe, E. D. Hughes and C. K. Ingold, *J. Chem. Soc.*, (1962) 2418.
46  D. V. Banthorpe, *J. Chem. Soc.*, (1962) 2407.
47  R. O. C. Norman and R. Taylor, *Electrophilic Substitution in Benzenoid Compounds*, Elsevier, Amsterdam, 1965, pp. 29, 30, etc.
48  Ref. 25, pp. 29–44.
49  E. D. Hughes and C. K. Ingold, *J. Chem. Soc.*, (1941) 608; (1950) 1638.

50  Ref. 42, p. 2872.
51  Ref. 25, pp. 7–13; ref. 2 pp. 154–162.
52  V. STERBA AND M. VECERA, *Collection Czech. Chem. Commun.*, 31 (1966) 3486.
53  M. O'SULLIVAN, The acid-catalysed rearrangements of hydrazobenzenes. *M. Sc. Thesis*, London, 1966.
54  Ref. 2, p. 165.
    W. N. WHITE AND E. E. MOORE, *J. Am. Chem. Soc.*, 90 (1968) 526.
55  Ref. 25, pp. 44, 45.
56  V. O. LUKASHEVICH AND L. G. KROLIK, *Dokl. Akad. Nauk SSSR*, 129 (1959) 117.
57  M. VECERA AND J. PETRANEK, *Collection Czech. Chem. Commun.*, 23 (1958) 249.
58  M. VECERA, L. SYNEK AND V. STERBA, *Collection Czech. Chem. Commun.*, 25 (1960) 1992.
59  D. V. BANTHORPE, R. BRAMLEY AND J. A. THOMAS, *J. Chem. Soc.*, (1964) 2900.
60  G. WITTIG, R. BÖRZEL, F. NEUMANN AND G. KLAR, *Ann. Chem.*, 691 (1966) 109.
    H. J. SHINE AND J. P. STANLEY, *J. Org. Chem.*, 32 (1967) 905.
61  D. V. BANTHORPE, *J. Chem. Soc.*, (1964) 2854.
62  H. J. SHINE, *J. Am. Chem. Soc.*, 78 (1956) 4807.
63  Ref. 25, p. 47.
64  E. BAMBERGER AND K. LANDSTEINER, *Chem. Ber.*, 26 (1893) 482.
    E. BAMBERGER, *Chem. Ber.*, 30 (1897) 1248.
65  K. J. P. ORTON AND C. PEARSON, *J. Chem. Soc.*, 93 (1908) 725.
66  E. D. HUGHES AND G. T. JONES, *J. Chem. Soc.*, (1950) 2678.
67  W. N. WHITE AND J. T. GOLDEN, *Chem. Ind. (London)*, (1962) 138.
68  S. BROWNSTEIN, C. A. BUNTON AND E. D. HUGHES, *J. Chem. Soc.*, (1958) 4354.
69  D. V. BANTHORPE, E. D. HUGHES AND D. L. H. WILLIAMS, *J. Chem. Soc.*, (1964) 5349.
70  D. V. BANTHORPE, J. A. THOMAS AND D. L. H. WILLIAMS, *J. Chem. Soc.*, (1965) 6135.
71  B. A. GELLER AND L. N. DUBROVA, *J. Gen. Chem. USSR*, 30 (1960) 2627.
72  D. V. BANTHORPE AND J. A. THOMAS, *J. Chem. Soc.*, (1965) 7149.
73  S. MARKINKIEWICZ, J. GREEN AND P. MAMALIS, *Tetrahedron*, 14 (1961) 208.
74  W. N. WHITE, J. R. KLINK, D. LAZDINS, C. HATHAWAY, J. T. GOLDEN AND H. S. WHITE, *J. Am. Chem. Soc.*, 83 (1961) 2024.
75  K. J. P. ORTON AND A. E. SMITH, *J. Chem. Soc.*, 87 (1905) 389.
    A. E. BRADFIELD AND K. J. P. ORTON, *J. Chem. Soc.*, (1929) 915.
76  W. N. WHITE, D. LAZDINS AND H. S. WHITE, *J. Am. Chem. Soc.*, 86 (1964) 1517.
77  D. V. BANTHORPE AND J. A. THOMAS, *J. Chem. Soc.*, (1965) 7138.
78  Ref. 1, p. 295.
79  D. W. RUSSELL, *J. Chem. Soc.*, (1963) 894.
80  O. FISCHER, *Chem. Ber.*, 32 (1899) 247.
81  T. D. B. MORGAN AND D. L. H. WILLIAMS, work to be published; reported in The Mechanisms of Reactions in Solution, Chemical Society Conference at the University of Kent at Canterbury, July 1970.
82  O. FISCHER AND E. HEPP, *Chem. Ber.*, 19 (1886) 2991.
83  O. FISCHER AND P. NEBER, *Chem. Ber.*, 45 (1912) 1093.
84  P. W. NEBER AND H. RAUSCHER, *Ann. Chem.*, 550 (1942) 182.
85  J. HOUBEN, *Chem. Ber.*, 46 (1913) 3984.
86  Ref. 2, p. 234.
87  J. H. BOYER, *The Chemistry of the Nitro and Nitroso Groups*, H. FEUER (Ed.)., Interscience, 1969, p. 223.
88  T. I. ASLAPOVSKAYA, E. Y. BELYAEV, V. P. KUMAREV AND B. A. PORAI-KOSHITS, *Reaktsii Sposobnost Org. Soedin.*, 5 (1968) 456.
89  E. A. GUGGENHEIM, *Phil. Mag.*, 2 (1926) 538.
90  G. STEEL AND D. L. H. WILLIAMS, *Chem. Commun.*, (1969) 975.
91  E. KALATZIS AND J. H. RIDD, *J. Chem. Soc.*, (1966) 529.
92  B. A. PORAI-KOSHITS, E. Y. BELYAEV AND J. SZADOWSKI, *Reaktsii Sposobnost Org. Soedin.*, 1 (1964) 10.
93  Z. VRBA AND Z. J. ALLAN, *Tetrahedron Letters*, (1968) 4507.
94  B. C. CHALLIS AND J. H. RIDD, *J. Chem. Soc.*, (1962) 5208.

95 E. BAMBERGER, *Ann. Chem.*, 424 (1921) 233, 297; 441 (1925) 207.
96 S. OKAZAKI AND M. OKUMURA, unpublished work, quoted by S. OAE, T. FUKUMOTO AND M. YAMAGAMI, *Bull. Chem. Soc. Japan*, 34 (1961) 1873; 36 (1963) 601.
97 H. E. HELLER, E. D. HUGHES AND C. K. INGOLD, *Nature*, 168 (1951) 909.
98 Y. YUKAWA, *J. Chem. Soc. Japan*, 71 (1950) 547, 603.
99 E. BOYLAND AND R. NERY, *J. Chem. Soc.*, (1962) 5217.
100 E. BOYLAND AND R. NERY, *Biochem. J.*, 91 (1964) 362.
101 P. GRIESS AND C. A. MARTIUS, *Z. Chem.*, 2 (1866) 132.
    A. KEKULÉ, *Z. Chem.*, 2 (1866) 689.
102 R. J. FRISWELL AND A. G. GREEN, *J. Chem. Soc.*, 47 (1885) 917.
103 E. NOELTING AND F. BINDER, *Chem. Ber.*, 20 (1887) 3004.
104 K. HEUMANN AND L. OECONOMIDES, *Chem. Ber.*, 20 (1887) 372.
105 H. GOLDSCHMIDT, S. JOHNSEN AND E. OVERWIEN, *Z. Physik. Chem.*, 110 (1924) 251.
106 E. D. HUGHES AND C. K. INGOLD, *Quart. Rev.*, 6 (1952) 34.
107 B. MILLER, *J. Am. Chem. Soc.*, 86 (1964) 1127.
108 T. DYL, H. LAHMER AND H. BEYER, *Chem. Ber.*, 94 (1961) 3217.
109 H. BEYER, H. J. HAASE AND W. WILDGRUBE, *Chem. Ber.*, 91 (1958) 247.
110 H. BEYER AND H. J. HAASE, *Chem. Ber.*, 90 (1957) 66.
111 O. WALLACH AND L. BELLI, *Chem. Ber.*, 13 (1880) 525.
112 E. BUNCEL, in *Mechanism of Molecular Rearrangements*, Vol. I, Interscience, New York, 1968, p. 61.
113 M. M. SHEMYAKIN, T. E. AGADZHANYAN, V. I. MAIMIND, R. V. KUDRYAVTSEV AND D. N. KURSANOV, *Proc. Acad. Sci. USSR*, 135 (1960) 1295.
    S. OAE, T. FUKOMOTO AND M. YAMAGAMI, *Bull. Chem. Soc. Japan*, 36 (1963) 601.
114 M. M. SHEMYAKIN, V. I. MAIMIND AND B. K. VAICHUNAITE, *Chem. Ind. (London)*, (1958) 755; *Bull. Acad. Sci. USSR, Div. Chem. Sci.*, (1960) 808.
115 E. BUNCEL AND B. T. LAWTON, *Can. J. Chem.*, 43 (1965) 862.
116 C. S. HAHN AND H. H. JAFFÉ, *J. Am. Chem. Soc.*, 84 (1962) 949.
117 C. S. HAHN, K. W. LEE AND H. H. JAFFÉ, *J. Am. Chem. Soc.*, 89 (1967) 4975.
118 L. CLAISEN, *Chem. Ber.*, 45 (1912) 3157.
119 Ref. 2, pp. 89–120.
120 S. J. RHOADS, in *Molecular Rearrangements*, Vol. 1, P. DE MAYO (Ed.), Interscience, New York, 1963, pp. 665–706.
    D. L. DALRYMPLE, T. L. KRUGER AND W. N. WHITE, in *The Chemistry of the Ether Linkage*, S. PATAI (Ed.), Interscience, London, 1967, pp. 635–660.
    A. JEFFERSON AND F. SCHEINMANN, *Quart. Rev.*, 22 (1968) 391.
121 C. D. HURD AND L. SCHMERLING, *J. Am. Chem. Soc.*, 59 (1937) 107.
122 E. R. ALEXANDER AND R. W. KLUIBER, *J. Am. Chem. Soc.*, 73 (1951) 4304.
123 L. CLAISEN AND E. TIETZE, *Chem. Ber.*, 58B (1925) 275.
124 H. CONROY AND R. A. FIRESTONE, *J. Am. Chem. Soc.*, 75 (1953) 2530; 78 (1956) 2290.
125 D. Y. CURTIN AND R. J. CRAWFORD, *J. Am. Chem. Soc.*, 79 (1957) 3156.
126 H. SCHMID AND K. SCHMID, *Helv. Chim. Acta*, 35 (1952) 1879; 36 (1953) 489.
    J. P. RYAN AND P. R. O'CONNOR, *J. Am. Chem. Soc.*, 74 (1952) 5866.
127 J. F. KINCAID AND D. S. TARBELL, *J. Am. Chem. Soc.*, 61 (1939) 3085.
    D. S. TARBELL AND J. F. KINCAID, *J. Am. Chem. Soc.*, 62 (1940) 728.
128 F. KALBERER, K. SCHMID AND H. SCHMID, *Helv. Chim. Acta*, 39 (1956) 555.
129 W. HAEGELE AND H. SCHMID, *Helv. Chim. Acta*, 41 (1958) 657.
130 H. L. GOERING AND R. R. JACOBSON, *J. Am. Chem. Soc.*, 80 (1958) 3277.
131 W. N. WHITE, D. GWYNN, R. SCHLITT, C. GIRARD AND W. K. FIFE, *J. Am. Chem. Soc.*, 80 (1958) 3271.
132 S. J. RHOADS AND R. L. CRECELIUS, *J. Am. Chem. Soc.*, 77 (1955) 5057.
133 F. W. SHULER AND G. W. MURPHY, *J. Am. Chem. Soc.*, 72 (1950) 3155.
    L. STEIN AND G. W. MURPHY, *J. Am. Chem. Soc.*, 74 (1952) 1041.
134 K. R. BROWER, *J. Am. Chem. Soc.*, 83 (1961) 4370.
135 W. N. WHITE AND W. K. FIFE, *J. Am. Chem. Soc.*, 83 (1961) 3846.
136 W. N. WHITE, C. D. SLATER AND W. K. FIFE, *J. Org. Chem.*, 26 (1961) 627.

W. N. WHITE AND C. D. SLATER, *J. Org. Chem.*, 26 (1961) 3631; 27 (1962) 2908.

137  R. HOFFMANN AND R. B. WOODWARD, *J. Am. Chem. Soc.*, 87 (1965) 4389.

138  W. N. WHITE AND B. E. NORCROSS, *J. Am. Chem. Soc.*, 83 (1961) 1968.
L. D. HUESTIS AND L. J. ANDREWS, *J. Am. Chem. Soc.*, 83 (1961) 1963.

139  W. M. LAUER AND W. F. FILBERT, *J. Am. Chem. Soc.*, 58 (1936) 1388.

140  E. N. MARVELL, D. R. ANDERSON AND J. ONG, *J. Org. Chem.*, 27 (1962) 1109.
R. M. ROBERTS AND R. G. LANDOLT, *J. Org. Chem.*, 31 (1966) 2699.
E. N. MARVELL AND B. SCHATZ, *Tetrahedron Letters* (1967) 67.

141  S. MARCINKIEWICZ, J. GREEN AND P. MAMALIS, *Chem. Ind. (London)* (1961) 438; *Tetrahedron*, 14 (1961) 208.

142  W. VON E. DOERING AND R. A. BRAGOLE, *Tetrahedron*, 22 (1966) 385.

143  H. KWART AND E. R. EVANS, *J. Org. Chem.*, 31 (1966) 413.

144  J. K. ELWOOD AND J. W. GATES, *J. Org. Chem.*, 32 (1967) 2956.

145  A. H. BLATT, *Organic Reactions*, Vol. 1, Wiley, New York, 1942, p. 342.

146  K. ROSENMUND AND W. SCHNURR, *Ann. Chem.*, 460 (1928) 56.

147  R. MARTIN AND J. M. BETOUX, *Bull. Soc. Chim. France*, (1969) 2079.

148  F. KRAUSZ AND R. MARTIN, *Bull. Soc. Chim. France*, (1965) 2192.
C. R. HAUSER AND E. H. MAN, *J. Org. Chem.*, 17 (1952) 390.

149  N. M. CULLINANE, R. A. WOOLHOUSE AND B. F. R. EDWARDS, *J. Chem. Soc.*, (1961) 3842 and earlier papers.

150  Y. OGATA AND H. TABUCHI, *Tetrahedron*, 20 (1964) 1661.

151  E. H. COX, *J. Am. Chem. Soc.*, 52 (1930) 352.

152  K. ROSENMUND AND W. SCHNURR, *Ann. Chem.*, 460 (1927) 96.

153  R. BALTZLY, W. S. IDE AND A. P. PHILLIPS, *J. Am. Chem. Soc.*, 77 (1955) 2522.

154  Ref. 1, p. 319.

155  R. A. SMITH, *J. Am. Chem. Soc.*, 55 (1933) 3718.

156  W. I. GILBERT AND E. S. WALLIS, *J. Org. Chem.*, 5 (1940) 184.

157  H. HART AND R. J. ELIA, *J. Am. Chem. Soc.*, 76 (1954) 3031.

158  M. J. S. DEWAR AND N. A. PUTTNAM. *J. Chem. Soc.*, (1960) 959 and earlier papers.

159  D. S. TARBELL AND J. C. PETROPOULOUS, *J. Am. Chem. Soc.*, 74 (1952) 244.

160  N. M. CULLINANE, R. A. WOOLHOUSE AND G. B. CARTER, *J. Chem. Soc.*, (1962) 2995.

161  P. A. SPANNINGER AND J. L. VON ROSENBERG, *Chem. Commun.*, (1970) 795.

162  H. STEINBERG AND F. L. J. SIXMA, *Rec. Trav. Chim.*, 81 (1962) 185.
V. A. KOPTYUG, I. S. ISAEV AND N. N. VOROZHTSOV, *Proc. Akad. Sci. USSR*, 149 (1963) 191.

163  See L. I. SMITH, *Org. Reactions*, 1 (1942) 370.

164  D. A. MCCAULAY AND A. P. LIEN, *J. Am. Chem. Soc.*, 74 (1952) 6246.

165  E. N. MARVELL AND B. M. GRAYBILL, *J. Org. Chem.*, 30 (1965) 4014.

166  G. A. OLAH AND S. J. KUHN, *J. Am. Chem. Soc.*, 80 (1958) 6541.

167  D. A. MCCAULAY AND A. P. LIEN, *J. Am. Chem. Soc.*, 73 (1951) 2013.

168  H. C. BROWN AND C. R. SMOOT, *J. Am. Chem. Soc.*, 78 (1956) 2176.

169  M. J. S. DEWAR, in ref. 1, p. 302.

170  H. SUZUKI AND Y. TAMURA, *Chem. Commun.*, (1969) 244.

171  R. H. ALLEN, *J. Am. Chem. Soc.*, 82 (1960) 4856.

172  K. S. PITZER AND D. W. SCOTT, *J. Am. Chem. Soc.*, 65 (1943) 803.
W. J. TAYLOR, D. D. WAGMAN, M. G. WILLIAMS, K. S. PITZER AND F. D. ROSSINI, *J. Res. Natl. Bur. Std.*, 37 (1946) 95.

173  S. EHRENSON, *J. Am. Chem. Soc.*, 83 (1961) 4493.

174  H. C. BROWN AND H. JUNGK, *J. Am. Chem. Soc.*, 77 (1955) 5579.

175  M. KILPATRICK, J. A. S. BETT AND M. L. KILPATRICK, *J. Am. Chem. Soc.*, 85 (1963) 1038.

176  G. A. OLAH, W. S. TOLGYESI AND R. E. A. DEAR, *J. Org. Chem.*, 27 (1962) 3449.

177  F. REVERDIN, *Chem. Ber.*, 29 (1896) 1000.

178  G. M. ROBINSON, *J. Chem. Soc.*, 109 (1916) 1078.

179  G. A. OLAH, W. S. TOLGYESI AND R. E. A. DEAR, *J. Org. Chem.*, 27 (1962) 3441.

180  J. W. CRUMP AND G. A. GORNOWICZ, *J. Org. Chem.*, 28 (1963) 949.

181  G. A. OLAH AND M. W. MEYER, *J. Org. Chem.*, 27 (1962) 3464.

182  Ref. 2, pp. 40–47.

486 AROMATIC REARRANGEMENTS

183  W. N. WHITE et al., *J. Org. Chem.*, 35 (1970) 737, 965, 1803, 2048.
184  B. T. BALIGA, *J. Org. Chem.*, 35 (1970) 2031.
185  T. D. B. MORGAN AND D. L. H. WILLIAMS, *Chem. Commun.*, (1970) 1671.

# Index

## A

acenaphthene, bromination of, 119, 406
—, hydrogen exchange with, 259
4-acetamidodiphenyl, chlorination of, 104
acetanilides, bromination of, 114, 115
—, chlorination of, 98–100, 104
—, iodination of, 137
—, mercuration of, 192
—, nitration of, 38
acetate ion, catalysis of nucleophilic substitution by, 411, 415
acetic acid, and acetoxylation, 56
—, and alkylation, 155–160
—, and $ArB(OH)_2$+halogens, 368–370
—, and $H_2O_2$+$PhB(OH)_2$, 364
—, and nitration, 30–34, 37, 38, 43, 44
—, and nitrodesilylation, 378
—, and nucleophilic substitution, 414
—, and Orton rearrangement, 434–436
—, and positive iodination, 93
—, and protodecarboxylation, 314, 316
—, ArCOOH+$Br_2$ in, 371, 373
—, ArH+HOBr in, 84–87
—, ArH+HOCl in, 89–91
—, ArH+$I_2$, ICl in, 136–139
—, bromination in, 114–120, 122, 125–129, 132, 133, 405
—, bromodeboronation in, 367, 369
—, chlorination in, 98–109
—, chloromethylation in, 164, 165
—, halodesilylation in, 378, 379
—, hydrogen exchange in, 203, 205–207, 209, 210, 215–219, 237, 239–241, 252–255, 260, 261, 263, 264
—, mercuration in, 187–194
—, mercuridesilylation in, 376, 377
—, nitrosation in, 47, 48
—, protodegermylation in, 341, 342
—, protodehalogenation in, 355
—, protodemercuration by, 279, 280
—, protodesilylation in, 325, 327, 332–337
—, protodesulphonation by, 352
—, rearrangement of quinamines in, 464
acetic anhydride, acetoxylation in, 56
—, acylation by, 172, 184, 185
—, and Fries rearrangement, 475
—, chlorination in, 100, 101
—, effect on nitration by $AcONO_2$, 42
—, nitration in, 35–40
—, nitrodesilylation in, 378

acetone, mercuridemercuration in, 357
—, $NO_2C_6H_4F$+piperidine in, 423
acetonitrile, alkylation in, 152
—, chlorination in, 100, 101
—, nitration in, 40, 46
—, $NO_2C_6H_4F$+piperidine in, 423
acetophenones, Claisen rearrangement in, 471
—, protodeacylation of, 317
acetoxymercury perchlorate, and mercuration, 188–190
acetylanthranilic acid, reaction+$Cl_2$, 99, 100
acetyl bromide (fluoride, iodide), acetylation by, 173
acetyl chloride, acetylation by, 168, 172, 173, 177, 180–184
N-acetyldiphenylamine, chlorination of, 106
4-acetylhydrindacene, protodeacylation of, 316, 317
S-acetylmercaptoethylamine, isomerisation of, 420
acetyl nitrate, acetoxylation by, 56
—, and nitration in $Ac_2O$, 35–40
—, nitration by, 42
acetyloctahydroanthracene, protodeacylation of, 316, 317
acid catalysis, in electrophilic substitution, 3–5, 7
acidity function, and ArH+$Br_2$, 120
—, and ArH+HOBr, 84, 86
—, and ArH+HOCl, 89
—, and benzidine rearrangement, 441, 442
—, and chloromethylation, 165
—, and cyclialkylation, 161, 162
—, and electrophilic substitution, 4, 5
—, and Fischer–Hepp rearrangement, 455–457, 461
—, and $H_2O_2$+$PhB(OH)_2$, 366
—, and hydrogen exchange, 196, 198, 200, 203–205, 207–209, 213, 214, 216, 221, 222, 224, 226–237, 241, 249, 253, 254, 257
—, and nitramine rearrangement, 449, 450, 452
—, and nitrosation, 49, 50
—, and Orton rearrangement, 436
—, and protodeacylation, 317
—, and protodealkylation, 323
—, and protodeboronation, 287–293, 301, 302
—, and protodeiodination, 356
—, and protodesilylation, 324, 325
—, and protodesulphonation, 351, 352, 354

—, —+ IC$_6$H$_3$(NO$_2$)$_2$, 423, 424
butylbenzene, acylation of, 175, 177, 181, 182
—, alkylation of, 140, 146, 149
—, chloromethylation of, 164
—, hydrogen exchange with, 197, 198, 207, 244, 251, 255–257, 260, 264, 266, 406
—, mercuration of, 192, 193
—, nitration of, 33, 34, 39, 45
—, protodealkylation of, 323
—, reaction+Br$_2$, 116, 133
—, —+ Cl$_2$, 99, 101, 104, 106, 375
—, —+ ClOAc, 108
—, —+ HOBr, 85, 375
—, —+ HOCl, 90
—, —+ SO$_2$Cl$_2$, 113
—, sulphonation of, 71, 72
—, sulphonylation of, 79
butylbenzene sulphonic acids, protodesulphonation of, 353
t-butylbenzoate, nitration of, 41
butyl bromide, alkylation by, 144, 151
butyl chloride, alkylation by, 145, 147, 152
butyl phenyl ethers, rearrangement of, 476, 477

C

cadmium bromide, catalysis of acylation by, 168
—, — of ArH+Br$_2$ by, 132
cadmium iodide, and demercuration, 361, 362
cadmium ion, catalysis of protodeboronation by, 298–301
caesium cyclohexylamine, and hydrogen exchange, 273–275
carbanions, in electrophilic substitution, 3
—, in protodestannylation, 347, 348
carbazoles, chlorination of, 106
—, hydrogen exchange with, 247
—, nitration of, 39
carbitol, Claisen rearrangement in, 469, 471
carbon dioxide, from protodecarboxylation, 303, 308
carbon disulphide, acylation in, 171, 173, 182
—, alkylation in, 145, 150
carbonium ions, and alkylation, 140, 144, 156, 158, 160
carbon monoxide, from protodeacylation, 319
carbon tetrachloride, alkylation in, 145, 147
—, bromination in, 115, 117, 123, 130, 131
—, bromodesilylation in, 381
—, chlorination in, 107, 109, 110
—, hydrogen exchange and, 247, 248, 257–259
—, iodination in, 137
—, iododestannylation in, 382
—, nitration in, 35, 40–42
chloramine-T, reaction+p-cresol, 437

chloride ion, and rearrangement of ArNClAc, 434–436
—, effect on ArH+HOBr, HOAc, 84, 128
—, — on ArH+HOCl, 87
—, — on ArNH$_2$+ICl, 95
—, — on ArOH+R$_2$NCl, 92
—, — on mercuration, 186
—, — on PhNHOH rearrangement, 462
—, — on PhSiMe$_3$+ICl, 381
—, — on protodeiodination, 356
—, — on protodemercuration, 279, 283, 285, 286
chlorine, and rearrangement of PhNClAc, 434
—, chlorination by, 98–107
—, reaction+PhSiMe$_3$, 379
chlorine acetate, and positive chlorination, 90, 91
—, reaction+ArH, 107, 108
N-chloroacetamide, reaction+PhOH, 92
N-chloroacetanilides, Orton rearrangement of, 433–437
chloroacetic acid, and H$_2$O$_2$+PhB(OH)$_2$, 364
—, and hydrogen exchange, 215, 216, 263
—, catalysis of nucleophilic substitution, 414
chlorobenzene, acylation of, 167–169, 175, 177, 180, 182, 183
—, alkylation of, 140, 141, 145, 146, 149–151
—, chlorination in, 101, 112, 113
—, hydrogen exchange with, 248, 256
—, mercuration of, 193
—, nitration of, 13, 31, 33, 36, 37, 39, 42, 45
—, Orton rearrangement in, 436
—, reaction+Br$_2$, 134
—, —+ Cl$_2$, 105, 111
—, —+ HOCl, 91
—, sulphonation of, 58, 59, 61, 75, 76
—, sulphonylation of, 77–79, 81–83
chlorobenzene boronic acids, protodeboronation of, 289, 297, 300
—, reaction+Br$_2$, 368, 369
m-chlorobenzoyl chloride, and Fries rearrangement, 475
chlorofluorobenzenes, isomerisation of, 481
chloroform, acylation in, 182, 183
—, bromination in, 115, 132
—, chlorination in, 99, 100
—, hydrogen exchange in, 249
—, nitration in, 40
—, nucleophilic substitution in, 410, 423, 424
chloromethyl methyl ether, chloromethylation by, 163, 164
N-chloromorpholine, reaction+PhOH, 92
chloronitrobenzenes, nitration of, 12, 13
4-chlorophenetole, reaction+Br$_2$, 114
—, —+ Cl$_2$, 99, 100
chlorophenyl ethers, bromination of, 114

stannic chloride, catalysis of acylation by, 168, 170, 183–185
—, — of alkylation by, 151, 152
—, — of bromination by, 130
—, — of chlorination by, 110, 111
—, — of Fries rearrangement by, 474
—, — of hydrogen exchange by, 206, 207, 238–242
stannous chloride, and protodehalogenation, 355
steric effects, in acylation, 166, 167, 172, 173, 181–183
—, in alkylation, 151, 158, 159
—, in bromination, 114, 124, 125, 405
—, in bromodecarboxylation, 374
—, in chlorination, 104, 105
—, in chlorodebutylation, 375
—, in Claisen rearrangement, 470
—, in diazonium coupling, 50, 53
—, in electrophilic substitution, 8
—, in hydrogen exchange, 207, 224, 262, 264, 272, 274
—, in iodination, 136, 137
—, in iododestannylation, 382, 384
—, in mercuridesilylation, 376, 377
—, in nitration, 10, 18
—, in protodeacylation, 316, 320, 321
—, in protodeboronation, 298
—, in protodecarboxylation, 306, 315
—, in protodesilylation, 327, 329, 335, 338
—, in protodestannylation, 343–347
—, in protodesulphonation, 352
—, in rearrangement of ArNHOH, 462
—, in sulphonation, 72, 75
stirred flow reactor, for iodination, 94
sulphamic acid, effect on Fischer–Hepp rearrangement, 455, 456
sulphanilic acid, diazonium coupling of, 50–52, 54
sulpholane, see tetramethylene sulphone
sulphur dioxide, alkylation in, 150
—, effect on ArH+SO$_2$Cl$_2$, 113
sulphuric acid, alkylation in, 151, 155, 156, 158, 161–163
—, catalyst for cycliacylation, 185
—, chloromethylation and, 165
—, effect on ArH+Br$_2$, 115, 116, 124
—, — on ArH+Cl$_2$, 99
—, — on ArH+HOBr, 84, 86, 87
—, — on ArH+HOCl, 87
—, — on H$_2$O$_2$+PhB(OH)$_2$, 365, 366
—, — on nitrodesilylation, 378
—, — on protodemercuration, 280, 285
—, Fischer–Hepp rearrangement and, 455, 457
—, hydrogen exchange and, 195–199, 203, 205, 207, 222–229, 232, 233, 235–238, 255–262

—, nitramine rearrangement in, 448, 449
—, nitration and, 12–30, 37, 38, 43, 46
—, nitrosation and, 49, 458
—, protodeacylation by, 317–321
—, protodealkylation by, 323
—, protodeboronation by, 287–290, 292, 294
—, protodecarboxylation by, 304, 306, 307, 310, 311, 315, 316
—, protodegermylation by, 341, 342
—, protodesilylation by, 327, 332–335
—, protodesulphonation by, 350–354
—, rearrangement of aromatic hydrocarbons with, 477, 478
—, — of azoxybenzene in, 465
—, — of PhNHOH in, 461
—, sulphonation by, 60–77
sulphur trioxide, sulphonation by, 58–60, 62, 64, 65, 67, 70, 76
sulphuryl chloride, chlorination by, 112, 113

T

tartaric acid, hydrogen exchange with PhCH$_2$B(OH)$_3$$^-$, 249
tellurium chlorides, catalysis of acylation by, 168
p-terphenyl, hydrogen exchange with, 265
—, nitration of, 39
tetrachloroethane, alkylation in, 145
tetraethylammonium hydrogensulphate (nitrate, picrate), effect on nitration, 41
tetrafluorobenzene, hydrogen exchange with, 275
tetrahydrofuran, nucleophilic substitution in, 423
—, protodemercuration in, 284
tetrahydropyran, effect on PhOH+Ph$_3$CCl, 148
tetralin, hydrogen exchange with, 244, 248, 251, 252
tetramethylammonium nitrate, effect on nitrosation, 48
tetramethylbenzenes, acylation of, 177, 182
—, chloromethylation of, 165
—, hydrogen exchange with, 251
—, isomerisation of, 477–479
—, mercuration of, 193
—, reaction+Br$_2$, 117, 122, 125, 132, 133
—, — +Cl$_2$, 102, 104, 107, 109, 110
—, — +ICl, 137
tetramethylbenzene sulphonic acids, protodesulphonation of, 352
tetramethylene sulphone, alkylation in, 150, 151
—, nitration in, 34, 44, 45, 47
—, nitrodesilylation in, 378

NEW HALL HALL
OVERLE LIBRARY
LONDON S.W.